Birkhäuser

Systems & Control: Foundations & Applications

Bing Sun • Bao-Zhu Guo • Zhen-Zhen Tao

Maximum Principle and Dynamic Programming Viscosity Solution Approach

From Open-Loop to Closed-Loop

Birkhäuser

Bing Sun
School of Mathematics and Statistics
and MIIT Key Laboratory of MTCIS
Beijing Institute of Technology
Beijing, China

Bao-Zhu Guo
Academy of Mathematics and Systems
Science
Chinese Academy of Sciences
Beijing, China

Zhen-Zhen Tao
School of Mathematics and Information
Science
Henan Polytechnic University
Jiaozuo, China

ISSN 2324-9749 ISSN 2324-9757 (electronic)
Systems & Control: Foundations & Applications
ISBN 978-981-96-5738-4 ISBN 978-981-96-5739-1 (eBook)
https://doi.org/10.1007/978-981-96-5739-1

Mathematics Subject Classification: 49L25, 49K15, 49N35, 49M41, 49K20, 65N35

This book is published under the imprint Birkhäuser, www.birkhauser-science.com by the registered company Springer Nature Singapore Pte Ltd.
The registered company address is: 152 Beach Road, #21-01/04 Gateway East, Singapore 189721, Singapore

To the Memory of My Mother (B. S.)

Preface

Control theory serves as a mathematical framework that steers automation engineering technology, with its cornerstone centered on the art of designing controls that empower systems to attain predetermined objectives. These objectives span across diverse domains, including output tracking, ensuring internal stability within the system, and efficiently mitigating or eliminating both internal and external disturbances. At the heart of control design lies the formulation of both feedforward and feedback control laws, with the latter occupying a pivotal role. Feedback control encompasses two main forms: state-based feedback and output feedback, the latter reliant on measurements. Intrinsic to feedback control is its inherent ability to overcome system uncertainties, a feature that underscores its paramount importance in various applications. Nevertheless, not all feedback mechanisms are equally beneficial; only those that are relatively optimized carry practical weight, thereby fostering a natural alliance between control and optimization principles.

The Hamiltonian principle in physics asserts that a physical system transitioning from one state to another will naturally select the path that extremizes (typically minimizes) the action. The quest to identify the extremum of a function was a pivotal impetus behind the inception of calculus in the seventeenth century. Shortly after Isaac Newton's passing in 1727, a new mathematical discipline, Calculus of Variations, emerged, complete with its distinctive challenges and methodologies. Newton's exploration of the optimal shape for a rotating surface to minimize resistance during axial motion at a constant speed exemplifies a classic variational problem, with profound implications in ship, submarine, and aircraft design. However, Newton's most celebrated achievement in 1696, in response to a challenge posed by John Bernoulli, was the resolution of the brachistochrone problem: determining the curve shape that enables a bead, starting from rest and solely accelerated by gravity (without friction), to slide from one point to another in the shortest possible time. The cycloid, the path traced by a point on the circumference of a rolling circle along a straight line, first studied and named by Galileo Galilei in 1599, emerged as the solution. In 1744, Leonhard Euler (1707–1783) published a treatise dedicated to uncovering curves characterized by maximum or minimum properties. A renowned quote from this treatise states, "nothing at all takes place in the universe in which some rule of maximum or minimum does not appear." This perspective is widely recognized as the formal origin of the Calculus of Variations [73, 202].

Optimal control dates back to the 1950s, where it aimed to find a time-varying control function that minimizes a performance criterion while adhering to specific constraints for systems governed by ordinary differential equations (ODEs) or partial differential equations (PDEs). Since its inception in avionics, optimal control has expanded its horizons, finding successful applications in diverse domains such as space exploration, military strategies, and industrial manufacturing. Its efficacy and adaptability have rendered it a captivating field of inquiry. Its roots are deeply intertwined with the Calculus of Variations, classical control theory, stochastic processes, and linear and nonlinear programming [35, 224]. The optimal control theory was significantly shaped by important figures and theories in the 1950s. One such cornerstone is Dynamic Programming, introduced by Richard Bellman in 1956, which laid a fundamental foundation for the field. Dynamic Programming is a mathematical technique that dissects a multi-stage process into a sequence of single-stage problems, leveraging the interdependencies between stages to iteratively optimize the entire process. Its distinctive strength lies in the feedback nature of the optimal control solutions it yields. Another important contribution is the Maximum Principle, introduced by L.S. Pontryagin in 1958, which occupies another important position within optimal control theory. The Maximum Principle excels in addressing scenarios with constrained control variables, furnishing necessary conditions for optimal control in such problems. It serves as a means to identify the extremum of functionals within a closed domain, although the controls derived from this principle are generally open-loop in nature. Notably, when control is unconstrained, the optimal control problem assumes the form of a variational problem, thus linking it intimately with the Calculus of Variations. Consequently, the Calculus of Variations can be considered as a method for finding the extremum of functionals on an open domain, capable of handling only optimal control problems with internal control constraints.

In 1963, A.Ya. Dubovitskiĭ and A.A. Milyutin, drawing parallels to the Karush-Kuhn-Tucker (KKT) method in mathematical optimization, offered a unified portrayal of the Maximum Principle through the lens of functional analysis. This approach not only streamlined the proofs but also imparted a deeper comprehension of the mathematical underpinnings. The second epoch-making advancement in Dynamic Programming emerged in the mid-1980s. Since Pontryagin's time, it has been acknowledged that the Dynamic Programming paradigm culminates in the emergence of the Hamilton-Jacobi-Bellman (HJB) equation, which governs the value function in optimal control problems. Solving the HJB equation yields a feedback control law, enabling the analytical synthesis of optimal control in a closed-loop configuration given knowledge of the value function and its gradient. Identifying the optimal feedback control law stands as a pivotal pursuit in optimal control, often regarded as the holy grail of control theory [116]. Nevertheless, despite the analytical coefficients of the HJB equation, a classical solution may not always exist, and even when it does, uniqueness is not guaranteed. This enduring limitation constrained Dynamic Programming's potential in deriving optimal controls. In the 1980s, M.G. Crandall and P.L. Lions pioneered the concept of viscosity solutions for the HJB equation, establishing that, in a broad context, the value function derived from optimal control uniquely corresponds to the viscosity solution of the HJB equation.

On the contrary, for problems beyond Linear Quadratic (LQ) in linear systems, analytically deriving the optimal control through either the Maximum Principle or Dynamic Programming is generally unfeasible. These constraints circumscribe the applicability of optimal control theory. Nevertheless, the emergence of digital computers in the mid-1950s revolutionized the field by enabling efficient numerical computation of optimal controls. Prior to this, engineers were confined to addressing relatively straightforward problems, thereby limiting the practical application of variational methods, optimal control, and nonlinear programming [35]. However, the swift advancements in computer technology and numerical algorithms have dramatically transformed this scenario.

This book delves into these issues of optimal control, encompassing both the theoretical underpinnings and numerical algorithms. It commences with an exploration of open-loop control and culminates in a comprehensive treatment of closed-loop control, seamlessly integrating Pontryagin's Maximum Principle and Bellman's Dynamic Programming Principle (DPP). Pontryagin's Maximum Principle paves the way for solving two-point boundary value problems, thereby facilitating numerical computations, including the renowned shooting method. Conversely, Dynamic Programming leads to the value function of the optimal control satisfying the HJB equation. If a numerical solution to the HJB equation can be procured, then the computation of the optimal control numerically becomes viable.

The book is organized into five chapters. Chapter 1 introduces essential mathematical concepts, encompassing cones and dual cones, C_0-semigroup theory, the calculus of variations and optimal control, Pontryagin's maximum principle, the dynamic programming approach for finite-dimensional systems, viscosity solution theory for PDEs, and numerical techniques for optimal control. The initial segment, encompassing Chaps. 2 and 3, centers on open-loop control. In Chap. 2, the Dobovitskiĭ-Milyutin functional analytical method is presented as a powerful tool for exploring optimal control problems subject to equality and inequality constraints. We delve into the optimal control challenges of an elastic beam and the sterilization of prepackaged food, deriving first-order optimality conditions for both fixed and free final horizon scenarios. Chapter 3 is dedicated to the spectral method, a crucial numerical approach for solving optimal control problems. Here, we apply the spectral method to analyze two PDE-constrained optimal control problems. After establishing the optimality conditions, we tackle the optimal control of a fourth-order PDE with an H^1-norm state constraint and the Stokes equation with an L^2-norm constraint on the control variable. Furthermore, we provide rigorous a priori and a posteriori error estimates for the spectral approximation problems. The second part of the book focuses on closed-loop control, constituting the core of this monograph. We integrate the concept of viscosity solutions for partial differential equations with the dynamic programming approach. Subsequently, the dynamic programming viscosity solution (DPVS) approach is employed to investigate the optimal control of diverse scenarios, including HIV/AIDS chemotherapy, the continuous casting process, and population dynamics. The final chapter, Chap. 5, offers a comprehensive array of algorithms for the DPVS approach. These methods include a direction derivative algorithm, an interpolation algorithm, a time difference method, the time-space approximation scheme, a local approximation algorithm, and an upwind finite-difference scheme. Some of these methods are accompanied by rigorous convergence proofs. A distinguishing feature of this book is the focus on dynamic systems primarily stemming from technical or biological backgrounds.

The completion of this book has been greatly facilitated by the numerous support and assistance we have received. We express our heartfelt gratitude for this invaluable help. Additionally, this work has been partially funded by the National Natural Science Foundation of China, with grant numbers 11471036, 12131008, and 12271034, and National Key R&D Program of China under grant no. 2024YFA1013101.

Beijing, China — Bing Sun
Beijing, China — Bao-Zhu Guo
Jiaozuo, Henan, China — Zhen-Zhen Tao
September 2024

Contents

Chapter 1
Mathematical Preliminary

In this chapter, we present pertinent results derived from functional analysis and control theory, which will be frequently referenced in subsequent chapters. For brevity and clarity, we may omit certain less relevant findings. For further details, we recommend consulting the provided references [16, 85, 89, 106, 181, 202, 245].

1.1 Cones and Dual Cones

1.1.1 Topological Linear Spaces

Definition 1.1.1 A set E of elements x over a field K is defined as a topological linear space if:

(i) E is a linear space, which means that the addition of elements and the multiplication of elements by real numbers (scalers) are defined and satisfy the natural conditions of commutativity, distributivity, and so on;
(ii) E is equipped with a topology;
(iii) The operations of addition and multiplication by scalers are continuous in the given topology.

A linear space, also known as a vector space, consists of elements known as vectors. Common fields within this context include the real number field $\mathbb{R}^1$ and the complex number field $\mathbb{C}^1$, often shortened to $\mathbb{R}$ and $\mathbb{C}$, respectively.

In a linear space, the semi-norm of a vector serves as a metric to measure its length. This semi-norm quantifies the "magnitude" or "size" of a vector and is formally defined in the subsequent Definition 1.1.2. It offers a mathematical tool for assessing the extent or scale of vectors within the vector space.

B. Sun et al., *Maximum Principle and Dynamic Programming Viscosity Solution Approach*, Systems & Control: Foundations & Applications,
https://doi.org/10.1007/978-981-96-5739-1_1

Definition 1.1.2 A real-valued function $p(x)$ defined in the linear space E over the field K is called a semi-norm on E if, for all $x, y \in E$ and $\alpha \in K$, the following conditions are satisfied:

(i) Subadditivity: $p(x + y) \leq p(x) + p(y)$.
(ii) Homogeneity: $p(\alpha x) = |\alpha| p(x)$.

Consequently, for a semi-norm $p(x)$, we have $p(0) = 0$, and

$$p(x_1 - x_2) \geq |p(x_1) - p(x_2)|$$

which implies that $p(x) \geq 0$.

Definition 1.1.3 A locally convex space is said to be a normed linear space, if its topology is defined by just one semi-norm.

Thus, a linear space E over the field K is said to be a normed linear space if, for all $x, y \in E$, $\alpha \in K$, there is an associated real number $\|x\|$, referred to as the norm of the vector x, that satisfies

$$\|x\| \geq 0 \text{ and } \|x\| = 0 \text{ if and only if } x = 0,$$

$$\|x + y\| \leq \|x\| + \|y\| \quad \text{(triangle inequality)},$$

$$\|\alpha x\| = |\alpha| \cdot \|x\|.$$

These conditions define the norm of a vector in a normed linear space.

It is important to note that, for a semi-norm, it is possible for $p(x) = 0$ even when x is a nonzero vector. However, a semi-norm becomes a norm only when the condition $p(x) = 0$ is equivalent to $x = 0$.

Using the norm, we can introduce a distance function that allows us to quantify the closeness between two vectors. Additionally, the norm enables us to characterize the convergence of sequences of vectors.

For two topological linear spaces E_1 and E_2, an operator $P(x)$ mapping E_2 into E_1 is considered continuous at a point $x \in E_2$ if, for any neighborhood V of $P(x)$, there exists a neighborhood U of x such that $P(U) \subset V$. Furthermore, if $P(x)$ is continuous at all points of E_2, it is considered continuous on E_2. Additionally, $P(x)$ is continuous if and only if the preimage of every open (closed) set is open (closed). If E_1 is the real line, meaning that the values of $P(x)$ are real numbers, then $P(x)$ is referred to as a functional.

Furthermore, in a normed linear space E, a linear functional F is considered bounded if there exists a constant $M > 0$ such that $|F(x)| \leq M\|x\|$ for all $x \in E$. The norm of F is defined as the supremum of $|F(x)|$ over all x with $\|x\| \leq 1$, denoted as

$$\|F\| = \sup_{\|x\| \leq 1} \{|F(x)|\}.$$

An operator $P(x)$ mapping E_2 into E_1 is termed as a linear operator if it possesses the properties of additivity and homogeneity, i.e.,

$$P(x + y) = P(x) + P(y), \quad P(\lambda x) = \lambda P(x),$$

where λ is a scalar. The collection of linear operators, equipped with the natural definitions of operator addition and scalar multiplication, forms a linear space. A noteworthy feature of linear operators is that continuity at a single point implies overall continuity. This is an inherent characteristic of linearity.

Furthermore, a linear operator is continuous if and only if it is bounded in the neighborhood of any point. A linear functional is a special linear operator if it is a real-valued linear function on E_2.

Theorem 1.1.4 *A linear functional F defined on the normed linear space E is bounded if and only it is continuous.*

Proof First, assume that the linear functional F defined on the normed linear space E is bounded. Then, for all $x, h \in E$, where h is nonzero, the inequality

$$|F(x + h) - F(x)| = |F(x) + F(h) - F(x)| = |F(h)| \leq M\|h\|$$

holds for some constant $M > 0$. This inequality implies the continuity of F at any point x by taking the limit as $h \to 0$.

Conversely, if F is continuous, then it is also continuous at 0. For any linear functional f, the property $f(0) = f(0 + 0) = f(0) + f(0)$ implies that $f(0) = 0$. As a result, there exists a $\delta > 0$ such that for all $h \in E$ with $\|h\| \leq \delta$, the inequality $|F(h)| = |F(h) - F(0)| \leq 1$ holds. Now, for any nonzero $x \in E$, we can write

$$|F(x)| = \left|\frac{\|x\|}{\delta} F\left(\delta \frac{x}{\|x\|}\right)\right| = \frac{\|x\|}{\delta}\left|F\left(\delta \frac{x}{\|x\|}\right)\right| \leq \frac{\|x\|}{\delta} \cdot 1 = \frac{\|x\|}{\delta}.$$

This shows that F is bounded, as the constant $\frac{1}{\delta}$ serves as an upper bound for F. □

The set of all continuous linear functionals on E_2 comprises a linear space, denoted as E_2^*. This space is referred to as the dual space of E_2. Linear functionals can be represented using either the notation $f(x)$ or (f, x).

Let P be a linear operator mapping E_2 into E_1. The adjoint operator of P, denoted as P^*, is a linear operator that maps E_1^* into E_2^*. It is defined by the relation

$$\left(P^* f, x\right) = (f, Px)$$

for any $f \in E_1^*$ and $x \in E_2$. If P is a continuous operator, then its adjoint operator P^* is also continuous.

Theorem 1.1.5 *A linear functional that is nonnegative on an open set is continuous.*

Proof If the function $f(\cdot) \geq 0$ on an open set Q, then there exists a neighborhood U of zero such that for some $x_0 \in Q$, the set $x_0 + U \subset Q$. Consequently, for all $x \in U$, the inequalities $f(x_0 + x) \geq 0$ and $f(x_0 - x) \geq 0$ hold. This implies that $|f(x)| \leq f(x_0)$ for all $x \in U$. Therefore, it demonstrates that $f(\cdot)$ is bounded and, as a result, continuous. □

A set L is a linear manifold if, for any two vectors $x \in L, y \in L$, the vector $x + \alpha(y - x)$ remains within L for any $\alpha \in \mathbb{R}$. Furthermore, the closure of a linear manifold retains its linear manifold properties. A linear manifold that includes the point 0 is referred to as a subspace. A linear manifold defined as $L = \{x \mid f(x) = \alpha\}$, where f is a nonzero linear functional, is termed a hyperplane. If such a functional $f(\cdot)$ is continuous, then the hyperplane is closed. Additionally, every hyperplane H in a topological linear space E is either closed or everywhere dense, meaning $\overline{H} = H$ or $\overline{H} = E$. A set S_1 contained on one side of a closed hyperplane, specially $S_1 = \{x \mid f(x) \leq \alpha\}$ with $f \in E^*$, is designated as a closed half-space. Similarly, the set $S_2 = \{x \mid f(x) < \alpha\}$ with $f \in E^*$ is referred to as an open half-space.

Theorem 1.1.6 mentioned below is well-known.

Theorem 1.1.6 *Let E be a topological linear space and E^* represent its dual. Let $f_1, f_2 \in E^*$, $Q_1 = \{x \mid f_1(x) = 0\}$, $Q_2 = \{x \mid f_2(x) = 0\}$ and $Q_1 \subset Q_2$. Then either $f_2 = 0$ (i.e., $Q_2 = E$) or $f_1 = \lambda f_2, \lambda \neq 0$ (i.e., $Q_1 = Q_2$).*

Among the vast array of topological linear spaces, the Banach space X stands out as a crucial linear space. It is endowed with a norm, denoted as $\|x\|$, that satisfies several key properties:

- $\|x\| = 0$ if and only if $x = 0$.
- $\|x + y\| \leq \|x\| + \|y\|, \ \forall\, x, y \in X$.
- $\|\alpha x\| = |\alpha|\|x\|, \ \forall\, x \in X, \alpha \in K$.
- X is complete in terms of the topology induced by the norm $\|\cdot\|$.

A Banach space H is designed as a Hilbert space if its norm adheres to the parallelogram law

$$\|x + y\|^2 + \|x - y\|^2 = 2\left(\|x\|^2 + \|y\|^2\right), \quad \forall\, x, y \in H.$$

When the parallelogram law holds true, one can define an inner product, denoted by $\langle x, y\rangle$, for every pair of elements x and y, such that

$$\langle x, x\rangle = \|x\|^2, \quad \forall\, x \in H.$$

The inner product satisfies the following properties:

- $\langle x, y\rangle = \overline{\langle y, x\rangle}$, the notation $\overline{\langle \cdot, \cdot\rangle}$ refers to the complex conjugate of the inner product.
- $\langle \alpha x, y\rangle = \alpha\langle x, y\rangle$ for all α in the field F and $x, y \in H$.
- $\langle x, x\rangle = 0$ if and only if $x = 0$.

The prime advantage of a Hilbert space lies in the inner product, which not only endows the space H with a topological structure but also a geometrical one. Specially, two elements x and y are considered perpendicular to each other, denoted as $x \perp y$, whenever their inner product $\langle x, y\rangle = 0$. Furthermore, a crucial property of the inner product is the Cauchy-Schwarz inequality, which is stated as follows:

$$|\langle x, y\rangle|^2 \leq \|x\|^2\|y\|^2, \quad \forall\, x, y \in H.$$

Additionally, in a Hilbert space H, the dual space H^* is identical to H itself, as established by the following Riesz representation theorem.

Theorem 1.1.7 (Riesz Representation Theorem) *Let T be a bounded linear functional on a Hilbert space H. Then, there exists $g \in H$ such that for every $f \in H$,*

$$T(f) = \langle f, g\rangle.$$

Moreover, $\|T\| = \|g\|$, where $\|T\|$ represents the operator norm of T and $\|g\|$ is the Hilbert space norm of g.

Proof Due to the continuity of T, it is evident that the set

$$M = \{x \in H \mid T(x) = 0\}$$

is closed. If $M = H$, then we simply choose $g = 0$. In the subsequent discussion, we presume that $M \neq H$. For any $x \in H$ such that $T(x) \neq 0$, we define

$$d = d(x, M) = \inf_{y\in M} \|x - y\|.$$

Suppose that $\{y_n\}$ is a minimal sequence, meaning $\{y_n\} \subset M$ and $\|y_n - x\| \to d$ as $n \to \infty$. By the parallelogram law,

$$2\left\|\frac{y_m - y_n}{2}\right\| = \|y_m - x\|^2 + \|y_n - x\|^2 - 2\left\|\frac{y_m + y_n}{2} - x\right\|^2.$$

Since $\frac{y_m+y_n}{2} \in M$, we deduce that

$$0 \leq 2\left\|\frac{y_m - y_n}{2}\right\| \leq \|y_m - x\|^2 + \|y_n - x\|^2 - 2d^2 \to 0 \quad \text{as } n, m \to \infty.$$

This demonstrates that $\{y_n\}$ is a Cauchy sequence. Consequently, there exists $y \in M$ such that $y_n \to y \in M$ as $n \to \infty$. Therefore

$$d = \|x - y\|.$$

For any given $\lambda \in F$ and $z \in M$, $z \neq 0$, it follows that $y + \lambda z \in M$. Therefore,

$$d^2 \leq \|x - y - \lambda z\|^2 = \|x - y\|^2 - 2\text{Re}\langle\overline{\lambda}(x - y), x\rangle.$$

Choosing $\lambda = \frac{\langle x-y,z\rangle}{\|z\|^2}$ leads to

$$d^2 \leq d^2 - \frac{|\langle x - y, z\rangle|^2}{\|z\|^2}$$

indicating that

$$\langle x - y, z\rangle = 0, \quad \forall\, z \in M.$$

Defining $y_0 = x - y$, we find that $y_0 \perp M,\ T(y_0) \neq 0$, meaning

$$\langle y_0, z\rangle = 0, \quad \forall\, z \in M.$$

Observe that for any $x \in H$,

$$T\left(x - T(x)\frac{y_0}{T(y_0)}\right) = 0.$$

This demonstrates that $x - T(x)\frac{y_0}{T(y_0)} \in M$. Therefore,

$$x = T(x)\frac{y_0}{T(y_0)} + z = T(x)x_0 + z, \quad z \in M, \quad x_0 = \frac{y_0}{T(y_0)} \neq 0,$$

from which we derive

$$\langle x, x_0\rangle = T(x)\|x_0\|.$$

Defining

$$g = \frac{x_0}{\|x_0\|},$$

we arrive at the conclusion that

$$T(x) = \langle x, g\rangle.$$

□

In what follows, we enumerate several important Banach spaces and Hilbert spaces.

First and foremost is the finite-dimensional Euclidean space, typically denoted as $\mathbb{R}^n$. This Euclidean space qualifies as a Hilbert space due to its inner product defined as $\langle x, y\rangle = \sum_{i=1}^{n} x_i y_i$. We refer to it as the n-dimensional vector space, and its elements are represented in the form of $x = \{x_1, x_2, \cdots, x_n\}$, where $x_i \in \mathbb{R}$ for all $i = 1, 2, \cdots, n$. The norm induced by the inner product is given by $|x| = \sqrt{\sum_{i=1}^{n} x_i^2}$.

The second space we consider is the Banach space $L^p(a, b)$, where $1 \le p < \infty$, or simply referred to as L^p. This space consists of n-tuples of measurable functions $x(t) = \{x_1(t), x_2(t), \cdots, x_n(t)\}$ defined over the interval $[a, b] \subset \mathbb{R}$ and possessing a summable p-th power. The norm in this space is given by

$$\|x\|_{L^p(a,b)} = \left(\int_a^b |x(t)|^p \, \mathrm{d}t\right)^{\frac{1}{p}} = \left(\int_a^b \left(\sum_{i=1}^{n} x_i^2\right)^{\frac{p}{2}} \mathrm{d}t\right)^{\frac{1}{p}}.$$

For $1 < p < \infty$, $\frac{1}{p} + \frac{1}{q} = 1$, the dual space of L^p is L^q. Consider T, any bounded linear functional on $L^p(a, b)$. There exists a unique function $g(t) \in L^q(a, b)$ such that for all $f(t) \in L^p(a, b)$,

$$T(f) = \int_a^b \langle f(t), g(t)\rangle \, \mathrm{d}t = \int_a^b \left(\sum_{i=1}^{n} f_i(t) g_i(t)\right) \mathrm{d}t.$$

Specially, $L^2(a, b)$ qualifies as a Hilbert space, equipped with the inner product

$$\langle x, y\rangle_{L^2(a,b)} = \int_a^b \langle x(t), y(t)\rangle \, \mathrm{d}t.$$

Furthermore, the Banach space $L^\infty(a, b)$, or simply referred to as L^∞, represents the space encompassing all n-tuples of essentially bounded functions on the interval $[a, b]$. The norm in this space is defined as

$$\|x\|_{L^\infty(a,b)} = \operatorname*{ess\,sup}_{t \in [a,b]} |x(t)|,$$

where ess sup signifies the smallest number λ such that $|x(t)| \le \lambda$ holds for almost all $t \in [a, b]$. It is worth noting that while L^∞ is the dual space of L^1, i.e., $L^\infty = (L^1)^*$, the converse is not true: $(L^\infty)^* \neq L^1$.

Next, let's introduce an intriguing Banach space, denoted as $C(a, b)$. This space comprises n-tuples of functions that are continuous over the interval $[a, b]$ and is equipped with the norm

$$\|x\|_{C(a,b)} = \max_{t \in [a,b]} |x(t)|.$$

The dual space $C(a, b)$ corresponds to the space of measures on $[a, b]$. In other words, for every f belonging to the dual space $C^*(a, b)$, there exists an n-dimensional measure denoted as $\mathrm{d}\mu(t)$ with support $[a, b]$, satisfying the following relationship:

$$f(x) = \int_a^b \langle x(t), \mathrm{d}\mu(t) \rangle = \int_a^b \sum_{i=1}^n x_i(t)\, \mathrm{d}\mu_i(t).$$

Now, let's introduce the concept of differentiation within function spaces.

Let Ω represent an open set of $\mathbb{R}^n$ possessing a smooth boundary, denoted as $\partial\Omega$, and let m signify a non-negative integer. Consider the vector $\alpha = \{\alpha_1, \alpha_2, \cdots, \alpha_n\}$, referred to as a multi-index. Here, each α_i (where $i = 1, 2, \cdots, n$) is a non-negative integer. We define the length of this multi-index as $|\alpha| = \sum_{i=1}^n \alpha_i$. Denote $D_j = \frac{\partial}{\partial x_j}$ and introduce the differential operator D^α of order $|\alpha|$ as follows:

$$D^\alpha = D_1^{\alpha_1} D_2^{\alpha_2} \cdots D_n^{\alpha_n}.$$

Subsequently, we have

$$D^\alpha f(x) = \frac{\partial^{|\alpha|}}{\partial x_1^{\alpha_1} \partial x_2^{\alpha_2} \cdots \partial x_n^{\alpha_n}} f(x).$$

Note that when α is the zero vector $\{0, 0, \cdots, 0\}$, $D^{\{0,0,\cdots,0\}} f(x)$ simplifies to $f(x)$.

Denote $C^m(\Omega)$ as the set of all functions defined over Ω that possess continuous partial derivatives of order $|\alpha|$ up to and including m. When $m = \infty$, we consider $|\alpha| < \infty$. Specially,

$$C^\infty(\Omega) = \bigcap_{m=0}^{\infty} C^m(\Omega)$$

and $C^0(\Omega)$ is equivalently denoted as $C(\Omega)$. Let $\overline{\Omega}$ represent the closure of Ω. The support set of a continuous function f is defined as

$$\mathrm{supp}(f) = \overline{\{x \in \Omega \mid f(x) \neq 0\}}.$$

Furthermore, $C_0^m(\Omega)$ signifies the set of all functions within $C^m(\Omega)$ that have compact support. The norm in $C^m(\overline{\Omega})$ is then given by

$$\|f\|_{C^m(\overline{\Omega})} = \max_{0 \le |\alpha| \le m} \max_{x \in \overline{\Omega}} |D^\alpha f(x)|.$$

The topology of $C_0^\infty(\Omega)$ is characterized as follows: a sequence $\{f_n\} \subset C_0^\infty(\Omega)$ converges to zero if:

(i) The support of each $f_n(\cdot)$ is contained within a predetermined compact set.
(ii) Both $f_n(\cdot)$ and all its partial derivatives converge uniformly to zero.

Equipped with this topology, the space $C_0^\infty(\Omega)$ is designated as the fundamental space, denoted by $\mathcal{D}(\Omega)$. Elements of $\mathcal{D}(\Omega)$ are referred to as fundamental functions or test functions. The dual space of $\mathcal{D}(\Omega)$, denoted by $\mathcal{D}^*(\Omega)$, is known as the space of generalized functions or Schwarz distributions on Ω.

For a distribution $f \in \mathcal{D}^*(\Omega)$, its weak or distributional partial derivative $D^\alpha f$ is defined as

$$\langle D^\alpha f, \varphi\rangle = (-1)^{|\alpha|}\langle f, D^\alpha \varphi\rangle, \quad \forall \varphi \in \mathcal{D}(\Omega).$$

In this expression, $\langle \cdot, \cdot\rangle$ signifies the duality inner product between $\mathcal{D}^*(\Omega)$ and $\mathcal{D}(\Omega)$. Specially,

$$\langle f, \varphi\rangle = f(\varphi)$$

represents the evaluation of $f(\cdot)$ at φ. Consequently, every distribution possesses derivatives of all orders, and the customary derivative rules apply.

Another crucial Banach space that holds a significant position in the study of partial differential equations is the Sobolev space.

Definition 1.1.8 Let Ω be an open set of $\mathbb{R}^n$ and k be a non-negative integer. For $1 \le p \le \infty$, the Sobolev space $W^{k,p}(\Omega)$ comprises all functions $f(x) = f(x_1, x_2, \cdots, x_n)$ defined in Ω such that both f and its distributional derivatives $D^\alpha f$ of order $|\alpha| = \sum_{i=1}^{n} \alpha_i \le k$ belong to $L^p(\Omega)$. When $p < \infty$, the norm of f in $W^{k,p}(\Omega)$ is given by

$$\|f\|_{k,p} = \left(\sum_{|\alpha|\le k} \int_\Omega |D^\alpha f(x)|^p \,\mathrm{d}x\right)^{\frac{1}{p}}, \quad \mathrm{d}x = \mathrm{d}x_1\,\mathrm{d}x_2 \cdots \mathrm{d}x_n,$$

and when $p = \infty$, the norm is defined as

$$\|f\|_{k,\infty} = \max_{|\alpha|\le k} \left\|D^{|\alpha|} f\right\|_{L^\infty(\Omega)}.$$

Clearly, $W^{0,p}(\Omega)$ coincides with $L^p(\Omega)$ for $1 \le p \le \infty$, and $W^{k,p}(\Omega)$ is a subspace of $L^p(\Omega)$ for the same range of p. In the special case where $p = 2$, $W^{k,2}(\Omega)$ is denoted specially as $H^k(\Omega)$ for $k = 0, 1, \cdots$. This space becomes a Hilbert space when equipped with the inner product

$$\langle f, g\rangle_{H^k(\Omega)} = \sum_{|\alpha|\le k} \langle D^\alpha f, D^\alpha g\rangle_{L^2(\Omega)}, \quad \forall f, g \in H^k(\Omega).$$

Moreover, it's worth noting that $H^0(\Omega)$ is identical to $L^2(\Omega)$. Additionally, $W_0^{k,p}(\Omega)$ represents the complete space of $C_0^\infty(\Omega)$ with respect to the norm defined in $W^{k,p}(\Omega)$.

Theorem 1.1.9 (Hölder's Inequality) *Let $p, q \in [1, \infty]$ and $\frac{1}{p} + \frac{1}{q} = 1$. For any functions $f \in L^p(a, b)$ and $g \in L^q(a, b)$, the following inequality holds:*

$$\int_a^b |f(t)g(t)|\, \mathrm{d}t \leq \left(\int_a^b |f(t)|^p\, \mathrm{d}t\right)^{\frac{1}{p}} \cdot \left(\int_a^b |g(t)|^q\, \mathrm{d}t\right)^{\frac{1}{q}}. \tag{1.1.1}$$

Specially, when $p = 2$, the inequality labeled as (1.1.1) *reduces to the well-known Cauchy-Schwarz inequality.*

Proof The case of $p = 1$ and $q = \infty$ becomes evident from the inequality

$$|f(t)g(t)| \leq \|g\|_{L^\infty}|f(t)|.$$

When either $f = 0$ or $g = 0$, the outcome is straightforward. For other scenarios, we can utilize the Young inequality. Actually, let $a < \infty, b < \infty$. Define the function $f(t) = \frac{t^p}{p} + \frac{1}{q} - t$ for all $t \in [0, \infty)$. Its derivative is given by

$$f'(t) = t^{p-1} - 1.$$

From this, we deduce that

$$f'(t) = 0 \Leftrightarrow t^{p-1} = 1 \Leftrightarrow t = 1.$$

This indicates that $f(\cdot)$ attains its minimum at $t = 1$ over the interval $[0, \infty)$ since $f'' \geq 0$. Given that $f(1) = \frac{1}{p} + \frac{1}{q} - 1 = 0$, we have

$$f(t) \geq f(1) \geq 0.$$

This leads to the conclusion that

$$\frac{t^p}{p} + \frac{1}{q} - t \geq 0 \Rightarrow \frac{t^p}{p} + \frac{1}{q} \geq t.$$

Assuming $t = ab^{\frac{1}{1-p}}$, we find that $t^p = a^p b^{\frac{p}{1-p}} = a^p b^{-q}$, as $\frac{p}{1-p} = -q$. Consequently,

$$ab^{\frac{1}{1-p}} \leq \frac{a^p b^{-q}}{p} + \frac{1}{q},$$

$$ab \leq \frac{a^p}{p} b^{\frac{p}{1-p}} b^{-\frac{p}{1-p}} + \frac{1}{q} b^{-\frac{p}{1-p}}.$$

Additionally, employing the identity $b^{1-\frac{1}{1-p}} = b^{-\frac{p}{1-p}}$, we arrive at the Young inequality

$$ab \leq \frac{a^p}{p} + \frac{b^q}{q}. \tag{1.1.2}$$

Let $A = \|f\|_{L^p(a,b)}$ and $B = \|g\|_{L^q(a,b)}$. Define $a = \frac{|f(t)|}{A}$, $b = \frac{|g(t)|}{B}$ and apply the Young inequality (1.1.2) to derive

$$ab = \frac{|f(t)g(t)|}{AB} \leq \frac{|f(t)|^p}{pA^p} + \frac{|g(t)|^q}{qB^q} = \frac{a^p}{p} + \frac{b^q}{q}.$$

Furthermore,

$$\frac{1}{AB}\int_a^b |f(t)g(t)|\,\mathrm{d}t \leq \frac{1}{pA^p}\int_a^b |f(t)|^p\,\mathrm{d}t + \frac{1}{qB^q}\int_a^b |g(t)|^q\,\mathrm{d}t. \tag{1.1.3}$$

Since $A^p = \int_a^b |f(t)|^p\,\mathrm{d}t$ and $B^q = \int_a^b |g(t)|^q\,\mathrm{d}t$, inequality (1.1.3) implies that

$$\frac{1}{\|f\|_{L^p(a,b)}\|g\|_{L^q(a,b)}}\|fg\|_{L^1(a,b)} \leq \frac{1}{p} + \frac{1}{q} = 1.$$

Therefore, we conclude that

$$\|fg\|_{L^1(a,b)} \leq \|f\|_{L^p(a,b)}\|g\|_{L^q(a,b)}.$$

□

Theorem 1.1.10 (Poincaré Inequality) *Let Ω be a bounded open set in $\mathbb{R}^n$ and let $1 \leq p < \infty$. Then there exists a constant K, dependent solely on Ω and p, such that for all functions f belonging to the Sobolev space $W_0^{1,p}(\Omega)$, the following inequality holds:*

$$\|f\|_{L^p(\Omega)} \leq K\|Df\|_{L^p(\Omega)} \tag{1.1.4}$$

Proof Due to density, it suffices to demonstrate the inequality solely for all functions $f \in C_0^\infty(\Omega)$.

Given that Ω is a bounded domain, for any index j ranging from 1 to n and for any function $f \in C_0^\infty(\Omega)$, the following holds.

$$\begin{aligned}\int_\Omega |f|^p\,\mathrm{d}x &= \int_\Omega \frac{\partial}{\partial x_j}(x_j)|f|^p\,\mathrm{d}x = -p\int_\Omega x_j|f|^{p-1}\operatorname{sgn}(f)f_{x_j}\,\mathrm{d}x \\ &\leq K\int_\Omega |f|^{p-1}|Df|\,\mathrm{d}x\end{aligned}$$

for some constant $K > 0$, in which $\operatorname{sgn}(\cdot)$ is the sign function, i.e., $\operatorname{sgn}(t) = 1$ for $t > 0$, $\operatorname{sgn}(t) = -1$ for $t < 0$ and $\operatorname{sgn}(t) = 0$ for $t = 0$. Since

$$\frac{1}{p} + \frac{1}{\frac{p}{p-1}} = 1,$$

by applying Hölder's inequality (referenced in Theorem 1.1.9), we obtain

$$\begin{aligned}\|f\|_{L^p(\Omega)}^p &\leq K \left(\int_\Omega \left(|f|^{p-1} \right)^{\frac{p}{p-1}} \mathrm{d}x \right)^{\frac{p-1}{p}} \left(\int_\Omega |Df|^p \,\mathrm{d}x \right)^{\frac{1}{p}} \\ &= \|f\|_{L^p(\Omega)}^{p-1} \|Df\|_{L^p(\Omega)},\end{aligned}$$

whence the desired inequality is derived. Leveraging the density of $C_0^\infty(\Omega)$ in $W_0^{1,p}(\Omega)$, we can extend this inequality to all functions $f \in W_0^{1,p}(\Omega)$. □

1.1.2 Convex Set

Definition 1.1.11 In a linear space E, a set A is convex if, for any two elements $x, y \in A$, the convex combination $\lambda x + (1 - \lambda)y$ also belongs to A, where λ is any real number in the closed interval $[0, 1]$.

The intersection of any quantity of convex sets remains convex. Given two convex sets, A and B, the set $\lambda_1 A + \lambda_2 B$ is convex for all $\lambda_1, \lambda_2 \in \mathbb{R}$.

Under linear mappings, both the image and the preimage of a convex set retain their convexity.

If A is a convex set within a topological linear space E, both its interior $\overset{\circ}{A}$ and its closure $\overline{A}$ are convex.

The convex hull of a set A, denoted as $\operatorname{conv}A$, represents the minimal convex set encompassing A. It is defined as the intersection of all convex sets that contain A. Alternatively, it comprises all possible convex combinations of points with A. Specially,

$$\operatorname{conv}A = \left\{ x \;\middle|\; x = \sum_{i=1}^n \lambda_i x_i,\ x_i \in A,\ \lambda_i \geq 0,\ \sum_{i=1}^n \lambda_i = 1,\ 1 \leq n < \infty \right\}.$$

Let A be a convex set. If x belongs to A and y belongs to $\overset{\circ}{A}$, then the convex combination $\lambda x + (1 - \lambda)y$ also lies within $\overset{\circ}{A}$ for any λ in the range $0 \leq \lambda < 1$. Moreover, provided that $\overset{\circ}{A} \neq \emptyset$, the following equalities hold

$$\overline{A} = \overline{\left(\overset{\circ}{A}\right)}, \quad \overset{\circ}{A} = \overset{\circ}{\left(\overline{A}\right)}.$$

A locally convex space, indicated by the term "locally convex space," refers to a topological linear space that boasts a basis around zero composed entirely of convex sets. Consequently, all standard function spaces, including Banach spaces, fall under the umbrella of locally convex spaces.

1.1.3 Hahn-Banach Theorem

Theorem 1.1.12 (Hahn-Banach Theorem) *Let $p(x)$ be a functional that exhibits both convexity and positive homogeneity on the real linear space E. Specially, it satisfies the following properties*

$$p(\lambda x + (1-\lambda)y) \le \lambda p(x) + (1-\lambda)p(y), \quad 0 \le \lambda \le 1,$$
$$p(\lambda x) = \lambda p(x), \quad \lambda \ge 0.$$

Define $f : E_0 \to \mathbb{R}$ as a linear functional acting on a subspace $E_0 \subset E$, subject to the condition that

$$f(x) \le p(x), \quad x \in E_0.$$

We assert that there exists an extension of the functional f to the entirety of the space E, preserving the aforementioned property. In other words, there exists a linear functional F operating on E that fulfills the following criteria:

$$F(x) = f(x), \quad x \in E_0,$$
$$F(x) \le p(x), \quad x \in E.$$

Citing [80], we will showcase the proof in two steps. Initially, we introduce Lemma 1.1.13, which is referred to as one step of the Hahn-Banach theorem.

Lemma 1.1.13 *Take E, f, E_0 as defined in Theorem* 1.1.12. *Consider a point $x_1 \notin E_0$, and let E_1 denote the subspace spanned by E_0 and x_1. Then, there exists a linear functional $F : E_1 \to \mathbb{R}$ that coincides with f in E_0 and fulfills the criterion $F(x) \le p(x)$ for all $x \in E$.*

Proof It is worth noting that the extension we seek must adhere to the following form:

$$F(x + \lambda x_1) = f(x) + \lambda F(x_1) = f(x) + \lambda a.$$

Our objective is to demonstrate that there exists a value of a that satisfies the condition

$$f(x) + \lambda a = F(x + \lambda x_1) \le p(x + \lambda x_1), \quad x \in E_0,\ \lambda \in \mathbb{R}. \tag{1.1.5}$$

In the case of $\lambda > 0$, the inequality labeled as (1.1.5) is valid if

$$a \leq p\left(\frac{x}{\lambda} + x_1\right) - f\left(\frac{x}{\lambda}\right), \quad x \in E_0,$$

or, equivalently, if

$$a \leq M = \inf_{x \in E_0} \{p(x + x_1) - f(x)\}.$$

For $\lambda = -\mu < 0$, the inequality (1.1.5) holds true if

$$a \geq -p\left(\frac{x}{\mu} - x_1\right) + f\left(\frac{x}{\mu}\right), \quad x \in E_0.$$

Namely, if

$$a \geq m = \sup_{x \in E_0} \{-p(x - x_1) + f(x)\}$$

so that the choice of a is possible whenever $m \leq M$. The inequality $m \leq M$ always holds because, for all $x, y \in E_0$,

$$\begin{aligned} f(x) + f(y) &= f(x + y) \\ &\leq p(x + y) = p(x + x_1 + y - x_1) \\ &\leq p(x + x_1) + p(y - x_1). \end{aligned}$$

Consequently,

$$f(y) - p(y - x_1) \leq -f(x) + p(x + x_1).$$

□

Next, we introduce a relation known as "order" for a set. A relation denoted by "$\leq$" in a set E is termed as an order if it satisfies the conditions that $u \leq u$ for every element $u \in E$, and whenever $u \leq v$ and $v \leq w$, then $u \leq w$ for u, v, w belonging to E. When the set E is endowed with such a relation $\leq$, it is referred to as an ordered set. In the set E, an element u is considered maximal if for every $v \in E$, whenever $u \leq v$, it necessarily implies that $v \leq u$. An upper bound for a subset $E_1 \subseteq E$ is defined as any element v such that $u \leq v$ for $u \in E_1$. Additionally, a subset $E_1 \subseteq E$ is said to be linearly ordered if and only if, for any arbitrary elements $u, v \in E$, either $u \leq v$ or $v \leq u$ holds true.

The Zorn lemma, which we will discuss next, is crucial for ordered sets. For further information, please refer to [70].

Lemma 1.1.14 (Zorn Lemma) *If every linearly ordered subset of E possesses an upper bound, then E necessarily contains a maximal element.*

Now, we present the proof of the Hahn-Banach theorem.

Proof *(Proof of Theorem* **1.1.12***)* By definition, an extension of f consists of a pair (G, g) where $G \supseteq E_0$ as a subspace of E and g is a linear functional that coincides with f in E_0 and satisfies $g(x) \leq p(x)$ for all $x \in G$. To introduce an order relation among the extensions (G, g) of (E_0, f), we define $(G_1, g_1) \leq (G_2, g_2)$ if $G_1 \subseteq G_2$ and g_2 coincides with g_1 in G_1. Let $\mathcal{A}$ be an arbitrary indexing set. For each $\alpha \in \mathcal{A}$, consider an arbitrary linearly ordered set of extensions (G_α, g_α). Then, (G, g) serves as an upper bound for this set, where $G = \bigcup_{\alpha \in \mathcal{A}} G_\alpha$ and g is a linear functional defined on G by setting $g(x) = g_\alpha(x)$ whenever $x \in G_\alpha$. Invoking Zorn's lemma, we conclude that the set of all possible extensions has a maximal element, denoted as (E', F). To complete the proof, we must show that $E' = E$. If not, we can apply the one-step lemma to construct an extension (E'', H) with E' being strictly contained in E'', contradicting the maximality of (E', F). □

Several useful corollaries can be derived from the general Hahn-Banach theorem ([89, 182]). For simplicity, we focus on normed vector spaces, which occupy a pivotal position in functional analysis. Specially, Banach spaces and Hilbert spaces serve as illustrations of normed vector spaces.

Theorem 1.1.15 *Let E be a normed linear space and L be its subspace. If f is a bounded linear functional defined on L with the norm $\|f\|_L$, then there exists a bounded linear extension F of f to E such that it satisfies the condition $\|F\| = \|f\|_L$. Furthermore, for any nonzero element $x_0 \in E$, there exists an $F \in E^*$ such that $\|F\| = 1$ and $F(x_0) = \|x_0\|$.*

Proof In the Hahn-Banach theorem (Theorem 1.1.12), let us define $p(x) = \|f\|_L \|x\|$. It is straightforward to verify that p is subadditive and positive homogeneous. Additionally, we observe that $f \leq p$ on L. Consequently, according to the Hahn-Banach theorem, there exists a linear extension F of f to E such that $F \leq p$. Utilizing the linearity of F and the norm properties, we establish the following chain of inequalities:

$$-p(x) = -p(-x) \leq -F(-x) = F(x) \leq p(x)$$

for all $x \in E$. Therefore,

$$-\|f\|_L \|x\| \leq F(x) \leq \|f\|_L \|x\|,$$

which implies $|F(x)| \leq \|f\|_L \|x\|$ for all $x \in E$, indicating that F is bounded. We can then conclude that $\|F\| \leq \|f\|_L$. Furthermore, since F is an extension of f, it cannot have a smaller norm. Hence $\|F\| \geq \|f\|_L$, leading to $\|F\| = \|f\|_L$.

Now, let $U = \text{span}\{x_0\} = \{\lambda x_0 \mid \lambda \in \mathbb{R}\}$ and define f on U by $f(\alpha x_0) = \alpha \|x_0\|$ for some $\alpha \in \mathbb{R}$. Clearly, f is a bounded linear functional with $\|f\|_U = 1$.

According to the proof above, there exists a norm-preserving extension F of f to E that is also a bounded linear functional (therefore, $\|F\| = 1$). Since $x_0 \in U$ and $F = f$ on U, it follows that $F(x_0) = f(x_0) = \|x_0\|$. □

Definition 1.1.16 Let L be a set in a real normed vector space E. The Minkowski functional of L is a functional $p : E \to \mathbb{R}$, defined as

$$p(x) = \inf_{x \in \lambda L} \{\lambda > 0\}.$$

It is evident that a Minkowski functional is inherently nonnegative, yet it is possible that it can be equal to ∞.

Lemma 1.1.17 *Let L be a non-empty convex subset of a real normed vector space E such that L contains an interior point, and let x_0 be an element not belonging to L. Then, there exists a continuous linear functional F defined on E that satisfies the condition $F(x) \leq F(x_0)$ for all $x \in L$.*

Proof Without loss of generality, we may assume that $0 \notin \overset{\circ}{L}$. Let $X = \text{span}\{x_0\} = \{\lambda x_0 \mid \lambda \in \mathbb{R}\}$ and define the linear functional f on X as $f(\lambda x_0) = \lambda$. Since $x_0 \notin L$, it follows that $\lambda x_0 \notin \lambda L$ for any $\lambda \neq 0$. Let p be the Minkowski functional of L. Then $p(\lambda x_0) \geq \lambda = f(\lambda x_0)$ for all $\lambda \in \mathbb{R}$, which implies that $f(x) \leq p(x)$ for all $x \in X$. By the Hahn-Banach theorem, there exists a linear extension F of f into E that satisfies $F(x) \leq p(x)$ for all $x \in E$. Now, for any $x \in L$, we have $F(x) \leq p(x) \leq 1 = F(x_0)$ (since the Minkowski functional $p(x) \leq 1$ for $x \in L$ by definition). Therefore, $F(x) \leq F(x_0)$ for all $x \in L$.

Now, suppose that $0 \in \overset{\circ}{L}$. Then, for any given $\varepsilon > 0$ and all y such that $\|y\|$ is sufficiently small, we have $\frac{y}{\varepsilon}, -\frac{y}{\varepsilon} \in L$. Since $F\left(\frac{y}{\varepsilon}\right) \leq 1$ and $F\left(-\frac{y}{\varepsilon}\right) \leq 1$, it follows that $-\varepsilon \leq F(y) \leq \varepsilon$. In other words, $|F(y)| \leq \varepsilon$ for sufficiently small $\|y\|$, i.e., $\|y\| \leq \delta$. By the linearity of F and setting $y = x - z$, we can deduce that $|F(x) - F(z)| = |F(x - z)| \leq \varepsilon$ for $\|x - z\| \leq \delta$. This establishes continuity of F. □

The geometric interpretation of Theorem 1.1.12 involves the separation of two convex sets, which holds significant importance in optimization techniques, such as the Dubovitskiĭ-Milyutin functional analysis approach discussed later in Chap. 2.

Let A, B be two sets in a topological linear space E. Consider a nonzero continuous linear functional f. If there exists a real number α such that for all $x \in A$, $f(x) \geq \alpha$, and for all $x \in B$, $f(x) \leq \alpha$, then we say that f separates A and B. Furthermore, when the above inequalities become strict, meaning f is strictly greater than α for all x in A and strictly less than α for all x in B, f is termed as strongly separating A and B. In this context, the closed hyperplane $H = \{x \mid f(x) = \alpha\}$ is referred to as a separating hyperplane. Additionally, the sets A and B are said to be separable. A fundamental theorem in this regard is the Theorem 1.1.18 that follows.

Theorem 1.1.18 *Let A and B be non-empty, convex subsets of a real normed vector space E. Suppose that A and B are disjoint, with A possessing an interior point. Then, there exists a hyperplane that separates A from B.*

Proof We consider the set $Z = B - A$, which is convex by the convexity of A and B. Since A and B are disjoint, $0 \notin Z$. Furthermore, Z must possess an interior point. The set Z with the point 0 fulfills the conditions of Lemma 1.1.17, and hence there exists a continuous linear functional F such that $F(z) \leq F(0) = 0$ for all $z \in Z$ (recall that for any linear functional f, we have that $f(0) = 0$). Now, since any $z \in Z$ can be written as $b - a$ for some $a \in A$ and $b \in B$, we have

$$F(z) = F(b - a) = F(b) - F(a) \leq 0,$$

or, equivalently, $F(a) \geq F(b)$ for all $a \in A$ and $b \in B$. This implies that $\inf_{\alpha \in A} F(a) \geq \sup_{b \in B} F(b)$. Additionally, $\inf_{a \in A} F(a) > -\infty$ since it is bounded below by $F(b_0)$, where b_0 can be any point in B (and $F(b_0) \neq -\infty$ due to the continuity of F). Similarly, we know that $\sup_{b \in B} F(b) < \infty$. Thus we can choose any real number c that lies between $\sup_{b \in B} F(b)$ and $\inf_{a \in A} F(a)$ to obtain $F(a) \geq c \geq F(b)$ for all $a \in A$ and $b \in B$. □

1.1.4 Cones and Its Duals

Definition 1.1.19 Let E be a linear space and K a set of E. If for any real number $\lambda > 0$ and any element $x \in K$ there exists an element λx that is also in K, then E is a cone with its apex at 0. Furthermore, the set $x_0 + K$ is called a cone with apex at x_0.

If a cone contains its apex, it is referred to as a sharp cone. Otherwise, it is classified as blunt. A cone K is considered proper or salient if, for any $x \in K$ and $x \neq 0$, there holds $-x \notin K$.

Any intersections of cones sharing a common apex result in a cone, as does any set of the form $\lambda_1 K_1 + \lambda_2 K_2$, where K_i is a cone and $\lambda_i > 0$ for $i = 1, 2$.

In a topological linear space, both the interior and closure of a cone are cones. A cone is designated as convex if it also possesses convex set characteristic. A convex cone with its apex at 0 is actually a set that comprises all points expressed as $\alpha x + \beta y$, where real numbers $\alpha > 0$, $\beta > 0$, and x, y are arbitrary elements of the set.

Theorem 1.1.20 *If K is a cone with its apex at x_0 and $f(x)$ is a linear functional satisfying $f(x) \geq \alpha$, $\alpha \in \mathbb{R}$ for all $x \in K$, then*

$$f(x) \geq f(x_0), \quad x \in K.$$

Proof Let $y \in K$ and $f(y) < f(x_0)$. According to the definition of a cone, the entire ray $x_0 + t(y - x_0),\ t > 0$, must lie in K. However, for sufficiently large t, we have $f(x_0 + t(y - x_0)) = f(x_0) + t(f(y) - f(x_0))$, which contradicts the assumption of the theorem. □

Theorem 1.1.20 referenced above geometrically implies that any cone situated on one side of a hyperplane must lie on one side of the parallel hyperplane through its apex. Moreover, Theorem 1.1.20 yields several important facts.

Firstly, if a cone K and a set A can be distinguished by a hyperplane, then they can also be separated by a parallel hyperplane that traverses the apex of the cone. In simpler terms, given that $f(x) \geq \alpha$ holds for $x \in K$ and $f(x) \leq \alpha$ holds for $x \in A$, it follows that $f(x) \geq f(x_0)$ for $x \in K$ and $f(x) \leq f(x_0)$ for $x \in A$, considering that $f(x) \geq \alpha$.

Secondly, it is feasible to discern a closed convex cone in a locally convex space from any ray emanating from its apex but avoiding the cone itself.

Thirdly, it is noteworthy that if two cones K_1 and K_2, both with their apexes at the origin 0, can be separated by a hyperplane, then that hyperplane must necessarily traverse the origin. Furthermore, if a cone with its apex at 0 and a subspace L can be separated by a hyperplane, it necessarily implies that L lies entirely within that hyperplane.

The theorem referenced as Theorem 1.1.21 characterizes the extensions of a positive linear functional.

Theorem 1.1.21 (Krein Theorem) *Let K be a convex cone with apex 0 and containing interior points, and let L be a subspace such that $\overset{\circ}{K} \cap L \neq \emptyset$. Let $\bar{f}(x)$ represent a linear functional on L that satisfies $\bar{f}(x) \geq 0$ on $K \cap L$. Then, there exists a continuous linear functional $f(x)$ defined on the linear space E that fulfills $f(x) = \bar{f}(x)$ for $x \in L,\ f(x) \geq 0$ for all $x \in K$.*

Proof If $\bar{f}(x) \equiv 0$ on L, we can safely take $f(x) \equiv 0$. Therefore, we assume that $\bar{f}$ is not identically zero. Let $Q_1 = \{x \in L \mid \bar{f}(x) = 0\}$. This set Q_1 is both convex and non-empty with $0 \in Q_1$. Furthermore, $Q_1 \cap \overset{\circ}{K} \neq \emptyset$, for if $x_0 \in Q_1 \cap \overset{\circ}{K}$, then $\bar{f}(x_0) = 0$. However, since $x_0 \in L \cap \overset{\circ}{K}$, there exists a point $x_1 \in L \cap K$ in the neighborhood of x_0 such that $\bar{f}(x_1) < 0$, which contradicts the fact that $\bar{f}(x)$ is non-negative on $L \cap K$. Therefore, a hyperplane exists that separates Q_1 and $\overset{\circ}{K}$. In other words, there exists a linear functional $g(x)$ (which can be assumed continuous; see Theorem 1.1.5) such that $g(x) > 0$ on $\overset{\circ}{K}$ and $g(x) = 0$ on Q_1 (this can be deduced from the third fact implied by the preceding Theorem 1.1.20). It is necessary to consider two linear functionals $\bar{f}(\cdot)$ and $g(\cdot)$ on the subspace L, as well as their zero sets $Q_1 = \{x \in L \mid \bar{f}(x) = 0\}$ and $Q_2 = \{x \in L \mid g(x) = 0\}$. Since $g(x) = 0$ on Q_1, $Q_1 \subset Q_2$. Thus there are only two possibilities: either $Q_1 = Q_2$ or $Q_2 = L$ by Theorem 1.1.6. However, the second scenario cannot occur since $g(x_0) > 0$ for any point $x_0 \in L \cap \overset{\circ}{K}$. According to Theorem 1.1.6, $\bar{f}(x) = \lambda g(x),\ \lambda \neq 0$ on L. Since $\bar{f}(x)$ and $g(x)$ have the same sign on $K \cap L$

and λ must be positive, $f(x) = \lambda g(x)$ is the required extension of the functional $\bar{f}(x)$. □

Definition 1.1.22 Let K be a cone with apex 0 in a topological linear space E. The dual cone $K^* \subset E^*$ is such a set that

$$K^* = \{f \mid f(x) \geq 0,\ x \in K\}.$$

Definition 1.1.22 tells us that K^* consists of all continuous nonnegative linear functionals on K. It is obvious that the dual cone K^* is a sharp convex cone with apex 0.

Theorem 1.1.23 *Let K be a cone. Then its dual* $K^* = \left(\overline{K}\right)^* = \left(\text{conv}\left(\overline{K}\right)\right)^*$ *where* $\overline{K}$ *is the weak closure of K.*

Actually, if $f(x) \geq 0$ on K, then since $f(x)$ is linear, it is also non-negative on $\text{conv}(K)$. Furthermore, due to its continuity, it remains non-negative on the weak closure $\text{conv}\left(\overline{K}\right)$.

Moreover, in the subsequent discussion, we shall refer to K_1 and K_2 as the cones having the origin, 0, as their apex in a locally convex space. Theorem 1.1.24 that follows is derived from the definition of the dual cone.

Theorem 1.1.24 *If* $K_1 \subset K_2$, *then* $K_1^* \supset K_2^*$.

Theorem 1.1.25 *Denote*

$$K^{**} = \left\{x \in E \mid f(x) \geq 0 \textit{ for all } f \in E^*\right\},$$

where E is a linear space. Then $K^{**} = \text{conv}\overline{K}$.

Proof Let $Q = \text{conv}\overline{K}$. By Theorem 1.1.23, $Q^* = K^*$. If $x \in Q$, then for every $f \in Q^*$, we have $f(x) \geq 0$. This, according to the definition of K^{**}, implies that $x \in K^{**}$. Therefore, $Q \subset K^{**}$.

Let $x_0 \in K^{**}, x_0 \notin Q$. According to the second fact generated by Theorem 1.1.20, the point x_0 can be separated from Q. In other words, there exists a functional $f \in E^*$ such that $f(x) \geq 0$ for all $x \in Q$ and $f(x_0) < 0$. This means that f belongs to Q^* and, consequently, to K^* as well. However, if $f(x_0) < 0$ and $f \in K^*$, then $x_0 \notin K^{**}$. This contradicts with our initial assumption. □

Furthermore, let Λ be an arbitrary index set, and let K_α, with $\alpha \in \Lambda$, represent a set of cones. Theorem 1.1.26 is a trivial fact.

Theorem 1.1.26

$$\left(\bigcup_{\alpha \in A} K_\alpha\right)^* = \bigcap_{\alpha \in A} K_\alpha^*.$$

Furthermore, according to the definition, we possess the following information.

Lemma 1.1.27

$$\left(\bigcap_{\alpha\in\Lambda} K_\alpha\right)^* \supset \sum_{\alpha\in\Lambda} K_\alpha^*. \tag{1.1.6}$$

Lemma 1.1.28 *If the cones K_α, $\alpha \in \Lambda$ are both weakly closed and convex, then we can deduce that the dual cone of their intersection is the weakly* closed convex hull of the sum of their dual cones. Namely,*

$$\left(\bigcap_{\alpha\in\Lambda} K_\alpha\right)^* = \overline{\text{conv}}\left(\sum_{\alpha\in\Lambda} K_\alpha^*\right).$$

Proof Let $Q = \overline{\text{conv}}\left(\sum_{\alpha\in\Lambda} K_\alpha^*\right)$. According to Theorem 1.1.23, we have

$$Q^* = \left(\text{conv}\left(\bigcup_{\alpha\in A} K_\alpha^*\right)\right)^* = \left(\bigcup_{\alpha\in A} K_\alpha^*\right)^*.$$

By applying Theorems 1.1.25 and 1.1.26, it follows that

$$\left(\bigcup_{\alpha\in A} K_\alpha^*\right)^* = \bigcap_{\alpha\in\Lambda} K_\alpha^{**} = \bigcap_{\alpha\in\Lambda} \overline{\text{conv} K_\alpha} = \bigcap_{\alpha\in\Lambda} K_\alpha.$$

Therefore, $Q^* = \bigcap_{\alpha\in\Lambda} K_\alpha$, and consequently $Q^{**} = \left(\bigcap_{\alpha\in\Lambda} K_\alpha\right)^*$. However, invoking Theorem 1.1.25 once again, we find that $Q^{**} = Q$, whence $Q = \left(\bigcap_{\alpha\in\Lambda} K_\alpha\right)^*$. □

Corollary 1.1.1 *If the cones K_α, $\alpha \in \Lambda$ are both weakly closed and convex, and if $\sum_{\alpha\in\Lambda} K_\alpha^*$ is weakly* closed, then*

$$\left(\bigcap_{\alpha\in\Lambda} K_\alpha\right)^* = \sum_{\alpha\in\Lambda} K_\alpha^*.$$

Theorem 1.1.29 *Let K be a convex cone with apex at 0, and let L be a subspace such that $\overset{\circ}{K} \cap L \neq \emptyset$. Then, $(K \cap L)^* = K^* + L^*$.*

Proof If $\overline{f} \in (K \cap L)^*$, then $\overline{f}(x) \geq 0$ on $K \cap L$. Furthermore, by Theorem 1.1.21, there exists a function $f \in K^*$ such that $\overline{f}(x) = f(x)$ on L. Therefore, we can write $\overline{f} = f + (\overline{f} - f)$ with $f \in K^*$ and $\overline{f} - f \in L^*$ (as $(f - \overline{f})(x) = 0$ for all $x \in L$). This implies that $\overline{f} \in K^* + L^*$. The inverse inclusion follows from Lemma 1.1.27. □

Furthermore, we proceed to present the subsequent general theory, namely, Theorem 1.1.30.

Theorem 1.1.30 *Let K_i, for $i = 1, 2, \cdots, n$, denote convex open cones, and assume that $\bigcap_{i=1}^{n} K_i \neq \emptyset$. Then,*

$$\left(\bigcap_{i=1}^{n} K_i\right)^* = \sum_{i=1}^{n} K_i^*.$$

Proof Let E represent a linear space. Consider the topological product $\tilde{E} = \overbrace{E \times E \times \cdots\cdots E}^{n}$, where points $\tilde{x} \in \tilde{E}$ are expressed as

$$\tilde{x} = (x_1, x_2, \cdots, x_n), \quad x_i \in E.$$

The set of functionals on $\tilde{E}$ is denoted as $\tilde{E}^* = \overbrace{E^* \times E^* \times \cdots \times E^*}^{n}$ and comprises elements of the form

$$\tilde{f} = (f_1, f_2, \cdots, f_n), \ \tilde{f} \in \tilde{E}^*, \ f_i \in E^*,$$

with $\tilde{f}(\tilde{x}) = \sum_{i=1}^{n} f_i(x_i)$ (see [89]).

In $\tilde{E}$, define the set $K = \{\tilde{x} \mid x_i \in K_i, \ i = 1, 2, \cdots, n\}$. This set constitutes an open convex cone since it's a direct product of open convex cones. Additionally, define the set L in $\tilde{E}$ as $L = \{\tilde{x} \mid x_1 = x_2 = \cdots = x_n = x \in E\}$. L is a linear subspace of $\tilde{E}$. The condition $\bigcap_{i=1}^{n} K_i \neq \emptyset$ ensures that $L \cap K \neq \emptyset$. A linear functional $\overline{f}(\tilde{x})$ is defined on the subspace L as $\overline{f}(\tilde{x}) = f(x)$, where $\tilde{x} = (x, x, \cdots, x) \in L$, $x \in E$, and $f \in \left(\bigcap_{i=1}^{n} K_i\right)^*$. Since $\overline{f}(\tilde{x}) = f(x)$, if $\tilde{x} = (x, x, \cdots, x) \in L \cap K$, then $x \in \bigcap_{i=1}^{n} K_i$ and $f \in \left(\bigcap_{i=1}^{n} K_i\right)^*$. Moreover, $\overline{f}$ is non-negative on $L \cap K$. Applying Theorem 1.1.21, there exists a non-negative function $\tilde{f} \in \tilde{E}^*$ on K that matches $\overline{f}$ on L. Let $\tilde{f} = (f_1, f_2, \cdots, f_n)$. On the subspace L, we have

$$\tilde{f}(\tilde{x}) = \sum_{i=1}^{n} f_i(x) = \overline{f}(\tilde{x}) = f(x).$$

Therefore, $f = \sum_{i=1}^{n} f_i$, where $f \in \left(\bigcap_{i=1}^{n} K_i\right)^*$. Since $\tilde{f}(\tilde{x}) \geq 0$ for $\tilde{x} \in K$, it follows that $\sum_{i=1}^{n} f_i(x_i) \geq 0$ for $x_i \in K_i$, $i = 1, 2, \cdots, n$. This implies that $f_i(x_i) \geq 0$ for $x_i \in K_i$, i.e., $f_i \in K_i^*$, $i = 1, 2, \cdots, n$. Hence, we conclude that if $f \in \left(\bigcap_{i=1}^{n} K_i\right)^*$, then there exist $f_i \in K_i^*$ such that $f = \sum_{i=1}^{n} f_i$. In other words, $\left(\bigcap_{i=1}^{n} K_i\right)^* \subset \sum_{i=1}^{n} K_i^*$. The inverse inclusion is derived from Lemma 1.1.27. □

Theorem 1.1.31 *In a topological linear space E, let K_i for $i = 1, 2, \cdots, n, n+1$ be convex cones with apex 0, where K_i for $i = 1, 2, \cdots, n$ are open. Then the intersection $\bigcap_{i=1}^{n+1} K_i = \emptyset$ if and only if there exist linear functionals $f_i \in K_i^*$ for $i = 1, 2, \cdots, n+1$, not all zero, such that*

$$f_1 + f_2 + \cdots + f_n + f_{n+1} = 0.$$

Proof *(*Necessity*)* Let $K = \bigcap_{i=1}^{n} K_i \neq \emptyset$ and $K_{n+1} \cap K = \emptyset$. Since K is open, the combination of Theorems 1.1.18 and 1.1.20 guarantees the existence of a non-zero $f \in E^*$ such that $f(x) \geq 0$ for all $x \in K$ and $f(x) \leq 0$ for all $x \in K_{n+1}$. Here, E^* represents the dual space of E. According to Theorem 1.1.30, since $f \in K^*$, we can express f as

$$f = \sum_{i=1}^{n} f_i, \quad f_i \in K_i^*, \quad i = 1, 2, \cdots, n.$$

By setting $f_{n+1} = -f_n$, we obtain $f_{n+1} \in K_{n+1}^*$ with $f_{n+1} \neq 0$ and $\sum_{i=1}^{n+1} f_i = 0$.

On the other hand, if $\bigcap_{i=1}^{n} K_i = \emptyset$, then there exists an integer s such that $1 \leq s < n$ and

$$K = \bigcap_{i=1}^{s} K_i \neq \emptyset, \quad \bigcap_{i=1}^{s+1} K_i = \emptyset.$$

By applying the previous result with n replaced by s, we obtain functionals $f_i \in K_i^*$ for $i = 1, 2, \cdots, s+1$ with $f_{s+1} \neq 0$ such that $\sum_{i=1}^{s+1} f_i = 0$. By setting $f_{s+2} = f_{s+3} = \cdots = f_{n+1} = 0$, we obtain the desired functionals $f_1, f_2, \cdots, f_{n+1}$.

(Sufficiency). Let $\sum_{i=1}^{n+1} f_i = 0$ with $f_i \neq 0$ and $\bigcap_{i=1}^{n+1} K_i \neq \emptyset$. There exists $x_0 \in K_i$ for $i = 1, 2, \cdots, n+1$. Now, for some $1 \leq j \leq n$, there must be a non-zero functional f_j. If not, we would have $f_{n+1} = -\sum_{i=1}^{n} f_i = 0$, which is a contradiction. Since K_j is open, it must satisfy $f_j(x_0) > 0$. However, this leads to the following contradiction:

$$0 = \sum_{i=1}^{n+1} f_i(x_0) \geq f_j(x_0) > 0.$$

Therefore, the theorem is proven. □

1.2 Semigroup Theory

The continuous semigroup of bounded linear operators plays a crucial role in the theoretical framework of functional analysis. It holds a pivotal position in elucidating the solutions of linear ODEs and PDEs.

Let $T > 0$. We consider the following Cauchy problem for linear ODE over the interval $[0, T]$,

$$\begin{cases} \dfrac{d}{dt}x(t) = ax(t), \\ x(0) = x_0, \end{cases} \tag{1.2.1}$$

where $a \in \mathbb{R}$ with $a \neq 0$, $x(\cdot) \in \mathbb{R}$, and x_0 is the initial value. Obviously, the solution to (1.2.1) is given by $x(t) = x_0 e^{at}$.

Furthermore, if $x(\cdot) \in \mathbb{R}^n$, then Eq. (1.2.1) can be rewritten as the following linear ODE:

$$\begin{cases} \dfrac{d}{dt}x(t) = Ax(t), \\ x(0) = x_0, \end{cases} \tag{1.2.2}$$

where the matrix $A \in \mathbb{R}^{n\times n}$. Denote

$$e^{tA} = \sum_{k=0}^{\infty} \frac{t^k A^k}{k!}.$$

It is similar to Eq. (1.2.1) that $x(t) = x_0 e^{tA}$ is the solution to Eq. (1.2.2).

Generally, in Eq. (1.2.2), if A is a closed operator acting on a Banach space X and $x_0 \in X$, we can establish an abstract Cauchy problem. Based on the preceding analysis of Eqs. (1.2.1) and (1.2.2), we anticipate that this problem admits a solution $x(t) = x_0 e^{tA}$. This gives rise to the exponential of tA for the operator A. The semigroup approach is generally used to deal with linear PDEs as ODE on an infinite-dimensional function space.

For instance, we are concerned with the following initial-boundary value problem for the heat equation on the spatial interval $[0, 1] \subset \mathbb{R}$ and times $t \geq 0$ [237],

$$\begin{cases} \frac{\partial}{\partial t} u(t, x) = \frac{\partial^2}{\partial x^2} u(t, x), \ x \in (0, 1), \ t > 0, \\ u(t, 0) = u(t, 1) = 0, \ t > 0, \\ u(0, x) = u_0(x), \ x \in (0, 1). \end{cases} \tag{1.2.3}$$

Let $X = L^2(0, 1; \mathbb{R})$ and define A as the second derivative operator with domain

$$D(A) = \left\{ u \in H^2(0, 1; \mathbb{R}) \mid u(0) = u(1) = 0 \right\}.$$

Then, Eq. (1.2.3) can be interpreted as an abstract Cauchy problem on the space X,

$$\begin{cases} \frac{d}{dt} u(t) = Au(t), \\ u(0) = u_0. \end{cases} \tag{1.2.4}$$

Again, we conjecture that the solution to this problem is $u(t) = e^{tA} u_0$.

In fact, as a function of t, e^{tA} represents the semigroup of operators from X to itself, mapping the initial state u_0 at time $t = 0$ to the state $u(t) = e^{tA} u_0$ at time t. The operator A is referred to as the infinitesimal generator of this semigroup.

Next, we delve into the concept of a continuous semigroup of linear bounded operators and explore its significance in the realm of PDEs. For a comprehensive and rigorous treatment of this topic, we recommend the monograph [181].

Definition 1.2.1 Let X be a Banach space. A one-parameter family $T(t)$ of bounded linear operators from X into X, parameterized by $t \in [0, \infty)$, is considered a semigroup of bounded linear operators if it satisfies the following two conditions:

(i) $T(0)$ is the identity operator on X, denoted by I.
(ii) $T(t + s) = T(t)T(s)$ holds for any $t, s \in [0, \infty)$.

A semigroup of linear bounded operators $T(t)$ is uniformly continuous if it satisfies the condition

$$\lim_{t \to 0+} \|T(t) - I\| = 0.$$

When this holds, it necessarily implies that

$$\lim_{s \to t} \|T(s) - T(t)\| = 0.$$

The linear operator A serves as the infinitesimal generator of the continuous semigroup $T(t)$ of linear bounded operators. Its domain $D(A)$ is given by

$$D(A) = \left\{ x \in X \;\middle|\; \lim_{t \to 0+} \frac{T(t)x - x}{t} \text{ exists} \right\},$$

and for $x \in D(A)$, we have

$$Ax = \lim_{t \to 0+} \frac{T(t)x - x}{t} = \left. \frac{d^+T(t)x}{dt} \right|_{t=0}.$$

According to Definition 1.2.1, a continuous semigroup $T(t)$ of linear bounded operators possesses a unique infinitesimal generator.

Theorem 1.2.2 *Let A be a linear operator. Then, A serves as the infinitesimal generator of a uniformly continuous semigroup of linear bounded operators if and only if A is itself bounded.*

Proof Let A be a bounded linear operator on X, and define

$$T(t) = e^{tA} = \sum_{n=0}^{\infty} \frac{(tA)^n}{n!}. \tag{1.2.5}$$

For all $t \geq 0$, the right-hand side of (1.2.5) converges in norm, thus defining a bounded linear operator $T(t)$. Clearly, $T(0) = I$. A simple power series calculation demonstrates that $T(t + s) = T(t)T(s)$. By estimating the power series, we obtain

$$\|T(t) - I\| \leq t\|A\|e^{t\|A\|},$$

and

$$\left\| \frac{T(t) - I}{t} - A \right\| \leq \|A\| \cdot \max_{0 \leq s \leq t} \|T(s) - I\|.$$

Consequently, $T(t)$ forms a uniformly continuous semigroup of bounded linear operators on X, with A as its infinitesimal generator.

Let $T(t)$ represent a uniformly continuous semigroup of linear bounded operators acting on X. Choose a value of $\rho > 0$ small enough such that

$$\left\| I - \rho^{-1} \int_0^{\rho} T(s)\, \mathrm{d}s \right\| < 1.$$

This ensures that $\rho^{-1}\int_0^\rho T(s)\,\mathrm{d}s$ is invertible, and thus $\int_0^\rho T(s)\,\mathrm{d}s$ is also invertible. Now,

$$\begin{aligned} h^{-1}(T(h)-I)\int_0^\rho T(s)\,\mathrm{d}s &= h^{-1}\left(\int_0^\rho T(s+h)\,\mathrm{d}s - \int_0^\rho T(s)\,\mathrm{d}s\right) \\ &= h^{-1}\left(\int_\rho^{\rho+h} T(s)\,\mathrm{d}s - \int_0^h T(s)\,\mathrm{d}s\right). \end{aligned}$$

This implies that

$$h^{-1}(T(h)-I) = \left(h^{-1}\int_\rho^{\rho+h} T(s)\,\mathrm{d}s - h^{-1}\int_0^h T(s)\,\mathrm{d}s\right)\left(\int_0^\rho T(s)\,\mathrm{d}s\right)^{-1}. \tag{1.2.6}$$

Taking the limit as $h \to 0+$ in (1.2.6) results in $h^{-1}(T(h)-I)$ converging strongly in norm to the bounded linear operator $(T(\rho)-I)\left(\int_0^\rho T(s)\,\mathrm{d}s\right)^{-1}$. This operator serves as the infinitesimal generator of $T(t)$. □

Theorem 1.2.3 *Let $T_1(t)$ and $T_2(t)$ be uniformly continuous semigroups of linear bounded operators. If*

$$\lim_{t\to 0+}\frac{T_1(t)-I}{t} = A = \lim_{t\to 0+}\frac{T_2(t)-I}{t}, \tag{1.2.7}$$

then it follows that $T_1(t) = T_2(t)$ for all $t \geq 0$.

Proof We demonstrate that $T_2(t) = T_1(t)$ for $0 \leq t \leq T$ with a given $T > 0$. Since $t \to \|T_1(t)\|$ and $t \to \|T_2(t)\|$ are continuous, there exists a constant C such that $\|T_1(t)\|\|T_2(s)\| \leq C$ for $0 \leq s, t \leq T$. For any given $\varepsilon > 0$, (1.2.7) guarantees the existence of a $\delta > 0$ such that

$$h^{-1}\|T_1(h) - T_2(h)\| < \frac{\varepsilon}{TC} \tag{1.2.8}$$

holds for $0 \leq h \leq \delta$. Now, let $0 \leq t \leq T$ and select $n \geq 1$ such that $\frac{t}{n} < \delta$. Applying the semigroup property and (1.2.8), we obtain

$$\begin{aligned} &\|T_1(t) - T_2(t)\| = \left\|T_1\left(n\frac{t}{n}\right) - T_2\left(n\frac{t}{n}\right)\right\| \\ &\leq \sum_{k=0}^{n-1}\left\|T_1\left((n-k)\frac{t}{n}\right)T_2\left(\frac{kt}{n}\right) - T_1\left((n-k-1)\frac{t}{n}\right)T_2\left(\frac{(k+1)t}{n}\right)\right\| \\ &\leq \sum_{k=0}^{n-1}\left\|T_1\left((n-k-1)\frac{t}{n}\right)\right\|\left\|T_1\left(\frac{t}{n}\right) - T_2\left(\frac{t}{n}\right)\right\|\left\|T_2\left(\frac{kt}{n}\right)\right\| \\ &\leq Cn\frac{\varepsilon}{TC}\frac{t}{n} \leq \varepsilon. \end{aligned} \tag{1.2.9}$$

Since $\varepsilon > 0$ is arbitrary, we conclude that $T_1(t) = T_2(t)$ for $0 \le t \le T$. □

Consequently, for any bounded linear operator A, there exists a unique, uniformly continuous semigroup $T(t)$ of linear bounded operators, of which A serves as the infinitesimal generator.

Theorem 1.2.4 *If $T(t)$ is a uniformly continuous semigroup of linear bounded operators, then the following properties hold:*

(i) *There exists a constant $w \ge 0$ such that $\|T(t)\| \le e^{\omega t}$ for all $t \ge 0$.*
(ii) *There exists a unique bounded linear operator A that produces $T(t) = e^{tA}$.*
(iii) *The operator A mentioned in* (ii) *serves as the infinitesimal generator of $T(t)$.*
(iv) *The mapping $t \to T(t)$ is differentiable in norm, and its derivative is given by*

$$\frac{dT(t)}{dt} = AT(t) = T(t)A.$$

Proof All the assertions in Theorem 1.2.4 can be readily derived from part (ii). To establish (ii), it is crucial to observe that A, as a bounded linear operator, generates $T(t)$ and, consequently, also generates e^{tA} according to (1.2.5). Consequently, by virtue of Theorem 1.2.3, we conclude that $T(t) = e^{tA}$. □

Then, we introduce a new definition of a semigroup, namely, the strongly continuous semigroup of bounded linear operators, henceforth referred to as the C_0-semigroup.

Definition 1.2.5 Let X be a Banach space. For $t \in [0, \infty)$, the semigroup $T(t)$ of bounded linear operators acting on X is a strongly continuous semigroup of bounded linear operators if it satisfies the condition

$$\lim_{t \to 0+} T(t)x = x, \quad \forall\, x \in X. \tag{1.2.10}$$

Indeed, the C_0-semigroup serves as a generalization of the exponential function. Similar to how exponential functions provide solutions to scalar linear ODEs with constant coefficients, C_0-semigroups generate solutions to linear ODEs with constant coefficients in Banach spaces. These differential equations in Banach spaces often arise from various mathematical models, including delay differential equations and PDEs [238].

Theorem 1.2.6 *Let $T(t)$ be a C_0-semigroup. There exist constants $\omega \ge 0$ and $M \ge 1$ such that for all $t \in [0, \infty)$,*

$$\|T(t)\| \le Me^{\omega t}.$$

Additionally, for every x in the Banach space X, the mapping $t \to T(t)x$ defines a continuous function from the non-negative real line $\mathbb{R}_0^+$ into X.

Proof First, we show that there exists an $\eta > 0$ such that $\|T(t)\|$ is bounded for $0 \le t \le \eta$. Suppose, for contradiction, that such an η does not exist. Then, there exists a sequence $\{t_n\}$ with $t_n \ge 0$, $\lim_{n\to\infty} t_n = 0$, and $\|T(t_n)\| \ge n$. However, by the uniform boundedness theorem, there exists an $x \in X$ such that $\|T(t_n)x\|$ is unbounded, which contradicts with (1.2.10). Hence, for $0 \le t \le \eta$, we have $\|T(t)\| \le M$. Since $\|T(0)\| = 1$ and $M \ge 1$, we can define $\omega = \eta^{-1} \log M \ge 0$. Now, given $t \ge 0$, we can write $t = n\eta + \delta$ where $0 \le \delta < \eta$. Using the semigroup property, we obtain

$$\|T(t)\| = \left\|T(\delta)T(\eta)^n\right\| \le M^{n+1} \le MM^{\frac{t}{\eta}} = Me^{\omega t}.$$

Furthermore, for $t, h \ge 0$, the continuity of $t \to T(t)x$ can be demonstrated by

$$\|T(t+h)x - T(t)x\| \le \|T(t)\|\|T(h)x - x\| \le Me^{\omega t}\|T(h)x - x\|.$$

Similarly, for $t \ge h \ge 0$, we observe continuity through

$$\|T(t-h)x - T(t)x\| \le \|T(t-h)\|\|x - T(h)x\| \le Me^{\omega t}\|x - T(h)x\|.$$

□

In Theorem 1.2.6, if $\omega = 0$, $T(t)$ is referred to as a uniformly bounded semigroup. Furthermore, if $M = 1$ and $\omega = 0$, it is called a C_0-semigroup of contractions.

Now, consider a Banach space X and let $T(t)$ be a C_0-semigroup with A as its infinitesimal generator. Since $t \to T(t)x$ is continuous, we can define

$$T(t) = \int_0^t T(s)x \, \mathrm{d}s$$

which is differentiable, and its derivative is given by $\frac{d}{dt}F(t) = T(t)x$. Specially,

$$\frac{d}{dt}F(t) = \lim_{h\to 0+} \frac{F(t+h) - F(t)}{h} = \lim_{h\to 0+} \frac{1}{h}\int_t^{t+h} T(s)x \, \mathrm{d}s.$$

From this, we obtain

$$\lim_{h\to 0+} \frac{1}{h}\int_t^{t+h} T(s)x \, \mathrm{d}s = T(t)x, \tag{1.2.11}$$

for any $x \in X$ and $t \ge 0$.

Furthermore, for any $x \in X$ and $t \geq 0$, we have

$$\int_0^t T(s)x\, ds \in D(A), \quad A\int_0^t T(s)x\, ds = T(t)x - x. \tag{1.2.12}$$

Actually, since for $h > 0$,

$$\begin{aligned}\lim_{h\to 0+} \frac{T(h)-I}{h}\int_0^t T(s)x\, ds &= \lim_{h\to 0+}\frac{1}{h}\int_t^{t+h}[T(h+s)x - T(s)x]\, ds\\ &= \lim_{h\to 0+}\frac{1}{h}\int_t^{t+h} T(s)x\, ds - \lim_{h\to 0+}\frac{1}{h}\int_0^h T(s)x\, ds,\end{aligned}$$

which demonstrates the validity of (1.2.12).

For any $x \in D(A)$, we note that $T(t)x \in D(A)$. Specially, we observe that

$$\frac{d}{dt}T(t)x = AT(t)x = T(t)Ax, \tag{1.2.13}$$

and further,

$$T(t)x - T(s)x = \int_s^t T(\tau)Ax\, d\tau = \int_s^t AT(\tau)x\, d\tau. \tag{1.2.14}$$

Moreover, if A serves as the infinitesimal generator of a C_0-semigroup $T(t)$, then $D(A)$ is dense in X, and A is a closed linear operator, as stated in [181].

Theorem 1.2.7 *Let $T(t)$ and $S(t)$ be C_0-semigroups with their corresponding infinitesimal generators be A and B. If $A = B$, then $T(t) = S(t)$ for all $t \in [0, \infty)$.*

Proof Let $x \in D(A) = D(B)$. Equation (1.2.13) implies that the function $s \to T(t-s)S(s)x$ is differentiable, and

$$\begin{aligned}\frac{d}{ds}T(t-s)S(s)x &= -AT(t-s)S(s)x + T(t-s)BS(s)x\\ &= -T(t-s)AS(s)x + T(t-s)BS(s)x = 0.\end{aligned}$$

Therefore, the function $s \to T(t-s)S(s)x$ is constant. By setting $s = 0$ and $s = t$, we obtain $T(t)x = S(t)x$ for every $x \in D(A)$. Additionally, since A generates a C_0-semigroup, $D(A)$ is dense in X. Since $T(t)$ and $S(t)$ are bounded, it follows that $T(t)x = S(t)x$ for all $x \in X$. □

Theorem 1.2.8 *Let $T(t)$ be a C_0-semigroup with the property that $\|T(t)\| \leq M$ for all $t \geq 0$, and let A be its infinitesimal generator. If $x \in D(A^2)$, then it follows that*

$$\|Ax\|^2 \leq 4M^2 \left\|A^2x\right\| \|x\|. \tag{1.2.15}$$

Proof Using (1.2.14), for $x \in D(A^2)$,

$$T(t)x - x = tAx + \int_0^t (t-s)T(s)A^2x \, \mathrm{d}s.$$

From this,

$$\begin{aligned} \|Ax\| &\le t^{-1}(\|T(t)x\| + \|x\|) + t^{-1}\int_0^t (t-s)\left\|T(s)A^2x\right\| \mathrm{d}s \\ &\le \frac{2M}{t}\|x\| + \frac{Mt}{2}\left\|A^2x\right\|, \end{aligned} \tag{1.2.16}$$

where we have used the fact that $\|T(t)\| \le M$ for all $t \ge 0$. Assuming $M \ge 1$ (which implies $\|T(0)\| = 1$ since $T(0)$ is the identity operator), we proceed with the analysis. If $A^2x = 0$, then by (1.2.16), $Ax = 0$, and thus (1.2.15) is satisfied. When $A^2x \neq 0$, we substitute $t = 2\|x\|^{\frac{1}{2}}\left\|A^2x\right\|^{-\frac{1}{2}}$ into (1.2.16) to obtain (1.2.15). □

Let A be a linear (not necessarily bounded) operator acting on the Banach space X. The resolvent set of A, denoted by $\rho(A)$, comprises all complex numbers λ such that $\lambda I - A$ is invertible, i.e., $(\lambda I - A)^{-1}$ exists and is a bounded linear operator in X. The resolvent of A is defined as the family of bounded linear operators $R(\lambda : A) = (\lambda I - A)^{-1}$ for all $\lambda \in \rho(A)$. For a C_0-semigroup of contractions, the following Hille-Yosida theorem plays a crucial role in characterizing the generation of this C_0-semigroup.

Theorem 1.2.9 (Hille-Yosida Theorem) *Let X be a Banach space. A linear (unbounded) operator A is the infinitesimal generator of a C_0-semigroup of contractions $T(t)$, $t \ge 0$ if and only if*

(i) *A is closed and $\overline{D(A)} = X$.*
(ii) *The resolvent set $\rho(A)$ of A contains $\mathbb{R}^+$ and*

$$\|R(\lambda : A)\| \le \frac{1}{\lambda}, \quad \forall\, \lambda > 0. \tag{1.2.17}$$

Proof (Necessity) We observe that if A is the infinitesimal generator of a C_0-semigroup $T(t)$, then $D(A)$ is dense in X. Consequently, $\overline{D(A)} = X$ and A is a closed linear operator, as established in [181]. For $\lambda > 0$ and $x \in X$, we define

$$R(\lambda)x = \int_0^\infty e^{-\lambda t}T(t)x \, \mathrm{d}t. \tag{1.2.18}$$

Since the mapping $t \to T(t)x$ is continuous and uniformly bounded, the integral exists as an improper Riemann integral and defines a bounded linear operator $R(\lambda)$. Therefore,

$$\|R(\lambda)x\| \leq \int_0^\infty e^{-\lambda t}\|T(t)x\| \, dt \leq \frac{1}{\lambda}\|x\|. \tag{1.2.19}$$

For $h > 0$,

$$\begin{aligned}\frac{T(h)-I}{h}R(\lambda)x &= \frac{1}{h}\int_0^\infty e^{-\lambda t}(T(t+h)x - T(t)x) \, dt \\ &= \frac{e^{\lambda h}-1}{h}\int_0^\infty e^{-\lambda t}T(t)x \, dt - \frac{e^{\lambda h}}{h}\int_0^h e^{-\lambda t}T(t)x \, dt.\end{aligned} \tag{1.2.20}$$

As $h \to 0+$, the right-side of Eq. (1.2.20) converges to $\lambda R(\lambda)x - x$. This implies that for all $x \in X$ and $\lambda > 0$, $R(\lambda)x \in D(A)$ and $AR(\lambda) = \lambda R(\lambda) - I$, or, equivalently,

$$(\lambda I - A)R(\lambda) = I. \tag{1.2.21}$$

Whenever $x \in D(A)$,

$$\begin{aligned}R(\lambda)Ax &= \int_0^\infty e^{-\lambda t}T(t)Ax \, dt = \int_0^\infty e^{-\lambda t}AT(t)x \, dt \\ &= A\left(\int_0^\infty e^{-\lambda t}T(t)x \, dt\right) = AR(\lambda)x,\end{aligned} \tag{1.2.22}$$

where we have used Eq. (1.2.13) and the closedness of A. By applying (1.2.21) and (1.2.22), we conclude that for $x \in D(A)$,

$$R(\lambda)(\lambda I - A)x = x. \tag{1.2.23}$$

Therefore, $R(\lambda)$ is the inverse of $\lambda I - A$ for all $\lambda > 0$ and satisfies the estimate (1.2.17). This demonstrates the necessity of conditions (i) and (ii).

(Sufficiency). Initially, we present three useful results.

Firstly, let A satisfy conditions (i) and (ii) and denote $R(\lambda : A) = (\lambda I - A)^{-1}$. We assert that for any $x \in X$,

$$\lim_{\lambda \to \infty} \lambda R(\lambda : A)x = x. \tag{1.2.24}$$

In fact, for $x \in D(A)$,

$$\begin{aligned}\|\lambda R(\lambda : A)x - x\| &= \|AR(\lambda : A)x\| = \|R(\lambda : A)Ax\| \\ &\leq \frac{1}{\lambda}\|Ax\| \to 0, \qquad \text{as } \lambda \to \infty.\end{aligned}$$

Furthermore, since $D(A)$ is dense in X and $\|\lambda R(\lambda : A)\| \leq 1$, $\lambda R(\lambda : A)x \to x$ as $\lambda \to \infty$ for every $x \in X$.

We then define the Yosida approximation of A by

$$A_\lambda = \lambda A R(\lambda : A) = \lambda^2 R(\lambda : A) - \lambda I, \tag{1.2.25}$$

for every $\lambda > 0$. In this context, if A satisfies conditions (i) and (ii), then for $x \in D(A)$,

$$\lim_{\lambda \to \infty} A_\lambda x = Ax. \tag{1.2.26}$$

Actually, for $x \in D(A)$, we observe from (1.2.24) and (1.2.25) that

$$\lim_{\lambda \to \infty} A_\lambda x = \lim_{\lambda \to \infty} \lambda R(\lambda : A)Ax = Ax.$$

Thirdly, let A satisfy conditions (i) and (ii). If A_λ is the Yosida approximation of A, then A_λ is the infinitesimal generator of a uniformly continuous semigroup e^{tA_λ} of contractions. Moreover, for all $x \in X$, $\lambda, \mu > 0$,

$$\left\| e^{tA_\lambda} x - e^{tA_\mu} x \right\| \leq t \|A_\lambda x - A_\mu x\|. \tag{1.2.27}$$

We now prove the third result. It is apparent from (1.2.25) that A_λ is a bounded linear operator, and consequently, A is the infinitesimal generator of a uniformly continuous semigroup e^{tA_λ} of bounded linear operators (refer to, for instance, Theorem 1.2.2). Additionally, e^{tA_λ} is a semigroup of contractions because

$$\left\| e^{tA_\lambda} \right\| = e^{-t\lambda} \left\| e^{t\lambda^2 R(\lambda:A)} \right\| \leq e^{-t\lambda} e^{t\lambda^2 \|R(\lambda:A)\|} \leq 1. \tag{1.2.28}$$

Since e^{tA_λ}, e^{tA_μ}, A_λ and A_μ are commutable by definition, a simple calculations shows that

$$\left\| e^{tA_\lambda} x - e^{tA_\mu} x \right\| = \left\| \int_0^1 \frac{d}{ds} \left(e^{tsA_\lambda} e^{t(1-s)A_\mu} x \right) \mathrm{d}s \right\|$$

$$\leq \int_0^1 t \left\| e^{tsA_\lambda} e^{t(1-s)A_\mu} \left(A_\lambda x - A_\mu x \right) \right\| \mathrm{d}s \leq t \left\| A_\lambda x - A_\mu x \right\|,$$

which is equivalent to (1.2.27).

To establish the sufficiency, let $x \in D(A)$. Then,

$$\|e^{tA_\lambda} x - e^{tA_\mu} x\| \leq t\|A_\lambda x - A_\mu x\| \leq t\|A_\lambda x - Ax\| + t\|Ax - A_\mu x\|. \tag{1.2.29}$$

Using (1.2.26) and (1.2.29), we deduce that for $x \in D(A)$, $e^{tA_\lambda} x$ converges uniformly over bounded intervals as $\lambda \to \infty$. Since $D(A)$ is dense in X and

$\|e^{tA_\lambda}\| \le 1$, we conclude that

$$\lim_{\lambda\to\infty} e^{tA_\lambda}x = T(t)x, \tag{1.2.30}$$

for every $x \in X$. Moreover, this limit is again uniformly over bounded intervals. It is apparent from (1.2.30) that the limit $T(t)$ inherits the semigroup property, $T(0) = I$, and $\|T(t)\| \le 1$. Additionally, for $t \ge 0$, the mapping $t \to T(t)x$ is continuous as a uniform limit of the continuous functions $t \to e^{tA_\lambda}x$. Therefore, $T(t)$ is a C_0-semigroup of contractions on X. Finally, to complete the proof, it suffices to show that A is the infinitesimal generator of $T(t)$. Let $x \in D(A)$. Then, using (1.2.30) and Theorem 2.4 in [181], we can observe that

$$\begin{aligned} T(t)x - x &= \lim_{\lambda\to\infty}\left(e^{tA_\lambda}x - x\right) \\ &= \lim_{\lambda\to\infty}\int_0^t e^{sA_\lambda}A_\lambda x \, ds = \int_0^t T(s)Ax \, ds. \end{aligned} \tag{1.2.31}$$

The final equality is a consequence of the uniform convergence of $e^{tA_\lambda}A_\lambda x$ to $T(t)Ax$ on bounded intervals. Let B denote the infinitesimal generator of $T(t)$, and let $x \in D(A)$. Dividing (1.2.31) by $t > 0$ and taking the limit as $t \to 0+$ indicates that $x \in D(B)$ and $Bx = Ax$. Hence, $B \supseteq A$. Since B is assumed to be the infinitesimal generator of $T(t)$, it follows from the necessary conditions that $1 \in \rho(B)$. Furthermore, from (ii), we have $1 \in \rho(A)$ as well. Since $B \supseteq A$, it follows that $(I - B)D(A) = (I - A)D(A) = X$. Consequently, $D(B) = (I - B)^{-1}X = D(A)$, and thus, $A = B$. □

For every $\lambda > 0$, the Yosida approximation of the operator A is defined as

$$A_\lambda = \lambda AR(\lambda : A) = \lambda^2 R(\lambda : A) - \lambda I. \tag{1.2.32}$$

Furthermore, if operator A fulfills conditions (i) and (ii) of Theorem 1.2.9, then A_λ approximates A in the sense that

$$\lim_{\lambda\to\infty} A_\lambda x = Ax, \quad \forall\, x \in D(A).$$

Corollary 1.2.1 *Let X be a Banach space, and let $T(t)$ be a C_0-semigroup of contractions with the infinitesimal generator A. For the Yosida approximation of A,*

$$T(t)x = \lim_{\lambda\to\infty} e^{tA_\lambda}x, \quad \forall\, x \in X. \tag{1.2.33}$$

Moreover, the resolvent set of A encompasses the open right half-plane, specially, $\rho(A) \supseteq \{\lambda \mid \mathrm{Re}\lambda > 0\}$ *and for such* λ,

$$\|R(\lambda : A)\| \leq \frac{1}{\mathrm{Re}\lambda}. \tag{1.2.34}$$

Proof Based on the proof of Theorem 1.2.9, it is deduced that the right-hand side of (1.2.33) characterizes a C_0-semigroup of contractions $S(t)$ whose infinitesimal generator is A. Combining this with Theorem 1.2.7, we arrive at the conclusion that $T(t) = S(t)$.

Therefore, the operator $R(\lambda)x = \int_0^\infty e^{-\lambda t} T(t)x \, dt$ is well defined for λ satisfying $\mathrm{Re}\lambda > 0$. The proof of the necessary part of Theorem 1.2.9 demonstrates that $R(\lambda) = (\lambda I - A)^{-1}$, indicating that $\rho(A) \supseteq \{\lambda \mid \mathrm{Re}\lambda > 0\}$. The estimate for $R(\lambda)$ in Eq. (1.2.34) can be readily derived. □

Let $T(t)$ denote a C_0-semigroup such that $\|T(t)\| \leq e^{\omega t}$ for some $\omega \geq 0$. We consider the transformation $S(t) = e^{-\omega t} T(t)$, which is evidently a C_0-semigroup of contractions. If A is the infinitesimal generator of $T(t)$, then $A - \omega I$ serves as the infinitesimal generator of $S(t)$. Conversely, if A is the infinitesimal generator of a C_0-semigroup of contractions $S(t)$, then $A + \omega I$ becomes the infinitesimal generator of a C_0-semigroup $T(t)$ that satisfies $\|T(t)\| \leq e^{\omega t}$ and is expressed as $T(t) = e^{\omega t} S(t)$. These observations enable us to characterize the infinitesimal generators of C_0-semigroups that adhere to the criterion $\|T(t)\| \leq e^{\omega t}$.

Corollary 1.2.2 *Let X be a Banach space. A linear operator A is the infinitesimal generator of a* C_0*-semigroup* $T(t)$ *such that* $\|T(t)\| \leq e^{\omega t}$ *if and only if*

(i) *A is closed and* $\overline{D(A)} = X$.
(ii) *The resolvent set* $\rho(A)$ *of A contains the ray* $\{\lambda \mid \mathrm{Im}\lambda = 0, \lambda > \omega\}$, *for which*

$$\|R(\lambda : A)\| \leq \frac{1}{\lambda - \omega}.$$

Furthermore, in addition to the Hille-Yosida theorem (Theorem 1.2.9), which characterizes the infinitesimal generator of a C_0-semigroup of contractions, there exists another fundamental result known as the Lumer-Phillips theorem. We will present this theorem later.

Let X be a Banach space and X^* its dual. We denote the value of $x^* \in X^*$ at $x \in X$ by $\langle x^*, x \rangle$ or $\langle x, x^* \rangle$. If A is a linear operator in X, its numerical range $S(A)$ is defined as the set

$$\begin{aligned} S(A) = \big\{ \langle x^*, Ax \rangle \mid x \in D(A), \ \|x\| = 1, \\ x^* \in X^*, \ \|x^*\| = 1, \ \langle x^*, x \rangle = 1 \big\}. \end{aligned} \tag{1.2.35}$$

Additionally, for every $x \in X$, the duality set $F(x) \subseteq X^*$ is given by

$$F(x) = \left\{ x^* \;\middle|\; x^* \in X^*, \langle x^*, x \rangle = \|x\|^2 = \left\|x^*\right\|^2 \right\}.$$

By virtue of the Hahn-Banach theorem (Theorem 1.1.12), $F(x) \neq \emptyset$ for every $x \in X$.

Definition 1.2.10 A linear operator A is dissipative if, for each $x \in D(A)$, there exists an $x^* \in F(x)$ such that $\text{Re}\langle Ax, x^* \rangle \leq 0$.

Theorem 1.2.11 *The linear operator A is dissipative if and only if it satisfies*

$$\|(\lambda I - A)x\| \geq \lambda\|x\|, \quad \forall\, x \in D(A),\ \lambda > 0. \tag{1.2.36}$$

Proof Let A be dissipative, $\lambda > 0$, and $x \in D(A)$. If $x^* \in F(x)$ and $\text{Re}\langle Ax, x^* \rangle \leq 0$, then

$$\|\lambda x - Ax\|\,\|x\| \geq |\langle \lambda x - Ax, x^* \rangle| \geq \text{Re}\langle \lambda x - Ax, x^* \rangle \geq \lambda\|x\|^2.$$

Thus (1.2.36) follows immediately.

Conversely, suppose that $x \in D(A)$ and $\lambda\|x\| \leq \|\lambda x - Ax\|$ for all $\lambda > 0$. For each $\lambda > 0$, let $y_\lambda^* \in F(\lambda x - Ax)$ and define $z_\lambda^* = \frac{y_\lambda^*}{\|y_\lambda^*\|}$ so that $\|z_\lambda^*\| = 1$. Then,

$$\begin{aligned} \lambda\|x\| \leq \|\lambda x - Ax\| &= \langle \lambda x - Ax, z_\lambda^* \rangle \\ &= \lambda \text{Re}\langle x, z_\lambda^* \rangle - \text{Re}\langle Ax, z_\lambda^* \rangle \leq \lambda\|x\| - \text{Re}\langle Ax, z_\lambda^* \rangle, \end{aligned}$$

which implies,

$$\text{Re}\langle Ax, z_\lambda^* \rangle \leq 0, \quad \text{and} \quad \text{Re}\langle x, z_\lambda^* \rangle \geq \|x\| - \frac{1}{\lambda}\|Ax\|. \tag{1.2.37}$$

Since the unit ball of X^* is compact in the weak-star topology of X^*, the net z_λ^*, $\lambda \to 0$, has a weak-star cluster point $z^* \in X^*$ with $\|z^*\| \leq 1$. From (1.2.37), it follows that $\text{Re}\langle Ax, z^* \rangle \leq 0$ and $\text{Re}\langle x, z^* \rangle \geq \|x\|$. However, since $\text{Re}\langle x, z^* \rangle \leq |\langle x, z^* \rangle| \leq \|x\|$, we have $\langle x, z^* \rangle = \|x\|$. Taking $x^* = \|x\| z^*$ gives $x^* \in F(x)$ and $\text{Re}\langle Ax, x^* \rangle \leq 0$. Therefore, for every $x \in D(A)$, there exists an $x^* \in F(x)$ such that $\text{Re}\langle Ax, x^* \rangle \leq 0$, proving that A is dissipative. □

Theorem 1.2.12 (Lumer-Phillips Theorem) *Let A be a linear operator with a dense domain D(A) in a Banach space X.*

(i) *If A is dissipative and there exists a* $\lambda_0 > 0$ *such that the range* $R(\lambda_0 I - A)$ *of* $\lambda_0 I - A$ *is X, then A is the infinitesimal generator of a* C_0*-semigroup of contractions on X.*

(ii) *If A is the infinitesimal generator of a C_0-semigroup of contractions on X, then $R(AI - A) = X$ for all $\lambda > 0$ and A is dissipative. Moreover, for every $x \in D(A)$ and every $x^* \in F(x)$,* $\mathrm{Re}\langle Ax, x^*\rangle \leq 0$.

Proof Given $\lambda > 0$, Theorem 1.2.11 asserts that the dissipativeness of A guarantees

$$\|\lambda x - Ax\| \geq \lambda\|x\| \quad \text{for every } \lambda > 0 \text{ and } x \in D(A). \tag{1.2.38}$$

Since $R(\lambda_0 I - A) = X$, it follows from (1.2.38) with $\lambda = \lambda_0$ that $(\lambda_0 I - A)^{-1}$ is a bounded linear operator, thereby implying it is closed. Consequently, $\lambda_0 I - A$ is also closed, and thus A is closed. If $R(\lambda I - A) = X$ for all $\lambda > 0$, then $\rho(A) \supseteq (0, \infty)$, and $\|R(\lambda : A)\| \leq \lambda^{-1}$ due to (1.2.38). Subsequently, the Hille-Yosida theorem (Theorem 1.2.9) implies that A is the infinitesimal generator of a C_0-semigroup of contractions on X.

To finalize the proof of (i), it is necessary to demonstrate that $R(\lambda I - A) = X$ for all $\lambda > 0$. Consider the set

$$\Lambda = \{\lambda \mid 0 < \lambda < \infty \text{ and } R(\lambda I - A) = X\}.$$

Let $\lambda \in \Lambda$. By Eq. (1.2.38), we have $\lambda \in \rho(A)$. Since $\rho(A)$ is an open set, there exists a neighborhood of λ contained in $\rho(A)$. The intersection of this neighborhood with the real line is a subset of Λ, implying that Λ is also an open set. On the other hand, let $\lambda_n \in \Lambda$ with $\lambda_n \to \lambda > 0$. For any $y \in X$, there exists $x_n \in D(A)$ such that

$$\lambda_n x_n - Ax_n = y. \tag{1.2.39}$$

From (1.2.38), we have $\|x_n\| \leq \lambda_n^{-1}\|y\| \leq C$ for some $C > 0$. Now,

$$\begin{aligned} \lambda_m\|x_n - x_m\| &\leq \|\lambda_m(x_n - x_m) - A(x_n - x_m)\| \\ &= |\lambda_n - \lambda_m|\,\|x_n\| \leq C|\lambda_n - \lambda_m|. \end{aligned} \tag{1.2.40}$$

This shows that $\{x_n\}$ is a Cauchy sequence. Suppose $x_n \to x$. By (1.2.39), $Ax_n \to \lambda x - y$. Since A is closed, $x \in D(A)$ and $\lambda x - Ax = y$. Therefore, $R(\lambda I - A) = X$ and $\lambda \in \Lambda$. This proves that Λ is closed in $(0, \infty)$. Since the assumption $\lambda_0 \in \Lambda$ ensures that $\Lambda \neq \emptyset$, we conclude that $\Lambda = (0, \infty)$. This completes the proof of assertion (i).

If A is the infinitesimal generator of a C_0-semigroup of contractions $T(t)$ on X, then the Hille-Yosida theorem (Theorem 1.2.9) implies that $\rho(A) \supseteq (0, \infty)$, thereby ensuring that $R(\lambda I - A) = X$ for all $\lambda > 0$. In addition, if $x \in D(A)$ and $x^* \in F(x)$, then

$$\left|\langle T(t)x, x^*\rangle\right| \leq \|T(t)x\|\,\|x^*\| \leq \|x\|^2,$$

which implies

$$\operatorname{Re}\left\langle T(t)x - x, x^*\right\rangle = \operatorname{Re}\left\langle T(t)x, x^*\right\rangle - \|x\|^2 \leq 0. \tag{1.2.41}$$

Dividing (1.2.41) by $t > 0$ and passing the limit as $t \to 0+$ yields

$$\operatorname{Re}\left\langle Ax, x^*\right\rangle \leq 0. \tag{1.2.42}$$

This shows that the desired conclusion holds for every $x^* \in F(x)$. □

Recall that a linear operator A mapping its domain $D(A)$ to a Hilbert space $\mathcal{H}$ is said to be closable if there exists a closed extension $B : D(B) \to \mathcal{H}$ such that $D(A) \subset D(B)$.

Theorem 1.2.13 *Let A be a dissipative operator defined in a Banach space X. Then,*

(i) *If $R(\lambda_0 I - A) = X$ for some $\lambda_0 > 0$, $R(\lambda I - A) = X$, $\forall\, \lambda > 0$.*
(ii) *If A is closable, its closure $\overline{A}$ is also dissipative.*
(iii) *If $\overline{D(A)} = X$, then A is closable.*

Proof The proof of assertion (i) in Theorem 1.2.12 directly establishes assertion (i).

To prove (ii), assume $x \in D\left(\overline{A}\right)$ and let $y = \overline{A}x$. Then, there exists a sequence $\{x_n\}$ with $x_n \in D(A)$ such that $x_n \to x$ and $Ax_n \to y = \overline{A}x$. By Theorem 1.2.11, we obtain $\|\lambda x_n - Ax_n\| \geq \lambda \|x_n\|$ for $\lambda > 0$. Taking the limit as $n \to \infty$,

$$\left\|\lambda x - \overline{A}x_n\right\| \geq \lambda \|x\| \quad \text{for } \lambda > 0. \tag{1.2.43}$$

Since (1.2.43) holds for every $x \in D\left(\overline{A}\right)$, it follows from Theorem 1.2.11 that $\overline{A}$ is dissipative.

Finally, to prove (iii), assume that A is not closable. Then, there exists a sequence $\{x_n\}$ such that $x_n \in D(A)$, $x_n \to 0$, and $Ax_n \to y$ with $\|y\| = 1$. By Theorem 1.2.11, for every $t > 0$ and $x \in D(A)$,

$$\left\|\left(x + t^{-1}x_n\right) - tA\left(x + t^{-1}x_n\right)\right\| \geq \left\|x + t^{-1}x_n\right\|.$$

Taking the limit as $n \to \infty$ and then $t \to 0$ yields $\|x - y\| \geq \|x\|$ for every $x \in D(A)$. However, this is a contradiction if $D(A)$ is dense in X; thus, A must be closable. □

The Hille-Yosida theorem (Theorem 1.2.9) and the Lumer-Phillips theorem (Theorem 1.2.12) play pivotal roles in characterizing the infinitesimal generators of C_0-semigroups of contractions. These theorems provide fundamental insights into the properties of these semigroups and their generators. To extend these characterizations to general C_0-semigroups, it is necessary to delve deeper into the infinitesimal generators of uniformly bounded C_0-semigroups. In this pursuit, the renorming theorem becomes a crucial tool. By appropriately renorming the Banach space X, we can gain a new perspective on the structure and properties of the semigroup and its generator. This renorming process allows us to leverage the existing characterizations of contractions and apply them to a more general setting.

Theorem 1.2.14 *Suppose that A is a linear operator such that* $\rho(A) \supset (0, \infty)$. *If*

$$\left\|\lambda^n R(\lambda : A)^n\right\| \leq M \quad \textit{for}\ n = 1, 2, \cdots, \ \lambda > 0, \tag{1.2.44}$$

then there is a norm $|\cdot|$ *on* X *that is equivalent to the original norm* $\|\cdot\|$ *on* X *satisfying*

$$\|x\| \leq |x| \leq M\|x\| \quad \textit{for}\ x \in X \tag{1.2.45}$$

and

$$|\lambda R(\lambda : A)x| \leq |x| \quad \textit{for}\ x \in X,\ \lambda > 0. \tag{1.2.46}$$

Proof Let $\mu > 0$ and define the norm

$$\|x\|_\mu = \sup_{n \geq 0} \left\|\mu^n R(\mu : A)^n x\right\|. \tag{1.2.47}$$

Then, it is clear that

$$\|x\| \leq \|x\|_\mu \leq M\|x\| \tag{1.2.48}$$

where M is a constant, and

$$\|\mu R(\mu : A)\|_\mu \leq 1. \tag{1.2.49}$$

In addition,

$$\|\lambda R(\lambda : A)\|_\mu \leq 1 \quad \text{for } 0 < \lambda \leq \mu. \tag{1.2.50}$$

Evidently, if $y = R(\lambda : A)x$, then $y = R(\mu : A)(x + (\mu - \lambda)y)$. By (1.2.49),

$$\|y\|_\mu \leq \frac{1}{\mu}\|x\|_\mu + \left(1 - \frac{\lambda}{\mu}\right)\|y\|_\mu,$$

which implies $\lambda\|y\|_\mu \leq \|x\|_\mu$ as claimed. Combining (1.2.48) and (1.2.50), we have

$$\left\|\lambda^n R(\lambda : A)^n x\right\| \leq \left\|\lambda^n R(\lambda : A)^n x\right\|_\mu \leq \|x\|_\mu \quad \text{for } 0 < \lambda \leq \mu. \tag{1.2.51}$$

Taking the supremum over $n \geq 0$ on the left-hand side of (1.2.51) yields $\|x\|_\lambda \leq \|x\|_\mu$ for $0 < \lambda \leq \mu$. Finally, we define the limit norm

$$|x| = \lim_{\mu \to \infty} \|x\|_\mu. \tag{1.2.52}$$

Next, (1.2.45) is guaranteed by (1.2.48). Setting $n = 1$ in (1.2.51), noting

$$\|\lambda R(\lambda : A)x\|_\mu \leq \|x\|_\mu,$$

and letting $\mu \to \infty$, we obtain (1.2.46). □

Theorem 1.2.15 *Let X be a Banach space. The linear operator A is the infinitesimal generator of a C_0-semigroup $T(t)$ that satisfies $\|T(t)\| \leq M$ for some $M \geq 1$ if and only if*

(i) *A is closed, and its domain $D(A)$ is dense in X.*
(ii) *The resolvent set $\rho(A)$ of A contains $\mathbb{R}^+$ and*

$$\|R(\lambda : A)^n\| \leq \frac{M}{\lambda^n}, \quad \text{for } \lambda > 0, \ n = 1, 2, \cdots.$$

Proof Let $T(t)$ be a C_0-semigroup acting on the space X, with A as its infinitesimal generator. If an equivalent norm is introduced on X, $T(t)$ continues to be a C_0-semigroup under this new norm. Additionally, the infinitesimal generator A remains closed and densely defined, as the topological properties are invariant under equivalent norm.

Given that A is the infinitesimal generator of a C_0-semigroup satisfying $\|T(t)\| \leq M$, we define a new norm $|\cdot|$ on X as

$$|x| = \sup_{t \geq 0} \|T(t)x\|. \tag{1.2.53}$$

This leads to the inequality

$$\|x\| \leq |x| \leq M\|x\|, \tag{1.2.54}$$

indicating that $|\cdot|$ is equivalent to the original norm $\|\cdot\|$ on X. Moreover,

$$|T(t)x| = \sup_{s \geq 0} \|T(s)T(t)x\| \leq \sup_{s \geq 0} \|T(s)x\| = |x|, \tag{1.2.55}$$

confirming that $T(t)$ is a C_0-semigroup of contractions on X under the $|\cdot|$ norm. By virtue of the Hille-Yosida theorem (Theorem 1.2.9) and the initial remarks in the proof, we know that A is closed and densely defined, and $|R(\lambda : A)| \leq \lambda^{-1}$ for $\lambda > 0$. Applying (1.2.54) and (1.2.55)

$$\left\| R(\lambda : A)^n x \right\| \leq \left| R(\lambda : A)^n x \right| \leq \lambda^{-n}|x| \leq M\lambda^{-n}\|x\|,$$

thereby verifying the necessity of conditions (i) and (ii).

Assuming that conditions (i) and (ii) are satisfied, Theorem 1.2.14 guarantees the existence of a norm $| \cdot |$ on X that fulfills (1.2.45) and (1.2.46). When X is equipped with this norm, A emerges as a closed, densely defined operator with $\rho(A) \supset (0, \infty)$ and $|R(\lambda : A)| \leq \lambda^{-1}$ for $\lambda > 0$. Consequently, the Hille-Yosida theorem establishes that A serves as the infinitesimal generator of a C_0-semigroup of contractions on X under the $|\cdot|$ norm. Returning to the original norm, we observe that A remains the infinitesimal generator of $T(t)$. Furthermore,

$$\|T(t)x\| \leq |T(t)x| \leq |x| \leq M\|x\|,$$

which consequently implies that $\|T(t)\| \leq M$, as desired. Thus, conditions (i) and (ii) are not only necessary but also sufficient. □

If $T(t)$ is a general C_0-semigroup acting on X, then according to Theorem 1.2.6, there exist constants $M \geq 1$ and $\omega \geq 0$ such that

$$\|T(t)\| \leq Me^{\omega t}.$$

Now, consider the C_0-semigroup $S(t) = e^{-\omega t}T(t)$. It then follows that $\|S(t)\| \leq M$ for all t. Furthermore, A is the infinitesimal generator of $T(t)$ if and only if $A - \omega I$ is the infinitesimal generator of $S(t)$. Combining these observations with Theorem 1.2.15, we obtain the following result.

Theorem 1.2.16 *Let X be a Banach space. A linear operator A serves as the infinitesimal generator of a C_0-semigroup $T(t)$ that satisfies $\|T(t)\| \leq Me^{\omega t}$ for all $t \in [0, \infty)$, where $M \geq 1$ and $\omega \geq 0$, if and only if*

(i) *A is closed and $D(A)$ is dense in X.*
(ii) *The resolvent set $\rho(A)$ of A contains the ray (ω, ∞) and*

$$\|R(\lambda : A)^n\| \leq \frac{M}{(\lambda - \omega)^n}, \quad \text{for } \lambda > \omega, \ n = 1, 2, \cdots. \tag{1.2.56}$$

It should be noted that the requirement that all real λ, $\lambda > \omega$, exist in the resolvent set of A along with the estimate (1.2.56), leading to the conclusion that all complex λ that satisfy $\mathrm{Re}\lambda > \omega$ also exist in the resolvent set of A and

$$\|R(\lambda : A)^n\| \leq \frac{M}{(\mathrm{Re}\lambda - \omega)^n}, \quad \text{for } \mathrm{Re}\lambda > \omega, \ n = 1, 2, \cdots. \tag{1.2.57}$$

In the following, we shall proceed to prove this remark. Define

$$R(\lambda)x = \int_0^\infty e^{-\lambda t} T(t)x \, dt.$$

Since $\|T(t)\| \le Me^{\omega t}$, $R(\lambda)$ is well defined for every λ with $\mathrm{Re}\lambda > \omega$. An argument used in the proof of the Hille-Yosida theorem (Theorem 1.2.9) shows that $R(\lambda) = R(\lambda : A)$. In order to prove (1.2.57), we assume that $\mathrm{Re}\lambda > \omega$. Then,

$$\frac{d}{d\lambda} R(\lambda : A)x = \frac{d}{d\lambda} \int_0^\infty e^{-\lambda t} T(t)x \, dt = -\int_0^\infty te^{-\lambda t} T(t)x \, dt.$$

By induction, we can derive

$$\frac{d^n}{d\lambda^n} R(\lambda : A)x = (-1)^n \int_0^\infty t^n e^{-\lambda t} T(t)x \, dt. \tag{1.2.58}$$

Additionally, we can deduce from the resolvent identity

$$R(\lambda : A) - R(\mu : A) = (\mu - \lambda) R(\lambda : A) R(\mu : A)$$

that for every $\lambda \in \rho(A)$, $\lambda \to R(\lambda : A)$ is holomorphic and

$$\frac{d}{d\lambda} R(\lambda : A) = -R(\lambda : A)^2. \tag{1.2.59}$$

Further, by induction, we find

$$\frac{d^n}{d\lambda^n} R(\lambda : A) = (-1)^n n! R(\lambda : A)^{n+1}. \tag{1.2.60}$$

Comparing (1.2.58) and (1.2.60) leads to

$$R(\lambda : A)^n x = \frac{1}{(n-1)!} \int_0^\infty t^{n-1} e^{-\lambda t} T(t)x \, dt. \tag{1.2.61}$$

Consequently,

$$\left\| R(\lambda : A)^n x \right\| \le \frac{M}{(n-1)!} \int_0^\infty t^{n-1} e^{(\omega - \mathrm{Re}\lambda)t} \|x\| \, dt = \frac{M}{(\mathrm{Re}\lambda - \omega)^n} \|x\|,$$

which completes the proof.

We then have the representation formula of Corollary 1.2.1 in the general case.

Theorem 1.2.17 *Let A be the infinitesimal generator of a C_0-semigroup $T(t)$ on a Banach space X. If A_λ denotes the Yosida approximation of A, defined as $A_\lambda = \lambda AR(\lambda : A)$, then*

$$T(t)x = \lim_{\lambda\to\infty} e^{tA_\lambda}x, \quad \forall x \in X. \tag{1.2.62}$$

Proof We begin with the scenario in which $\|T(t)\| \le M$. In the proof of Theorem 1.2.15, we introduced a norm $|||\cdot|||$ on X that is equivalent to the original norm $\|\cdot\|$ on X and under which $T(t)$ is a C_0-semigroup of contractions. Following Corollary 1.2.1, it then follows that $|||\, e^{tA_\lambda}x - T(t)x\,||| \to 0$ as $\lambda \to \infty$ for every $x \in X$. Since $|||\cdot|||$ is equivalent to $\|\cdot\|$, (1.2.62) holds in X. In the general case where $\|T(t)\| \le Me^{\omega t}$ we have $\|T(t)\| \le M$ for $\omega \le 0$. By previous proof, the result holds. It now remains to show the result for $\omega > 0$. Take $\omega > 0$ and observe that $\lambda \to \|e^{tA_\lambda}\|$ is bounded for $\lambda > 2\omega$. Indeed, we have

$$\begin{aligned}\left\|e^{tA_\lambda}\right\| &= e^{-\lambda t}\left\|e^{\lambda^2 R(\lambda:A)t}\right\| \\ &\le e^{-\lambda t}\sum_{k=0}^{\infty}\frac{\lambda^{2k}t^k\left\|R(\lambda : A)^k\right\|}{k!} \le Me^{\frac{\lambda\omega}{\lambda-\omega}t} \le Me^{2\omega t}.\end{aligned} \tag{1.2.63}$$

Furthermore, we focus on the uniformly bounded semigroup $S(t) = e^{-\omega t}T(t)$ with $A - \omega I$ as the infinitesimal generator. As per the first part of the proof,

$$T(t)x = \lim_{\lambda\to\infty} e^{t(A-\omega I)_\lambda + \omega t}x \quad \text{for } x \in X. \tag{1.2.64}$$

A straightforward computation shows that $(A-\omega I)_\lambda + \omega I = A_{\lambda+\omega} + H(\lambda)$, where

$$\begin{aligned}H(\lambda) &= 2\omega I - \omega(\omega + 2\lambda)R(\lambda + \omega : A) \\ &= \omega\left[\omega R(\lambda + \omega : A) - 2AR(\lambda + \omega : A)\right].\end{aligned}$$

We can easily verify that $\|H(\lambda)\| \le 2\omega + (2\omega + \lambda^{-1}\omega^2)M$ and that for $x \in D(A)$, $\|H(\lambda)x\| \le M\lambda^{-1}\left(\omega^2\|x\| + 2\omega\|Ax\|\right) \to 0$ as $\lambda \to \infty$. Therefore $H(\lambda)x \to 0$ as $\lambda \to \infty$ for every $x \in X$. Since

$$\left\|e^{tH(\lambda)}x - x\right\| \le te^{t\|H(\lambda)\|}\|H(\lambda)x\|,$$

it follows that

$$\lim_{\lambda\to\infty} e^{tH(\lambda)}x = x \quad \text{for } x \in X. \tag{1.2.65}$$

Finally, since $H(\lambda)$ and $A_{\lambda+\omega}$ commute, there holds

$$\begin{aligned} &\left\| e^{tA_\lambda}x - T(t)x \right\| \\ &\leq \left\| e^{tA_\lambda + tH(\lambda-\omega)}x - T(t)x \right\| + \left\| e^{tA_\lambda} \right\| \left\| e^{tH(\lambda-\omega)}x - x \right\|. \end{aligned} \tag{1.2.66}$$

As $\lambda \to \infty$, the first term on the right-hand side tends to zero according to (1.2.64), while the second term tends to zero according to (1.2.63) and (1.2.65). Consequently, for $x \in X$,

$$\lim_{\lambda\to\infty} e^{tA_\lambda}x = T(t)x$$

and this completes the proof of the theorem. □

Let A be a closed operator on a Banach space X. Consider the abstract Cauchy problem

$$\begin{cases} \dfrac{d}{dt}u(t) = Au(t), \\ u(0) = x, \end{cases} \tag{1.2.67}$$

in which $x \in X$ and $A : D(A)(\subset X) \to X$. A continuously differential function $u : [0, \infty) \to X$ is called a classical solution of the Cauchy problem if $u(\cdot) \in D(A)$ satisfies the problem (1.2.67) for all $t > 0$.

Theorem 1.2.18 *Suppose that A is a densely defined linear operator with $\rho(A) \neq \emptyset$ in a Banach space X. The Cauchy problem* (1.2.67) *admits a unique solution for $x \in D(A)$, which is continuously differentiable for $t \geq 0$ if and only if A generates a C_0-semigroup $T(t)$ on X.*

Proof If A generates a C_0-semigroup $T(t)$, then according to (1.2.13), for $x \in D(A)$, $u(t) = T(t)x$ is a solution of (1.2.67) that is continuously differentiable for $t \geq 0$. Let us suppose the existence of another solution $y(t)$. Then the difference $e(t) = u(t) - y(t)$ satisfies (1.2.67) with $e(0) = 0$. For a given $t > 0$, we define $z(s) = T(t-s)e(s)$ for $0 \leq s \leq t$. Then $z(s)$ is strongly differentiable for $0 \leq s \leq t$, and for any $0 < \delta < t$,

$$z(t) - z(\delta) = \int_\delta^t \frac{dz(s)}{ds}\, ds.$$

However,

$$\frac{dz(s)}{ds} = -T(t-s)Ae(s) + T(t-s)Ae(s) = 0.$$

Consequently, $z(t) = e(t) = z(\delta) = T(t-\delta)e(\delta)$. Since $T(t)$ exhibits strong continuity and $e(0) = 0$, if we take the limit as $\delta \to 0$, the result would indicate that $e(t) = 0$ and $u(t) = y(t)$ for $t > 0$.

Conversely, suppose that $u(t, x_0)$ is the unique solution for any $x_0 \in D(A)$ in Eq. (1.2.67). For any $x \in D(A)$, we recall the graph norm $|||\cdot|||$ on $D(A)$ as $|||x||| = \|x\| + \|Ax\|$. Due to $\rho(A) \neq \emptyset$, A is a closed operator. Therefore, $D(A)$ with the given graph norm forms a Banach space, which is denoted by $[D(A)]$. A becomes a bounded operator on $[D(A)]$ as well as stated in [129]. For $t_0 > 0$, let $C(I, D(A))$ be the set of continuous functions from $I = [0, t_0]$ to $[D(A)]$. This set is also a Banach space, with the usual supremum norm. The mapping $S : [D(A)] \to C(I, D(A))$ can be defined as $S_x = u(t, x)$ for $t \in [0, t_0]$. Given that (1.2.67) is linear and has a unique solution, S is a linear operator. To demonstrate that S is closed, assume $x_n \to x$ in $[D(A)]$ and $S_{x_n} \to v$ in $C(I, D(A))$. Since $S_{x_n} = u(t, x_n)$ yields the solution of (1.2.67) with $u(0) = x_n$, it follows that

$$u(t, x_n) = x_n + \int_0^t Au(\tau, x_n)\,\mathrm{d}\tau.$$

As A is closed,

$$v(t) = x + \int_0^t Av(\tau)\,\mathrm{d}\tau,$$

which shows that $v(t) = u(t, x) = S_x$. Hence S is closed. Using the closed graph theorem, S is a linear operator and is bounded as well. Define a mapping $T(t) : [D(A)] \to [D(A)]$ by $T(t)x = u(t, x) = S_x$. With the help of the uniqueness, we can deduce that $T(t)$ satisfies the semigroup properties. In addition, given that $u(\cdot, x)$ is continuous, it is clear that $T(t)$ is uniformly bounded on $[0, t_0]$. Following the proof of Theorem 1.2.6, it is shown that $T(t)$ can be extended to a semigroup on $[D(A)]$, which satisfies $|||T(t)x||| \leq Me^{\omega t} |||x|||$ for $t > 0$. Furthermore, we demonstrate that $T(t)$ can be further extended to X. Assuming that $x \in D\left(A^2\right)$, define

$$z(t) = x + \int_0^t T(s)Ax\,\mathrm{d}s. \tag{1.2.68}$$

Upon differentiation of (1.2.68), we obtain

$$\frac{dz(t)}{dt} = T(t)Ax = Ax + A\int_0^t T(s)Ax\,\mathrm{d}s = Az(t), \tag{1.2.69}$$

where (1.2.12) and the fact that A is closed were used. Since $z(0) = x$, thanks to the uniqueness assumption, we deduce that $z(t) = u(t, x) = T(t)x$. It then follows from Eq. (1.2.69) that

$$T(t)Ax = AT(t)x, \quad \forall\, x \in D\left(A^2\right). \tag{1.2.70}$$

Let $z \in D(A)$ and $\lambda \in \rho(A)$. The equation $z = (\lambda I - A)x$ has a unique solution $x = R(\lambda : A)z \in D\left(A^2\right)$. Therefore,

$$\|T(t)z\| \leq |\lambda|\|T(t)x\| + \|T(t)Ax\| \leq (1 + |\lambda|) \, |||\, T(t)x \,||| \leq M_1 e^{\omega t} \, |||\, x \,|||, \tag{1.2.71}$$

where $M_1 = (1 + |\lambda|)M$ and (1.2.70) was used. As $R(\lambda : A)$ and $AR(\lambda : A)$ are bounded, it follows that $|||\, x \,||| \leq C\|z\|$ for some $C > 0$. Consequently,

$$\|T(t)z\| \leq M_2 e^{\omega t}\|z\|, \quad \forall\, z \in D(A), \tag{1.2.72}$$

in which $M_2 = CM_1$. Hence, $T(t)$ can be extended to a C_0-semigroup on X. The proof will be accomplished if A is the infinitesimal generator of $T(t)$. Let A_1 be the infinitesimal generator of $T(t)$ and let $z \in D(A)$. As $u(t) = T(t)z$ is the solution of (1 2.67), by evaluating (1.2.67) at $t = 0$, we obtain $A_z = A_1 z$. Therefore, $A \subset A_1$ when A_1 is restricted to $D(A)$. Using (1.2.70), we can determine that

$$e^{-\lambda t} AT(t)x = e^{-\lambda t} T(t)Ax = e^{-\lambda t} T(t)A_1 x, \quad x \in D\left(A^2\right), \tag{1.2.73}$$

where $\lambda > \omega$. Integrating (1.2.73) on $[0, \infty)$ and using (1.2.61) as well as the fact that A is closed, we can obtain

$$AR(\lambda : A_1)x = R(\lambda : A_1)A_1 x = A_1 R(\lambda : A_1)x. \tag{1.2.74}$$

Since $A_1 R(\lambda : A_1)$ is uniformly bounded, A is closed, and $D(A^2)$ is dense in X, it follows that (1.2.74) holds for $x \in X$. This implies that $D(A_1) = \text{Range}R(\lambda : A_1) \subset D(A)$, and so $A_1 z = Az$ for $z \in D(A_1)$. This proves that $A_1 \subset A$, and hence $A = A_1$. □

It is noted that if $u_0 \notin D(A)$, the Cauchy problem (1.2.67) may not have a classical solution. We provide the definition of the mild solution.

Let $T(t)$ be a C_0-semigroup on a Banach space X and let A be its infinitesimal generator. For any $u_0 \in X$, $u(t) = T(t)u_0$ is defined as a mild solution of Eq. (1.2.67).

In the following, we present some special classes of semigroups, except for the uniformly continuous semigroups that have been discussed previously.

Definition 1.2.19 Let $T(t)$ be a C_0-semigroup on a Banach space X. The semigroup $T(t)$ is called differentiable for $t > t_0$ if for every $x \in X$, $t \to T(t)x$ is differentiable for $t > t_0$. Furthermore, $T(t)$ is called differentiable if it is differentiable for all $t > 0$.

It is observed that if $t \to T(t)x$ is differentiable for every $x \in X$ and $t \geq 0$, then $D(A) = X$, and since A is closed, it is necessarily bounded. Furthermore, a strongly continuous semigroup $T(t)$ is called eventually differentiable if there exists a $t_0 > 0$

such that $T(t_0)X \subset D(A)$ or, equivalently, $T(t)X \subset D(A)$ for all $t \geq t_0$, and $T(t)$ is immediately differentiable if $T(t)X \subset D(A)$ for all $t > 0$.

Theorem 1.2.20 *If A generates a differentiable C_0-semigroup $T(t)$ on a Banach space X, then Eq.* (1.2.67) *possesses a unique solution $u(t)$ for every initial value $x \in X$, which is equal to the mild solution, namely, $u(t) = T(t)x$.*

Proof From the assumptions, it follows that $T(t)x$ is differentiable for $x \in X$ and that

$$\frac{d}{dt}T(t)x = AT(t)x.$$

Since A is closed and $T(t)$ is bounded, $AT(t)$ is also closed and defined for all $x \in X$. Therefore, according to the closed graph theorem, $AT(t)$ is bounded and thus is Lipschitz for $t > 0$. Consequently, (1.2.67) has a unique solution, and since $u(t) = T(t)x$ is also a solution, it is the unique solution. □

Another kind of important semigroups is the analytic semigroup, which extends the domain of the parameter to regions in the complex plane that include the nonnegative real axis.

Definition 1.2.21 Let Δ be a subset of the complex plane. If $J(\lambda)$, $\lambda \in \Delta$, is a family of bounded linear operators on X that meets the criteria of

$$J(\lambda) - J(\mu) = (\mu - \lambda)J(\lambda)J(\mu) \tag{1.2.75}$$

for $\lambda, \mu \in \Delta$, then it is called a pseudo resolvent on Δ.

It is apparent from (1.2.75) that $J(\lambda)$ and $J(\mu)$ commute for $\lambda, \mu \in \Delta$. Additionally, when rewriting (1.2.75) as

$$J(\lambda) = J(\mu)[I + (\mu - \lambda)J(\lambda)],$$

it becomes evident that $R(J(\mu)) \supset R(J(\lambda))$. By symmetry,

$$J(\lambda)J(\mu) = J(\mu)J(\lambda).$$

Consequently, $N(J(\lambda)) = N(J(\mu))$. Additionally, the closedness of $N(J(\lambda))$ is readily apparent. Based on these observations, we can state the following Lemma 1.2.22.

Lemma 1.2.22 *Let Δ be a subset of the complex plane $\mathbb{C}$. If $J(\lambda)$ is a pseudo resolvent on Δ, then $J(\lambda)J(\mu) = J(\mu)J(\lambda)$ for all $\lambda, \mu \in \Delta$. Furthermore, the null space $N(J(\lambda))$ and the range $R(J(\lambda))$ are both independent of the choice of $\lambda \in \Delta$. Specially, $N(J(\lambda))$ is a closed subspace of X.*

Definition 1.2.23 Let X be a Banach space and $\Xi = \{z \mid \varphi_1 < \arg z < \varphi_2,\ \varphi_1 < 0 < \varphi_2\}$. For $z \in \Xi$ denote $T(z)$ as a bounded linear operator. The family $T(z),\ z \in \Xi$ is an analytic semigroup in Ξ if:

(i) $z \to T(z)$ is analytic in Ξ.
(ii) $T(0) = I$ and $\lim_{\substack{z\to 0\\ z\in\Xi}} T(z)x = x$ for every $x \in X$.
(iii) $T(z_1 + z_2) = T(z_1)T(z_2)$ for $z_1,\ z_2 \in \Xi$.

A semigroup $T(t)$ is called analytic if it is analytic in some sector Ξ containing the nonnegative real axis.

By the definition provided, it is indeed true that every analytic semigroup possesses immediate differentiability. This property ensures a smooth and continuous evolution of the semigroup over time, reflecting the underlying regularity of the system it represents.

Now, turning to the Trotter approximation theorem, this fundamental result establishes a profound connection between a semigroup $T(t)$ and its infinitesimal generator A. In essence, the theorem asserts that there is a continuous interdependence between the semigroup and its generator. More specifically, it states that in a precise mathematical sense, the convergence of a sequence of infinitesimal generators is equivalent to the convergence of the corresponding semigroups.

This equivalence is remarkable because it bridges the gap between two seemingly distinct mathematical objects: the semigroup, which represents the evolution of a system over time, and the infinitesimal generator, which captures the instantaneous behavior of that system. The Trotter approximation theorem thus provides a powerful tool for analyzing and understanding the behavior of complex systems described by semigroups and their generators.

In practical applications, this theorem can be used to approximate the behavior of a system by considering a sequence of simpler systems (or "approximations") whose infinitesimal generators converge to the generator of the original system. By studying the convergence of these semigroups, one can gain insights into the behavior of the original, potentially more complex system.

Lemma 1.2.24 *Let A and B be the infinitesimal generators of C_0-semigroups $T(t)$ and $S(t)$, respectively. For any $x \in X$ and $\lambda \in \rho(A) \cap \rho(B)$, there holds*

$$\begin{aligned} &R(\lambda : B)[T(t) - S(t)]R(\lambda : A)x \\ &= \int_0^t S(t-s)[R(\lambda : A) - R(\lambda : B)]T(s)x \, ds. \end{aligned} \tag{1.2.76}$$

Proof The differentiability of the X-valued function

$$s \to S(t-s)R(\lambda : B)T(s)R(\lambda : A)x$$

is established for every $x \in X$ and $\lambda \in \rho(A) \cap \rho(B)$. A straightforward calculation shows that

$$\begin{aligned}&\frac{d}{dx}[S(t-s)R(\lambda : B)T(s)R(\lambda : A)x]\\&= S(t-s)[-BR(\lambda : B)T(s) + R(\lambda : B)T(s)A]R(\lambda : A)x\\&= S(t-s)[R(\lambda : A) - R(\lambda : B)]T(s)x,\end{aligned} \tag{1.2.77}$$

where $R(\lambda : A)T(s)x = T(s)R(\lambda : A)x$ was used. The fact (1.2.76) is then obtained by integrating both sides of Eq. (1.2.77) from 0 to t. □

For simplicity, denote $A \in G(M, \omega)$ as an operator A, which is the infinitesimal generator of a C_0-semigroup $T(t)$ satisfying $\|T(t)\| \leq Me^{\omega t}$.

Theorem 1.2.25 *Let A, $A_n \in G(M, \omega)$ be operators, and let $T(t)$ and $T_n(t)$ represent the semigroups generated by A and A_n, respectively. Then, the following two statements are equivalent:*

(i) *For every $x \in X$ and λ with $\mathrm{Re}\lambda > \omega$, $R(\lambda : A_n)x \to R(\lambda : A)x$ as $n \to \infty$.*
(ii) *For every $x \in X$ and $t \geq 0$, $T_n(t)x \to T(t)x$ as $n \to \infty$.*

Moreover, in part (ii)*, there is uniform convergence on bounded t-intervals.*

Proof We begin by proving that (i) ⇒ (ii). Fix $x \in X$ and an interval $0 \leq t \leq T$. Consider

$$\begin{aligned}\|(T_n(t) - T(t))R(\lambda : A)x\| \leq &\|T_n(t)\,(R(\lambda : A) - R(\lambda : A_n))\,x\|\\&+\|R(\lambda : A_n)\,(T_n(t) - T(t))\,x\|\\&+\|\,(R(\lambda : A_n) - R(\lambda : A))\,T(t)x\|\\=&D_1 + D_2 + D_3.\end{aligned} \tag{1.2.78}$$

Since $\|T_n(t)\| \leq Me^{\omega T}$ for $0 \leq t \leq T$, it follows from (i) that $D_1 \to 0$ as $n \to \infty$, uniformly over $[0, T]$. Additionally, since $t \to T(t)x$ is continuous, the set $\{T(t)x \mid 0 \leq t \leq T\}$ is compact in X, and therefore $D_3 \to 0$ as $n \to \infty$, uniformly on $[0, T]$. Finally, applying Lemma 1.2.24 with $B = A_n$, we obtain

$$\begin{aligned}&\|R(\lambda : A_n)(T_n(t) - T(t))R(\lambda : A)x\|\\&\leq \int_0^t \|T_n(t-s)\|\,\|(R(\lambda : A) - R(\lambda : A_n))\,T(s)x\|\,ds.\end{aligned} \tag{1.2.79}$$

The integrand on the right-hand side of (1.2.79) is bounded by

$$2M^3e^{2\omega t}(\mathrm{Re}\lambda - \omega)^{-1}\|x\|,$$

and it tends to zero as $n \to \infty$. By Lebesgue's bounded convergence theorem, the right-hand side of (1.2.79) tends to 0. Therefore,

$$\lim_{n\to\infty} \|R(\lambda : A_n)(T_n(t) - T(t))R(\lambda : A)x\| = 0 \tag{1.2.80}$$

and the limit in (1.2.80) is uniform on $[0, T]$. Since every $x \in D(A)$ can be written as $x = R(\lambda : A)z$ for some $z \in X$, (1.2.80) indicates that for $x \in D(A)$, $D_2 \to 0$ as $n \to \infty$ uniformly on $[0, T]$. From Eq. (1.2.78), it follows that for all $x \in D(A^2)$,

$$\lim_{n\to\infty} \|(T_n(t) - T(t))x\| = 0. \tag{1.2.81}$$

Additionally, the limit in (1.2.81) is uniform on $[0, T]$. Since $\|T_n(t) - T(t)\|$ are uniformly bounded on $[0, T]$ and $D(A^2)$ is dense in X, Eq. (1.2.81) holds uniformly on $[0, T]$ for every $x \in X$. Therefore, (i) $\Rightarrow$ (ii).

Now assume that (ii) holds and that $\mathrm{Re}\lambda > \omega$. Then,

$$\|R(\lambda : A_n)x - R(\lambda : A)x\| \le \int_0^\infty e^{-\mathrm{Re}\lambda t} \|(T_n(t) - T(t))x\|\, dt. \tag{1.2.82}$$

The right-hand side of (1.2.82) tends to zero as $n \to \infty$ due to (ii) and Lebesgue's dominated convergence theorem, and hence (ii) $\Rightarrow$ (i). □

Theorem 1.2.26 *Let $A_n \in G(M, \omega)$. If there is a λ_0 with $\mathrm{Re}\lambda_0 > \omega$ and*

(i) *for each $x \in X$, $R(\lambda_0 : A_n)x \to R(\lambda_0)x$ as $n \to \infty$, and*
(ii) *the range of $R(\lambda_0)$ is dense in X,*

then there exists a unique operator $A \in G(M, \omega)$ such that $R(\lambda_0) = R(\lambda_0 : A)$.

Proof We assume without loss of generality that $\omega = 0$ and prove that $R(\lambda : A_n)x$ converges as $n \to \infty$ for all λ with $\mathrm{Re}\lambda > 0$. Let $S = \{\lambda \mid \mathrm{Re}\lambda > 0, R(\lambda : A_n)x$ converges as $n \to \infty\}$. We claim that S is open. Actually, we can expand $R(\lambda : A_n)$ in a Taylor series around a point μ in which $R(\mu : A_n)x$ converges as $n \to \infty$. Hence,

$$R(\lambda : A_n) = \sum_{k=0}^{\infty} (\mu - \lambda)^k R(\mu : A_n)^{k+1}. \tag{1.2.83}$$

Based on (1.2.57), the inequality $\|R(\mu : A_n)^k\| \le M(\mathrm{Re}\mu)^{-k}$ holds. Consequently, the series (1.2.83) converges in the uniform operator topology for all λ values such that $|\mu - \lambda|(\mathrm{Re}\mu)^{-1} < 1$. Additionally, such convergence is uniformly guaranteed for all λ satisfying $|\mu - \lambda|(\mathrm{Re}\mu)^{-1} \le \vartheta < 1$ where the series of constants

$$\sum_{k=0}^{\infty} M\vartheta^{k+1}$$

is a majorant of the series

$$\sum_{k=0}^{\infty} |\mu - \lambda|^k \|R(\mu : A_n)^{k+1}\|.$$

As a result, $R(\lambda : A_n)x$ converges as $n \to \infty$ for all λ satisfying $|\mu - \lambda|(\mathrm{Re}\mu)^{-1} \leq \vartheta < 1$. This shows that the set S is open. Now, let λ be a cluster point of S with $\mathrm{Re}\lambda > 0$. If $0 < \vartheta < 1$, there is a point $\mu \in S$ for which $|\mu - \lambda|(\mathrm{Re}\mu)^{-1} \leq \vartheta < 1$. Consequently, by the first part of the proof, $R(\lambda : A_n)x$ converges as $n \to \infty$, implying that $\lambda \in S$. Therefore, S is relatively closed in $\mathrm{Re}\lambda > 0$. As $\lambda_0 \in S$ by assumption, we conclude that $S = \{\lambda \mid \mathrm{Re}\lambda > 0\}$.

For each λ with $\mathrm{Re}\lambda > 0$, we define a linear operator $R(\lambda)$ by

$$R(\lambda)x = \lim_{n\to\infty} R(\lambda : A_n)x. \tag{1.2.84}$$

It is evident that

$$R(\lambda) - R(\mu) = (\mu - \lambda)R(\lambda)R(\mu) \tag{1.2.85}$$

for $\mathrm{Re}\lambda > 0$ and $\mathrm{Re}\mu > 0$. Consider $R(\lambda)$ as a pseudo resolvent on $\mathrm{Re}\lambda > 0$. Since the range of $R(\lambda)$ is independent of λ for a pseudo resolvent by Lemma 1.2.22, we can infer from (ii) that the range of $R(\lambda)$ is dense in X. Moreover, it is clear from the definition of $R(\lambda)$ that

$$\left\| R(\lambda)^k \right\| \leq M(\mathrm{Re}\lambda)^{-k} \tag{1.2.86}$$

for $\mathrm{Re}\lambda > 0,\ k = 1, 2, \cdots$. In particular, for real $\lambda > 0$,

$$\|\lambda R(\lambda)\| \leq M, \tag{1.2.87}$$

for all $\lambda > 0$. By Theorem 1.9.4 of [181], there is a unique closed densely defined linear operator A such that $R(\lambda) = R(\lambda : A)$. Further, it follows from (1.2.86) and Theorem 1.2.16 that $A \in G(M, 0)$. This completes the proof of the theorem. □

Theorem 1.2.27 below directly follows from Theorems 1.2.25 and 1.2.26.

Theorem 1.2.27 (Trotter-Kato Theorem) *Let X be a Banach space. Let $A_n \in G(M, \omega)$ and $T_n(t)$ be the semigroup whose infinitesimal generator is A_n. If for some λ_0 with* $\mathrm{Re}\lambda_0 > \omega$*, there is*

(i) $R(\lambda_0 : A_n)x \to R(\lambda_0)x$ *for all $x \in X$ as $n \to \infty$;*
(ii) *the range of $R(\lambda_0)$ is dense in X;*

then there exists a unique operator $A \in G(M, \omega)$ such that $R(\lambda_0) = R(\lambda_0 : A)$. If $T(t)$ is the C_0-semigroup generated by A, then $T_n(t)x \to T(t)x$ for all $t \geq 0$ and $x \in X$ as $n \to \infty$, and the limit is uniform for t in bounded intervals.

1.3 Calculus of Variations and Optimal Control

The universality of extremum problems has led to the establishment of numerous necessary and sufficient conditions in calculus to aid in finding solutions. However, contrary to classical calculus, there are instances where the minimization of a function over a set in an infinite-dimensional space is considered.

Johann Bernoulli, a Swiss mathematician, posed an intriguing problem on the brachistochrone curve back in 1696. Consider the following scenario: On a plane, there are two points, A and B, where B is lower and not directly below A. A particle or bead, influenced by a uniform gravitational field, slides frictionlessly from A to the endpoint B in the shortest possible time. Johann Bernoulli published this problem in the journal *Acta Eruditorum*, inviting mathematicians worldwide to find such a curve [35, 90, 202, 231].

In the end, five mathematicians, respectively, Isaac Newton, Jakob Bernoulli, Gottfried Leibniz, Ehrenfried Walther von Tschirnhaus, and Guillaume de l'Hôpital, responded with solutions. Apart from l'Hôpital's, the remaining four solutions were published in the same edition of the journal as Johann Bernoulli's. It was revealed that the solution is a segment of a cycloid. Interestingly, Jakob Bernoulli formulated a more challenging version of the brachistochrone problem in his paper. His efforts in solving it led to the development of new methods, which were later refined by Leonhard Euler and became known as the calculus of variations. Joseph-Louis Lagrange's further contributions eventually culminated in the modern infinitesimal calculus.

The brachistochrone problem stands as one of the earliest challenges in the field of the calculus of variations. Essentially, it boils down to finding a curve, $y(t)$, that minimizes a specific functional:

$$J(y) = \int_a^b \sqrt{\frac{1 + (y')^2}{2gy}}\, dt,$$

where y' is the derivative of y, g is the gravitational acceleration, and a and b are the abscissas of points A and B. Actually, the classical variational problem is essentially the quest to identify the optimal path from a set of permissible paths that minimizes the value of a specific functional. This involves exploring the infinite possibilities to find the one that satisfies the given conditions and achieves the desired minimum. Illustrative examples abound in this field. One such example is Dido's isoperimetric problem, which challenges us to determine the shape of a region with a fixed perimeter that maximizes its area. Another fascinating example is the light reflection and refraction problem, where we investigate the optimal path of light rays as they interact with different media. Lastly, the catenary curve problem explores the shape of a hanging chain or rope under the influence of gravity. In general, a problem of the calculus of variations can be formulated as follows. Find a curve $y(t)$ over the interval $a \leq t \leq b$, which minimize a cost functional

$$J(y) = \int_a^b L\left(t, y(t), y'(t)\right)\, dt,$$

subject to

$$y(a) = y_0, \ \ y(b) = y_1,$$

in which $L : \mathbb{R} \times \mathbb{R} \times \mathbb{R} \to \mathbb{R}$ is a given functional. The Euler necessary condition of optimality of the curve $y(\cdot)$ is

$$\frac{d}{dt} L_{y'}\left(t, y(t), y'(t)\right) = L_y\left(t, y(t), y'(t)\right),$$

where subscript in y or y' denote the partial derivative with respect to y or y'. Utilizing perturbations or first-order variations of the optimal curve, Lagrange employed an analytical approach to derive the Euler necessary condition, which is now widely recognized as the Euler-Lagrange equation. Subsequently, the Legendre-Clebsch condition, a second-order necessary condition of optimality derived from the second variation, was formulated as follows:

$$L_{y'y'}\left(t, y(t), y'(t)\right) \geq 0, \quad t \in [a, b].$$

Moreover, as the investigation deepens, the optimal control problem emerges and typically assumes the following form.

Given a time interval (a, b) and for every t within this interval, the objective is to identify a parameter $u(\cdot)$ or control within a fixed subset $U \subset \mathbb{R}^m$ that minimizes the cost functional

$$J(y) = \int_a^b L(t, y(t), y'(t))\, dt + \psi(y(b)) \tag{1.3.1}$$

subject to the state equation

$$\begin{cases} y'(t) = f(t, y(t), u(t)), & t \in (a, b), \\ y(a) = y_0, \ \ y(b) = y_1. \end{cases} \tag{1.3.2}$$

Here, the mapping $f : [a, b] \times \mathbb{R}^n \times U \to \mathbb{R}^n$, the output $y(\cdot) \in \mathbb{R}^n$ denotes the state of system (1.3.2), the input $u(\cdot) \in U$ is the admissible control, and U denotes the admissible control set. For $u(\cdot) \in U$, the solution $y(\cdot)$ to (1.3.2) is said to be the state trajectory, and $(y(\cdot), u(\cdot))$ is the state-control or trajectory-control pair.

In (1.3.1), the functional $L : [a, b] \times \mathbb{R}^n \times \mathbb{R}^n \to \mathbb{R}$ is the running cost and the functional $\psi : \mathbb{R}^n \to \mathbb{R}$ is the endpoint cost or the terminal cost. In practical applications, it is common to construct diverse cost functionals to cater to specific needs. For instance, if our sole focus is on the final state at time b, we can set $L \equiv 0$ and take

$$J(y) = \psi(y(b)),$$

which is of the Mayer form. Certainly, there are instances where our concern lies solely with the state and control within the interval $[a, b]$. For instance, we may desire the state to remain close to a prescribed trajectory throughout this interval. In such cases, we set $\psi \equiv 0$ and define

$$J(y) = \int_a^b L(t, y(t), u(t))\, dt,$$

which is known as the Lagrange form. Finally, combining the two forms above, we call the cost functional of the form (1.3.1) as the Bolza form.

Moreover, let X be a set and $\mathbb{R}^+$ be the set of nonnegative reals. A metric ρ in X is a mapping $\rho : X \times X \to \mathbb{R}^+$, which has the following properties:

$$\rho(x, y) = 0 \Rightarrow x = y;$$

$$\rho(x, y) = \rho(y, x);$$

$$\rho(x, z) \le \rho(x, y) + \rho(y, z) \quad \text{(the triangle inequality)}.$$

The set X equipped with a metric ρ is called a metric space and is denoted by (X, ρ).

For an optimal control problem, both the state and the control are usually equipped with some constraints such as

$$F(y(\cdot), u(\cdot)) \in \Gamma,$$

in which Γ is a subset of some metric space Γ_0 and $F : \mathbb{R}^n \times U \to \Gamma_0$ is a given mapping. Analogous to the various forms of cost functionals, there exist classical constraints such as equality constraints and inequality constraints. These constraints can be applied to the entire interval $[a, b]$ or solely at the initial moment a and the terminal moment b.

When dealing with an optimal control problem, the existence and regularity theorems for optimal controls are crucial elements of research. Furthermore, it is essential to establish various conditions, including necessary and sufficient conditions, to solve such problems. Broadly speaking, the Pontryagin maximum principle presents a necessary condition for optimality and furnishes the optimal control in an open-loop form. Conversely, the dynamic programming approach constructs a sufficient condition for optimality and offers the optimal control in a closed-loop form. Both of these topics constitute significant components of optimal control theory, and we delve into them in the subsequent two sections.

In control theory, a system that lacks any feedback connections is referred to as an open-loop system. Such systems are independent of their output, meaning that the output is not utilized as a control variable and has no influence on the input. Consequently, in open-loop systems, if the output is perturbed by external disturbances, there is no automatic mechanism to correct these disturbances as they lack feedback. While this is a significant disadvantage, open-loop systems are generally simpler and exhibit a certain degree of stability. Contrastingly, a closed-loop control system utilizes feedback signals to generate its output. In these systems, the feedback signal from the output is used to adjust the input, enabling the correction of any errors that occur. Although more complex, closed-loop systems offer high accuracy and are less susceptible to noise.

Indeed, the realm of optimal control encompasses a diverse array of challenges, among which linear quadratic regulator problems (LQRPs) occupy a prominent position. In LQRPs, the system dynamics adhere to a linear structure, while the cost functional assumes a quadratic form. This particular structure greatly simplifies the analytical complexity of the problem, rendering it more tractable and amenable to various solution techniques [151]. The simplicity and elegance of LQRPs have made them a cornerstone in control theory, with widespread applications across numerous fields.

Consider a linear time-varying system

$$\frac{d}{dt}x(t) = A(t)x(t) + B(t)u(t), \quad x(t_0) = x_0, \tag{1.3.3}$$

in which $x(\cdot) \in \mathbb{R}^n$ and $u(\cdot) \in \mathbb{R}^m$. To be simple, we assume that the control is unconstrained. Let $t_1 > 0$. Take the cost functional as follows:

$$J(y) = \int_{t_0}^{t_1} \left(x^\top(t)Q(t)x(t) + u^\top(t)R(t)u(t) \right) \mathrm{d}t + x^\top(t_1)Mx(t_1), \tag{1.3.4}$$

where for all $t \in [t_0, t_1]$, $Q(\cdot)$, M are positive semidefinite, and $R(\cdot)$ are positive definite symmetric matrices with appropriate dimensions. $f^\top$ denotes the transpose of f. Furthermore, we assume that $R(t)$ is invertible for all t. For this LQR problem, by the necessary condition of optimality provided by the maximum principle, an optimal control in linear state feedback law form can be established. For the sake of brevity, we only give the main results here, which were described in [151].

Introduce the Lagrange multiplier $p(\cdot) \in \mathbb{R}^n$, and define the Hamiltonian by

$$H(t, x, u, p) = p^\top A(t)x + p^\top B(t)u - x^\top Q(t)x - u^\top R(t)u. \tag{1.3.5}$$

Its gradient with respect to u is

$$H_u = B^\top(t)p - 2R(t)u.$$

Denote the terminal cost $K = K(x(t_1)) = x^\top(t_1)Mx(t_1)$. Along an optimal trajectory, H_u must vanish. Therefore, if the optimal control u^* exists, it must have

$$u^*(t) = \frac{1}{2}R^{-1}(t)B^\top(t)p^*(t). \tag{1.3.6}$$

Here, p^* is the costate corresponding to an optimal control-trajectory pair (u^*, x^*) and satisfies the adjoint equation

$$\begin{cases} \dfrac{d}{dt}p^* = -H_x\left(t, x^*, u^*, p^*\right) = 2Q(t)x^* - A^\top(t)p^*, \\ p^*(t_1) = -K_x(x^*(t_1)) = -2Mx^*(t_1). \end{cases} \tag{1.3.7}$$

Further computation gives

$$p^*(t) = -2P(t)x^*(t), \quad \forall\, t \in [t_0, t_1], \tag{1.3.8}$$

where $P(\cdot)$ is some matrix-valued function to be determined and its boundary condition $P(t_1) = M$ can be derived from the second equation of (1.3.7) and (1.3.8).

Combining (1.3.3), (1.3.6), and (1.3.7), we have the equations in closed-loop form:

$$\begin{pmatrix} \dfrac{d}{dt}x^* \\ \dfrac{d}{dt}p^* \end{pmatrix} = \begin{pmatrix} A(t) & \dfrac{1}{2}B(t)R^{-1}(t)B^\top(t) \\ 2Q(t) & -A^\top(t) \end{pmatrix} \begin{pmatrix} x^* \\ p^* \end{pmatrix}. \tag{1.3.9}$$

Denote the transition matrix for the system (1.3.9) by $\Phi(\cdot, \cdot)$. As already known, the state transition matrix relates the state of the system at $t = \tau_0$ to its state at a subsequent time t when the input $u(t) = 0$. Hence,

$$\begin{pmatrix} x^*(t) \\ p^*(t) \end{pmatrix} = \Phi(t, t_1) \begin{pmatrix} x^*(t_1) \\ p^*(t_1) \end{pmatrix}, \tag{1.3.10}$$

where $\Phi(t, t_1) = \Phi^{-1}(t_1, t)$. Partitioning Φ into $n \times n$ blocks as

$$\Phi = \begin{pmatrix} \Phi_{11} & \Phi_{12} \\ \Phi_{21} & \Phi_{22} \end{pmatrix}$$

changes the Eqs. (1.3.10) into

$$\begin{pmatrix} x^*(t) \\ p^*(t) \end{pmatrix} = \begin{pmatrix} \Phi_{11}(t,t_1) & \Phi_{12}(t,t_1) \\ \Phi_{21}(t,t_1) & \Phi_{22}(t,t_1) \end{pmatrix} \begin{pmatrix} x^*(t_1) \\ p^*(t_1) \end{pmatrix},$$

which, by the terminal condition in (1.3.7), can be written as

$$x^*(t) = (\Phi_{11}(t,t_1) - 2\Phi_{12}(t,t_1)M)\, x^*(t_1), \tag{1.3.11}$$

$$p^*(t) = (\Phi_{21}(t,t_1) - 2\Phi_{22}(t,t_1)M)\, x^*(t_1). \tag{1.3.12}$$

Solving (1.3.11) for $x^*(t_1)$ and by virtue of (1.3.12) gives

$$p^*(t) = (\Phi_{21}(t,t_1) - 2\Phi_{22}(t,t_1)M)\,(\Phi_{11}(t,t_1) - 2\Phi_{12}(t,t_1)M)^{-1}\, x^*(t).$$

Moreover, by (1.3.8), there is

$$P(t) = -\frac{1}{2}\,(\Phi_{21}(t,t_1) - 2\Phi_{22}(t,t_1)M)\,(\Phi_{11}(t,t_1) - 2\Phi_{12}(t,t_1)M)^{-1}. \tag{1.3.13}$$

Combining (1.3.6) and (1.3.8), we obtain the optimal linear state feedback control

$$u^*(t) = -R^{-1}(t)B^\top(t)P(t)x^*(t), \tag{1.3.14}$$

in which $P(\cdot)$ is from (1.3.13).

We will derive a differential equation that the matrix $P(\cdot)$ defined in (1.3.13) must satisfy. Initially, we differentiate both sides of Eq. (1.3.8), resulting in

$$\frac{d}{dt}p^*(t) = -2\frac{dP(t)}{dt}x^*(t) - 2P(t)\frac{dx^*(t)}{dt}.$$

We then expand $\frac{d}{dt}p^*(t)$ and $\frac{d}{dt}x^*(t)$ by using the canonical equations (1.3.9) to obtain

$$\begin{aligned} &2Q(t)x^*(t) - A^\top(t)p^*(t) \\ &= -2\frac{d}{dt}P(t)x^*(t) - 2P(t)A(t)x^*(t) - P(t)B(t)R^{-1}(t)B^\top(t)p^*(t). \end{aligned}$$

Apply (1.3.8) to omit p^* and divide by 2. Then, the equation

$$
\begin{aligned}
&Q(t)x^*(t) + A^\top(t)P(t)x^*(t) \\
&= -\frac{d}{dt}P(t)x^*(t) - P(t)A(t)x^*(t) + P(t)B(t)R^{-1}(t)B^\top(t)P(t)x^*(t)
\end{aligned}
$$

must be valid for all t where $P(t)$ is defined. Since the initial state x_0 is arbitrary and x^* is the state of the linear time-varying system given by (1.3.3) and (1.3.14) with a nonsingular transition matrix, $x^*(t)$ can be arbitrary. It follows that P must solve the matrix differential equation of the following

$$
\frac{d}{dt}P(t) = -P(t)A(t) - A^\top(t)P(t) - Q(t) + P(t)B(t)R^{-1}(t)B^\top(t)P(t), \tag{1.3.15}
$$

which is known as the Riccati differential equation (RDE). Its boundary condition is specified by $P(t_1) = M$. To propagate the solution $P(t)$, it is to be solved backward in time from $t = t_1$.

Finally, we introduce a crucial inequality that can be employed to derive numerous estimations within the realm of differential equation theory. Additionally, this inequality offers a comparison theorem that validates the uniqueness of a solution to the initial value problem, as outlined in [80].

Lemma 1.3.1 *For a given $\alpha(\cdot)$, $\eta(\cdot) \in L^\infty(0,T)$, and $\beta(\cdot) \in L^1(0,T)$, where $\alpha(t)$, $\eta(t)$, and $\beta(t)$ are non-negative almost everywhere, assume that*

$$
\eta(t) \le \alpha(t) + \int_s^T \beta(\tau)\eta(\tau)\,\mathrm{d}\tau, \quad s \le t \le T. \tag{1.3.16}
$$

Then,

$$
\eta(t) \le \alpha(t) + \frac{1}{B(t)}\int_s^t \alpha(\tau)\beta(\tau)B(\tau)\,\mathrm{d}\tau, \tag{1.3.17}
$$

for $s \le t \le T$, where

$$
B(t) = \exp\left(-\int_s^t \beta(\tau)\,\mathrm{d}\tau\right).
$$

Proof By setting

$$
N(t) = \int_s^T \beta(\tau)\eta(\tau)\,\mathrm{d}\tau,
$$

Equation (1.3.16) can be rewritten as $N'(\tau) \le \alpha(\tau)\beta(\tau) + \beta(\tau)N(\tau)$. Therefore, $B(\tau)N'(\tau) - \beta(\tau)B(\tau)N(\tau) \le \alpha(\tau)\beta(\tau)B(\tau)$, or $(B(\tau)N(\tau))' \le \alpha(\tau)\beta(\tau)B(\tau)$. Integrating over $s \le \tau \le t$, we obtain

$$N(t) \le \frac{1}{B(t)} \int_s^t \alpha(\tau)\beta(\tau)B(\tau)\, \mathrm{d}\tau.$$

This, together with (1.3.16), leads to (1.3.17). □

Corollary 1.3.1 (Gronwall) *Assume* $\alpha(t) = \alpha$ *in* (1.3.16). *Then*

$$\eta(t) \le \alpha \exp\left(\int_s^t \beta(\tau)\, \mathrm{d}\tau\right), \qquad s \le t \le T. \tag{1.3.18}$$

Proof This can be verified by noting that the equation $B'(\tau) = -\beta(\tau)B(\tau)$ solves the integral on the right-hand side of (1.3.17). □

Broadly speaking, there exist two forms of the Gronwall inequality. One is in differential form, while the other is in integral form. The Gronwall lemma encompasses numerous variants and generalizations, as detailed in [179]. For informational purposes, we present one such variant directly below.

For a continuous function $\eta = \eta(t)$ in the interval $[0, T]$, which is differentiable on $(0, T)$, if there exists a constant $\alpha \in \mathbb{R}$ and a continuous function $g(t)$ such that for $0 < t < T$, η satisfies the inequality

$$\eta'(t) \le \alpha\eta(t) + g(t)$$

(or, equivalently,

$$\eta(t) \le \eta(0) + \int_0^t [\alpha\eta(s) + g(s)]\, \mathrm{d}s),$$

then η satisfies the inequality

$$\eta(t) \le e^{\alpha t}\eta(0) + \int_0^t g(s)e^{\alpha(t-s)}\, \mathrm{d}s.$$

1.4 Pontryagin's Maximum Principle

Pontryagin's maximum principle, established by Lev Semenovich Pontryagin and his team in 1956 [186], is a cornerstone in optimal control theory. Derived from classical calculus of variations, this principle explores the optimal control of dynamic systems by examining the behavior of systems under slight perturbations. By considering the first-order term of a Taylor expansion with respect to these perturbations and then letting the perturbation tend to zero, a variational inequality emerges, leading to the maximum principle.

In the context of optimal control, Pontryagin's maximum principle serves as a vital tool for finding the most effective control strategies to steer a dynamic system from one state to another, especially in scenarios involving state or input constraints. It poses a necessary condition that must be satisfied by any optimal trajectory, making it invaluable both as a computational and analytical technique. This principle is not limited to finite-dimensional systems; it finds applications in infinite-dimensional systems as well.

Due to its versatility, Pontryagin's maximum principle has found widespread use in various fields, including mathematics, engineering, economics, and agriculture, successfully addressing numerous practical challenges. As the fundamental and essential necessary optimality condition, it stipulates that the optimal trajectory must adhere to the Euler-Lagrange equation. Within this equation, the optimal control ensures that the Hamiltonian is maximized at every point within its bounded domain along the trajectory. Concurrently, this variational problem transforms into a nonlinear programming problem at each point along the trajectory.

To illustrate the principle's application, we present below two examples of optimal control problems in finite dimensions, highlighting their respective Pontryagin maximum principles. These examples provide a basic understanding of the principle, noting that its form may vary depending on the specific problem. One example is presented without a detailed proof, while the other is accompanied by its corresponding proof. For a deeper understanding and further applications, the reader is encouraged to consult relevant literature for the maximum principle of optimal control problems in finite dimensions [231] and in infinite dimensions [150].

Let $T > 0$. Given a control system [98]

$$\begin{cases} \frac{d}{dt}x = f(x, u), \\ x(t_0) = x_0, \end{cases} \tag{1.4.1}$$

subject to the terminal constraints

$$\begin{cases} g_1(x(T)) = 0, \\ g_2(x(T)) \geq 0, \end{cases} \tag{1.4.2}$$

where $x \in \mathbb{R}^n$, $u \in \mathbb{R}^r$,

$$\begin{aligned} f^\top(x, u) &= (f_1(x, u), f_2(x, u), \cdots, f_n(x, u)), \\ g_1^\top(x) &= \left(g_{11}(x), g_{12}(x), \cdots, g_{1p}(x)\right), \\ g_2^\top(x) &= \left(g_{21}(x), g_{22}(x), \cdots, g_{2q}(x)\right), \end{aligned}$$

and $p \leq n$. Here and thereafter, whenever we mention $g_2(x) \geq 0$, it is implicitly understood that for all indices $j = 1, 2, \cdots, q$, the corresponding element $g_{2j}(x) \geq 0$.

In this section, the objective of the optimal control problem is to identify an admissible control, denoted as $u^*(t) \in U$ within the interval $t \in [t_0, T^*]$, along with a specific time instant T^*. The goal is to ensure that, under the influence of the control function $u(t)$, the trajectory $x(t)$ satisfies the constraints stated in (1.4.2). Specially, the objective is to minimize the functional

$$J(u(\cdot)) = S(x(T)). \tag{1.4.3}$$

Here $(u^*(t), T^*)$ is the optimal solution to the optimal control problem, and $(u^*(t), x^*(t))$ is the optimal control-trajectory pair.

As is well established, the Pontryagin maximum principle serves as the necessary optimality condition satisfied by the optimal solution $(u^*(t), T^*)$. To proceed, we make the following assumptions. The functions

$$f_i(x, u), \ \frac{\partial f_i(x, u)}{\partial x_j}, \ g_{ik}(x), \ \frac{\partial g_{ik}(x)}{\partial x_j}, \ S(x), \ \frac{\partial S(x)}{\partial x_j},$$

are assumed to be continuous in their respective domains. For any bounded region $D \subset \mathbb{R}^n$ contained within its defined domain, the function $f(x, u)$ satisfies the Lipschitz condition with respect to the variable x. That is to say, for any $x_1, x_2 \in D$, there holds

$$|f(x_1, u) - f(x_2, u)| \leq a|x_1 - x_2|, \quad a > 0.$$

Introduce the following Hamiltonian of the system (1.4.1):

$$H(x, \psi, u) = \sum_{i=1}^{n} \psi_i f_i(x, u),$$

where $\psi = (\psi_1, \psi_2, \cdots, \psi_n)$ is the adjoint variable of the state variable x.

Theorem 1.4.1 (Pontryagin's Maximum Principle [98]) *Let* $(u^*(t), T^*)$ *with* $t_0 \leq t \leq T^*$ *solve the optimal control problem* (1.4.1)–(1.4.3). *The optimal control-trajectory pair is* $(u^*(t), x^*(t))$. *Then, there exists a nonzero vector-valued function* $\psi^*(t)$, $t_0 \leq t \leq T^*$, *such that* $u^*(t), x^*(t), \psi^*(t)$, *and* T^* *satisfy the following conditions:*

(i)

$$\frac{d}{dt}x_i^*(t) = \frac{\partial H(x^*(t), \psi^*(t), u^*(t))}{\partial \psi_i} = f_i(x^*(t), u^*(t)),$$

$$x_i^*(t_0) = x_{i0}, \quad i = 1, 2, \cdots, n,$$

$$\frac{d}{dt}\psi_j^*(t) = -\frac{\partial H(x^*(t), \psi^*(t), u^*(t))}{\partial x_j} = -\sum_{i=1}^{n} \psi_i^*(t)\frac{\partial f_i(x^*(t), u^*(t))}{\partial x_j},$$

$$j = 1, 2, \cdots, n.$$

(ii)

$$\psi_j^*(T) = -\mu_0 \frac{\partial S(x^*(T))}{\partial x_j} - \sum_{i=1}^{p} \mu_i \frac{\partial g_{1j}(x^*(T))}{\partial x_j}\frac{\partial g_{1i}(x^*(T))}{\partial x_j}$$

$$-\sum_{i=1}^{q} \nu_i \frac{\partial g_{2j}(x^*(T))}{\partial x_j}\frac{\partial g_{2i}(x^*(T))}{\partial x_j}, \quad j = 1, 2, \cdots, n,$$

where $\mu_0 \geq 0,\ \nu_j \geq 0,\ \nu_j g_{2j}(x^*(T^*)) = 0,\ j = 1, 2, \cdots, q$ *and* μ_0, μ_i, ν_j, $i = 1, 2, \cdots, p$, $j = 1, 2, \cdots, q$, *are constants not all identical to zero.*

(iii) $H(x^*(t), \psi^*(t), u^*(t)) = \max_{u \in U} H(x^*(t), \psi^*(t), u)$.

(iv) $H(x^*(t), \psi^*(t), u^*(t)) \equiv \text{Constant} = 0$.

In the case of a fixed final horizon, the constant mentioned in condition (iv) does not necessarily have to be zero. Instead, under such circumstances, the corresponding condition takes the following form: (iv′).

$$H(x^*(t), \psi^*(t), u^*(t)) \equiv \text{Constant}.$$

The Pontryagin maximum principle is named so due to condition (iii), which stipulates that if $u(t)$ is the optimal control, it must maximize the Hamiltonian along the optimal trajectory. This condition ensures that the control chosen is optimal in terms of maximizing the desired objective function. Meanwhile, condition (ii) is commonly referred to as the transversality condition in the context of optimal control problems. It represents a specific requirement satisfied by $\psi^*(T)$ in problems involving terminal constraints. This condition ensures that the optimal trajectory satisfies the necessary boundary conditions at the final time T.

Indeed, the determination of whether the extreme value is a maximum or a minimum hinges on the sign convention adopted for defining the Hamiltonian. It is noteworthy that L. S. Pontryagin employed the classical concept of the Hamiltonian, which differs in sign from the contemporary form we utilize today. This historical convention gave rise to the term "maximum principle." Nowadays, however, it is more commonly known simply as Pontryagin's principle, without the qualifiers of maximum or minimum.

Furthermore, Pontryagin's principle stands as one of the three milestones in control theory. The other two are the Bellman DPP and the Kalman linear quadratic regulator theory. These three principles have significantly contributed to the advancement and understanding of control systems, enabling engineers and scientists to design and analyze complex systems with greater precision and efficiency.

Next, we delve into exploring the maximum principle of a newly formulated optimal control problem and providing the corresponding proof [124]. The objective is to minimize the cost functional

$$\ell_0(x(0), x(T)), \tag{1.4.4}$$

subject to the following state equation and its constraints:

$$\begin{cases} \dfrac{d}{dt}x = f(t, x, u), \quad u \in U, \\ \ell_j(x(0), x(T)) \le 0, \quad j = 1, 2, \cdots, l, \\ \ell_j(x(0), x(T)) = 0, \quad j = l+1, l+2, \cdots, r. \end{cases} \tag{1.4.5}$$

Here, x and u are elements of Euclidean spaces, specially $\mathbb{R}^n$ and $\mathbb{R}^m$. A pair $(x(\cdot), u(\cdot))$ is said to be feasible in (1.4.4)–(1.4.5) if $x(\cdot)$ is absolutely continuous, $u(\cdot)$ is bounded and measurable, $x(\cdot)$ and $u(\cdot)$ satisfy the equation almost everywhere, and $x(\cdot)$ satisfies the terminal point conditions of (1.4.4) and (1.4.5). Throughout this section, we fix a feasible pair $(\overline{x}(\cdot), \overline{u}(\cdot))$, and assume that ℓ_j are continuously differentiable in a neighborhood of $(\overline{x}(0), \overline{x}(T))$; furthermore, U is a closed set. Additionally, assume that

(H$'$) The functions $f(t, x, u)$ and $f_x'(t, x, u)$ are measurable in t and continuous in (x, u). For any $N > 0$, there exists a summable function $c_N(\cdot)$ on $[0, T]$ and an $\varepsilon_N > 0$ such that the condition $\|f(t, x, u)\| \le c_N(t)$ and $\|f_x'(t, x, u)\| \le c_N(t)$ is met whenever $\|x - \overline{x}(t)\| \le \varepsilon_N, \ u \in U, \ \|u\| \le N$.

It is generally considered that $(\overline{x}(\cdot), \overline{u}(\cdot))$ is either an optimal process or a strong minimum in (1.4.4) and (1.4.5) if there is an $\varepsilon > 0$ guaranteeing $\ell_0(x(0), x(T)) \ge \ell_0(\overline{x}(0), \overline{x}(T))$ for any feasible $(x(\cdot), u(\cdot))$ such that $\|x(t) - \overline{x}(t)\| \le \varepsilon$ for all t. We will utilize the usual inner product $\langle \cdot, \cdot \rangle$ in $\mathbb{R}^n$ to define

$$H(t, x, p, u) = \langle p, f(t, x, u) \rangle, \quad \mathcal{H}(t, x, p) = \max_{u \in U} H(t, x, p, u).$$

To thoroughly examine and furnish a rigorous proof for the maximum principle of the optimal control problem defined by Eqs. (1.4.4) through (1.4.5), it is imperative to establish a robust theoretical foundation. We first introduce a crucial component in this endeavor, the Lyusternik-Graves theorem, which plays a pivotal role in the subsequent reduction process of the proof [124].

Let X and Y be Banach spaces and let B_X and B_Y be the unit balls in X and Y, respectively. The Banach constant of a linear continuous operator $A : X \to Y$ is defined as

$$C(A) = \sup\{r \geq 0 \mid rB_Y \subset A(B_X)\}.$$

Theorem 1.4.2 *Let $F : X \to Y$ be a C^1-mapping defined in a neighborhood of a $\overline{x} \in X$ and set $\overline{y} = F(\overline{x})$. If the Fréchet derivative $F'(\overline{x})$ is a linear operator onto Y, then for any $K > 0$ with $KC\left(F'(\overline{x})\right) > 1$, there exists a neighborhood of $(\overline{x}, \overline{y})$ such that the distance estimate*

$$d\left(x, F^{-1}(y)\right) \leq K\|y - F(x)\|$$

holds for all (x, y) of the neighborhood.

This result is equivalent to the Lyusternik-Graves theorem, which states that under the assumptions, for any $r < C\left(F'(\overline{x})\right)$, there exists an $\varepsilon > 0$ such that $\overline{B}(y, rt) \cap \overline{B}(\overline{y}, \varepsilon) \subset F\left(\overline{B}(\overline{x}, t)\right)$ if $\|y - \overline{y}\| < \varepsilon$ and $0 \leq t < \varepsilon$.

This equivalence is explicitly stated in [68]. The notation $B(x, r)$ represents an open ball centered at x with a radius of r, whereas $\overline{B}(x, r)$ denotes a closed ball. For detailed historical comments and a straightforward proof of the theorem, please refer to [123]. The theorem asserts that if y is sufficiently close to $\overline{y}$, there exists a solution to the equation $F(x) = y$ such that $\|x - \overline{x}\| \leq K\|y - \overline{y}\|$.

Furthermore, let Z be another Banach space. The theorem presented below is a direct consequence of the Lyusternik-Graves theorem [123, 124].

Theorem 1.4.3 *Consider a function $F : X \times Z \to Y$ that is both defined and continuously differentiable in a neighborhood of $(\overline{x}, \overline{z})$. Let $\overline{y} = F(\overline{x}, \overline{z})$ and define $S(z, y)$ as the solution set for the equation $F(x, z) = y$. Assuming that $F_x'(\overline{x}, \overline{z})$ maps X onto Y, there exist positive constants K, ε, along with a neighborhood of $(\overline{z}, \overline{y})$ such that for any two pairs (z, y) and (z', y') within this neighborhood, we have*

$$S(z', y') \cap \overline{B}(\overline{x}, \varepsilon) \subset S(z, y) + K\left(\|z - z'\|B_Z \times \|y - y'\|B_Y\right).$$

Specially, $S(z, y) \neq \emptyset$ for (z, y) sufficiently close to $(\overline{z}, \overline{y})$.

We now apply the theorems discussed above to the differential equation $\frac{d}{dt}x = f(t, x)$, where $x \in \mathbb{R}^n$, $t \in [0, T]$. Consider a $\overline{x}(\cdot) \in W^{1,1}$. Assume that there exists an $\varepsilon > 0$ such that f is continuously differentiable with respect to x in the ε-ball around $\overline{x}(t)$ for almost every t. Additionally, suppose there is a summable $c(t)$ such that both $\|f(t,x)\| \le c(t)$ and $\|f'_x(t,x)\| \le c(t)$ for any x within this ball and for almost every t. Then the mapping F from $W^{1,1}$ into L^1 defined by

$$F(x(\cdot))(t) = \frac{d}{dt}x(t) - f(t, x(t))$$

is continuously differentiable near $\overline{x}(\cdot)$.

The derivative $F'(x(\cdot))$ transforms any $h(\cdot) \in W^{1,1}$ into $\frac{d}{dt}h(t) - f'_x(t, x(t))h(t)$. The linear equation $\frac{d}{dt}h - f'_x(t, x(t))h = z(t)$ admits a solution for any $z(\cdot) \in L^1$. Consequently, for any $x(\cdot) \in W^{1,1}$ in proximity to $\overline{x}(\cdot)$, $F'(x(\cdot))$ maps $W^{1,1}$ onto the entirety of L^1.

Let $S = F^{-1}(0) \subset W^{1,1}$ be the set of solutions to the equation. According to Theorem 1.4.2, there exists a constant $K > 0$ such that for all such $x(\cdot) \in W^{1,1}$, $d(x(\cdot), S) < K \int_0^T \left\| \frac{d}{dt}x(t) - f(t, x(t)) \right\| \, dt$. In simpler terms, for any $x(\cdot)$ in close proximity to $\overline{x}(\cdot)$, there exists a solution $\hat{x}(\cdot)$ of the equation that fulfills the condition

$$\|x(\cdot) - \hat{x}(\cdot)\| \le K \int_0^1 \left\| \frac{d}{dt}x(t) - f(t, x(t)) \right\| \, dt.$$

Assume that f now depends on a parameter p residing in some Euclidean space, and let $\overline{x}(\cdot)$ be a solution of the equation $\frac{d}{dt}x = f(t, x, \overline{p})$. Additionally, suppose there exist constants $\varepsilon > 0$ and a summable $c(t)$ such that, for almost every t, the norms of $f(t, x, p)$, $f'_x(t, x, p)$ and $f'_p(t, x, p)$ do not exceed $c(t)$ provided x and p are ε-close to $(\overline{x}(t), \overline{p})$. Define $S(p)$ as the set of solutions to the equation associated with a given p. By invoking Theorem 1.4.3 (with $y = y' = 0,\ z = p,\ z' = q$), we deduce the existence of a constant $K > 0$ such that, whenever p, q are in close proximity to $\overline{p}$ and $\omega(\cdot)$ is a solution of $\frac{d}{dt}x = f(t, x, q)$ that is sufficiently close to $\overline{x}(\cdot)$, there exists a solution $x(\cdot)$ of the equation $\frac{d}{dt}x = f(t, x, p)$ satisfying

$$\|x(\cdot) - \omega(\cdot)\| \le K\|p - q\|.$$

In fact, the Lyusternik-Graves theorem is a cornerstone of calculus of variations. It builds upon two distinct results, independently discovered by L. A. Lyusternik in 1934 and L. M. Graves in 1950 [123]. This theorem occupies a middle ground, being slightly weaker than the Graves theorem and slightly stronger than the Lyusternik theorem. The Graves theorem is outlined below, while the Lyusternik theorem will be discussed at a later stage.

Theorem 1.4.4 (Graves Theorem) *Let F be a continuous mapping that maps a neighborhood of $\overline{x} \in X$ into Y. Suppose that there exist a linear bounded operator $A : X \to Y$ and positive numbers $\delta > 0,\ \gamma > 0,\ \varepsilon > 0$ such that $C(A) > \delta + \gamma$ and*

$$\left\| F(x') - F(x) - A(x' - x) \right\| < \delta \left\| x' - x \right\|, \tag{1.4.6}$$

whenever x and x' belong to the open ε-ball around $\overline{x}$. Then, for all $t \in (0, \varepsilon)$,

$$\overline{B}\,(F(\overline{x}), \gamma t) \subset F\left(\overline{B}\,(\overline{x}, t)\right).$$

Proof We suppose $F(\overline{x}) = 0$ for convenience and take $K > 0$ such that $KC(A) > 1 > K(\delta + \gamma)$. Let $\|y\| < \gamma t$ for some $t < \varepsilon$, and set $x_0 = \overline{x},\ y_0 = y$. Define recursively first x_n and then y_n as follows:

$$\begin{aligned} y_{n-1} &= A(x_n - x_{n-1}), \quad \|x_n - x_{n-1}\| \le K\|y_{n-1}\|, \\ y_n &= A(x_n - x_{n-1}) - (F(x_n) - F(x_{n-1})). \end{aligned} \tag{1.4.7}$$

By subtracting the second equality from the first one, we obtain $y_{n-1} - y_n = F(x_n) - F(x_{n-1})$. Therefore,

$$y - y_n = \sum_{i=1}^{n} (y_{i-1} - y_i) = \sum_{i=1}^{n} (F(x_i) - F(x_{i-1})) = F(x_n).$$

Thus, if the process can be extended infinitely and x_n converges to some x, we would have $F(x) = y$.

To confirm this and determine the distance between x and $\overline{x}$, we observe that $\|y_n\| \le \delta\|x_n - x_{n-1}\|$ by (1.4.6) and the second equation in (1.4.7). This, together with the inequality in (1.4.7), gives $\|y_n\| \le K\delta\|y_{n-1}\|$. Therefore,

$$\|x_n - x_{n-1}\| \le (K\delta)^{n-1} K\|y\|, \quad \|y_n\| \le (K\delta)^n \|y\|. \tag{1.4.8}$$

As a result,

$$\|x_n - \overline{x}\| \le \frac{K}{1 - K\delta}\|y\| \le \gamma^{-1}\|y\| < t.$$

This shows that the process can be continued and, by (1.4.8), $\{x_n\}$ is a Cauchy sequence. Finally, if $x = \lim_{n\to\infty} x_n$, then $\|x - \overline{x}\| \le t$ as claimed. □

Below, we present a corollary, which is sometimes referred to as the Graves theorem at nearby points.

Corollary 1.4.1 *Under the assumptions of the theorem, it follows that*

$$\overline{B}(F(x), \gamma t) \subset F\left(\overline{B}(x, t)\right),$$

whenever $\|x - \overline{x}\| + t < \varepsilon,\ t > 0$.

Proof Replacing $\overline{x}$ by x and ε by $\varepsilon - \|x\|$ leads to the required result. □

Additionally, a normed space is a metric space with a metric $\rho(x, y) = \|x - y\|$. Given a metric space (X, ρ), a point $x \in X$, and a set $C \subset X$, the quantity

$$d(x, C) = \inf_{y \in C} \rho(x, y)$$

is called the distance from x to C. Moreover, we can present another corollary as follows.

Corollary 1.4.2 *Under the assumptions of Graves' theorem, for all* (x, y) *satisfying* $\|x - \overline{x}\| + \gamma^{-1}\|y - F(x)\| < \varepsilon$, *there holds the inequality*

$$d\left(x, F^{-1}(y)\right) \leq \frac{1}{\gamma}\|y - F(x)\|.$$

Proof Given γ, we can set $t = \gamma^{-1}\|y - F(x)\|$. This allows $\|x - \overline{x}\| + t < \varepsilon$ and $y \in \overline{B}(F(x), \gamma t)$. By applying Corollary 1.4.1, we can find a $u \in X$ such that $F(u) = y$ and $\|x - u\| \leq \gamma^{-1}t$. As a result,

$$d\left(x, F^{-1}(y)\right) \leq \|x - u\| \leq \gamma^{-1}\|y - F(x)\|.$$

□

The second step involves presenting the fact regarding measurable set-valued mappings, as outlined in [191]. We will simply enumerate the fundamental facts as follows.

If $f(t, u)$ is a Carathéodory function on $[0, T] \times \mathbb{R}^n$, then the set-valued mapping $t \to \{u \in U(t) \mid f(t, u) \leq \gamma(t)\}$ is measurable for any measurable $\gamma(t)$. Additionally,

$$\inf_{u(\cdot) \in \mathcal{U}} \int_0^T f(t, u(t))\, \mathrm{d}t = \int_0^T \inf_{u \in U(t)} f(t, u)\, \mathrm{d}t, \tag{1.4.9}$$

where $\mathcal{U}$ is the collection of measurable selections of $U(\cdot)$.

The third step entails introducing the relaxation theorem. Herein, we present its simplified version. Define Ξ as the standard $(k-1)$-simplex in $\mathbb{R}^k$, where

$$\Xi = \left\{ a = (\alpha_1, \alpha_2, \cdots, \alpha_k) \in \mathbb{R}^k \;\middle|\; \alpha_i \geq 0,\ \sum_{i=1}^{k} \alpha_i = 1 \right\}.$$

Let $f_i : [0, T] \times \mathbb{R}^n \to \mathbb{R}^n$ be Carathéodory mappings. Consider the differential inclusion

$$\frac{d}{dt}x \in \{f_1(t, x), f_2(t, x), \cdots, f_k(t, x)\} \tag{1.4.10}$$

and the equation

$$\frac{d}{dt}x = \alpha_1 f_1(t,x) + \alpha_2 f_2(t,x) + \cdots \alpha_k f_k(t,x), \quad (\alpha_1, \alpha_2, \cdots, \alpha_k) \in \Xi. \tag{1.4.11}$$

Let $x(\cdot)$ be a solution of (1.4.11) for some $(\alpha_1, \alpha_2, \cdots, \alpha_k) \in \Xi$ (which certainly suggests that $f_i(\cdot, x(\cdot))$ are summable).

Theorem 1.4.5 *Suppose that*

$$\left\| f_i(t,x) - f_i(t,x') \right\| \le k(t) \left\| x - x' \right\|, \quad x, x' \in \overline{B}(x(t), \varepsilon),$$

for certain $\varepsilon > 0$ *and a summable* $k(\cdot)$. *Then there is a sequence of* $\{x_s(\cdot)\}$ *that solves the Eq.* (1.4.10) *and converges uniformly to* $x(\cdot)$.

Actually, the proof of this theorem is straightforward. We only need to examine a sequence $\{\alpha_{1s}(t), \alpha_{2s}(t), \cdots, \alpha_{ks}(t)\}$ with each $\alpha_{is}(t)$ being either 0 or 1, and $\sum_{i=1}^{k} \alpha_{is}(t) = 1$ a.e. for all $s = 1, 2, \cdots$, which converges in weak*-topology to a function that is precisely $(\alpha_1, \alpha_2, \cdots, \alpha_k)$. The solutions for the desired $x_s(\cdot)$ are derived from the equation $\frac{d}{dt}x = \sum_{i=1}^{k} \alpha_{is}(t) f_i(t,x)$, with $x_s(0) = x(0)$ as the initial conditions.

Furthermore, let us consider X as a complete metric space, with $\overline{x}$ representing a local minimum for the problem of minimizing a function f on a subset $M \subset X$. Additionally, it is assumed that f satisfies the Lipschitz condition in a neighborhood of $\overline{x}$. Furthermore, we have a non-negative lower semicontinuous function φ defined on X, which vanishes at $\overline{x}$. Consequently, this gives rise to optimality alternatives, which can be categorized as either the regular case or the singular case.

In the regular case, there is a $\rho > 0$ such that the function $\rho f + \varphi$ achieves an unconditional local minimum at $\overline{x}$.

In the singular case, there exists a sequence of $\{x_m\}$ converging to $\overline{x}$, and for any m, the function $\varphi + m^{-1}d(\cdot, x_m)$ attains global minimum on X at x_m and $d(x_m, M) > 0$.

In fact, the regular case arises when there exists an $N > 0$ such that $d(x, M) \le N\varphi(x)$ holds for x in a neighborhood of $\overline{x}$. Any positive $\rho > 0$ satisfying ρ^{-1} being greater than N times the Lipschitz constant of f can be selected. Consequently, if X is a Banach space and $M = \{x \mid F(x) = 0\}$, where F is a mapping into another Banach space Y and F is continuously differentiable at x with $\mathrm{Im}F'(\overline{x}) = Y$, then, according to the Lyusternik-Graves theorem, x is an unconditional local minimum of $f(x) + K\|F(x)\|$ for a sufficiently large K. For further details, we refer to [123, 124].

We present the maximum principle as outlined in Theorem 1.4.6.

Theorem 1.4.6 *Let* $(\overline{x}(\cdot), \overline{u}(\cdot))$ *be an optimal process in* (1.4.4) *and* (1.4.5). *Then, there exist constants* $\lambda_i,\ i = 0, 1, \cdots, r$ *and an absolutely continuous function* $p(\cdot)$ *such that* $\lambda_j \geq 0,\ j = 0, 1, \cdots, l,\ \lambda_j \ell_j(\overline{x}(0), \overline{x}(T)) = 0,\ j = 1, 2, \cdots, l$, *and the following conditions hold:*

(i) $\sum_{j=0}^{l} \lambda_j + \sum_{j=l+1}^{r} |\lambda_j| > 0$ *(nontriviality).*

(ii) $(p(0), -p(T)) = \sum_{j=0}^{r} \lambda_j \ell_j'(\overline{x}(0), \overline{x}(T))$ *(transversality).*

(iii) $-\frac{d}{dt} p = H_x'(t, \overline{x}(t), p, \overline{u}(t))$ a.e. *(adjoint equation).*

(iv) $H(t, \overline{x}(t), p(t), \overline{u}(t)) = \mathcal{H}(t, \overline{x}(t), p(t))$ a.e. *(maximum principle).*

Proof We may assume that $\ell_0(\overline{x}(0), \overline{x}(T)) = 0$. This leads to $(\overline{x}(\cdot), \overline{u}(\cdot))$ being a strong local minimum in the problem that minimize

$$\max_{0 \leq j \leq l} \ell_j(x(0), x(T)), \tag{1.4.12}$$

subject to

$$\begin{cases} \frac{d}{dt} x = f(t, x, u), \quad u \in U, \\ \ell_j(x(0), x(T)) = 0, \quad j = l+1, l+2, \cdots, r. \end{cases} \tag{1.4.13}$$

The collection of all summable functions $u(\cdot)$ on $[0, T]$ taking values in U is denoted as $\mathcal{U}$. We consider $\mathcal{U}$ with the L^1-metric because U is a closed set and thus turns $\mathcal{U}$ into a complete metric space.

First, we introduce an auxiliary problem. Choose a finite collection of bounded elements of $\mathcal{U}$ denoted as $u_1(\cdot), u_2(\cdot), \cdots, u_k(\cdot)$. Set $U_k(t) = \{\overline{u}(t), u_1(t), \cdots, u_k(t)\}$. Consider the problem ($\mathrm{OC}_k$) as stated in (1.4.12) and (1.4.13), with the only difference that in (OC_k), it is required that $u(t) \in U_k(t)$ almost everywhere for any feasible pair $(x(\cdot), u(\cdot))$. The collection of feasible pairs in (OC_k) is denoted as M_k.

Let $\mathcal{U}_k \subset \mathcal{U}$ be the collection of measurable $u(\cdot)$ that takes values in $U_k(t)$ almost everywhere (which is necessarily bounded). X_k consists of all pairs $(x(\cdot), u(\cdot))$ where $u(\cdot) \in \mathcal{U}_k$ satisfies the equation $\frac{d}{dt} x = f(t, x, u(t))$. The metric for X_k is

$$\begin{aligned} & d\left((x(\cdot), u(\cdot)), \left(x'(\cdot), u'(\cdot)\right)\right) \\ & = \left\|x(0) - x'(0)\right\| + \int_0^T \left(\left\|x(t) - x'(t)\right\| + \left\|u(t) - u'(t)\right\|\right) dt. \end{aligned}$$

It is easy to see that (X_k, d) is a complete metric space. For X_k, the topology that is defined by this metric and the $W^{1,1} \times L^1$ topology are equivalent.

We then proceed to the first reduction step. By applying the optimality alternative to (OC_k) and considering a problem defined on X_k with

$$\varphi(x(\cdot), u(\cdot)) = \sum_{i=l+1}^{r} |\ell_i(x(0), x(T))|,$$

we conclude that either there is a $\rho > 0$ such that $(\overline{x}(\cdot), \overline{u}(\cdot))$ is a local minimum on X_k of the functional

$$\mathcal{J}_0(x(\cdot)) = \rho \max_{0\le i\le 1} \ell_i(x(0), x(T)) + \sum_{i=l+1}^{r} |\ell_j(x(0), x(T))|$$

(in the regular case), or there exists a sequence

$$\{\overline{x}_s(\cdot), \overline{u}_s(\cdot)\} \in X_k$$

with $d((\overline{x}_s(\cdot), \overline{u}_s(\cdot)), M_k) > 0$ converging to $(\overline{x}(\cdot), \overline{u}(\cdot))$ and such that for any s, the functional

$$\mathcal{J}_s(x(\cdot), u(\cdot)) = \sum_{i=l+1}^{r} |\ell_i(x(0), x(T))| + s^{-1} d((x(\cdot), u(\cdot)), (\overline{x}_s(\cdot), \overline{u}_s(\cdot)))$$

reaches its global minimum on X_k at $(\overline{x}_s(\cdot), \overline{u}_s(\cdot))$ (in the singular case).

We can write both problems in a unified way as minimizing on X_k of the function

$$\begin{aligned} \mathcal{J}(x(\cdot), u(\cdot)) &= \rho \max_{0\le i\le l} \ell_i(x(0), x(T)) + \sum_{i=l+1}^{r} |\ell_j(x(0), x(T))| \\ &+\varepsilon\left(\int_0^T (\|x(t) - x_*(t)\| + \|u(t) - u_*(t)\|)\, dt + \|x(0) - x_*(0)\|\right), \end{aligned} \tag{1.4.14}$$

where $\rho \ge 0$ and $\varepsilon \ge 0$ satisfy $\rho + \varepsilon > 0$, $\rho\varepsilon = 0$ and $(x_*(\cdot), u_*(\cdot))$ stands for either $(\overline{x}(\cdot), \overline{u}(\cdot))$ (if $\rho > 0$) or $(\overline{x}_s(\cdot), \overline{u}_s(\cdot))$ (if $\rho = 0$ and $s = 1, 2, \cdots$).

Next, we move on to the second reduction. Set

$$v_i(t) = \begin{cases} u_i(t), & u_i(t) \ne u_*(t), \\ \overline{u}(t), & u_i(t) = u_*(t), \end{cases}$$

and examine the following equation:

$$\frac{d}{dt}x = f(t, x, u_*(t)) + \sum_{i=1}^{k} \alpha_i \left(f(t, x, v_i(t)) - f(t, x, u_*(t))\right). \tag{1.4.15}$$

For $\alpha \in \mathbb{R}$, denote $\alpha^+ = \max\{\alpha, 0\}$, $\alpha^- = (-\alpha)^+$. If $\alpha_i \geq 0$, $\sum_{i=1}^{k} \alpha_i \leq 1$ and $x(\cdot)$ is a solution of the equation, which is close enough to $\overline{x}(\cdot)$, then by the relaxation theorem, we can find a sequence $\{x_j(\cdot), u_j(\cdot)\} \in X_k$ with $x_j(\cdot)$ converging uniformly to $x(\cdot)$. As stated after the statement of the relaxation theorem,

$$w_j(t) = u_*(t) + \sum_{i=1}^{k} \alpha_{ij}(t)(v_i(t) - u_*(t)),$$

where $\alpha_{ij}(t) \in \{0, 1\}$, and only one of $\alpha_{ij}(t)$, $i = 1, 2, \cdots, k$, can be equal to 1. $\alpha_{ij}(\cdot)$ weak* converges to α_i in L^∞. Then,

$$\int_0^T \|w_j(t) - u_*(t)\| \, \mathrm{d}t = \sum_{i=1}^{k} \int_0^T \alpha_{ij}(t)\|v_j(t) - u_*(t)\| \, \mathrm{d}t$$

$$\to \sum_{i=1}^{k} \alpha_i \int_0^T \|v_j(t) - u_*(t)\| \, \mathrm{d}t.$$

Since all $u_i(\cdot)$ and $\overline{u}(\cdot)$ are bounded, so are $v_i(\cdot)$ and $u_*(\cdot)$. Consequently, a $K^+ > 0$ exists such that

$$\int_0^T \|v_i(t) - u_*(t)\| \, \mathrm{d}t \leq K^+.$$

To summarize the results of the previous paragraph, we conclude that $(x_*(\cdot), 0, \cdots, 0)$ is a local minimum, based on (1.4.15) and $\alpha_i \geq 0$, of the following functional on $W^{1,1} \times \mathbb{R}^n$:

$$\mathcal{J}_*(x(\cdot), u(\cdot), \alpha_1, \cdots, \alpha_k) = \rho \max_{0 \leq i \leq l} \ell_i((x(0), x(T)) + \sum_{i=l+1}^{r} |\ell_j(x(0), x(T))|$$

$$+\varepsilon \left(\int_0^T \|x(t) - x_*(t)\| \, \mathrm{d}t + K^+ \sum_{i=1}^{k} \alpha_i + \|x(0) - x_*(0)\| \right).$$

Let $S(\alpha_1, \alpha_2, \cdots, \alpha_k)$ be the set of solutions of (1.4.15), which are defined on $[0, T]$ and have the given $\alpha_i \in \mathbb{R}$. Our subsequent discussion after stating Theorem 1.4.3 has revealed that there exists $K_0 > 0$ such that whenever $w(\cdot) \in S(\alpha_1, \alpha_2, \cdots, \alpha_k)$ is sufficiently close to $\overline{x}(\cdot)$ and α_i is sufficiently close to zero, there is a $x(\cdot) \in S(\alpha_1^+, \alpha_2^+, \cdots, \alpha_k^+)$ such that

$$\|x(\cdot) - w(\cdot)\| \leq K_0 \sum_{i=1}^{k} \alpha_i^-.$$

Since $\mathcal{J}_*$ is a Lipschitz function of $x(\cdot)$, it can be deduced that there is a $K^- > 0$ such that $(x_*(\cdot), 0, \cdots, 0)$ is a local minimum in minimizing the function

$$\mathcal{J}_*(x(\cdot), \alpha_1, \cdots, \alpha_k) + K^- \sum_{i=0}^{k} \alpha_i^-$$

on the set of $(x(\cdot), \alpha_1, \cdots, \alpha_k) \in W^{1,1} \times \mathbb{R}^k$ satisfying (1.4.15).

An operator $\Phi : W^{1,1} \times \mathbb{R}^n \to L^1$ associated with (1.4.15) can be used to transform any $(x(\cdot), \alpha_1, \cdots, \alpha_k)$ into

$$\frac{d}{dt}x(t) - f(t, x(t), u_*(t)) - \sum_{i=1}^{k} \alpha_i \left(f(t, x(t), v_i(t)) - f(t, x(t), u_*(t))\right).$$

According to (H′), Φ is Fréchet differentiable at $(x_*(\cdot), 0, \cdots, 0)$ and

$$\Phi'(x_*(\cdot), \alpha_1, \cdots, 0)(h(\cdot), \beta_1, \cdots, \beta_k)(t) = \frac{d}{dt}h(t) - A(t)h(t) - \sum_{i=1}^{k} \beta_i \psi_i(t),$$

where $A(t) = f_x'(t, x_*(t), u_*(t))$ and

$$\psi_i(t) = f(t, x_*(t), v_i(t)) - f(t, x_*(t), u_*(t)).$$

The equation

$$\frac{d}{dt}h - A(t)h = z(t)$$

has a solution for each $z(\cdot) \in L^1$ (e.g., with $h(0) = 0$). Thus $\Phi'(x_*(\cdot), 0, \cdots, 0)$ is a surjective operator. By consecutively applying the Lyusternik-Graves theorem and the discussion on the regular part of the optimality alternative provided immediately prior to Theorem 1.4.6, we can obtain an $N > 0$ such that $(x_*(\cdot), 0, \cdots, 0)$ represents a local minimum of the functional

$$\begin{aligned}
&\mathcal{I}_*(x(\cdot), \alpha_1, \cdots, \alpha_k) \\
&= \mathcal{J}_*(x(\cdot), \alpha_1, \cdots, \alpha_k) + K^- \sum_{i=1}^{k} \alpha_i^- \\
&\quad + N \int_0^T \left\| \frac{d}{dt}x(t) - f(t, x, u_*(t)) - \sum_{i=1}^{k} \alpha_i \left(f(t, x, v_i(t)) - f(t, x, u_*(t))\right) \right\| dt.
\end{aligned}$$

Furthermore, we conduct the following analyses. The directional derivative

$$\mathcal{I}'_*((x_*(\cdot), 0, \cdots, 0); (h(\cdot), \beta_1, \cdots, \beta_k))$$

of $\mathcal{I}_*$ at $(x_*(\cdot), 0, \cdots, 0)$ along $(h(\cdot), \beta_1, \cdots, \beta_k)$ is given by

$$\rho \max_{i\in I}\{\langle a_i, h(0)\rangle + \langle b_i, h(T)\rangle\} + \sum_{i=l+1}^{r} \mu_i(\langle a_i, h(0)\rangle + \langle b_i, h(T)\rangle)$$
$$+K^- \sum_{i=1}^{k} \beta_i^- + \varepsilon\left(\int_0^T \|h(t)\|\, dt + K^+ \sum_{i=1}^{k} \beta_i^+ + \|h(0)\|\right)$$
$$+N \int_0^T \left\| \frac{d}{dt}h(t) - A(t)h(t) - \sum_{i=1}^{k} \beta_i \psi_i(t)\right\| dt,$$

where $I = \{i \in \{0, 1, \cdots, l\} \mid \ell_i(\overline{x}(0), \overline{x}(T)) = 0\}$ if $\rho > 0$, a_i and b_i are equal, respectively, to the derivatives of $\ell_i(\cdot, \cdot)$ at $(x_*(0), x_*(T))$ with respect to the first and second variable, and $A(\cdot)$ and $\psi(\cdot)$ remain unchanged as before, while $\mu_i(t) = |t|$ if $\ell_i(\overline{x}(0), \overline{x}(T)) = 0$ and μ_i sign $\ell_i(\overline{x}(0), \overline{x}(T))$ otherwise. Clearly, this is a nonnegative convex function on $W^{1,1} \times \mathbb{R}^k$. Therefore, zero is included in its subdifferential. By applying standard rules of convex subdifferential calculus, we can find constants λ_i, where $i = 0, 1, \cdots, r$, such that $\lambda_i \geq 0$, $\lambda_i \ell_i(\overline{x}(0), \overline{x}(T)) = 0$ for $i = 0, 1, \cdots, l$ and $\lambda_0 + \lambda_1 + \cdots + \lambda_l = \rho$. Additionally, $|\lambda_i| \leq 1$ and $\lambda_i \ell_i(x_*(0), x_*(T)) = |\ell_i(x_*(0), x_*(T))|$ for $i = l+1, l+2, \cdots, r$. We can also find a vector $\theta \in \mathbb{R}^n$ with $\|\theta\| \leq \varepsilon$, two bounded measurable $\mathbb{R}^n$-valued functions $q(t)$ and $p(t)$ such that $\|q(t)\| \leq \varepsilon$ and $\|p(t)\| \leq N$ and two more sets of real numbers $\xi_i^- \in [-K, 0]$ and $\xi_i^+ \in [0, \varepsilon K^+]$, $i = 1, 2, \cdots, k$ such that for all $h(\cdot) \in W^{1,1}$,

$$\left\langle \theta + \sum_{i=0}^{r} a_i, h(0)\right\rangle + \left\langle \sum_{i=0}^{k} b_i, h(T)\right\rangle$$
$$+ \int_0^T \left\langle p(t), \frac{d}{dt}h(t) - A(t)h(t)\right\rangle dt + \int_0^T \langle q(t), h(t)\rangle\, dt = 0$$

and

$$\xi_i^- + \xi_i^+ - \int_0^T \langle p(t), \psi_i(t)\rangle\, dt = 0, \quad i = 1, 2, \cdots, k.$$

The last equality directly implies that

$$\int_0^T (H(t, x_*(t), p(t), v_j(t)) - H(t, x_*(t), p(t), u_*(t)))\,\mathrm{d}t \le \varepsilon K^+, \qquad i = 1, 2, \cdots, k. \tag{1.4.16}$$

The first equality necessitates a slightly more intricate analysis. By integrating

$$\langle p(t), A(t)h(t)\rangle + \langle q(t), h(t)\rangle$$

by parts and applying the result with $h(\cdot)$ vanishing at the endpoints of $[0, T]$, we find that

$$\int_0^T \left\langle p(t) + \int_0^t \left(A^\top(\tau)p(\tau) - q(\tau)\right)\,\mathrm{d}\tau,\ \frac{d}{dt}h(t)\right\rangle\,\mathrm{d}t = 0$$

for all $h(\cdot) \in W^{1,1}$ with $h(0) = h(T) = 0$. Therefore, the equation

$$p(t) + \int_0^T \left(A^\top(\tau)p(\tau) + q(\tau)\right)\,\mathrm{d}\tau \equiv c = \text{const}$$

and

$$c = \theta + \sum_{i=0}^{r} a_i = -\sum_{i=0}^{k} b_i.$$

It can be deduced that

$$\frac{d}{dt}p(t) = -H_x'(t, x_*(t), p(t), u_*(t)) + q(t). \tag{1.4.17}$$

The initial and final values of $p(\cdot)$ are, respectively,

$$\begin{aligned} p(0) &= \theta + \sum_{i=1}^{k} \lambda_i \ell_{i0}'(x_*(0), x_*(T)),\\ p(T) &= \int_0^T q(t)\,\mathrm{d}t - \sum_{i=1}^{k} \lambda_i \ell_{iT}'(x_*(0), x_*(T)). \end{aligned} \tag{1.4.18}$$

Now, we are on the verge of completing the proof. Let us first consider the regular case where $\rho > 0$, $\varepsilon = 0$, resulting in $\theta = 0$ and $q(t) \equiv 0$. In this scenario, Eqs. (1.4.17) and (1.4.18) reduce to the adjoint equation and the transversality conditions, with zero emerging on the right-hand side of inequality

(1.4.16). Additionally, the nontriviality condition is also satisfied due to

$$\sum_{i=0}^{l} \lambda_i = \rho > 0.$$

Let us consider the singular case where $\rho = 0$, $\varepsilon = s^{-1}$, $x_*(\cdot) = \overline{x}_s(\cdot)$, and $u_*(\cdot) = \overline{u}_s(\cdot)$. To ensure clarity, we will use λ_{is}, $p_s(\cdot)$ to denote the multipliers λ_i and $p(\cdot)$ for different s. So do θ, $\xi_i^{\pm}$ and $q(\cdot)$. Additionally, we note that K^+, K^-, and N can be chosen independently of s.

To observe the outcomes as $s \to \infty$, it is important to note that $p_s(\cdot)$ remain uniformly bounded. As per (H′), the norms of matrices $A_s(t) = f_x'(t, \overline{x}_s(t), \overline{u}_s(t))$ are also uniformly bounded since $\overline{x}_s(t) \to x(t)$ uniformly. Therefore, the derivatives $\frac{d}{dt} p_s(t)$ are also uniformly bounded. We can assume (taking a subsequence, if necessary) that $p_s(\cdot)$ converge uniformly to some $p(\cdot) \in W^{1,\infty}$ and λ_{is} also converge to some $\lambda_i \in [-1, 1]$ for all $i = l+1, l+2, \cdots, r$. As $\theta_s \to 0$ and $q_s(t) \to 0$ uniformly, the limiting $p(\cdot)$ satisfies the adjoint equation and transversality conditions.

Next, it is noted that for any s, at least one of λ_{is}, $l+1 \le i \le r$ is equal to 1 or -1. This follows from the fact that $(\overline{x}_s(\cdot), \overline{u}(\cdot)) \notin M_k$, which means that $\ell_i(\overline{x}_s(0), \overline{x}_s(T)) \neq 0$ for some i. Therefore, $\partial|\cdot|$ at $\xi_i = \ell_i(\overline{x}_s(0), \overline{x}_s(T))$ consists of a single element, which is either 1 or -1. Hence, the nontriviality condition is satisfied in the singular case as well. As the relations (1.4.16)–(1.4.18) are homogeneous functions of multipliers $(\lambda_0, \lambda_1, \cdots, \lambda_r, p(\cdot))$ with ε, θ and $q(\cdot)$ equal to zeros, the nontriviality condition can be expressed as

$$\lambda_0 \ge 0,\ \lambda_1 \ge 0, \cdots, \lambda_l \ge 0; \quad \sum_{i=0}^{l} \lambda_i + \sum_{i=l+1}^{r} |\lambda_i| = 1. \tag{1.4.19}$$

In summary, for any finite collection $u = \{u_1(\cdot), u_2(\cdot), \cdots, u_k(\cdot)\}$, the set $\Lambda(u)$ of multipliers $(\lambda_0, \lambda_1, \cdots, \lambda_r, p(\cdot))$ satisfying the adjoint equation, the transversality condition, the non-triviality condition (1.4.19), and

$$\int_0^T (H(t, \overline{x}(t), p(t), u_i(t)) - H(t, \overline{x}(t), p(t), \overline{u}(t)))\, \mathrm{d}t \le 0, \quad i = 1, 2, \cdots, k, \tag{1.4.20}$$

is non-empty.

In order to finalize the proof, it is imperative to establish that the intersection of all sets $\Lambda(u)$ are non-empty. Furthermore, for any set of multipliers belonging to this intersection, the maximum principle (iv) must hold true. To demonstrate the non-emptiness of the intersection, we observe that the space $W^{1,\infty}$ (comprising Lipschitz functions) is isomorphic to the adjoint of $W^{1,1}$. Notably, the weak*-topology in $W^{1,\infty}$ simplifies to the topology of weak*-convergence L^∞ for the derivatives, coupled with convergence of the functions at least at a single point. Each set $\Lambda(u)$ is compact in the weak*-topology of $\mathbb{R}^{r+1} \times W^{1,\infty}$. Additionally, $\Lambda(u) \subset \Lambda(u')$ if u' belongs to u. Therefore, the intersection Λ of all $\Lambda(u)$ is indeed non-empty.

Let $(\lambda_0, \lambda_1, \cdots, \lambda_r, p(\cdot)) \in \Lambda$. This implies, by (1.4.20), that

$$\int_0^T (H(t, \overline{x}(t), p(t), u(t)) - H(t, \overline{x}(t), p(t), \overline{u}(t)))\, dt \leq 0,$$

for any bounded $u(\cdot) \in \mathcal{U}$. As previously demonstrated in (1.4.9), this implies that $H(t, \overline{x}(t), p(t), u) \leq H(t, \overline{x}(t), p(t), \overline{u}(t))$ for all $u \in U$ almost everywhere, and thus (iv) follows. This completes the proof of Theorem 1.4.6. □

Furthermore, it is noteworthy that constraints of the form $g_j(t, x) \leq 0$ can be seamlessly integrated into the proof by replacing the function $\max_i \ell_i(x(0), x(T))$ with

$$\max \left\{ \max_i \ell_i(x(0), x(T)),\ \max_j \max_t g_j(t, x(t)) \right\}.$$

Finally, Pontryagin's maximum principle serves as a fundamental tool in optimal control theory, enabling the identification of the optimal control strategy for guiding a dynamical system from one state to another, particularly in scenarios involving state or input constraints. The principle asserts that any optimal control, along with the corresponding optimal state trajectory, must satisfy the Hamiltonian system, which constitutes a two-point boundary value problem, in addition to a maximum condition pertaining to the control Hamiltonian. Notably, under certain convexity assumptions regarding the objective and constraint functions, these necessary conditions become sufficient. In certain instances, Pontryagin's maximum principle can facilitate the synthesis of optimal feedback controls. A notable example is the renowned time-optimal control problem, which has been extensively studied in the literature [81, 235].

Generally speaking, there are several limitations to Pontryagin's maximum principle. Firstly, the derived optimal control is typically of an open-loop nature, meaning it lacks adaptability to changing conditions. Secondly, due to the nonlinearity of the equations involved, it is practically impossible to obtain an analytical solution to the optimal control problem. Thirdly, even from a numerical perspective, solving a boundary value problem is significantly more complex than tackling an initial value problem.

In contrast, the DPVS approach, which integrates the dynamic programming methodology with the viscosity solution theory of PDEs, offers substantial advantages. The optimal control is generally closed loop and formulated as feedback, enabling it to respond dynamically to changes in the system. Furthermore, in numerical analysis, the task is simplified to solving an initial value problem for a nonlinear partial differential equation, which is generally less complex than dealing with boundary value problems.

1.5 Dynamic Programming Approach and Viscosity Solution

As a crucial component of modern control theory, the dynamic programming approach, when applied to the optimal control problem, centers on treating the initial value of the system as a parameter. Leveraging the properties of the cost functional, specifically the value function, it derives the dynamic programming equation that this function satisfies. This is the core essence of the dynamic programming method [244]. In contrast to Pontryagin's maximum principle, which focuses on a single extremal at a time, the dynamic programming method addresses families of extremals. Once the HJB equation satisfied by the value function is established, the feedback control law can be obtained by solving this first-order nonlinear partial differential equation. This approach offers a comprehensive framework for addressing optimal control problems, particularly in scenarios where multiple extremals or families of extremals need to be considered.

Consider a controlled system [15, 16]

$$\begin{cases} y'(t) = f(y(t), a(t)), & t > 0, \\ y(0) = x, \end{cases} \tag{1.5.1}$$

where $a(\cdot) \in \mathcal{A} := \{a \mid [0, +\infty) \to A \text{ measurable}\}$, A is a given compact metric space, and $x \in \mathbb{R}^N$,

$$f : \mathbb{R}^N \times A \to \mathbb{R}^N \text{ continuous,}$$

$$|f(y_1, a) - f(y_2, a)| \le L|y_1 - y_2|, \quad \forall\, y_1,\ y_2 \in \mathbb{R}^N,\ a \in A.$$

In (1.5.1), $y(\cdot)$ and $a(\cdot)$ are, respectively, the state and control in open loop of the system. Denote $y_x(\cdot, a) = y_x(\cdot)$ to be the solution to the system (1.5.1).

For the system (1.5.1), there exist various forms of cost functionals that can be utilized to establish the optimal control problem. These encompass the infinite horizon case, the finite horizon case, the minimum time case, the discounted minimum time case, and numerous others.

In this section, we first focus on the minimum time problem with the cost functional given by

$$t_x(a) = \begin{cases} \min\{s \mid y_x(s, a) \in \mathcal{T}\}, & \{s \mid y_x(s, a) \in \mathcal{T}\} \neq \emptyset, \\ +\infty, & \text{otherwise}, \end{cases} \tag{1.5.2}$$

in which the given $\mathcal{T} \subseteq \mathbb{R}^N$ is the closed target.

Our objective is to minimize the cost functional (1.5.2) subject to the constraint $a(\cdot) \in \mathcal{A}$. In this context, (1.5.1) and (1.5.2), along with the given constraint, jointly constitute an optimal control problem. If the minimum of this problem is achieved at a particular control $a^*(\cdot)$, then such a control is designated as the optimal control corresponding to the initial state x.

The dynamic programming approach delves into the optimal control problem mentioned above by initially introducing the value function of the problem. Specially, define the value function

$$T(x) = \inf_{a(\cdot)\in \mathcal{A}} t_x(a). \tag{1.5.3}$$

Central to the dynamic programming approach, the value function represents the optimal cost associated with a given state of the system described by (1.5.1) and (1.5.2). By analyzing this function, we can derive the functional equation it satisfies, which is known as the DPP outlined below.

Lemma 1.5.1 *For all $s > 0$,*

$$T(x) = \inf_{a(\cdot)\in \mathcal{A}} \{s + T(y_x(s, a))\}, \tag{1.5.4}$$

if $s \leq T(x) < +\infty$.

Proof Firstly, it is noteworthy that $y_x(\cdot, a)$, which represents the solution to the system (1.5.1), possesses the semigroup property. In other words,

$$y_x(s + t, a) = y_{y_x(s,a)}(t, a(\cdot + s)).$$

For the admissible control, if $a(\cdot) \in \mathcal{A},\ t > 0$, then $a(\cdot + t) \in \mathcal{A}$. Furthermore, if $a_1(\cdot),\ a_2(\cdot) \in \mathcal{A},\ t > 0$ and

$$a(s) = \begin{cases} a_1(s),\ s \leq t, \\ a_2(s),\ s > t, \end{cases}$$

then $a(\cdot) \in \mathcal{A}$.

Next, for all $a(\cdot) \in \mathcal{A}$,

$$t_x(a) = s + t_{y_x(s,a)}(a(\cdot + s)) \geq s + T(y_x(s, a)),$$

so

$$T(x) = \inf_{a(\cdot)\in\mathcal{A}} t_x(a) \geq \inf_{a(\cdot)\in\mathcal{A}} \{s + T(y_x(s, a))\}.$$

To establish the inverse inequality, we fix an arbitrary control $a \in \mathcal{A}$ and define $z = y_x(s, a)$ as the state reached by the system at time s under the control a. For simplicity, we assume the existence of a control $a_1(\cdot) \in \mathcal{A}$ such that the time taken by the system to reach a certain target state from z is $T(z) = t_z(a_1)$. Define

$$\overline{a}(t) = \begin{cases} a(t), & t \leq s, \\ a_1(t-s), & t > s. \end{cases}$$

Then

$$T(x) \leq t_x(\overline{a}) = s + t_z(a_1) = s + T(y_x(s,a)).$$

Since $a(\cdot)$ is arbitrary, we obtain

$$T(x) \leq \inf_{a(\cdot)\in\mathcal{A}} \{s + T(y_x(s,a))\}.$$

□

Furthermore, there exists another formulation of the DPP, which aligns more closely with the optimality principle originally expressed by Richard Ernest Bellman. This alternative form is captured in Lemma 1.5.2.

Lemma 1.5.2 *For all $a(\cdot) \in \mathcal{A}$, the function*

$$s \to s + T(y_x(s,a)), \quad s \in [0, t_x(a)], \text{ if } T(x) < +\infty, \tag{1.5.5}$$

is nondecreasing. Moreover, it is constant if and only if the control $a(\cdot)$ is optimal for the initial state x.

Proof For any $a(\cdot) \in \mathcal{A}$, the DPP with initial position $y_x(s,a)$ satisfies

$$T(y_x(s,a)) \leq \varepsilon + T(y_x(s+\varepsilon,a))$$

for $\varepsilon > 0$ small enough, proving the first statement.

If $h(s) = s + T(y_x(s,a))$ is constant, then $h(s) \equiv h(0) = T(x)$. Hence $0 \leq T(x) < +\infty$ implies that $t_x(a) < +\infty$ and $h(t_x(a)) = t_x(a)$ because $T \equiv 0$ on the target $\mathcal{T}$. Therefore, $T(x) = t_x(a)$.

Finally, if $a(\cdot) \in \mathcal{A}$ is optimal for x, then

$$h(0) = T(x) = t_x(a),$$

while the proof of Lemma 1.5.1 shows

$$t_x(a) \geq h(s),$$

thus $h(0) = h(s)$ because h is nondecreasing. □

Subsequently, it can be demonstrated that we arrive at the following HJB equation (1.5.6), which represents the infinitesimal form of the DPP. This equation provides a crucial tool for analyzing and solving optimal control problems, allowing us to determine the optimal control strategy by minimizing a certain cost functional.

Lemma 1.5.3 *Suppose that the value function $T(x)$ is continuously differentiable to the first order in some neighborhood of x. Then it satisfies the following equation:*

$$\max_{a\in A}\{-f(x,a)\cdot DT(x)\} = 1, \quad x \notin \mathcal{T},\ T(x) < +\infty, \tag{1.5.6}$$

where $DT(x)$ is the gradient of $T(x)$ at the point x.

Proof To commence, let us fix a constant control $\overline{a}(t) \equiv a_0$ and let $y(t) = y_x(t, \overline{a})$. The DPP indicates that

$$T(x) - T(y(s)) \le s \quad \text{for } 0 \le s < T(x).$$

Dividing by $s > 0$ and letting $s \to 0$ yields

$$-DT(x)\cdot y'(0) \le 1.$$

Since $y'(0) = f(x, a_0)$ and $a_0 \in A$ is arbitrary, we obtain

$$\max_{a\in A}\{-f(x,a)\cdot DT(x)\} \le 1.$$

Additionally, for small $\varepsilon,\ s > 0$, the DPP implies that there exists $\overline{a} \in \mathcal{A}$ such that

$$T(x) \ge s + T(y(s)) - \varepsilon s,$$

where $y(s) = y_x(s, \overline{a})$. Then,

$$\begin{aligned}
1-\varepsilon &\le \frac{T(x)-T(y(s))}{s} \\
&= -\frac{1}{s}\int_0^s \frac{d}{ds}T(y(s))\,\mathrm{ds} = -\frac{1}{s}\int_0^s DT(y(s))\cdot y'(s)\,\mathrm{ds} \\
&= -\frac{1}{s}\int_0^s DT(x)\cdot f(x,\overline{a}(s))\,\mathrm{ds} + o(1) \quad (\text{as } s\to 0) \\
&\le \max_{a\in A}\{-DT(x)\cdot f(x,a)\} + o(1).
\end{aligned}$$

By letting s and ε approach zero, we obtain $1 \le \max_{a\in A}\{-f(x,a)\cdot DT(x)\}$. □

Equation (1.5.6) can be augmented with a natural boundary condition, thereby giving rise to the following boundary value problem

$$\begin{cases} \max_{a\in A}\{-f(x,a)\cdot Du\}=1, & \text{in } \Omega\setminus\mathcal{T}, \\ u=0, & \text{on } \partial\mathcal{T}, \\ u(x)\to+\infty, & \text{as } x\to\partial\Omega, \end{cases} \tag{1.5.7}$$

where $\Omega\supseteq\mathcal{T}$ is an open set to be determined. The value function is the candidate solution for this problem. The boundary value problem for the minimum time problem is a free boundary problem [17]. Denote

$$\mathcal{R}=\{x\mid T(x)<+\infty\}. \tag{1.5.8}$$

It is expected that $\Omega=\mathcal{R}$, and it can be proven that $\mathcal{R}$ is open and $T(x)\to+\infty$ as $x\to\partial\mathcal{R}$, under a controllability assumption on the system near $\mathcal{T}$ [15].

Furthermore, the boundary value problem associated with the discounted minimum time problem corresponds to the Dirichlet problem. Comparable results for the minimum time problem can also be established for the discounted minimum time problem, as detailed in references [15] and [16].

In fact, for the discounted minimum time problem, we can present its cost functional as

$$J(x,a)=\begin{cases} \displaystyle\int_0^{t_x(a)} e^{-s}\,\mathrm{d}s=1-e^{-t_x(a)}, & t_x(a)<+\infty, \\ 1, & \text{otherwise}, \end{cases} \tag{1.5.9}$$

where $t_x(a)$ is defined in (1.5.2). In this case, take $T(x)$ in (1.5.3) and set

$$V(x)=\inf_{a(\cdot)\in\mathcal{A}} J(x,a)=1-e^{-T(x)},$$

as the corresponding value function. Its DPP is

$$V(x)=\inf_{a(\cdot)\in\mathcal{A}}\left\{\int_0^s e^{-t}\,\mathrm{d}t+V(y_x(s,a))e^{-s}\right\},\quad \text{if } s\le T(x). \tag{1.5.10}$$

Similarly, we can prove that for all $a(\cdot)\in\mathcal{A}$, the function

$$s\to\int_0^s e^{-t}\,\mathrm{d}t+V(y_x(s,a))e^{-s},\quad s\in[0,t_x(a)],$$

is non-decreasing. Moreover, it is constant if and only if the control $a(\cdot)$ is optimal for the initial position x. Then, the value function satisfies the following HJB equation:

$$V(x) + H(x, DV(x)) = 1, \ x \notin \mathcal{T},$$

provided it is C^1 in a neighborhood of x, where

$$H(x, p) = \max_{a \in A}\{-f(x, a) \cdot p\}.$$

After being coupled with a natural boundary condition, the problem can be formulated as the following boundary value problem:

$$\begin{cases} u + H(x, Du) = 1, & \text{in } \mathbb{R}^N \setminus \mathcal{T}, \\ u = 0, & \text{on } \partial\mathcal{T}, \end{cases}$$

whose candidate solution is the value function.

For the discounted minimum time problem (1.5.1) and (1.5.9), define a classical verification function as a bounded function $u \in C(\mathbb{R}^N) \cap C^1(\mathbb{R}^N \setminus \mathcal{T})$ such that

$$\begin{cases} u + H(x, Du) \leq 1, & \text{in } \mathbb{R}^N \setminus \mathcal{T}, \\ u \leq 0, & \text{on } \partial\mathcal{T}. \end{cases} \tag{1.5.11}$$

The verification theorem outlined below establishes the sufficient conditions for optimality.

Theorem 1.5.4 *Suppose that u is a classical verification function for the discounted minimum time problem, where $x \notin \mathcal{T}$, $a^*(\cdot) \in \mathcal{A}$. Then the following conditions hold:*

(i) *If $u(x) \geq J(x, a^*)$, then $a^*(\cdot)$ is optimal for x.*
(ii) *If*

$$\begin{cases} u(y^*(t)) - f(y^*(t), a^*(t)) \cdot Du(y^*(t)) = 1, & \textit{for } \text{a.e. } t \leq t_x(a^*), \\ u = 0, & \textit{on } \partial\mathcal{T}, \end{cases} \tag{1.5.12}$$

where $y^*(\cdot) = y_x(\cdot, a^*)$, then $a^*(\cdot)$ is optimal for x.

Proof (i) First, we prove that $u(x) \leq V(x)$, where V is the value function, and hence $V(x) = J(x, a^*)$, and $a^*(\cdot)$ is optimal for x. To this purpose, we take any $a(\cdot) \in \mathcal{A}$, $y(\cdot) = y_x(\cdot, a)$, and use (1.5.11) to obtain

$$\frac{d}{dt}\left[-e^{-t}u(y(t))\right] = e^{-t}\left[u(y(t)) - Du(y(t)) \cdot f(y(t), a(t))\right] \leq e^{-t}$$

for a.e. $t \le t_x(a)$. Integrate both sides to obtain

$$u(x) - e^{-t}u(y(t)) \le 1 - e^{-t}.$$

By letting $t \to t_x(a)$ and using the boundedness of u if $t_x(a) = +\infty$, and $u \le 0$ on $\partial\mathcal{T}$ if $t_x(a) < +\infty$, we obtain

$$J(x, a) = 1 - e^{-t_x(a)} \ge u(x).$$

Since a is arbitrary, we conclude that $V(x) \ge u(x)$.
(ii) Let

$$h(s) = \int_0^s e^{-t}\, \mathrm{d}t + u(y^*(s))e^{-s}. \tag{1.5.13}$$

We claim that h is non-increasing, and indeed, it is actually a constant. Since $u = 0$ on $\partial\mathcal{T}$, we have

$$u(x) = h(0) \ge h(t_x(a^*)) = 1 - e^{-t_x(a^*)} = J(x, a^*),$$

and we can conclude by part (i). As h is C^1, the claim is proven by

$$h'(s) = e^{-s}[1 - u(y^*(s)) + Du(y^*(s)) \cdot f(y^*(s), a^*(s))] = 0,$$

where the last equality follows from (1.5.12).

□

Moreover, if the verification function u is a solution of the HJB equation

$$u + H(x, Du) = 1 \quad \text{in } \mathbb{R}^N \setminus \mathcal{T},$$

then the sufficient condition of optimality (1.5.12) is equivalent to

$$\begin{aligned} -f(y^*(s), a^*(s)) \cdot Du(y^*(s)) &= \max_{a \in A}\{-f(y^*(s), a) \cdot Du(y^*(s))\} \\ &= H(y^*(s), Du(y^*(s))) \end{aligned}$$

for almost every $0 < s < t_x(a^*)$.

Furthermore, if the value function V is smooth and employed as a verification function, then the sufficient condition of optimality (1.5.12) is also necessary. In fact, according to the DPP (1.5.10), if a^* is optimal for x, then the function h, defined by (1.5.13) with $u = V$, is constant. The equation

$$0 = h'(s) = e^{-s}\left[1 - V(y^*(s)) + DV(y^*(s)) \cdot f(y^*(s), a^*(s))\right],$$

and therefore (1.5.12) holds with $u = V$.

The final stage of the classical dynamic programming approach involves designing an optimal control in feedback form utilizing the value function. Let us illustrate this with the discounted minimum time problem as an example, subject to certain assumptions. Naturally, analogous results can be derived for other types of optimal control problems as well.

Assuming that the value function V is smooth, we consider the following subset of A, denoted by $S(z)$, which is defined as

$$\begin{aligned} S(z) &= \arg\min_{a \in A}\{f(z, a) \cdot DV(z)\} \\ &= \{a \in A \mid H(z, DV(z)) = -f(z, a) \cdot DV(z)\}. \end{aligned}$$

These controls facilitate the steepest descent of the value function V along the direction of the corresponding vector f. Let us refer to an admissible feedback as a map $\Phi : \mathbb{R}^N \to A$ such that, for all $x \in \mathbb{R}^N$, there exists a unique solution to the system

$$\begin{cases} y' = f(y, \Phi(y)), & t > 0, \\ y(0) = x, \end{cases}$$

and we say that Φ is optimal for x if $\Phi(y(\cdot)) \in \mathcal{A}$ is an optimal open-loop control for x. Given that if V is smooth and serves as a verification function, the sufficient condition of optimality is also necessary. Furthermore, based on Theorem 1.5.4, we can derive the following conclusion. A control $a^*(\cdot) \in \mathcal{A}$ is optimal for x if and only if

$$a^*(t) \in S(y_x(t, a^*))$$

for almost every $t > 0$. Therefore, an admissible feedback that satisfies

$$\Phi(z) \in S(z) \quad \text{for all } z \in \mathbb{R}^N,$$

is optimal for all initial points $x \in \mathbb{R}^N$.

Next, we will explore the linear quadratic regulator problem (LQRP), a prevalent problem in engineering, as an illustrative example. Through this analysis, we will delve deeper into the strategy employed by the dynamic programming approach to obtain the optimal feedback control. This approach has been extensively studied in the literature, as referenced in [85].

Let $x(s) \in \mathbb{R}^n$, $u(s) \in \mathbb{R}^m$ satisfy

$$x'(s) = A(s)x(s) + B(s)u(s), \quad s \in [t, t_1]. \tag{1.5.14}$$

Here, $A(s)$, $B(s)$ are matrices of dimensions $n \times n$, $n \times m$, respectively. The $n \times n$ matrices $M(s)$ and D are symmetric and nonnegative definite. The invertible $m \times m$ matrix $N(s)$ is symmetric and positive definite. The initial condition is given as $x(t) = x$.

The LQRP problem is to choose $u(\cdot) \in U$ such that

$$\int_t^{t_1} [x(s) \cdot M(s)x(s) + u(s) \cdot N(s)u(s)] \, ds + x(t_1) \cdot Dx(t_1) \tag{1.5.15}$$

attains the minimum. Here, the control space U is a closed subspace in $\mathbb{R}^m$ and "$\cdot$" denotes the inner product of the vectors. Let $t_0 < t_1$ and consider initial times t in the finite interval $[t_0, t_1)$. By the dynamic programming approach, one obtains the following HJB equation that the value function of the optimal control problem satisfies.

$$\begin{cases} -V_t(t,x) + H(t,x,D_x V(t,x)) = 0, & t_0 \le t < t_1, \\ V(t_1,x) = x \cdot Dx, & x \in \mathbb{R}^n, \end{cases} \tag{1.5.16}$$

in which the Hamiltonian is given by

$$\begin{aligned} H(t,x,p) &= \sup_u \{-(A(t)x + B(t)u) \cdot p - x \cdot M(t)x - u \cdot N(t)u\} \\ &= \frac{1}{4} N^{-1}(t)B^\top(t)p \cdot B^\top(t)p - A(t)x \cdot p - x \cdot M(t)x. \end{aligned} \tag{1.5.17}$$

It is observed that the unique maximizer of (1.5.17) is

$$u^* = -\frac{1}{2} N^{-1}(t)B^\top(t)p. \tag{1.5.18}$$

Let the solution to (1.5.16) be a quadratic function of x. Denote $V(t,x) = x \cdot P(t)x$, in which $P(t)$ is a symmetric matrix. Substituting $V(t,x)$ into (1.5.16), we obtain the following matrix Riccati equation satisfied by $P(t)$.

$$\begin{cases} P'(t) = P(t)B(t)N^{-1}(t)B^\top(t)P(t) \\ \qquad\qquad -A(t)P(t) - P(t)A^\top(t) - M(t), \quad t \in [t_0, t_1). \\ P(t_1) = D. \end{cases} \tag{1.5.19}$$

By the classical verification theorem for this problem, we get the optimal state feedback control

$$\begin{aligned} u^*(s) &= -\frac{1}{2} N^{-1}(s)B^\top(s)D_x V(s, x^*(s)) \\ &= -N^{-1}(s)B^\top(s)P(s)x^*(s), \end{aligned} \tag{1.5.20}$$

in which the optimal trajectory $x^*(s)$ satisfies

$$\begin{cases} x'(s) = \Big(A(s) - B(s)N^{-1}(s)B^\top(s)P(s)\Big)\, x(s), \\ x(t) = x, \quad s \in [t, t_1]. \end{cases} \tag{1.5.21}$$

In other words, (1.5.20) is an optimal feedback control law.

In engineering, it is always desirable for the optimal control of the problem under consideration to be expressed as state feedback. The dynamic programming approach outlined earlier, along with the HJB equation, offers a formal method for discovering such optimal feedback control.

However, as previously discussed, this conclusion heavily relies on the sufficient smoothness of the value function. Unfortunately, with rare exceptions, the value function is not continuously differentiable. As a result, L. S. Pontryagin emphasized early on the challenges in asserting that the value function is indeed a solution to the HJB equation.

Even under conditions of exceptional smoothness, the HJB equation derived from the dynamic programming approach often lacks a classical solution. This long-standing issue has hampered the rigorous mathematical foundation of the dynamic programming approach. However, in the 1980s, American mathematician M.G. Crandall and French mathematician P.L. Lions [61] introduced the concept of viscosity solutions for nonlinear partial differential equations, providing a solid mathematical basis for the dynamic programming approach. Crucially, the value function is shown to be the unique viscosity solution of the corresponding HJB equation. This breakthrough effectively tackles the challenges encountered by the classical dynamic programming approach in dealing with optimal control problems.

Now, let us review some fundamental concepts of viscosity solution theory.

Definition 1.5.5 Given $E \subseteq \mathbb{R}^M$, denote

$$\mathrm{USC}(E) = \{\text{the upper semicontinuous mapping } u \mid E \to \mathbb{R}\},$$
$$\mathrm{LSC}(E) = \{\text{the lower semicontinuous mapping } u \mid E \to \mathbb{R}\}.$$

Let $F : E \times \mathbb{R} \times \mathbb{R}^M \to \mathbb{R}$ be a continuous function. A function $u \in \mathrm{USC}(E)$ (or $u \in \mathrm{LSC}(E)$) is a viscosity subsolution (or a viscosity supersolution) to the equation below

$$F(x, u, Du) = 0, \quad \forall\, x \in E, \tag{1.5.22}$$

if for any test function $\phi \in C^1(E)$ and any local maximizer (or minimizer) x of $u - \phi$, there is

$$F(x, u(x), D\phi(x)) \le 0 \quad (\text{or } \ge 0).$$

Moreover, u is the viscosity solution to Eq. (1.5.22), if it is both a viscosity subsolution and a viscosity supersolution.

Subsequently, we arrive at the following theorem concerning the HJB equation (1.5.7) in the context of the minimum time problem. It is worth noting that similar results also apply to other HJB equations, such as those encountered in scenarios with infinite horizons, whether or not state constraints are present, as well as in cases involving finite horizons, discounted minimum time, and so forth.

Theorem 1.5.6 *Assume that the value function* $T(x)$ *is continuous. Then it is a viscosity solution to the free boundary problem* (1.5.7) *with* $\Omega = \mathcal{R}$ *in* (1.5.8).

Proof Denote

$$H(x, Du) = \max_{a \in A} \{-f(x, a) \cdot Du\}$$

and $H(x, Du(x)) = 1$ is the HJB equation.

In the proof of Lemma 1.5.3, we showed that $T \in C^1$, $T(x) - T(y_x(s)) \le s$ for s small, for all trajectories $y_x(\cdot)$ of the system, implies that $H(x, DT(x)) \le 1$. If we only know $T \in C\left(\overline{\mathbb{R}^N \setminus \mathcal{T}}\right)$ and $\phi \in C^1$ is such that $T - \phi$ has a local maximum at x, then for small s,

$$\phi(x) - \phi(y_x(s)) \le T(x) - T(y_x(s)) \le s.$$

Replace T with ϕ in the previous proof to get

$$H(x, D\phi(x)) \le 1,$$

which shows that T is a viscosity subsolution.

Similarly, in the proof of Lemma 1.5.3, we showed that $T \in C^1$, $T(x) - T(y_x(s)) \ge s(1 - \varepsilon)$ for s and small ε, for some trajectories $y_x(\cdot)$ of the system, implies that $H(x, DT(x)) \le 1$. Now, if $\phi \in C^1$ so that $T - \phi$ has a local minimum at x, then for s and small ε,

$$\phi(x) - \phi(y_x(s)) \ge T(x) - T(y_x(s)) \ge s(1 - \varepsilon).$$

Once again, the previous proof illustrates that

$$H(x, D\phi(x)) \ge 1,$$

which shows that T is also a viscosity supersolution. □

The employment of viscosity solutions for partial differential equations offers the advantage of guaranteeing a unique solution. This ensures that the value function is unambiguously characterized as the sole viscosity solution to the appropriate boundary value problem.

Theorem 1.5.7 *Assuming that the value function is both continuous and bounded, and considering that Ω represents a smooth domain in the context of the infinite horizon problem with state constraints, we can draw the following conclusions. The value function serves as the minimal supersolution and the maximal subsolution, respectively, for the infinite horizon problem, the finite horizon problem, the discounted minimum time problem among bounded functions, and the infinite horizon problem with state constraints among bounded and continuous functions. Notably, it is the unique bounded and continuous viscosity solution.*

We recommend readers to consult [16] for a detailed proof of this result. The concept of viscosity solution for PDEs provides a solid theoretical foundation for the dynamic programming approach. It is worth mentioning that the theory of viscosity solutions extends beyond the optimal control of lumped parameter systems described earlier and can also be applied to the optimal control of distributed parameter systems. For a comprehensive understanding of the viscosity solution theory pertaining to infinite-dimensional HJB equations, we suggest referring to [19, 39, 96, 136, 150, 244] and the references cited within these works.

Indeed, there are various challenges in seeking optimal control through the viscosity solution of the HJB equation satisfied by the value function of the optimal control problem. A prominent issue is the need for both the value function and its gradient to derive the optimal feedback control. However, the existence of the gradient is not guaranteed as the viscosity solution may not be differentiable. Nevertheless, for certain large inertial systems, it has been observed that employing proper differences as approximations for the gradient of the value function can still yield effective numerical solutions for the optimal feedback control. As evidenced in [85], certain simple difference schemes have demonstrated their ability to numerically approximate the viscosity solution.

More recently, there have been intriguing efforts to numerically solve the HJB equation for finite-dimensional control problems. These efforts can be found in works such as [44, 121, 233, 234]. These studies not only highlight the progress made in numerical methods for HJB equations but also underscore the importance of such efforts in addressing real-world control problems.

1.6 Numerical Solution of Optimal Control

Compared to the intricate and sophisticated theoretical advancements, the historical development of numerical solution techniques in optimal control has been relatively straightforward. Fundamentally, there are three primary numerical methods employed to solve optimal control problems [202]. The first approach involves solving a two-point boundary value problem using the necessary conditions of optimal control, such as the Pontryagin maximum principle. This method leverages the theoretical framework of optimal control to directly address the problem's core characteristics. The second method involves discretizing the original optimal control problem entirely, converting it into a finite-dimensional nonlinear programming problem. This approach simplifies the problem by reducing it to a more manageable form, suitable for numerical techniques. The third method involves finitely parameterizing the control trajectory. Again, this transforms the original problem into a nonlinear programming problem. However, in this case, the cost functional and constraints are derived by integrating the system equations, and their gradients with respect to the control parameters are obtained by integrating the adjoint or sensitivity equations. This method offers a more flexible approach to handle different types of optimal control problems. Each of these methods has its own advantages and limitations, depending on the specific characteristics of the optimal control problem at hand. The choice of the appropriate method often depends on factors such as the complexity of the system equations, the nature of the constraints, and the desired accuracy of the solution.

Early numerical methods for dealing with control problems without control constraints or terminal constraints often involve solving a two-point boundary value problem. This problem consists of the initial value problem of the system equation combined with the terminal value problem of the adjoint system. The shooting method is typically employed to tackle such issues. In the shooting method, one begins by making an initial guess for the unknown initial values of the adjoint variables. The original system, along with its adjoint system, is then combined to form a system of differential equations. A range of classical numerical methods, such as the Runge-Kutta method, can be used to find a numerical solution to this initial value problem. After obtaining a solution, the initial guess is refined based on the mismatch at the terminal point. This iterative process continues until the desired target value is achieved. To improve upon the basic shooting method, the multiple-shooting method has been developed. In this approach, the time interval is subdivided, and initial values are re-estimated for each subinterval based on the mismatches. The multiple-shooting method addresses the extreme sensitivity and potential instability of the solution to the initial guess, marking a significant advancement in the shooting method. Another approach is the linearization method. Here, the system equation is linearized with respect to the current trajectory and then integrated forward in time. The integration interval must span the unknown initial value. Linear combinations of solutions are then computed to satisfy both the initial and terminal conditions. The process is repeated, linearizing the system with respect to the updated trajectory, until a satisfactory solution is obtained. These methods provide a foundation for numerical solutions to control problems without constraints. However, as control problems become more complex, incorporating constraints and considering additional factors such as optimality criteria, more advanced techniques and algorithms may be required.

The multiple-shooting method is a sophisticated numerical technique for solving boundary value problems, particularly those arising in control theory. Its core idea is to transform a complex boundary value problem into a series of smaller, more manageable initial value problems. This approach offers several advantages but also presents some challenges. One of the primary strengths of the multiple-shooting method is its ability to handle high-precision requirements and incorporate multiple necessary conditions, including inequality constraints and interior point constraints. This makes it particularly suitable for solving complex optimal control problems where accuracy and optimality are paramount. However, the method's dependency on an initial guess can be a significant limitation. The initial guess must be carefully chosen to ensure the convergence of the iterative process. This requires a deep understanding of both the calculus of variations and the physical nature of the problem being solved. Moreover, the practical implementation of the multiple-shooting method can be quite complicated. It often involves the use of the Newton method for iteration, which has a limited convergence domain and can make finding a numerical solution challenging. The lack of information about the adjoint variable further complicates the process. Additionally, determining the appropriate switch structure, specially how to divide the optimal trajectory into different subarcs, such as bang-bang type, singular types, and constrained and unconstrained types, is a difficult task. This often requires the use of homotopy methods, which can add further complexity to the solution process. In summary, while the multiple-shooting method offers powerful capabilities for solving boundary value problems in control theory, its successful application requires a deep understanding of the problem, careful choice of initial guesses, and sophisticated numerical techniques. Despite these challenges, it remains a valuable tool for addressing complex control problems where high precision and optimality are desired. So broadly speaking, the difficulties of the multiple shooting in applications come from two aspects. One is the limitation of the method itself, which holds a strong dependence on the initial guess. The other is the complexity of the optimal control problem, which makes it harder to transform the boundary value problem. Such a difficulty is mainly seen in the guess of the switch structure.

When dealing with optimal control problems, one approach is to transform the functional extremum problem into an extremum problem related to the Hamiltonian. This transformation forms the basis for classifying numerical methods into indirect and direct methods. If there is no such transformation process involved, the method is considered a direct method. Otherwise, it falls into the category of indirect methods. After adopting the indirect approach and transforming the functional extremum problem into an extremum problem of the Hamiltonian, there are two primary ways of thinking. The first approach utilizes the maximum principle to establish expressions for control variables that include state variables and adjoint variables. This leads to a two-point boundary value problem that can then be solved. The second approach stems from the extremum method of functions of several variables. Here, an iterative search method is employed to find a solution, starting from an initial guess. Gradually, through iterative steps, the process aims to reach an extremum. It is crucial to note that, instead of maximizing a function of several variables, the search space in this context becomes a function space. Specific methods that can be employed include the gradient method, the conjugate gradient method, the Newton method, the penalty function method, and a range of other techniques. Each of these methods has its own strengths and weaknesses, and the choice of the most appropriate method often depends on the specific characteristics of the optimal control problem at hand.

The direct method and the indirect method indeed exhibit significant differences in their approach to solving optimal control problems. The direct method focuses on transforming the optimal control problem into a static optimization problem through parameterization and discretization. Parameterization typically involves representing control variables as parameterized functions, which can further enable the parameterization of other variables. Different parameterization techniques, such as using power series, Fourier series, Chebyshev series, or interpolation functions, are available. This approach converts the original problem into a more tractable form that can be solved using standard optimization techniques. The multiple-shooting method can be employed as a direct method, where it is used to divide the original problem into smaller subproblems, each solved independently, and then combined to obtain the overall solution. However, it's worth noting that the multiple-shooting method can also be used as an indirect method, depending on how it is incorporated into the solution process. On the other hand, the indirect method involves transforming the functional extremum problem into an extremum problem related to the Hamiltonian. Additionally, there is a direct method known as direct transcription, which transforms the optimal control problem into a nonlinear programming problem. This approach allows the application of powerful nonlinear programming solvers to find the optimal control trajectory. In summary, the direct method focuses on discretizing and parameterizing the problem to convert it into a static optimization problem, while the indirect method utilizes transformations related to the Hamiltonian to solve a boundary value problem.

Based on the different systems that the optimal control problems belong to, the theory and methods of optimal control can be categorized into two primary directions: optimal control of lumped parameter systems and optimal control of distributed parameter systems. Lumped parameter systems refer to physical systems with a finite number of degrees of freedom. These systems are typically described by ODEs. On the other hand, distributed parameter systems possess an infinite number of degrees of freedom. Their behavior is generally captured by PDEs, partial differential integral equations, or functional differential equations. Many crucial phenomena in the physical world are accurately represented by PDEs. The control problems related to these systems often have a profound practical significance. Distributed parameter systems find widespread application in various engineering fields, including thermal engineering, chemical engineering, missile technology, aerospace, aviation, nuclear fission and fusion, as well as ecological, environmental, and social systems. Since the 1960s, there has been a significant growth in research on distributed parameter systems. The utilization of modern PDE and functional analysis theories has provided a rigorous theoretical foundation for distributed parameter systems and has furnished powerful research tools. Similar results to those obtained for lumped parameter systems have been achieved in areas such as stabilization, optimal control, controllability, observability, identification, and filtering of distributed parameter systems. These advancements can be viewed as extensions of the corresponding results from lumped parameter systems. Nevertheless, the study of lumped parameter systems often offers valuable insights, particularly for systems with strong practical backgrounds. For instance, in the context of biomedical applications, the numerical solution of optimal control problems for ODEs plays a crucial role. These ODEs might describe the dynamics of biological systems, drug delivery mechanisms, or physiological processes. By employing optimal control techniques, researchers can design strategies to optimize therapeutic outcomes, minimize side effects, or enhance the overall efficiency of medical treatments. In summary, both lumped and distributed parameter systems have their unique challenges and applications within the realm of optimal control. The study of these systems, particularly in the context of practical problems like those found in biomedical applications, continues to be an active and exciting area of research.

Next, we will illustrate the numerical solution of optimal control problems through a specific example. Consider an optimal control problem related to a two-drug therapy for a HIV model. This model aims to capture the dynamics of the interaction between the HIV virus and the immune system of the human body while incorporating the effects of two therapeutic drugs. Let T denote the concentration of the uninfected CD4$^+$ T cells and T^*, T^{**} the concentrations of latently infected and actively infected CD4$^+$ T cells, respectively. The concentration of free infectious virus particles is presented by V. Select two control variables $u_1(\cdot),\ u_2(\cdot)$ to denote two different treatment strategies. And their functions are, respectively, to decrease the infection of the T cell and to lower the number of the HIV virus. Therefore, we can get the following equations to model the dynamics of HIV/AIDS:

$$\begin{cases} \dfrac{dT}{dt} = \dfrac{s}{1+V} - \mu_T T + rT\left(1 - \dfrac{T+T^*+T^{**}}{T_{\max}}\right) - (1-u_1(t))\, K_1 V T, \\ \dfrac{dT^*}{dt} = (1-u_1(t))\, K_1 V T - \mu_T T^* - K_2 T^*, \\ \dfrac{dT^{**}}{dt} = K_2 T^* - \mu_b T^{**}, \\ \dfrac{dV}{dt} = (1-u_2(t))\, N \mu_b T^{**} - K_1 V T - \mu_V V, \end{cases} \tag{1.6.1}$$

satisfying the initial conditions $T(t_0) = T_0,\ T^*(t_0) = T_0^*,\ T^{**}(t_0) = T_0^{**},\ V(t_0) = V_0$ [184]. Here, $u(\cdot) = (u_1(\cdot), u_2(\cdot))$ represents the immune boosting and viral-suppressing drugs, respectively, with constraints $0 \le u_i(t) \le 1,\ i = 1, 2$. Additionally, μ_T denotes the death rate of uninfected and latently infected CD4$^+$ T cells, μ_b represents the death rate of actively infected CD4$^+$ T cell population, and μ_V is the death rate of free virus. K_1 is the rate constant for CD4$^+$ T cells becoming infected by free virus, K_2 is the rate latently infected cells convert to actively infected, and r is the growth rate of the CD4$^+$ T cell population. Furthermore, N is the number of free virus produced by lysing a CD4$^+$ T cell, $T_{\max}$ is the maximum CD4$^+$ T cell population level, and s is the rate of supply of CD4$^+$ T cells from precursors.

Define the cost functional as

$$J(u_1, u_2) = \int_{t_0}^{t_1} \left[T - \frac{1}{2} \left(A_1 u_1^2(t) + A_2 u_2^2(t) \right) \right] \mathrm{d}t, \tag{1.6.2}$$

in which the parameters A_1, A_2 represent the weight factors of the cost functional, indicating the severity of the drug side effects, respectively. Our objective is to minimize these side effects as much as possible while maximizing the system's harvest, which depends on the number of T cells. Simultaneously, we aim to minimize the cost associated with the chemotherapy side effects. Consequently, our goal can be concisely stated as finding an optimal control pair (u_1^*, u_2^*) such that

$$J(u_1^*, u_2^*) = \max \{ J(u_1, u_2) \mid (u_1, u_2) \in U \}, \tag{1.6.3}$$

where $U = \{(u_1, u_2) \mid u_i \in L^\infty(t_0, t_1),\ 0 \le u_i \le 1,\ t \in [t_0, t_1],\ i = 1, 2\}$ is the admissible control set.

For the optimal control problem (1.6.3) mentioned above, we derive the necessary first-order optimality condition using standard arguments.

Theorem 1.6.1 *Suppose that* (u_1^*, u_2^*) *is an optimal control pair to the optimal control problem* (1.6.3). *Then there exist the adjoint variables* λ_i, *satisfying*

$$\begin{cases} \lambda_1' = -1 + \lambda_1 \mu_T + \lambda_1 r \left(1 - \dfrac{2T(t) + T^*(t) + T^{**}(t)}{T_{\max}} \right) \\ \qquad + [(\lambda_1 - \lambda_2)(1 - u_1(t)) - \lambda_4] K_1 V(t), \\ \lambda_2' = \dfrac{\lambda_1 r T(t)}{T_{\max}} + \lambda_2(\mu_T - K_2) - \lambda_3 K_2, \\ \lambda_3' = \dfrac{\lambda_1 r T(t)}{T_{\max}} + \lambda_3 \mu_b - \lambda_4 (1 - u_2(t)) N \mu_b, \\ \lambda_4' = \dfrac{\lambda_1 s}{(1 + V(t))^2} + [(\lambda_1 - \lambda_2)(1 - u_1(t)) - \lambda_4] K_1 T(t) + \lambda_4 \mu_V, \end{cases} \tag{1.6.4}$$

with the transversality condition $\lambda_i(t_1) = 0,\ i = 1, 2, 3, 4$. *In addition,*

$$\begin{aligned} u_1^*(t) &= \min \left(1, \left(\frac{1}{A_1} (\lambda_1 - \lambda_2) K_1 T(t) V(t) \right)^+ \right), \\ u_2^*(t) &= \min \left(1, \left(-\frac{1}{A_2} \lambda_4 N \mu_b T^{**}(t) \right)^+ \right), \end{aligned} \tag{1.6.5}$$

where $(\cdot)^+ = \max(0, \cdot)$.

Furthermore, we have the following uniqueness theorem of the optimality system, which is derived from [127].

Theorem 1.6.2 *For t_1 sufficiently small, the bounded solutions to the optimality system are unique.*

By MATLAB 7.9, we proceed the numerical simulation on the HP Compaq 2210b computer. Here, we select three groups of different parameters, $A_1 = A_2 = 30$, $A_1 = A_2 = 150$, and $A_1 = A_2 = 300$, and obtain the numerical solutions to optimal control-trajectory pair.

During the solving process, the iterative min-H algorithm, which incorporates the modified Euler method, is employed [93, 180]. Initially, we utilize the initial conditions of the state to solve the state equation forward in time. Subsequently, the obtained state results are then used to solve the adjoint equation, which involves a backward computation in time. This methodology generates a solution for the control system, specifically, a pair of control functions. Using these control functions, we compute the corresponding cost. The computation continues until the cost of the optimal control problem converges, typically requiring between 10 and 20 iterations.

By analyzing the numerical solutions of the optimal control-trajectory pair, we obtain interesting insights into the optimal control problem (1.6.3). Regardless of the parameter values, the control u_2 always maintains its maximum value during the initial stages of treatment. Subsequently, it gradually decays to zero. In contrast, the trajectory of u_1 exhibits a completely different pattern. Initially, it quickly converges to zero. As u_2 begins to decrease, u_1 bounces within a certain level before gradually decaying as well. As the weight factors increase from 30 to 300, the point at which u_1 begins to decrease occurs earlier. However, the magnitude of u_1 remains relatively low throughout the treatment period. When $A_1 = A_2 = 300$, the peak value of u_1 is higher than when $A_1 = A_2 = 30$. Furthermore, as the parameters A_1 and A_2 increase, the point at which u_2 begins to decay occurs earlier. From a mathematical perspective, this behavior can be attributed to the reciprocal relationship between the chemotherapy administration dose and the weight factors. As the cost functional increases for the patient, the duration of chemotherapy with the maximum dose must decrease. This observation suggests that when two types of drugs are administered simultaneously, the HIV-proteinase-suppressing drug u_2 plays a crucial role. In contrast, without u_2, the trajectory of u_1 closely resembles that of u_2. Figures 1.1 and 1.2 provide further details on these trajectories. These findings have important implications for the design of optimal treatment strategies. By understanding how the control variables u_1 and u_2 evolve over time, we can refine the dosing schedule to minimize side effects and maximize the effectiveness of the treatment.

The pictures of the state variables are plotted in Figs. 1.3 and 1.4. When the optimal control is applied to the state system, the concentration of T cells begins to increase. However, the rate of this increase gradually tapers off. As the HIV-proteinase-suppressing drug u_2 decreases, the concentration of T cells also diminishes. As the cost functional of the system increases, which is reflected in the increase of the weight factors, the point at which the concentration of T cells begins to decrease occurs earlier. This trend is also observed in the decay of u_2. During

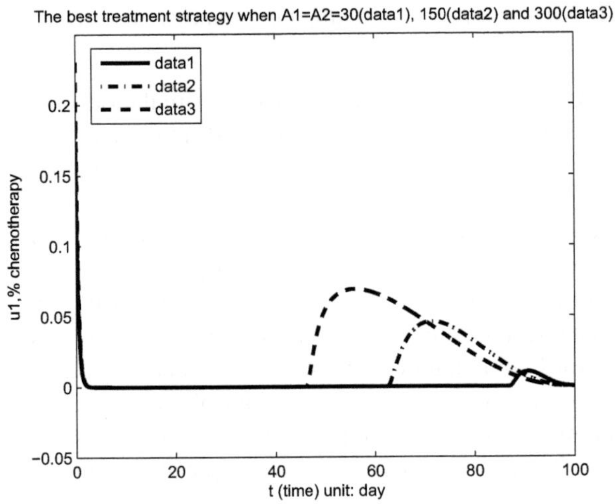

Fig. 1.1 The optimal control u_1^*

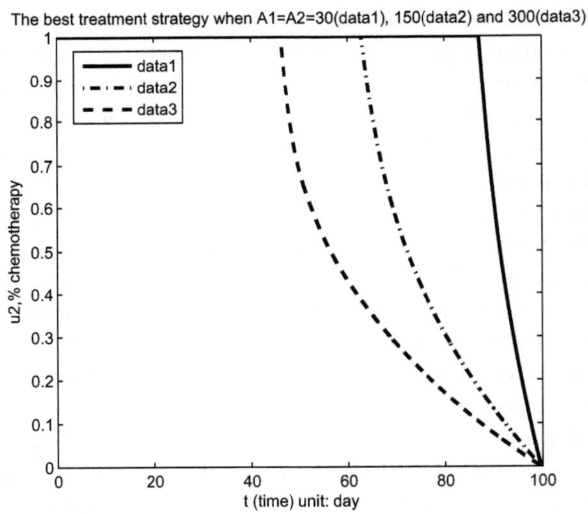

Fig. 1.2 The optimal control u_2^*

the first 50 days of treatment, the profiles of T cells under the optimal controls look similar. However, if the cost of the system is too high, the body may not be able to effectively control the concentration of the virus. This observation can be corroborated by examining the figure of the virus concentration (Fig. 1.4). As the weight factors increase, the point at which the virus population begins to increase occurs earlier. Therefore, under high system costs, the optimal control pair may result in a lower T cell count and a higher virus concentration. Initially, the virus population decreases to a low value during the first days of treatment. However,

Fig. 1.3 The optimal state $T(t)$

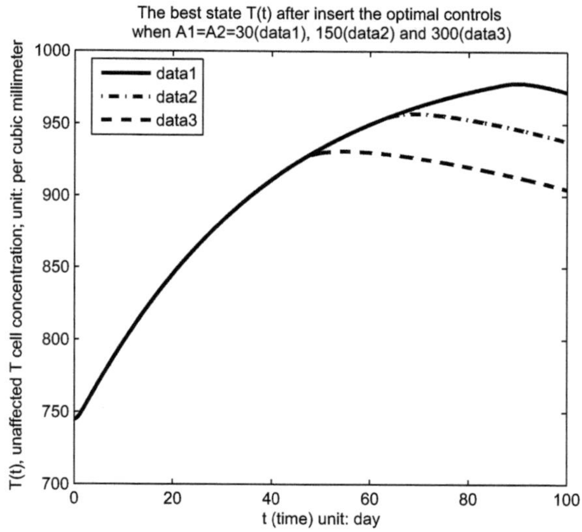

Fig. 1.4 The optimal state $V(t)$

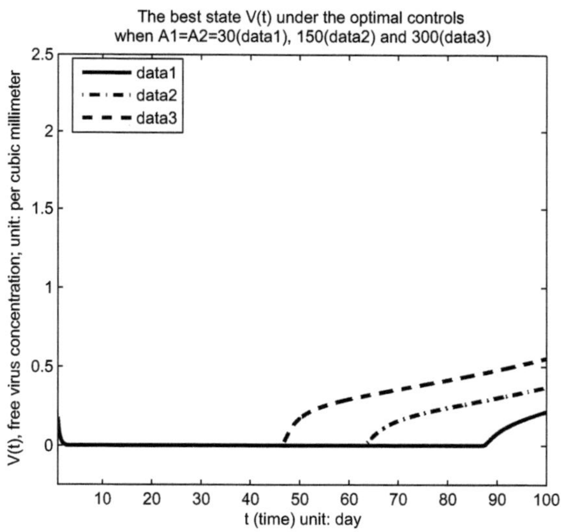

after approximately 50 days, it rebounds. These findings underscore the importance of thoroughly evaluating the cost functional and its corresponding weight factors in designing effective treatment strategies that strike a balance between controlling virus concentration and minimizing the burden on the immune system.

Part I
Open-Loop Control

Chapter 2
Dubovitskiĭ-Milyutin Approach

This chapter is dedicated to exploring optimal control problems for both finite- and infinite-dimensional systems, employing the Dubovitskiĭ-Milyutin functional analytical approach. Initially, we delve into the analysis of extremum problems in finite dimensions. Subsequently, in the realm of infinite dimensions, we apply this methodology to investigate two distinct optimal control problems, each with unique technical backgrounds. The first problem pertains to the optimal control of vibrations in an elastic beam, whereas the second problem focuses on the optimal control of sterilization processes for prepackaged food.

2.1 Necessary Optimality Condition

Using the Dubovitskiĭ-Milyutin approach, optimal control problems, which are broadly referred to as extremum problems, are first reformulated as the task of finding the extremum of a specific functional subject to inequality and equality constraints [89].

In a locally convex topological linear space E, consider a functional $F(x)$ defined in a neighborhood of a point x_0. The point x is subject to a set of constraints, encompassing both equality and inequality constraints. In one type of constraint, $x \in \Omega_i,\ i = 1, 2, \cdots, n$, where each Ω_i is defined by inequality constraints and possesses a nonempty interior. In the other type, $x \in \Omega_{n+1}$, a set defined by a system of equality constraints and lacking interior points. The extremum problem in this context involves determining the minimum of the functional $F(x)$ over the

B. Sun et al., *Maximum Principle and Dynamic Programming Viscosity Solution Approach*, Systems & Control: Foundations & Applications,
https://doi.org/10.1007/978-981-96-5739-1_2

intersection of all these sets, denoted as $\Omega = \bigcap_{i=1}^{n+1} \Omega_i$. In essence, our objective is to identify a point x_0 within Ω that minimizes the functional $F(x)$, as expressed in

$$F(x_0) = \min_{x \in \Omega} F(x). \tag{2.1.1}$$

Occasionally, the focus shifts to a local minimum problem, and the corresponding formulation can be established as follows:

$$F(x_0) = \min_{x \in \Omega \cap U} F(x),$$

where U is some neighborhood of x_0.

Regarding the functional to be optimized, we introduce the following definition of a direction of decrease.

Definition 2.1.1 A vector h in a topological linear space E is said to be a direction of decrease of the functional $F(x)$ at the point x_0 if there exists a neighborhood U of h, a number $\varepsilon_0 > 0$, and a number $\alpha = \alpha(F, x_0, h) < 0$ that depends on $F,\ x_0,\ h$ such that for all $\varepsilon \in (0, \varepsilon_0)$ and any $\bar{h} \in U$,

$$F\left(x_0 + \varepsilon \bar{h}\right) \leq F(x_0) + \varepsilon\alpha. \tag{2.1.2}$$

It is noteworthy that the directions of decrease for the functional $F(x)$ at the point x_0 form an open cone K having its apex at 0. In other words,

$$K = \{\lambda h \mid h \text{ is the direction of decrease of } F(x) \text{ at } x_0,\ \lambda > 0\}.$$

Furthermore, if the set of directions of decrease of $F(x)$ at x_0 is convex, $F(x)$ is said to be a regularly decreasing functional.

Regarding the given constraints, let us start with the inequality constraint Q_1 that possesses a nonempty interior. Here, we introduce the concept of a feasible direction.

Definition 2.1.2 A vector h in a topological linear space E is said to be a feasible direction for Q_1 at a point x_0 if there exists a neighborhood U of the vector h and a number $\varepsilon_0 > 0$ such that, for all $\varepsilon \in (0, \varepsilon_0)$ and all $\bar{h} \in U$, the vectors $x_0 + \varepsilon\bar{h} \in Q_1$.

Similarly, it is observed that the feasible directions generate an open cone K with its apex at 0. If the cone of feasible directions for an inequality constraint Q_1 at a point x_0 is convex, then Q_1 is deemed regular at x_0.

Moving on to the second type of constraint, the equality constraint Q_2 lacks interior points. Here, we introduce the concept of a tangent direction.

Definition 2.1.3 A vector h in a topological linear space E is said to be a tangent direction for Q_2 at a point x_0 if given a number $\varepsilon_0 > 0$, for all $\varepsilon \in (0, \varepsilon_0)$, there exists a point $x(\varepsilon) \in Q_2$ such that, if we set $x(\varepsilon) = x_0 + \varepsilon h + r(\varepsilon)$, then the vector

$r(\varepsilon) \in E$ has the property that for any neighborhood U of 0,

$$\frac{1}{\varepsilon} r(\varepsilon) \in U \tag{2.1.3}$$

for all small $\varepsilon > 0$.

It is worth noting that in a Banach space, (2.1.3) is replaced by $\|r(\varepsilon)\| = o(\varepsilon)$.

Furthermore, we observe that tangent directions also form a cone K with its apex at 0. Analogously, an equality constraint Q_2 is considered regular at a point x_0 if the cone of tangent directions to Q_2 at x_0 exhibits convexity.

In this manner, we establish the necessary condition for optimality in the given problem (2.1.1). Let K_i, $i = 1, 2, \cdots, n$, be the cones of feasible directions corresponding to the inequality constraints Ω_i. Denote the cone of tangent directions by K_{n+1} for the equality constraint Ω_{n+1}. Let K_0 be the cone of the directions of decrease. A necessary condition such that $F(x)$ attains a minimum at x_0 is that

$$\bigcap_{i=0}^{n+1} K_i = \emptyset. \tag{2.1.4}$$

Namely, the necessary condition for optimality is the absence of a common point among the cones K_i, $i = 0, 1, \cdots, n, n+1$. This assertion is rigorously proven in Theorem 2.1.4 presented later.

Indeed, Eq. (2.1.4) can be reformulated as a condition involving continuous linear functionals in the given space. This condition draws parallels to the Lagrange multipliers rule in classical analysis and the Euler equation in the calculus of variations. Consequently, this condition is also known as the Euler-Lagrange equation. We now present this condition in the form of a theorem.

Theorem 2.1.4 (Dubovitskiĭ-Milyutin Theorem) *Assume that the functional $F(x)$ attains a local minimum at the point x_0 in $\Omega = \bigcap_{i=0}^{n+1} \Omega_i$. Assume that $F(x)$ is regularly decreasing at x_0 with the cone of directions of decrease K_0 and the inequality constraint Ω_i, $i = 1, 2, \cdots, n$, are regular at x_0 with the cone of feasible directions K_i and that the equality constraint Ω_{n+1} is also regular at x_0 with the cone of tangent directions K_{n+1}. Then there exist continuous linear functionals f_i, not all identically zero, such that $f_i \in K_i^*$, the dual cone of K_i, $i = 0, 1, \cdots, n, n+1$, which satisfy the condition*

$$f_0 + f_1 + \cdots + f_n + f_{n+1} = 0.$$

Proof Firstly, we prove that a necessary condition for the functional to have a minimum at x_0 is $\bigcap_{i=0}^{n+1} K_i = \emptyset$. In other words, no direction of decrease of the functional $F(x)$ can be feasible for all constraints. Actually, this argument is exactly condition (2.1.4).

Suppose that this is false so that there exists a vector $h \in K_i$, $i = 0, 1, \cdots, n+1$. By the definition of K_i, $i = 0, 1, \cdots, n$, there exists a neighborhood U of h such that, whenever $0 < \varepsilon < \varepsilon_0$, any vector $x_0 + \varepsilon\bar{h}$, $\bar{h} \in U$, lies in $\bigcap_{i=1}^{n} \Omega_i$ and satisfies the inequality (2.1.2). Now consider the vector $x(\varepsilon) = x_0 + \varepsilon h + r(\varepsilon) \in \Omega_{n+1}$ as in the definition of tangent directions, and let ε_1 be such that $\frac{1}{\varepsilon}r(\varepsilon) \in U - h$, or $\bar{h}(\varepsilon) = h + \frac{1}{\varepsilon}r(\varepsilon) \in U$, for $0 < \varepsilon < \varepsilon_1$. Then, whenever $0 < \varepsilon < \min\{\varepsilon_0, \varepsilon_1\}$, on the one hand, the vectors $x(\varepsilon) = x_0 + \varepsilon\bar{h}(\varepsilon)$ lie in $\bigcap_{i=1}^{n} \Omega_i$ and, on the other hand, lie in Ω_{n+1}. In other words, these vectors satisfy all the constraints but they also satisfy the inequality

$$F\left(x_0 + \varepsilon\bar{h}(\varepsilon)\right) \le F(x_0) + \varepsilon\alpha < F(x_0),$$

which contradicts the assumption that x_0 is a minimum point. Thus, $\bigcap_{i=0}^{n+1} K_i = \emptyset$.

By definition, K_i, $i = 0, 1, \cdots, n$, are convex open cones, and K_{n+1} is a convex cone. Utilizing Theorem 1.1.31, we conclude the proof of this theorem. □

2.2 Determination of Cones

In order to effectively apply Theorem 2.1.4 to a specific optimal control problem, it is crucial to identify all relevant cones: the cone of directions of decrease, the cone of feasible directions, the cone of tangent directions, and their respective dual cones. This is the primary focus of this section. For brevity and clarity, we will only summarize the key findings and omit the detailed proofs. For a comprehensive understanding, readers can refer to [89, 98].

2.2.1 Directions of Decrease

Definition 2.2.1 Let $F(x)$ be a functional in a topological linear space E. Its directional derivative $F'(x_0, h)$ at a point x_0 in the direction $h \in E$ is given by

$$F'(x_0, h) = \lim_{\varepsilon \to 0+} \frac{F(x_0 + \varepsilon h) - F(x_0)}{\varepsilon}.$$

Theorem 2.2.2 *Let K_d be the cone of directions of decrease for the functional $F(x)$ at the point x_0. If $h \in K_d$ and $F'(x_0, h)$ exists, then $F'(x_0, h) < 0$.*

In fact, by (2.1.2),

$$F(x_0 + \varepsilon h) \le F(x_0) + \varepsilon\alpha.$$

This gives

$$\lim_{\varepsilon \to 0+} \frac{F(x_0 + \varepsilon h) - F(x_0)}{\varepsilon} \le \alpha < 0.$$

Conversely, for certain specific functionals, we observe the following facts.

Theorem 2.2.3 *Let E be a Banach space. Let the functional $F(x)$ satisfy a Lipschitz condition in a neighborhood of the point x_0. Namely, there exist $\varepsilon_0 > 0$ and $\beta > 0$ such that for all $\|x_1 - x_0\| \le \varepsilon_0$, $\|x_2 - x_0\| \le \varepsilon_0$, the following inequality holds:*

$$|F(x_1) - F(x_2)| \le \beta \|x_1 - x_2\|.$$

Hence, for $F'(x_0, h) < 0$, there must exist a vector $h \in K_d$ that is the cone of directions of decrease for $F(x)$ at x_0.

Proof Let $F'(x_0, h) = -\delta < 0$. It then follows from Definition 2.2.1 that

$$F(x_0 + \varepsilon h) \le F(x_0) - \frac{\varepsilon \delta}{2}, \quad 0 < \varepsilon < \varepsilon_1.$$

Let $\bar{h}$ be an arbitrary vector such that $\|h - \bar{h}\| \le \frac{\delta}{4}$. Then, whenever $0 < \varepsilon < \min\{\varepsilon_0, \varepsilon_1\}$, we have

$$\begin{aligned} F\left(x_0 + \varepsilon \bar{h}\right) &\le F(x_0 + \varepsilon h) + \beta \left\|\varepsilon \bar{h} - \varepsilon h\right\| \\ &\le F(x_0) - \frac{\varepsilon \delta}{2} + \frac{\varepsilon \delta}{4} = F(x_0) - \frac{\varepsilon \delta}{4}. \end{aligned}$$

Namely, (2.1.2) holds with $\alpha = -\frac{\delta}{4}$ so that $h \in K_d$. □

Definition 2.2.4 Let $F(x)$ be a functional in a topological linear space E. If for any $\lambda \in [0, 1]$, there holds

$$F(\lambda x + (1 - \lambda) y) \le \lambda F(x) + (1 - \lambda) F(y), \quad \forall\, x, y \in E,$$

then $F(x)$ is said to be a convex functional.

Obviously, for the convex functionals $F_i(x)$, $i = 1, 2, \cdots, n$, any linear combination with nonnegative coefficients

$$\sum_{i=1}^{n} \alpha_i F_i(x), \quad \alpha_i \ge 0$$

will also be convex. Moreover, let $F_s(x)$, $s \in S$ be convex functionals, where S is an arbitrary index set. Then, $\sup\limits_{s \in S} F_s(x)$, the supremum of $F_s(x)$ over S, is also a convex functional. If $F(x)$ is a convex functional, then for any given λ, the sets defined as

$$\{x \mid F(x) < \lambda\}, \quad \{x \mid F(x) \leq \lambda\}, \tag{2.2.1}$$

are both convex.

Theorem 2.2.5 *Let the functional $F(x)$ satisfy the Lipschitz condition in a neighborhood of the point x_0 in a Banach space E. Assume that $F(x)$ is differentiable at x_0 in any direction and that $F'(x_0, h)$ is convex as a functional of h. Then $F(x)$ is regularly decreasing at x_0, and its cone of directions of decrease*

$$K_d = \{h \mid F'(x_0, h) < 0\}. \tag{2.2.2}$$

Proof By Theorems 2.2.2 and 2.2.3, we have (2.2.2). Moreover, by the fact on the convex functional with (2.2.1), $K_d = \{h \mid F'(x_0, h) < 0\}$ is convex. □

Alternatively, we also have the following variant pertaining to regularly decreasing functionals.

Theorem 2.2.6 *Let $F(x)$ be a convex continuous functional in a topological linear space E. Then $F(x)$ is differentiable at any point x_0 in any direction h and*

$$F(x_0 + h) \geq F(x_0) + F'(x_0, h).$$

Moreover, $F(x)$ is regularly decreasing at any point, and its cone of directions of decrease K_d is as shown in (2.2.2).

Proof Let $x_0 \in E$, $h \in E$ be fixed. The function of one variable $\varphi(\varepsilon) = F(x_0 + \varepsilon h)$ is convex and continuous since $F(x)$ is convex and continuous. Thus, for any $0 < \lambda < 1$ we have $\varphi(\lambda\varepsilon) \leq \lambda\varphi(\varepsilon) + (1 - \lambda)\varphi(0)$. Namely, for $\varepsilon > 0$,

$$\frac{\varphi(\lambda\varepsilon) - \varphi(0)}{\lambda\varepsilon} \leq \frac{\varphi(\varepsilon) - \varphi(0)}{\varepsilon}.$$

Therefore, $\frac{\varphi(\varepsilon)-\varphi(0)}{\varepsilon}$ is monotone decreasing for $\varepsilon \to 0+$. Moreover, for $\varepsilon > 0$,

$$\varphi(0) \leq \frac{\varepsilon}{\varepsilon + 1}\varphi(-1) + \frac{1}{\varepsilon + 1}\varphi(\varepsilon), \quad \frac{\varphi(\varepsilon) - \varphi(0)}{\varepsilon} \geq \varphi(-1) - \varphi(0),$$

so that $\frac{\varphi(\varepsilon)-\varphi(0)}{\varepsilon}$ is bounded below for $\varepsilon > 0$. Thus, $\lim\limits_{\varepsilon \to 0+} \frac{\varphi(\varepsilon)-\varphi(0)}{\varepsilon}$ exists. By the definition of the directional derivative that

$$F'(x_0, h) = \lim_{\varepsilon \to 0+} \frac{\varphi(\varepsilon) - \varphi(0)}{\varepsilon},$$

exists at each point in an arbitrary direction. Moreover, the fact that $\frac{\varphi(\varepsilon)-\varphi(0)}{\varepsilon}$ is monotone decreasing implies that

$$\varphi(1)-\varphi(0) \geq \lim_{\varepsilon\to 0+} \frac{\varphi(\varepsilon)-\varphi(0)}{\varepsilon}.$$

This again means that $F(x_0+h) \geq F(x_0)+F'(x_0,h)$.

Assuming now that $F'(x_0,h)<0$, let us prove that $h \in K_d$. By Definition 2.2.1 of the directional derivative, there exists a number $\varepsilon_0>0$ such that $F(x_0+\varepsilon_0 h) < F(x_0)$. Set $\delta = F(x_0)-F(x_0+\varepsilon_0 h) > 0$. Since $F(x)$ is continuous, let U be a neighborhood of the point h such that for all $\bar{h} \in U$,

$$\left|F\left(x_0+\varepsilon_0\bar{h}\right)-F(x_0+\varepsilon_0 h)\right| \leq \frac{\delta}{2}.$$

Then $F\left(x_0+\varepsilon_0\bar{h}\right) \leq F(x_0)-\frac{\delta}{2}$. By the condition of convexity of $F(x)$, we have for $0<\varepsilon<\varepsilon_0$ that

$$F\left(x_0+\varepsilon_0\bar{h}\right) \leq \left(1-\frac{\varepsilon}{\varepsilon_0}\right)F(x_0)+\frac{\varepsilon}{\varepsilon_0}F\left(x_0+\varepsilon_0\bar{h}\right) \leq F(x_0)-\frac{\varepsilon\delta}{2\varepsilon_0}.$$

Namely, condition (2.1.2) holds with $\alpha = -\frac{\delta}{2\varepsilon_0}$, by which $h \in K_d$. This, together with Theorem 2.2.2, leads to $K_d = \{h \mid F'(x_0,h)<0\}$.

Finally, we proceed to establish that K_d is convex. Let $h_1,\ h_2 \in K_d$ and $h = \lambda h_1+(1-\lambda)h_2$ for $0<\lambda<1$. We have

$$\begin{aligned}&\frac{F(x_0+\varepsilon h)-F(x_0)}{\varepsilon}\\ \leq\ &\frac{\lambda[F(x_0+\varepsilon h_1)-F(x_0)]+(1-\lambda)[F(x_0+\varepsilon h_2)-F(x_0)]}{\varepsilon},\end{aligned}$$

so that

$$F'(x_0,h) \leq \lambda F'(x_0,h_1)+(1-\lambda)F'(x_0,h_2).$$

Since $h_1,\ h_2 \in K_d$, $F'(x_0,h_1)<0,\ F'(x_0,h_2)<0$. Thus, $F'(x_0,h)<0$ and hence $h \in K_d$. This completes the proof of the theorem. □

Definition 2.2.7 A functional $F(x)$ on a Banach space E is said to be Fréchet differentiable (or simply differentiable) at a point x_0 if there exists a linear functional $f \in E^*$ such that, for all $h \in E$,

$$F(x_0+h) = F(x_0)+\langle f,h\rangle+\circ(\|h\|). \tag{2.2.3}$$

Here, the linear functional f can be denoted by $F'(x_0)$ that is called the derivative, Fréchet derivative, or gradient of $F(x)$ at x_0.

It is obvious that if functionals $F_i(x),\ i = 1, 2, \cdots, n$, are differentiable at x_0, then $F(x) = \sum_{i=1}^{n} \alpha_i F_i(x)$, $\alpha_i \in \mathbb{R}$, is also differentiable at x_0 and

$$F'(x_0) = \sum_{i=1}^{n} \alpha_i F_i'(x_0).$$

Furthermore, in relation to the directional derivative, we have the following observation: If $F(x)$ is Fréchet differentiable at the point x_0, then at x_0 in any direction h, there exists the directional derivative $F'(x_0, h)$ and

$$F'(x_0, h) = \langle F'(x_0), h\rangle.$$

Therefore, by Theorem 2.2.6, if $F(x)$ is a differentiable convex functional, then

$$F(x_0 + h) \geq F(x_0) + \langle F'(x_0), h\rangle.$$

In view of the results presented above and Theorems 2.2.2 and 2.2.3, we have $K_d = \{h \mid \langle F'(x_0), h\rangle < 0\}$, which is convex since it is either a half-space or the empty set. Thus, $F(x)$ is regularly decreasing. We establish Theorem 2.2.8 below.

Theorem 2.2.8 *If the functional $F(x)$ is differentiable at the point x_0, then it is regularly decreasing at x_0 and its cone of directions of decrease*

$$K_d = \{h \mid \langle F'(x_0), h\rangle < 0\}.$$

2.2.2 *Feasible Directions*

Throughout this subsection, let K_f denote the cone of feasible directions at a point x_0 for a set Q in a topological linear space E. If $x_0 \in \overset{\circ}{Q}$, by Definition 2.1.2, any direction is feasible, so in this case, $K_f = E$. Therefore, we only need to consider the case that x_0 is a boundary point of Q. In addition, we always assume here that $\overset{\circ}{Q} \neq \emptyset$ Because otherwise, K_f would be an empty set, which would render the discussion trivial.

We first investigate a case where Q is defined by a specific functional $F(x)$, namely,

$$Q = \{x \mid F(x) \leq F(x_0)\}. \tag{2.2.4}$$

Let K_d be the cone of directions of decrease for $F(x)$ at x_0. Then $K_d \subset K_f$. In fact, if $h \in K_d$, it follows from (2.1.2) that

$$F\left(x_0 + \varepsilon\bar{h}\right) \le F(x_0) + \varepsilon\alpha < F(x_0),$$

for all $\bar{h}$ in some neighborhood U of the vector h and all $0 < \varepsilon < \varepsilon_0$. Thus, $x_0 + \varepsilon\bar{h} \in Q$ for all such $\bar{h}$ and ε, but this is precisely the definition of a feasible direction.

Furthermore, we have the following results.

Theorem 2.2.9 *Let the functional $F(x)$ be differentiable at the point x_0 in any direction and its directional derivative $F'(x_0, h)$ at x_0 in the direction h be convex in h. Assume that there exists $\tilde{h}$ such that $F'\left(x_0, \tilde{h}\right) < 0$, and then*

$$K_f \subset \{h \mid F'(x_0, h) < 0\}.$$

Proof Let $h \in K_f$. Then $x_0 + \varepsilon h \in Q$ for $0 < \varepsilon < \varepsilon_0$. Namely, $F(x_0 + \varepsilon h) \le F(x_0)$. Thus, $F'(x_0, h) \le 0$. Moreover, since the feasible directions generate an open cone with apex at 0, it follows that the cone K_f is open. Therefore, there exists a neighborhood U of the vector h such that $U \in K_f$.

Take $\gamma > 0$ so that $h_\gamma = h - \gamma\left(h - \tilde{h}\right) \in U$. Then $F'(x_0, h_\gamma) \le 0$ since $h_\gamma \in K_f$. By the convexity of $F'(x_0, h)$ in h,

$$\begin{aligned} F'(x_0, h) &= F'\left(x_0, \frac{1}{1+\gamma}h_\gamma + \frac{\gamma}{1+\gamma}\tilde{h}\right) \\ &\le \frac{1}{1+\gamma}F'(x_0, h_\gamma) + \frac{\gamma}{1+\gamma}F'\left(x_0, \tilde{h}\right) \\ &\le \frac{\gamma}{1+\gamma}F'\left(x_0, \tilde{h}\right) < 0. \end{aligned}$$

This completes the proof. □

Moreover, by Theorems 2.2.5, 2.2.6, and 2.2.8, in some cases, K_f can be directly determined by K_d from Theorem 2.2.9 and the fact that $K_d \subset K_f$. The results are listed below.

Theorem 2.2.10 *Let E be a Banach space. Suppose that any one of the following three conditions is satisfied.*

(i) *The functional $F(x)$ satisfies the Lipschitz condition in a neighborhood of the point x_0; $F(x)$ is differentiable at x_0 in any direction and $F'(x_0, h)$ is convex in h; there exists h_1 such that $F'(x_0, h_1) < 0$.*
(ii) *$F(x)$ is a convex continuous functional and there exists x_1 such that $F(x_1) < F(x_0)$.*
(iii) *$F(x)$ is differentiable at x_0 and $F'(x_0) \ne 0$.*

Then

$$K_f = K_d = \{h \mid F'(x_0, h) < 0\}.$$

Finally, in another scenario, the set Q is not defined by a functional, which differs from (2.2.4). In this instance, assuming Q is a convex set, the cone of feasible directions K_f can be straightforwardly determined using its definition.

Theorem 2.2.11 *If Q is a convex set, then at x_0,*

$$\begin{aligned} K_f &= \left\{\lambda\left(\overset{\circ}{Q} - x_0\right) \,\middle|\, \lambda > 0\right\} \\ &= \left\{h \,\middle|\, h = \lambda(x - x_0),\ x \in \overset{\circ}{Q},\ \lambda > 0\right\}. \end{aligned}$$

2.2.3 Tangent Directions

Definition 2.2.12 Let $E_1,\ E_2$ be two Banach spaces. And $P(x)$ is an operator with domain in E_1 and range in E_2. Then $P(x)$ is said to be differentiable at a point $x_0 \in E_1$ if there exists a continuous linear operator A mapping E_1 into E_2, such that for all $h \in E_1$,

$$P(x_0 + h) = P(x_0) + Ah + r(x_0, h),$$

where $\|r(x_0, h)\| = o(\|h\|)$. The operator A is called the (Fréchet) derivative of the operator $P(x)$ and denoted by $A = P'(x_0)$.

Specifically, if $P(x)$ is a functional (in this context, where $E_2 = \mathbb{R}$), this definition aligns with Definition 2.2.7. Moreover, the derivative of an operator possesses the standard properties associated with the differentiation of sums, composite functions, and so on.

Subsequently, we introduce a fundamental theorem in extremal theory, namely, the Lyusternik theorem. This theorem serves as a powerful tool for characterizing tangent directions. In particular, it provides a crucial means for deriving first-order necessary conditions for optimality in extremum problems involving equality constraints.

Theorem 2.2.13 (Lyusternik Theorem [164]) *Let $E_1,\ E_2$ be Banach spaces. Let $P(x)$ be an operator mapping E_1 into E_2 and differentiable in a neighborhood of a point $x_0,\ P(x_0) = 0$. The Fréchet derivative $P'(x)$ is continuous in a neighborhood* *of x_0, and assume that $P'(x_0)$ maps E_1 onto E_2. Then the set of tangent directions K_t to the set $Q = \{x \mid P(x) = 0\}$ at x_0 is the subspace*

$$K_t = \{h \mid P'(x_0)h = 0\}.$$

It is worth noting that in this context, the Fréchet derivative $P'(x_0)$ must be onto, which is an essential requirement. In fact, if the Fréchet derivative $P'(x_0)$ is not onto, the corresponding conditions would take on a trivial or degenerate form, as discussed in [145]. Furthermore, in variational analysis, the tangent cone and the Lyusternik theorem are defined as follows.

Definition 2.2.14 Let X be a Banach space. A vector $h \in X$ is called a tangent to a set $S \subset X$ at $x \in S$ if the distance from $x + th$ to S,

$$d(x + th, S) = o(t), \quad t \to 0+.$$

The collection of all tangent vectors to S at x, denoted by $T(S, x)$, forms a cone that always contains the zero vector. If this set is a subspace, it is called the tangent space to S at x.

Theorem 2.2.15 (Lyusternik Theorem) *Suppose that F is strictly differentiable and regular at $\overline{x}$. The collection of vectors tangent to the level set $M = \{x \mid F(x) = F(\overline{x})\}$ at $\overline{x}$ is a subspace that coincides with* $\ker F'(\overline{x})$, *the kernel of the operator F'.*

Proof It is evident that $T(M, \overline{x}) \subset \ker F'(\overline{x})$. This is because if h is a tangent vector to M at x, then for any t, there exists an $h(t)$ such that $F(x + th(t)) \in M$ and $| h - h(t)\| = o(t)$. This implies that

$$F'(x)h = \lim_{t\to 0} F'(\overline{x})h(t) = \lim_{t\to 0} \frac{F(\overline{x} + th(t)) - F(\overline{x}) + o(\|th(t)\|)}{t} = 0.$$

This implication holds regardless of any assumption made regarding the range of $F'(\overline{x})$.

Alternatively, if we set $A = F'(\overline{x})$, then F and A satisfy the assumptions of Theorem 1.4.4 for any positive δ. Therefore, given an $h \in \ker A$, for any small δ (e.g., all $\delta < \frac{C(A)}{2}$), by Corollary 1.4.2,

$$d(\overline{x} + th, M) \leq \frac{C(A)}{2} \|F(\overline{x}) - F(\overline{x} + th)\| = o(t).$$

□

For further information on Lyusternik's theorem, one can refer to the following sources for detailed discussions: [68, 123, 164].

For various sets Q at a point x_0, the cone of tangent directions K_t can be determined by the Lyusternik theorem mentioned above. For example, for $n,\ m \in \mathbb{N}$, let

$$Q = \left\{x \in \mathbb{R}^m \mid G_i(x) = 0,\ i = 1, 2, \cdots, n\right\},$$

in which $G_i(x)$ are functions that are continuously differentiable at the point x_0, $G_i(x_0) = 0,\ i = 1, 2, \cdots, n$, and the vectors $G_i'(x_0),\ i = 1, 2, \cdots, n$, are linearly independent. Then,

$$K_t = \left\{h \in \mathbb{R}^n \mid \left\langle G_i'(x_0), h\right\rangle = 0,\ i = 1, 2, \cdots, n\right\}.$$

Then, here is the second example.

Example 2.2.16 Let $T > 0$ and define the space $E_1 = C(0, T) \times L^{\infty}(0, T)$. The elements $x(t) \in \mathbb{R}^n$ and $u(t) \in \mathbb{R}^m$, $n,\ m \in \mathbb{N}$, satisfy the differential equation

$$\frac{d}{dt}x(t) = \varphi(x(t), u(t), t), \quad x(0) = c.$$

Set

$$Q = \left\{(x, u) \in E_1 \,\middle|\, x(t) = c + \int_0^t \varphi(x(\tau), u(\tau), \tau)\, \mathrm{d}\tau\right\}.$$

We now proceed to determine the tangent subspace $K_t = \{(\bar{x}(t), \bar{u}(t))\}$ subject to this constraint. Introduce the operator

$$P(x, u) = x(t) - c - \int_0^t \varphi(x(\tau), u(\tau), \tau)\, \mathrm{d}\tau$$

which maps E_1 into $C(0, T)$. By this formulation, we can rewrite

$$Q = \{(x, u) \in E_1 \mid P(x, u) = 0\}.$$

To utilize the Lyusternik theorem (refer to Theorem 2.2.13), certain assumptions are indeed necessary. For all bounded $x,\ u$ and all $t \in [0, T]$, assume that the derivatives $\varphi_x(x, u, t)$, $\varphi_u(x, u, t)$ exist, continuous in $x,\ u$, measurable in t, and bounded. Then, the following equality holds:

$$\begin{aligned}&P\left(x + \bar{x}, u + \bar{u}\right) - P(x, u)\\ &= \bar{x}(t) - \int_0^t \left[\varphi_x(x, u, \tau)\bar{x}(\tau) + \varphi_u(x, u, \tau)\bar{u}(\tau)\right]\, \mathrm{d}\tau + \delta,\end{aligned}$$

where

$$\delta = \text{o}\left(\sqrt{\|\bar{x}\|^2_{C(0,T)} + \|\bar{u}\|^2_{L^\infty(0,T)}}\right)$$

is a remainder term and the first term on the right-hand side is a linear operator of $\bar{x}$, $\bar{u}$, defined in $C(0, T)$. In this manner, $P(x, u)$ is differentiable, and

$$P'(x, u)(\bar{x}, \bar{u}) = \bar{x}(t) - \int_0^t [\varphi_x(x, u, \tau)\bar{x}(\tau) + \varphi_u(x, u, \tau)\bar{u}(\tau)]\, d\tau.$$

Moreover, $P'(x, u)$ is continuous in a neighborhood of x, u.

Next, we show that $P'(x, u)$ maps E_1 onto $C(0, T)$. Namely, the equation

$$\bar{x}(t) - \int_0^t [\varphi_x(x, u, \tau)\bar{x}(\tau) + \varphi_u(x, u, \tau)\bar{u}(\tau)]\, d\tau = a(t)$$

has a solution $\bar{x}$, $\bar{u}$ for any $a(t) \in C(0, T)$. Set $\bar{u}(t) \equiv 0$. Then the equation becomes

$$\bar{x}(t) = a(t) + \int_0^t \varphi_x(x, u, \tau)\bar{x}(\tau)\, d\tau.$$

This integral equation is a linear Volterra equation of the second kind, which has a solution $\bar{x}(t) \in C(0, T)$ for any $a(t) \in C(0, T)$.

By the Lyusternik theorem (Theorem 2.2.13), the tangent subspace K_t consists of all pairs $\bar{x}(t)$, $\bar{u}(t)$ satisfying the integral equation

$$\bar{x}(t) = \int_0^t [\varphi_x(x, u, \tau)\bar{x}(\tau) + \varphi_u(x, u, \tau)\bar{u}(\tau)]\, d\tau,$$

or the linear differential equation that follows:

$$\frac{d}{dt}\bar{x}(t) = \varphi_x(x, u, t)\bar{x}(t) + \varphi_u(x, u, t)\bar{u}(t), \quad \bar{x}(0) = 0,$$

which is also called as the variational equation.

To characterize a more intricate example, two conditions are imperative. Herein, we enumerate them in the form of a definition.

Definition 2.2.17 Let $D \subset \mathbb{R}^n$ be the set of all vectors $\bar{x}(T)$, where $\bar{x}(t)$ satisfies the differentiable equation

$$\frac{d}{dt}\bar{x}(t) = A(t)\bar{x}(t) + B(t)\bar{u}(t), \quad \bar{x}(0) = 0, \tag{2.2.5}$$

with $\bar{u}(t)$ ranging over the entire space $L^\infty(0, T)$. Here $A(t),\ B(t)$ are $n \times n$ and $n \times r$ matrices, $\bar{u}(t) \in \mathbb{R}^r,\ \bar{x}(t) \in \mathbb{R}^n$, respectively. Equation (2.2.5) is nondegenerate if $D = \mathbb{R}^n$. Moreover, a system of differential equations satisfying this condition is sometimes said to be completely controllable.

Alternatively, the definition can be expressed in the following manner.

Definition 2.2.18 System (2.2.5) is nondegenerate if any nonzero solution $\psi(t)$ of the equation

$$\frac{d}{dt}\psi(t) = -A^\top(t)\psi(t) \tag{2.2.6}$$

satisfies the condition $B^\top(t)\psi(t) \not\equiv 0$. More precisely, $B^\top(t)\psi(t)$ is nonzero on a set of positive measure.

Definitely, the nondegeneracy condition in Definition 2.2.18 implies the condition in Definition 2.2.17 [89].

Subsequently, another example is provided by incorporating the terminal constraint into Example 2.2.16.

Example 2.2.19 In Example 2.2.16, we add the constraint $x(T) = d \in \mathbb{R}^n$ so that

$$Q = \left\{ (x, u) \in E_1 \;\middle|\; x(t) = c + \int_0^t \varphi(x(\tau), u(\tau), \tau)\, \mathrm{d}\tau,\ t \in [0, T],\ x(T) = d \right\}.$$

Similarly, we are intrigued by the tangent subspace of Q. Once again, let us introduce the operator

$$P(x, u) = \left(x(t) - c - \int_0^t \varphi(x(\tau), u(\tau), \tau)\, \mathrm{d}\tau,\ x(T) - d \right)$$

which maps E_1 into $E_2 = C(0, T) \times \mathbb{R}^n$.

Using similar arguments as in the previous example, we can deduce that $P(x, u)$ is continuously differentiable, and

$$P'(x, u)\,(\bar{x}, \bar{u}) = \left(\bar{x}(t) - \int_0^t [\varphi_x(x, u, \tau)\bar{x}(\tau) + \varphi_u(x, u, \tau)\bar{u}(\tau)]\, \mathrm{d}\tau,\ \bar{x}(T) \right).$$

Then, let $(a(t), b) \in E_2,\ A(t) = \varphi_x(x(t), u(t), t),\ B(t) = \varphi_u(x(t), u(t), t)$. Take $z(t) \in C(0, T)$ such that

$$z(t) = a(t) + \int_0^t A(\tau)z(\tau)\, \mathrm{d}\tau, \quad 0 \le t \le T.$$

Just like in Example 2.2.16, if the nondegeneracy condition holds for system (2.2.5), then $P'(x, u)$ maps E_1 onto E_2. Since the system is nondegenerate, we can deduce that

$$\bar{u}(t) \in L^{\infty}(0,T), \quad y(t) \in C(0,T),$$

such that

$$\frac{d}{dt}y(t) = \varphi_x(x,u,t)y(t) + \varphi_u(x,u,t)\bar{u}(t), \; y(0)=0, \; y(T) = b - z(T).$$

Set $\bar{z}(t) = y(t) + z(t)$. Then

$$P'(x,u)(\bar{x},\bar{u}) = \left(\bar{x}(t) - \int_0^t [\varphi_x(x,u,\tau)\bar{x}(\tau) + \varphi_u(x,u,\tau)\bar{u}(\tau)]\, d\tau, \; \bar{x}(T)\right)$$

$$= \left(z(t) + y(t) - \int_0^t \varphi_x(x,u,\tau)z(\tau)\, d\tau - y(t), b\right) = (a(t), b).$$

Under the assumptions of Example 2.2.16, let the system

$$\frac{d}{dt}\bar{x}(t) = \varphi_x(x,u,t)\bar{x}(t) + \varphi_u(x,u,t)\bar{u}(t), \quad \bar{x}(0) = 0$$

be nondegenerate. In this manner,

$$K_t = \left\{(\bar{x},\bar{u}) \;\middle|\; \bar{x}(t) = \int_0^t [\varphi_x(x,u,\tau)\bar{x}(\tau) + \varphi_u(x,u,\tau)\bar{u}(\tau)]\, d\tau, \; \bar{x}(T) = 0\right\}.$$

Alternatively, in the context of differential equations,

$$K_t = \left\{(\bar{x},\bar{u}) \;\middle|\; \frac{d}{dt}\bar{x}(t) = \varphi_x(x,u,t)\bar{x}(t) + \varphi_u(x,u,t)\bar{u}(t), \bar{x}(0) = \bar{x}(T) = 0\right\}.$$

2.2.4 Dual Cones

We proceed to determine the dual cones for various cases and enumerate the corresponding results.

Theorem 2.2.20 *Let E be a topological linear space. If K is a subspace, then*

$$K^* = \{f \in E^* \mid f(x) = 0, \; \forall\, x \in K\}.$$

Moreover, let $f \in E^*$ *and*

$$K_1 = \{x \in E \mid f(x) = 0\},$$
$$K_2 = \{x \in E \mid f(x) \geq 0\},$$
$$K_3 = \{x \in E \mid f(x) > 0\}.$$

Then

$$K_1^* = \{\lambda f \mid -\infty < \lambda < \infty\},$$
$$K_2^* = \{\lambda f \mid 0 \leq \lambda < \infty\},$$
$$K_3^* = \begin{cases} E^*, & f = 0, \\ K_2^*, & f \neq 0. \end{cases}$$

Proof Suppose for some $x \in K$, $f \in K^*$ we have $f(x) > 0$. Since $-x \in K$, $f(-x) = -f(x) < 0$, this contradicts with $f \in K^*$. This proves the first assertion.

Moreover, by the first assertion, if $f_1(x) \in K_1^*$, then $f_1(x) = 0$ on K_1. By Theorem 1.1.6, $f_1 = \lambda f$. Since $K_1 \subset K_2$, it follows that $K_2^* \subset K_1^*$ by Theorem 1.1.24. However, if $f_1 = \lambda f \in K_2^*$, then $f_1(x) = \lambda f(x) \geq 0$ for all $x \in K_2$ so that $\lambda \geq 0$.

Finally, considering two scenarios, if $f = 0$, then K_3 is empty, implying $K_3^* = E^*$. Conversely, if $f \neq 0$, then $\overline{K_3} = K_2$, and hence, according to Theorem 1.1.23, we have $K_3^* = K_2^*$. □

Then, let A^* denote the adjoint operator of operator A. We present the following facts.

Theorem 2.2.21 *Let* $E = E_1 \times E_2$, *where* E_1, E_2 *are two topological linear spaces. Let A be a linear operator mapping* E_1 *into* E_2. *Denote*

$$K = \{x = (x_1, x_2) \in E \mid Ax_1 = x_2\}.$$

Then,

$$K^* = \{f = (f_1, f_2) \in E^* \mid f_1 = -A^* f_2\}.$$

Proof If $f \in K^*$, then for all $x \in K$, by Theorem 2.2.20, we have

$$0 = \langle f, x\rangle = \langle f_1, x_1\rangle + \langle f_2, x_2\rangle = \langle f_1 + A^* f_2, x_1\rangle,$$

and hence $f_1 + A^* f_2 = 0$. On the contrary, if $f = (-A^* f_2, f_2)$, then for $x \in K$,

$$\langle f, x\rangle = \langle -A^* f_2, x_1\rangle + \langle f_2, x_2\rangle = -\langle f_2, Ax_1 - x_2\rangle = 0.$$

□

The Minkowski-Farkas theorem below outlines the form of dual cones for a wide range of cone types.

Theorem 2.2.22 (Minkowski-Farkas Theorem) *Let E_1, E_2 be two topological linear spaces and K_2 be a convex cone in E_2 with apex at 0, $K_1 = \{x_1 \in E_1 \mid Ax_1 \in K_2\}$. Denote $E = E_1 \times E_2$. Assume that either of the following conditions is satisfied:*

(i) *There exists $\tilde{x}_1 \in E_1$ such that $A\tilde{x}_1 \in \overset{\circ}{K}_2$.*
(ii) *Let E_1, E_2 be of finite-dimensional and K_2 be the positive octant of E_2. Namely, $K_1 = \{x \in \mathbb{R}^m \mid \langle a^i, x\rangle \geq 0,\ a^i \in \mathbb{R}^m, i = 1, 2, \cdots, n\}$. Alternatively, expressed in matrix notation, $K_1 = \{x \in \mathbb{R}^m \mid Ax \geq 0\}$, where A is an $n \times m$ matrix.*

Then $K_1^* = A^*K_2^*$. Namely, in Case (ii),

$$K_1^* = \left\{ \sum_{i=1}^{n} a^i y_i \;\middle|\; y_i \geq 0,\ i = 1, 2, \cdots, n \right\}$$

or $K_1^* = \{A^*y \mid y \geq 0\}$.

Proof Let us start with Case (i). Denote $K = \{x \in E \mid x = (x_1, x_2),\ x_2 \in K_2\}$ and $L = \{x \in E \mid Ax_1 = x_2\}$. It follows that K is a convex cone with apex at 0 and L is a subspace such that $\overset{\circ}{K} \cap L \neq \emptyset$. By Theorem 1.1.29, $(K \cap L)^* = K^* + L^*$. Let $\tilde{f}_1 \in K_1^*$. If $x \in K \cap L$, then $x = (x_1, Ax_1),\ x_1 \in K_1$, and therefore, $\tilde{f}(x) = \tilde{f}_1(x_1) \leq 0$. Hence, $\tilde{f} = \langle \tilde{f}_1, 0\rangle \in (K \cap L)^*$.

Consequently, it follows that $\tilde{f} \in K^* + L^*$. Furthermore,

$$K^* = \{f = (f_1, f_2) \mid f_1 = 0,\ f_2 \in K_2^*\},$$

and by Theorem 2.2.21,

$$L^* = \{f = (f_1, f_2) \mid f_1 = -A^* f_2\}.$$

Therefore, there exists $\tilde{f}_2 \in K_2^*$ such that $\tilde{f}_1 = A^*\tilde{f}_2$. Namely, $K_1^* \subset A^*K_2^*$. Furthermore, the inverse inclusion remains valid, which implies that if $f_1 = A^* f_2,\ f_2 \in K_2^*$, then

$$\langle f_1, x_1\rangle = \langle A^* f_2, x_1\rangle = \langle f_2, Ax_1\rangle \geq 0$$

holds true for all $x \in K_1$, or equivalently, for $Ax_1 \in K_2$.

Now, we proceed to Case (ii). Denote $Q_i = \{x \mid \langle a^i, x\rangle \geq 0\}$. Hence, we have $K_1 = \bigcap_{i=1}^{n} Q_i$. By Theorem 2.2.20, $Q_i^* = \{a^i y_i \mid y_i \geq 0\}$. Take

$$\sum_{i=1}^{n} Q_i^* = \left\{ \sum_{i=1}^{n} a^i y_i \;\middle|\; y_i \geq 0,\ i = 1, 2, \cdots, n \right\}.$$

By virtue of [246], it follows that any set of the type

$$\left\{ \sum_{i=1}^{n} a^i y_i \;\middle|\; y_i \geq 0,\ i = 1, 2, \cdots, n \right\}$$

is closed in $\mathbb{R}^m$. Since all topologies of $\mathbb{R}^m$ coincide, the set is also weakly* closed in $\mathbb{R}^m$. Therefore, by Lemma 1.1.28

$$\left(\bigcap_{i=1}^{n} Q_i \right)^* = \sum_{i=1}^{n} Q_i^*.$$

In other words,

$$K_1^* = \left\{ \sum_{i=1}^{n} a^i y_i \;\middle|\; y_i \geq 0,\ i = 1, 2, \cdots, n \right\}.$$

□

In the sequel, several crucial examples of dual cones are enumerated. Firstly, let $T > 0,\ E = C(0, T)$ and Ξ be a closed set in $[0, T]$. Denote

$$K_1 = \{x = (x_1(t), x_2(t), \cdots, x_n(t)) \in C(\Xi) \mid x(t) \geq 0,\ t \in \Xi\}.$$

By the Riesz representation theorem, for any $f \in K_1^*$, there exists a nonnegative vector measure $\mathrm{d}\mu(t)$ with support Ξ, such that

$$f(x) = \int_0^T \langle x(t), \mathrm{d}\mu(t)\rangle = \int_\Xi \langle x(t), \mathrm{d}\mu(t)\rangle.$$

Furthermore, let $h(t) \in C(0, T),\ h(t) \neq 0$ on Ξ. Denote

$$K_2 = \{x = (x_1(t), x_2(t), \cdots, x_n(t)) \in C(0, T) \mid \langle h(t), x(t)\rangle \geq 0,\ t \in \Xi\}.$$

Then for any $f \in K_2^*$, there exists a nonnegative scalar measure $\mathrm{d}\mu(t)$ with support Ξ, such that

$$f(x) = \int_0^T \langle h(t), x(t)\rangle \,\mathrm{d}\mu(t) = \int_\Xi \langle h(t), x(t)\rangle \,\mathrm{d}\mu(t).$$

Furthermore, the following fact holds true. Here, let Q be a set in a topological linear space E, $x_0 \in Q$, K_f the cone of feasible directions for Q at x_0 and K_t the cone of tangent directions for Q at the point x_0. In some cases, it turns out that the duals of these cones coincide with the set of linear functionals which are supports for Q at x_0. Denote the latter set by $Q^* = \{f \in E^* \mid f(x) \geq f(x_0),\ x \in Q\}$.

Theorem 2.2.23 *If Q is a closed convex set, then $K_t^* = Q^*$. Moreover, if the interior $\overset{\circ}{Q} \neq \emptyset$, then $K_f^* = Q^*$.*

Proof Let $f \in Q^*, h \in K_t$. Given $0 < \varepsilon < \varepsilon_0$, there exists $x(\varepsilon) \in Q$, $x(\varepsilon) = x_0 + \varepsilon h + r(\varepsilon)$, such that $\frac{1}{\varepsilon} r(\varepsilon) \in U$ for any neighborhood U of zero and sufficiently small ε. Since $x(\varepsilon) \in Q$ and $f \in Q^*$, it follows that $\langle f, x(\varepsilon)\rangle \geq \langle f, x_0\rangle$, and therefore,

$$\langle f, h\rangle = \frac{\langle f, x(\varepsilon)\rangle - \langle f, x_0\rangle}{\varepsilon} - \frac{\langle f, r(\varepsilon)\rangle}{\varepsilon} \geq -\left\langle f, \frac{r(\varepsilon)}{\varepsilon}\right\rangle.$$

Due to the fact f is continuous, for every $\delta > 0$ there exists a neighborhood U of zero such that $|\langle f, x\rangle| \leq \delta$ for $x \in U$, and hence, $\left|\left\langle f, \frac{r(\varepsilon)}{\varepsilon}\right\rangle\right| < \delta$ for sufficiently small $\varepsilon > 0$. In this manner,

$$\left\langle f, \frac{r(\varepsilon)}{\varepsilon}\right\rangle \to 0, \quad \text{as } \varepsilon \to 0+.$$

Therefore, $\langle f, h\rangle \geq 0$ or $f \in K_t^*$.

On the contrary, let $f \in K_t^*$, $x \in Q$. Then $h = x - x_0$ is a tangent direction for $x_0 + \varepsilon h = (1-\varepsilon)x_0 + \varepsilon x \in Q$ when $\varepsilon \in (0, 1)$ because Q is convex. Given that $f \in K_t^*$, it can be deduced that $\langle f, h\rangle \geq 0$. Namely, $\langle f, x\rangle \geq \langle f, x_0\rangle$. Hence, $f \in Q^*$.

At this juncture, we have successfully established that $Q^* = K_t^*$. If $\overset{\circ}{Q} \neq \emptyset$, by Theorem 2.2.11,

$$K_f = \left\{\lambda(x - x_0) \;\middle|\; \lambda \geq 0,\ x \in \overset{\circ}{Q}\right\}.$$

Therefore, if $f \in K_f^*$, then

$$\langle f, \lambda(x - x_0)\rangle \geq 0, \quad \lambda\langle f, x\rangle \geq \lambda\langle f, x_0\rangle,$$

for all $x \in \overset{\circ}{Q}$ and all $\lambda \geq 0$, in particular for $\lambda = 1$. But $\overline{\overset{\circ}{Q}} = \overline{Q}$ when $\overset{\circ}{Q} \neq \emptyset$ by the fact in Sect. 1.1.2. Hence, by the continuity of $\langle f, x\rangle$, $\langle f, x\rangle \geq \langle f, x_0\rangle$ for all $x \in \overset{\circ}{Q}$ implies that $\langle f, x\rangle \geq \langle f, x_0\rangle$ for all $x \in \overline{Q}$. In other words, $f \in Q^*$ and $K_f^* \subset K_t^* = Q^*$. On the other hand, $K_f \subset K_t$, and hence by Theorem 1.1.24, $K_f^* \supset K_t^* = Q^*$. Thus, $K_f^* = Q^*$. □

We now present additional examples as follows. Let Q be a polyhedron in $\mathbb{R}^m$ defined as

$$Q = \{x \mid \langle a_i, x\rangle \geq b_i,\ i = 1, 2, \cdots, n\},\ a_i \in \mathbb{R}^m,\ b_i \in \mathbb{R},\ i = 1, 2, \cdots, n.$$

One has

$$K_t^* = Q^* = \left\{ \sum_{i=1}^{n} \lambda_i a_i \;\middle|\; \lambda_i \geq 0,\ \lambda_i\left[\langle a_i, x_0\rangle - b_i\right] = 0 \right\}.$$

In matrix notation, this can be expressed as $Q = \{x \mid Ax \geq b\}$. Then,

$$K_t^* = Q^* = \left\{ A^* y \mid y \geq 0,\ y_i\,(Ax_0 - b)_i = 0,\ i = 1, 2, \cdots, n \right\}.$$

Lastly, we present another intriguing example here.

Example 2.2.24 Let $T > 0$ and

$$Q = \left\{ x \in L^\infty(0, T) \mid x(t) \in M \subset \mathbb{R}^r \text{ for almost all } t \in [0, T] \right\},\ x_0 \in Q.$$

Then, if the linear functional

$$f(x) = \int_0^T \langle a(t), x(t)\rangle\, dt, \quad a(t) \in \mathbb{R}^r, \quad a(t) \in L^1(0, T),$$

is a support to Q at the point x_0, then

$$\langle a(t), x - x_0(t)\rangle \geq 0$$

for all $x \in M$ and almost all $t \in [0, T]$. Namely, the vector $a(t) \in \mathbb{R}^r$ is a support to M at the point $x_0(t)$ for almost all $t \in [0, T]$.

Here, since the structure of the space $(L^\infty)^*$ is relatively intricate, it is challenging to determine the set of supporting functionals. Therefore, it is offset by providing a characterization of the general form of supporting functionals defined through integrals.

2.3 Finite-Dimensional System Control

In this section, we delve into an optimal control problem, guided by a finite-dimensional system. Utilizing the Dubovitskiĭ-Milyutin functional analytical framework, we derive the necessary optimality condition, specifically, the Pontryagin maximum principle [89, 98]. Our primary objective is to demonstrate the application of the Dubovitskiĭ-Milyutin approach in addressing optimal control problems. Additionally, this serves as a crucial prelude to our future exploration of optimal control problems involving infinite-dimensional systems.

2.3.1 Optimal Control Formulation

Let $T > 0$. Consider the following control system over $[0, T]$:

$$\begin{aligned} \frac{d}{dt}x(t) &= f(x(t), u(t), t), \\ x(0) &= c, \ x(T) = d, \end{aligned} \tag{2.3.1}$$

where the state $x \in \mathbb{R}^n$, the control $u \in \mathbb{R}^m$. Denote the cost functional as

$$J(x, u) = \int_0^T \Phi(x(t), u(t), t)\, \mathrm{d}t, \tag{2.3.2}$$

and define the control constraint as

$$u \in M, \quad t \in [0, T] \text{ a.e.} \tag{2.3.3}$$

Here, the vectors $c,\ d$ denote the initial and terminal condition, respectively. The set $M \subset \mathbb{R}^m$ is a closed convex set in $\mathbb{R}^m$ such that $\overset{\circ}{M} \neq \emptyset$.

Occasionally, rather than focusing on the differential equation in (2.3.1), an equivalent integral equation is taken into account:

$$x(t) = c + \int_0^t f(x(\tau), u(\tau), \tau)\, \mathrm{d}\tau.$$

Moreover, $f(\cdot, \cdot, \cdot)$ is a vector-valued function and $\Phi(\cdot, \cdot, \cdot)$ a scalar function.

The optimal control problem aims to identify $u \in M$ such that, along with the solution to system (2.3.1), it minimizes the cost functional J.

Define the state space as $C(0, T)$ and the control space as $L^\infty(0, T)$. Establish $E = C(0, T) \times L^\infty(0, T)$ as the initial space. Consider (2.3.3) as the inequality constraint and (2.3.1) as the equality constraint. Set these constraints accordingly as

$$\Omega_1 = \{(x, u) \in E \mid u(t) \in M,\ t \in [0, T] \text{ a.e.}\},$$
$$\Omega_2 = \left\{(x, u) \in E \;\middle|\; \frac{d}{dt}x(t) = f(x(t), u(t), t),\ x(0) = c,\ x(T) = d\right\}.$$

In this manner, the optimal control problem under investigation is equivalent to finding the optimal solution $(x^*, u^*) \in \Omega = \Omega_1 \cap \Omega_2$ such that

$$J\left(x^*, u^*\right) = \min_{(x,u)\in\Omega} J(x, u). \tag{2.3.4}$$

2.3.2 Cones and Their Duals

By virtue of the Dubovitskiĭ-Milyutin theorem (Theorem 2.1.4), it is imperative to establish the cone of directions of decrease K_0 pertaining to the functional $J(x, u)$, the cone of feasible directions K_1 corresponding to the inequality constraint Ω_1, the cone of tangent directions K_2 linked to the equality constraint Ω_2, and their respective dual cones.

Firstly, we consider the cone of directions of decrease. Regarding the cost functional, we assume that $\Phi(x, u, t)$ is continuous in x and u and that the partial derivatives $\Phi_x(x, u, t)$, $\Phi_u(x, u, t)$ exist for every (x, u). Additionally, we make the following assumptions that

$$\int_0^T \left|\frac{\partial\Phi(x, u, t)}{\partial x}\right| \mathrm{d}t, \quad \int_0^T \left|\frac{\partial\Phi(x, u, t)}{\partial u}\right| \mathrm{d}t,$$

are bounded for $t \in [0, T]$.

Therefore, $J(x, u)$ is differentiable at any point $\left(x^0, u^0\right)$ in a direction (x, u) and its directional derivative is

$$J'\left(x^0, u^0\right)(x, u) = \int_0^T \Big[\left\langle \Phi_x\left(x^0(t), u^0(t), t\right), x(t)\right\rangle + \left\langle \Phi_u\left(x^0(t), u^0(t), t\right), u(t)\right\rangle\Big] \mathrm{d}t.$$

Moreover, $J(x, u)$ is regularly decreasing at (x^*, u^*) and

$$K_0 = \{(x, u) \in E \mid J'(x^*, u^*)(x, u) < 0\}$$
$$= \left\{(x, u) \in E \,\middle|\, \int_0^T \Big[\langle \Phi_x(x^*(t), u^*(t), t), x(t)\rangle + \langle \Phi_u(x^*(t), u^*(t), t), u(t)\rangle\Big] \mathrm{d}t < 0\right\}.$$

If $K_0 \neq \emptyset$, then for any $f_0 \in K_0^*$, there exists a $\lambda_0 \geq 0$ such that

$$f_0(x, u) = -\lambda_0 \int_0^T \left[\langle \Phi_x\left(x^*(t), u^*(t), t\right), x(t)\rangle + \langle \Phi_u\left(x^*(t), u^*(t), t\right), u(t)\rangle\right] \mathrm{d}t. \tag{2.3.5}$$

Secondly, we derive the cone of feasible directions. Since $\Omega_1 = C(0, T) \times \Omega_1'$, where $\Omega_1' = \{u(t) \in L^\infty(0, T) \mid u(t) \in M,\ t \in [0, T] \text{ a.e.}\}$ By the assumption on M, Ω_1' is a closed convex set in $L^\infty(0, T)$ and its interior is not empty. Therefore, Ω_1 is also a closed convex set in E and possesses the nonempty interior. Denote the cone of feasible directions of Ω_1 at the point (x^*, u^*) by K_1. Then, if $f_1 \in K_1^*$, by Theorem 2.2.23, it follows that $f_1 = \langle 0, \tilde{f}_1\rangle$, where $\tilde{f}_1 \in (L^\infty)^*$ is a support to Ω_1' at the point u^*.

Thirdly, let us investigate the constraint Ω_2. We assume that $f_u^*(x^*, u^*, t)g(t) \not\equiv 0$ for any nonzero solution $g(t)$ of the system

$$\frac{d}{dt} g(t) = -f_x^\top(x^*, u^*, t)g(t).$$

Namely, the nondegeneracy condition in Definition 2.2.18 holds. In this manner, by Example 2.2.19, the tangent subspace K_2 consists of all pairs $\bar{x},\ \bar{u}$ such that

$$\frac{d}{dt}\bar{x}(t) = f_x(x^*, u^*, t)\bar{x}(t) + f_u(x^*, u^*, t)\bar{u}(t), \quad \bar{x}(0) = 0, \tag{2.3.6}$$

$$\bar{x}(T) = 0. \tag{2.3.7}$$

Next, denote the set of all pairs $\bar{x},\ \bar{u}$ satisfying Equation (2.3.6) by $L_1 \subset E$ and the set of pairs satisfying equation (2.3.7) by $L_2 \subset E$. Therefore, L_1 and L_2 are the subspaces and $K_2 = L_1 \cap L_2$. By Lemma 1.1.1, $K_2^* = L_1^* + L_2^*$. Given that L_1 is a subspace, by Theorem 2.2.20, for any $f_2 \in L_1^*$, we know that $f_2(\bar{x}, \bar{u}) = 0$ for all $\bar{x},\ \bar{u}$ satisfying (2.3.6). Moreover, if $f_3 \in L_2^*$, then $f_3(\bar{x}, \bar{u}) = \langle \bar{x}(T), a\rangle, a \in \mathbb{R}^n$.

2.3.3 Local Maximum Principle

Now, it is time to present the final result. According to the Dubovitskiĭ-Milyutin theorem (Theorem 2.1.4), there exist $f_i \in E^*$, $i = 0, 1, 2, 3$, not all zero, such that for all $(\bar{x}, \bar{u}) \in E$,

$$f_0(\bar{x}, \bar{u}) + f_1(\bar{x}, \bar{u}) + f_2(\bar{x}, \bar{u}) + f_3(\bar{x}, \bar{u}) = 0, \tag{2.3.8}$$

where $f_0(\bar{x}, \bar{u})$ is given by (2.3.5), $f_1(\bar{x}, \bar{u}) = \tilde{f}_1(\bar{u})$ is a support to Ω_1' at u^*, $f_2(\bar{x}, \bar{u}) = 0$ for $\bar{x}$, $\bar{u}$ satisfying equation (2.3.6), and $f_3(\bar{x}, \bar{u}) = \langle \bar{x}(T), a \rangle$. Then, let $\bar{u}$ be arbitrary and choose a solution $\bar{x} = \bar{x}(\bar{u})$ to Eq. (2.3.6). And Eq. (2.3.8) yields

$$\tilde{f}_1(\bar{u}) = \lambda_0 \int_0^T \left[\left\langle \Phi_x\left(x^*, u^*, t\right), \bar{x} \right\rangle + \left\langle \Phi_u\left(x^*, u^*, t\right), \bar{u} \right\rangle \right] \mathrm{d}t - \langle \bar{x}(T), a \rangle. \tag{2.3.9}$$

Let $g(t)$ be the solution to the system

$$\begin{aligned} \frac{d}{dt} g(t) &= -f_x^\top(x^*(t), u^*(t), t) g(t) + \lambda_0 \Phi_x(x^*(t), u^*(t), t), \\ g(T) &= a. \end{aligned} \tag{2.3.10}$$

Given that $\bar{x}$, $\bar{u}$ satisfy equation (2.3.6), we integrate by parts to derive

$$\begin{aligned} &\lambda_0 \int_0^T \left\langle \Phi_x\left(x^*, u^*, t\right), \bar{x} \right\rangle \mathrm{d}t - \langle \bar{x}(T), a \rangle \\ &= \int_0^T \left\langle \frac{dg}{dt} + f_x^\top(x^*, u^*, t) g(t), \bar{x} \right\rangle \mathrm{d}t - \langle \bar{x}(T), a \rangle \\ &= \langle g, \bar{x} \rangle |_0^T - \int_0^T \left\langle g, \frac{d\bar{x}}{dt} \right\rangle \mathrm{d}t + \int_0^T \left\langle g, f_x(x^*, u^*, t)\bar{x} \right\rangle \mathrm{d}t - \langle \bar{x}(T), a \rangle \\ &= -\int_0^T \left\langle g, f_u(x^*, u^*, t)\bar{u} \right\rangle \mathrm{d}t \\ &= -\int_0^T \left\langle f_u^\top(x^*, u^*, t) g, \bar{u} \right\rangle \mathrm{d}t. \end{aligned}$$

As a result, condition (2.3.8) or Eq. (2.3.9) can be rewritten as

$$\tilde{f}_1(\bar{u}) = \int_0^T \left\langle -f_u^\top(x^*, u^*, t) g + \lambda_0 \Phi_u(x^*, u^*, t), \bar{u} \right\rangle \mathrm{d}t,$$

where $\bar{u}$ is arbitrary and $\tilde{f}_1(\bar{u})$ is a support to Ω_1' at the point u^*. By Example 2.2.24, we obtain

$$\left\langle -f_u^{\top}(x^*, u^*, t)g + \lambda_0 \Phi_u(x^*, u^*, t), u - u^*(t) \right\rangle \geq 0 \tag{2.3.11}$$

for almost all $t \in [0, T]$ and all $u \in M$.

By now, we can present the following local maximum principle for the optimal control problem outlined in Sect. 2.3.1, albeit with the need to discuss certain cases.

Theorem 2.3.1 *Let $f(x, u, t)$ and $\Phi(x, u, t)$ be continuous in x and u, measurable in t, continuously differentiable with respect to x, u. Let $f_x(x, u, t)$, $f_u(x, u, t)$, $\Phi_x(x, u, t)$, $\Phi_u(x, u, t)$ be bounded for all bounded x, u. Then there exist a $\lambda_0 \geq 0$ and a function $g(t)$ satisfying equation* (2.3.10), *not identically zero, such that the maximum principle* (2.3.11) *holds true for almost all $t \in [0, T]$ and all $u \in M$.*

In fact, it must be emphasized that, under the assumptions stated in Theorem 2.3.1, the case of $\lambda_0 = 0$, $g(t) \equiv 0$ cannot occur. Otherwise, there are $f_0(\bar{x}, \bar{u}) \equiv 0$, $a = g(T) = 0$. Therefore, by (2.3.9), $\tilde{f}_1(\bar{u}) = 0$. That is to say, $f_0 = 0$, $f_1 = 0$, $f_3 = 0$. Given that (2.3.8), we have $f_2 = 0$, which contradicts the fact stated in Theorem 2.1.4 that these continuous linear functionals are not all identically zero.

Furthermore, in our preceding analysis, we had made two additional assumptions. Firstly, we assumed that $K_0 \neq \emptyset$. Secondly, we presumed that system (2.3.6) is nondegenerate. However, we now demonstrate that these assumptions are redundant and can be dispensed with. As a matter of fact, if $K_0 = \emptyset$, then

$$\int_0^T \left[\left\langle \Phi_x(x^*, u^*, t), \bar{x} \right\rangle + \left\langle \Phi_u(x^*, u^*, t), \bar{u} \right\rangle \right] \mathrm{d}t = 0 \tag{2.3.12}$$

for all $\bar{x}$, $\bar{u}$. Specially, let $\lambda_0 = 1$, $g(T) = 0$. Then, as before, by system (2.3.10), there is

$$\begin{aligned}
&\int_0^T \left\langle \Phi_x(x^*, u^*, t), \bar{x} \right\rangle \mathrm{d}t \\
&= \int_0^T \left\langle \frac{dg}{dt} + f_x^{\top}(x^*, u^*, t)g, \bar{x} \right\rangle \mathrm{d}t \\
&= \langle g(T), \bar{x}(T) \rangle - \langle g(0), \bar{x}(0) \rangle - \int_0^T \left\langle g, \frac{d\bar{x}}{dt} \right\rangle \mathrm{d}t \\
&\quad + \int_0^T \left\langle g, f_x(x^*, u^*, t)\bar{x} \right\rangle \mathrm{d}t \\
&= -\int_0^T \left\langle g, f_u(x^*, u^*, t)\bar{u} \right\rangle \mathrm{d}t
\end{aligned}$$

for any $\bar{x}(t)$ satisfying system (2.3.6). In this way, we see that Eq. (2.3.12) can be rewritten as

$$0 = \int_0^T \left\langle -f_u^\top(x^*, u^*, t)g + \Phi_u(x^*, u^*, t), \bar{u} \right\rangle \mathrm{d}t$$

for any $\bar{u}$. Therefore,

$$-f_u^\top(x^*, u^*, t)g + \Phi_u(x^*, u^*, t) = 0$$

for almost $t \in [0, T]$. That is to say, inequality (2.3.11) holds.

Then, if system (2.3.6) is degenerate, there exists a nonzero $g(t)$ that is a solution to Eq. (2.3.10) for $\lambda_0 = 0$ such that $-f_u^\top(x^*, u^*, t)g(t) \equiv 0$, which is once again inequality (2.3.11).

In conclusion, this section has delved into an optimal control problem within the context of finite dimensions, utilizing the Dubovitskiĭ-Milyutin functional analytical approach. Through this analysis, we have derived the local maximum principle for the problem. However, due to space constraints, we have omitted discussing additional optimal control examples, such as those involving free final time, constraints on phase coordinates, and minimax problems. For those interested in further exploring these topics, additional information can be found in [89, 98, 139].

2.4 Optimal Control of Vibrations of an Elastic Beam

In the preceding section, it was demonstrated through a finite-dimensional optimal control problem that the Dubovitskiĭ-Milyutin functional analytical approach is remarkably effective in addressing extremum problems that involve both inequality and equality constraints. In fact, as we will demonstrate in this and subsequent sections, this method can also be successfully applied to PDE-constrained optimal control problems. The initial application in this section focuses on an optimal control problem related to the vibration of an elastic beam, which is modeled using a PDE system. Utilizing the Dubovitskiĭ-Milyutin approach, we establish the corresponding Pontryagin maximum principle for this problem.

2.4.1 Background

The dynamical modeling and control of large space structures has been a focal point in aeronautics and astronautics. Real-world applications, such as controlling the vibration of space shuttle arms, have yielded numerous notable and satisfactory results. The need to construct these structures as lightweight as possible often leads to slender elements with minimal natural damping. The unique characteristics of such space structures necessitate the implementation of appropriate control measures to suppress system vibrations. Failure to do so can result in the eventual collapse of the system due to the increasing vibration. In fact, such applications can be found in various engineering fields, including the vibration research of robot system links. To achieve high-speed and precise endpoint positioning of flexible beams, controlling beam vibration is crucial. In the mathematical realm, such physical structures can be abstracted into beam equations represented by differential equations.

In the studies of beam equations, the focus has been primarily on stability and feedback stabilization of systems (see [101, 163, 241], to name a few). It is true that feedback control of dynamical systems offers numerous advantages compared to open-loop control. However, it cannot be denied that open-loop control also holds its own advantages in the exploration of infinite-dimensional systems. These advantages include the efficiency and accuracy of open-loop control algorithms, as well as the robustness of the systems under investigation [211]. Just as stated in "The No Free Lunch Theorem of Optimization (NFLT)" in [117], "a general-purpose universal optimization strategy is impossible." Therefore, the investigation of open-loop control for beam equation vibration is both necessary and intriguing. For instance, [38] studies an active vibration controller (incorporating time delay) for a flexible cantilever beam and proposed a control algorithm to address the time delay issue. On the other hand, [200] explores the problem of characterizing optimal controls for a class of damped distributed parameter systems governed by a variable-coefficient, linear PDE and derived Pontryagin's maximum principle for these systems. Furthermore, with regards to a wide range of linear PDEs, the Adali, Bruch, Sadek, and Sloss research group has accomplished numerous noteworthy works using the maximum principle approach. Some of their notable contributions include [20, 21, 33, 34, 72, 196–199, 201, 210, 212]. For instance, a problem involving the optimal control of a model for a rotating body beam, which is characterized by a linear state PDE, is considered in [154], and the necessary condition for the control was proposed. Additionally, [147, 148] investigates optimal control problems pertaining to planar networks of Euler-Bernoulli beams, as well as general networks of strings and/or Timoshenko beams, using the dynamic domain decomposition approach.

In this section, we delve into the optimal control problem of vibrations in elastic beams governed by a nonlinear PDE. Our focus is on a class of nonlinear beam models that incorporate linear damping and external time-dependent forcing within the Sobolev space H^2. This exploration is spurred by the aspiration to develop computational techniques for controlling smart material composites, particularly those that undergo significant deformations or adhere to nonlinear stress-strain laws [12]. In engineering scenarios, curved and twisted structures often experience large deflections even when strain levels remain low. Additionally, certain composites and elastomers exhibit nonlinear stress-strain relationships even under small deformations [162]. The models we investigate here encapsulate this nonlinear but monotone stress-strain behavior, which is pivotal in deriving the transverse bending equation for a beam. The external forces considered are distributed across the space-time domain and represent the kind of loads encountered in current "smart materials" structures. These structures, such as beams or plates, are often loaded with piezoceramic actuators and sensors [12]. Employing the Dubovitskiĭ and Milyutin functional approach, we delve into the Pontryagin maximum principle of the system. This approach allows us to derive the necessary condition for the optimal control problem in the fixed final horizon case. This condition serves as a crucial guide in designing effective control strategies for the nonlinear beam vibrations.

2.4.2 Model and Well-Posedness Revisited

Consider a nonlinear elastic beam equation that follows [1]:

$$\begin{cases} w_{tt}(t,x)+\kappa_1 w_{xxxx}(t,x)+\kappa_2 w_{txxxx}(t,x) \\ \quad =[g(w_{xx}(t,x))]_{xx}+f(t,x), \quad 0<x<\ell, \quad 0<t<T, \\ w_x(t,0)=w(t,0)=0, \quad w_x(t,\ell)=w(t,\ell)=0, \\ w(0,x)=w_0(x)\in H_0^2(0,\ell), \quad w_t(0,x)=w_1(x)\in L^2(0,\ell), \end{cases} \tag{2.4.1}$$

where the state variable $w(t,x)$ denotes the displacement of vibration of beam at time t and spatial position x, κ_1, κ_2 two constants. Two given functions g, f satisfy specific conditions outlined in the following text.

For the nonlinear PDE (2.4.1), [1] introduces a definition of weak solution and, under relaxed assumptions regarding the nonlinearity in the stiffness constitutive law, establishes the existence and uniqueness of such weak solutions. To proceed with our optimal control investigation, we require such a definition of weak solution. Therefore, we revisit this definition here.

Let $H=L^2(0,\ell)$ and $V=H_0^2(0,\ell)$ so the Gelfand triple $V\hookrightarrow H\hookrightarrow V^*$ with $V^*=H^{-2}(0,\ell)$. Here $\langle\cdot,\cdot\rangle$ is the inner product in H and $\langle\cdot,\cdot\rangle_{V^*,V}$ denotes the usual duality product. Denote $\|\cdot\|_S$ the norm of the space S. Define the space of solutions to be

$$\mathcal{S}(0,T)=\left\{p \;\middle|\; p\in L^2(0,T;V),\ p_t\in L^2(0,T;V),\ p_{tt}\in L^2(0,T;V^*)\right\}$$

with norm

$$\|p\|_{\mathcal{S}(0,T)}=\left(\|p\|^2_{L^2(0,T;V)}+\|p_t\|^2_{L^2(0,T;V)}+\|p_{tt}\|^2_{L^2(0,T;V^*)}\right)^{\frac{1}{2}}.$$

A function $w \in \mathcal{S}(0, T)$ is a weak solution of (2.4.1) if the following equation holds:

$$\begin{aligned}&\langle w_{tt}(t), \phi\rangle_{V^*,V} + \kappa_1\langle w_{xx}(t), \phi_{xx}\rangle + \kappa_2\langle w_{txx}(t), \phi_{xx}\rangle \\ &\quad = \langle g(w_{xx}(t)), \phi_{xx}\rangle + \langle f(t), \phi\rangle_{V^*,V}, \quad \forall\, \phi \in V\end{aligned} \tag{2.4.2}$$

and $w(0) = w_0 \in V,\ w_t(0) = w_1 \in H$.

Assume that the nonlinear function g satisfies the following local Lipschitz condition. Let $B_r(0)$ denote the ball of radius r centered at 0 in H, and for some constant L_{B_r}, we have

$$\|g(\xi) - g(\sigma)\|_H \le L_{B_r}\|\xi - \sigma\|_H, \quad \forall\, \xi,\ \sigma \in B_r(0). \tag{2.4.3}$$

With regard to the definition of weak solution (2.4.2) mentioned previously, [1] demonstrates that under the condition that the nonlinear function g satisfies the local Lipschitz property (2.4.3) and the forcing term $f \in L^2(0, T;\ V^*)$, there exists a time T^* for which Eq. (2.4.1) possesses a unique weak solution on the interval $[0, T^*]$.

Moreover, if the nonlinear function g not only satisfies the local Lipschitz condition (2.4.3) but also fulfills the boundedness condition

$$|g(\xi)|_H \le C_1|\xi|_H + C_2, \quad \text{for all } \xi \in H \tag{2.4.4}$$

with some constants $C_1, C_2 > 0$, then the weak solution to Eq. (2.4.1) is global.

2.4.3 Optimal Control Problem Formulation

Now, let us consider the optimal control issues related to the system described by (2.4.1). Additionally, within the context of optimal control, we make the assumption that $g \in W^{1,2}(0, T;\ H)$, the Sobolev space [216]. This ensures the global existence of a unique weak solution. Unless otherwise specified, when we refer to a solution of (2.4.1) in the following discussion, we are always referring to the weak solution in the sense defined by (2.4.2) above.

We are interested in the optimal control problem associated with system (2.4.1) on the interval $[0, T]$, specifically focusing on the case of a fixed final horizon. In this setup, we consider the forcing term $f(t, x)$ to be of the form $f(t, x) = q(t, x)\beta(t)$, where $q(t, x) \in L^2(0, T;\ V^*)$, the control $\beta(t) \in L^2(0, T)$ distributed over the space-time domain. To formulate the optimal control problem, we equip system (2.4.1) with a general cost functional

$$\min_{\beta(\cdot)\in U_{ad}} J(w, \beta) = \min_{\beta(\cdot)\in U_{ad}} \int_0^T \int_0^\ell L(w(t, x), \beta(t), t, x)\, \mathrm{d}x\, \mathrm{d}t \tag{2.4.5}$$

and the control constraint

$$U_{ad} = \left\{ \beta \in L^2(0,T) \,\middle|\, 0 \le \beta_0 \le \beta(t) \le \beta_1,\ t \in [0,T] \text{ a.e.} \right\} \tag{2.4.6}$$

The cost function L is designed to be quite general, encompassing a wide range of practically relevant cost functionals. One example of such a cost functional is the quadratic cost

$$J(w,\beta) = \int_0^T \int_0^\ell \alpha_1 \left| w(t,x) - w^\dagger(t,x) \right|^2 \mathrm{d}x\, \mathrm{d}t + \alpha_2 \int_0^T \left| \beta(t) - \beta^\dagger(t) \right|^2 \mathrm{d}t,$$

where $\alpha_i > 0,\ i = 1,2$ are constants, and $w^\dagger$, $\beta^\dagger$ are the optimal control goal for the displacement of vibrating beam and the optimal control force, respectively.

Take $w \in \mathcal{S}(0,T)$. The control space is $L^2(0,T)$ and the control constraint is $\beta_0 \le \beta(t) \le \beta_1$ in $(0,T)$.

The following assumptions are made regarding the cost functional.

(a) L is a functional defined on $V \times [\beta_0, \beta_1] \times [0,T] \times [0,\ell]$ and

$$\frac{\partial L(w(t,x),\beta(t),t,x)}{\partial w},\quad \frac{\partial L(w(t,x),\beta(t),t,x)}{\partial \beta}$$

exist for every $(w,\beta) \in V \times [\beta_0, \beta_1]$ and L is continuous in its variables.

(b)

$$\int_0^\ell \left| \frac{\partial L(w(t,x),\beta(t),t,x)}{\partial w} \right| \mathrm{d}x,\quad \int_0^\ell \left| \frac{\partial L(w(t,x),\beta(t),t,x)}{\partial \beta} \right| \mathrm{d}x$$

are bounded for $t \in [0,T]$.

Define $X_T = \mathcal{S}(0,T) \times L^2(0,T)$. Let (w^*, β^*) be the optimal solution to the control problem (2.4.5) subject to the state equation (2.4.1). Set

$$\begin{aligned}
\Omega_1 &= \{(w,\beta) \in X_T \mid \beta_0 \le \beta(t) \le \beta_1,\ t \in [0,T] \text{ a.e.}\},\\
\Omega_2 &= \{(w,\beta) \in X_T \mid w_{tt}(t,x) + \kappa_1 w_{xxxx}(t,x) + \kappa_2 w_{txxxx}(t,x)\\
&\qquad = [g(w_{xx}(t,x))]_{xx} + q(t,x)\beta(t),\\
&\qquad w_x(t,0) = w(t,0) = 0,\ w_x(t,\ell) = w(t,\ell) = 0,\\
&\qquad w(0,x) = w_0(x),\ w_t(0,x) = w_1(x),\\
&\qquad w(T,x) = w^*(T,x)\}.
\end{aligned} \tag{2.4.7}$$

Then the optimal control problem (2.4.5) is equivalent to seeking for $(w^*, \beta^*) \in \Omega = \Omega_1 \cap \Omega_2$ such that

$$J\left(w^*, \beta^*\right) = \min_{(w,\beta)\in\Omega} J(w, \beta). \tag{2.4.8}$$

It is evident that problem (2.4.8) involves finding an extremum under both inequality constraint Ω_1 and equality constraint Ω_2. In such scenarios, the functional approach proposed by Dubovitskiĭ and Milyutin has proven to be highly effective in solving this type of extremum problems (see, for instance, [50, 220]). The general theorem of Dubovitskiĭ and Milyutin pertaining to the extremum problem (2.4.8) can be succinctly expressed as Theorem 2.4.1, which is detailed below.

Theorem 2.4.1 (Dubovitskiĭ-Milyutin Theorem) *Suppose that the functional $J(w, \beta)$ achieves a minimum at the point (w^*, β^*) in Ω. Assume that $J(w, \beta)$ is regularly decreasing at (w^*, β^*) with the cone of directions of decrease K_0 and the inequality constraint is regular at (w^*, β^*) with the cone of feasible directions K_1 and that the equality constraint is also regular at (w^*, β^*) with the cone of tangent directions K_2. Then there exist continuous linear functionals f_0, f_1, f_2, not all identically zero, such that $f_i \in K_i^*$, the dual cone of K_i, $i = 0, 1, 2$, which satisfy the condition*

$$f_0 + f_1 + f_2 = 0.$$

2.4.4 *Pontryagin's Maximum Principle*

2.4.4.1 The Cone of Directions of Decrease K_0

In order to apply Theorem 2.4.1, it is necessary to identify all cones K_i, $i = 0, 1, 2$. Initially, let us determine the cone of directions of decrease K_0. By assumption, the functional $J(w, \beta)$ is differentiable at any point (w^0, β^0) along any direction (w, β), and its directional derivative is

$$\begin{aligned}
&J'(w^0, \beta^0; w, \beta)\\
&= \lim_{\varepsilon\to 0+} \frac{1}{\varepsilon}\left[J\left(w^0 + \varepsilon w, \beta^0 + \varepsilon\beta\right) - J\left(w^0, \beta^0\right)\right]\\
&= \lim_{\varepsilon\to 0+} \frac{1}{\varepsilon}\left\{\int_0^T \int_0^\ell \left[L\left(w^0 + \varepsilon w, \beta^0 + \varepsilon\beta, t, x\right) - L\left(w^0, \beta^0, t, x\right)\right] \mathrm{d}x\, \mathrm{d}t\right\}\\
&= \int_0^T \int_0^\ell \left[\frac{\partial L\left(w^0, \beta^0, t, x\right)}{\partial w} w + \frac{\partial L\left(w^0, \beta^0, t, x\right)}{\partial \beta}\beta\right] \mathrm{d}x\, \mathrm{d}t.
\end{aligned}$$

Hence, the cone of directions of decrease of the functional $J(w, \beta)$ at the point (w^*, β^*) is determined by

$$\begin{aligned} K_0 &= \left\{(w, \beta) \in X_T \,\middle|\, J'(w^*, \beta^*; w, \beta) < 0\right\} \\ &= \left\{(w, \beta) \in X_T \,\middle|\, \int_0^T \int_0^\ell \left[\frac{\partial L(w^*, \beta^*, t, x)}{\partial w} w \right.\right. \\ &\qquad \left.\left. + \frac{\partial L(w^*, \beta^*, t, x)}{\partial \beta} \beta\right] \mathrm{d}x\, \mathrm{d}t < 0\right\}. \end{aligned} \tag{2.4.9}$$

If $K_0 \neq \emptyset$, then for any $f_0 \in K_0^*$, there exists a $\kappa_0 \geq 0$ such that

$$f_0(w, \beta) = -\kappa_0 \int_0^T \int_0^\ell \left[\frac{\partial L(w^*, \beta^*, t, x)}{\partial w} w + \frac{\partial L(w^*, \beta^*, t, x)}{\partial \beta} \beta\right] \mathrm{d}x\, \mathrm{d}t. \tag{2.4.10}$$

2.4.4.2 The Cone of Feasible Directions K_1

Since $\Omega_1 = \mathcal{S}(0, T) \times \tilde{\Omega}_1$, in which

$$\tilde{\Omega}_1 = \left\{\beta \in L^2(0, T) \,\middle|\, \beta_0 \leq \beta(t) \leq \beta_1,\ t \in [0, T] \text{ a.e.}\right\}$$

is a closed convex subset of $L^2(0, T)$, the interior $\overset{\circ}{\Omega}_1$ of Ω_1 is not empty and at the point (w^*, β^*), the cone of feasible directions K_1 of Ω_1 is determined by

$$\begin{aligned} K_1 &= \left\{\kappa\left(\overset{\circ}{\Omega}_1 - (w^*, \beta^*)\right) \,\middle|\, \kappa > 0\right\} \\ &= \left\{h \,\middle|\, h = \kappa(w - w^*, \beta - \beta^*),\ (w, \beta) \in \overset{\circ}{\Omega}_1,\ \kappa > 0\right\}. \end{aligned} \tag{2.4.11}$$

Therefore, for an arbitrary $f_1 \in K_1^*$, if there is an $\bar{a}(t) \in L^2(0, T)$, such that the linear functional defined by

$$f_1(w, \beta) = \int_0^T \bar{a}(t)\beta(t)\, \mathrm{d}t \tag{2.4.12}$$

is a support to $\tilde{\Omega}_1$ at the point β^*, then

$$\bar{a}(t)\left[\beta(t) - \beta^*(t)\right] \geq 0, \quad \forall\, \beta(t) \in [\beta_0, \beta_1],\ t \in [0, T] \text{ a.e.} \tag{2.4.13}$$

2.4.4.3 The Cone of Tangent Directions K_2

Let

$$\chi(t,x) = w_{tt}(t,x) + \kappa_1 w_{xxxx}(t,x) + \kappa_2 w_{txxxx}(t,x) \\ -[g(w_{xx}(t,x))]_{xx} - q(t,x)\beta(t).$$

Define the operator

$$G : X_T \to L^2(0,T;V^*) \times \left(L^2(0,T)\right)^4 \times V \times H \times V$$

by

$$G(w,\beta) = (\chi(t,x), w_x(t,0), w(t,0), w_x(t,\ell), w(t,\ell), w(0,x) - w_0(x), \\ w_t(0,x) - w_1(x), w(T,x) - w^*(T,x)) . \tag{2.4.14}$$

Then

$$\Omega_2 = \{(w,\beta) \in X_T \mid G(w(t,x),\beta(t)) = 0\} \tag{2.4.15}$$

so the Fréchet derivative of the operator $G(w,\beta)$ is

$$G'(w,\beta)\left(\hat{w},\hat{\beta}\right) = (\hat{\chi}(t,x), \hat{w}_x(t,0), \hat{w}(t,0), \hat{w}_x(t,\ell), \hat{w}(t,\ell), \hat{w}(0,x), \\ \hat{w}_t(0,x), \hat{w}(T,x)) ,$$

in which

$$\hat{\chi}(t,x) = \hat{w}_{tt}(t,x) + \kappa_1 \hat{w}_{xxxx}(t,x) + \kappa_2 \hat{w}_{txxxx}(t,x) \\ -[g'(w_{xx}(t,x))\hat{w}_{xx}(t,x)]_{xx} - q(t,x)\hat{\beta}(t).$$

Since (w^*,β^*) is the solution to problem (2.4.5), it has $G(w^*,\beta^*) = 0$. Choosing arbitrary

$$(g_1, g_2, g_3, g_4, g_5, g_6, g_7, g_8) \in L^2(0,T;V^*) \times \left(L^2(0,T)\right)^4 \times V \times H \times V$$

and solving the equation

$$G'(w^*,\beta^*)\left(\hat{w},\hat{\beta}\right) = (g_1(t,x), g_2(t), g_3(t), g_4(t), g_5(t), g_6(x), g_7(x), g_8(x)),$$

we obtain

$$\begin{cases} \hat{w}_{tt}(t,x) + \kappa_1 \hat{w}_{xxxx}(t,x) + \kappa_2 \hat{w}_{txxxx}(t,x) \\ \quad = \left[g'(w^*_{xx}(t,x))\hat{w}_{xx}(t,x)\right]_{xx} + q(t,x)\hat{\beta}(t) + g_1(t,x), \\ \hat{w}_x(t,0) = g_2(t),\ \hat{w}(t,0) = g_3(t),\ \hat{w}_x(t,\ell) = g_4(t),\ \hat{w}(t,\ell) = g_5(t), \\ \hat{w}(0,x) = g_6(x),\ \hat{w}_t(0,x) = g_7(x),\ \hat{w}(T,x) = g_8(x). \end{cases} \tag{2.4.16}$$

Next, assume that the linearized system

$$\begin{cases} w_{tt}(t,x) + \kappa_1 w_{xxxx}(t,x) + \kappa_2 w_{txxxx}(t,x) \\ \quad = \left[g'(w^*_{xx}(t,x))w_{xx}(t,x)\right]_{xx} + q(t,x)\beta(t), \\ w_x(t,0) = w(t,0) = 0,\ w_x(t,\ell) = w(t,\ell) = 0, \\ w(0,x) = w_t(0,x) = 0 \end{cases} \tag{2.4.17}$$

is controllable. Then choose $\beta(t) = \hat{\beta}(t) \in L^2(0,T)$ such that $w(T,x) = g_8(x) - \eta(T,x)$ and let w be the solution to the linearized system (2.4.17). Choose $\hat{w}(t,x) = w(t,x) + \eta(t,x)$, where η satisfies

$$\begin{cases} \eta_{tt}(t,x) + \kappa_1 \eta_{xxxx}(t,x) + \kappa_2 \eta_{txxxx}(t,x) \\ \quad = \left[g'\left(w^*_{xx}(t,x)\right)\eta_{xx}(t,x)\right]_{xx} + g_1(t,x), \\ \eta_x(t,0) = g_2(t),\ \eta(t,0) = g_3(t),\ \eta_x(t,\ell) = g_4(t),\ \eta(t,\ell) = g_5(t), \\ \eta(0,x) = g_6(x),\ \eta_t(0,x) = g_7(x),\ \eta(T,x) = g_8(x). \end{cases}$$

In this manner, it suffices for $(\hat{w}, \hat{\beta})$ satisfying equation (2.4.16). Therefore, $G'(w^*, \beta^*)$ maps X_T onto $L^2(0,T;V^*) \times \left(L^2(0,T)\right)^4 \times V \times H \times V$. Moreover, the cone of tangent directions K_2 to the constraint Ω_2 at the point (w^*, β^*) consists of the kernel of $G'(w^*, \beta^*)$. Namely, in X_T, (w, β) satisfies

$$\begin{cases} w_{tt}(t,x) + \kappa_1 w_{xxxx}(t,x) + \kappa_2 w_{txxxx}(t,x) \\ \quad = \left[g'\left(w^*_{xx}(t,x)\right)w_{xx}(t,x)\right]_{xx} + q(t,x)\beta(t), \\ w_x(t,0) = w(t,0) = 0,\ w_x(t,\ell) = w(t,\ell) = 0, \\ w(0,x) = w_t(0,x) = 0 \end{cases} \tag{2.4.18}$$

and

$$w(T, x) = 0. \tag{2.4.19}$$

Define

$$K_{21} = \{(w, \beta) \in X_T \mid (w(t, x), \beta(t)) \text{ satisfies } (2.4.18)\},$$

$$K_{22} = \{(w, \beta) \in X_T \mid (w(t, x), \beta(t)) \text{ satisfies } (2.4.19)\}.$$

Then the cone of tangent directions $K_2 = K_{21} \bigcap K_{22}$. Hence,

$$K_2^* = K_{21}^* + K_{22}^*.$$

For any $f_2 \in K_2^*$, decompose $f_2 = f_{21} + f_{22}$, $f_{2i} \in K_{2i}^*$, the dual cone of K_{2i}, $i = 1, 2$. Then $f_{21}(w, \beta) = 0$ and for all $w(t, x) \in \mathcal{S}(0, T)$ satisfying $w(T, x) = 0$, there exists a $\varphi(x) \in V^*$ such that

$$f_{22}(w(t, x), \beta(t)) = \int_0^\ell w(T, x)\varphi(x)\, \mathrm{d}x.$$

It then follows from Theorem 2.4.1 that there exist continuous linear functionals, not all identically zero, such that

$$f_0 + f_1 + f_{21} + f_{22} = 0.$$

Therefore, when selecting (w, β) satisfying (2.4.18), $f_{21}(w, \beta) = 0$. Moreover,

$$\begin{aligned}
&f_1(w(t, x), \beta(t)) \\
&= -f_0(w(t, x), \beta(t)) - f_{22}(w(t, x), \beta(t)) \\
&= \kappa_0 \int_0^T \int_0^\ell \left[\frac{\partial L\,(w^*, \beta^*, t, x)}{\partial w} w(t, x) + \frac{\partial L\,(w^*, \beta^*, t, x)}{\partial \beta} \beta(t) \right] \mathrm{d}x\, \mathrm{d}t \\
&\quad - \int_0^\ell w(T, x)\varphi(x)\, \mathrm{d}x.
\end{aligned} \tag{2.4.20}$$

2.4.4.4 Maximum Principle of Problem (2.4.5)

Define the adjoint system of the linearized system (2.4.17) as

$$\begin{cases} \phi_{tt}(t,x) + \kappa_1 \phi_{xxxx}(t,x) - \kappa_2 \phi_{txxxx}(t,x) \\ \quad = \left[g'\left(w_{xx}^*(t,x)\right)\phi_{xx}(t,x)\right]_{xx} + \kappa_0 \dfrac{\partial L\left(w^*, \beta^*, t, x\right)}{\partial w}, \\ \phi_x(t,0) = \phi(t,0) = 0,\ \phi_x(t,\ell) = \phi(t,\ell) = 0, \\ \kappa_2 \phi(T,x) = \mu(x) \end{cases} \tag{2.4.21}$$

with

$$\mu(x) = \begin{cases} \displaystyle\int_0^x \int_0^{x_3} \int_0^{x_2} \int_0^{x_1} \varphi(s)\,\mathrm{d}s\,\mathrm{d}x_1\,\mathrm{d}x_2\,\mathrm{d}x_3, & x \neq \ell, \\ 0, & x = \ell. \end{cases} \tag{2.4.22}$$

Similarly to (2.4.1), the existence and uniqueness of a solution to (2.4.21) can be derived in a comparable manner.

Theorem 2.4.2 *The solution of system* (2.4.17) *and that of its adjoint system* (2.4.21) *has the following relation:*

$$\begin{aligned} &\kappa_0 \int_0^T \int_0^\ell \frac{\partial L\left(w^*, \beta^*, t, x\right)}{\partial w} w(t,x)\,\mathrm{d}x\,\mathrm{d}t - \int_0^\ell w(T,x)\varphi(x)\,\mathrm{d}x \\ &\quad = \int_0^T \int_0^\ell q(t,x)\phi(t,x)\beta(t)\,\mathrm{d}x\,\mathrm{d}t. \end{aligned} \tag{2.4.23}$$

Proof Multiply the first equation in (2.4.21) by $w(t,x)$ and integrate it by parts over $[0,T] \times [0,\ell]$ with respect to t and x, respectively. The proof then follows. □

Now, by virtue of Theorem 2.4.2, we can rewrite $f_1(w,\beta)$ as

$$f_1(w,\beta) = \int_0^T \left\{ \int_0^\ell \left[\kappa_0 \frac{\partial L\left(w^*, \beta^*, t, x\right)}{\partial \beta} + q(t,x)\phi(t,x) \right] \mathrm{d}x \right\} \beta(t)\,\mathrm{d}t. \tag{2.4.24}$$

Therefore,

$$\bar{a}(t) = \int_0^\ell \left[\kappa_0 \frac{\partial L\left(w^*, \beta^*, t, x\right)}{\partial \beta} + q(t,x)\phi(t,x) \right] \mathrm{d}x$$

and (2.4.13) then reads

$$\left\{\int_0^\ell \left[\kappa_0 \frac{\partial L\left(w^*, \beta^*, t, x\right)}{\partial \beta} + q(t,x)\phi(t,x)\right] \mathrm{d}x\right\} \cdot \left[\beta(t) - \beta^*(t)\right] \geq 0,$$
$$\forall\, \beta(t) \in [\beta_0, \beta_1],\ t \in [0, T] \text{ a.e.}, \tag{2.4.25}$$

where κ_0 and $\phi(\cdot,\cdot)$ are not identical to zero simultaneously. Since otherwise, there are definitely $f_0 = 0,\ f_1 = 0,\ f_{22} = 0$, and $f_{21} = 0$, which contradicts the fact in Theorem 2.4.1 that these continuous linear functionals are not all identically zero.

On the other hand, if K_0 is a null set, then there is

$$\int_0^T \int_0^\ell \left[\frac{\partial L\left(w^*, \beta^*, t, x\right)}{\partial w} w(t,x) + \frac{\partial L\left(w^*, \beta^*, t, x\right)}{\partial \beta}\beta(t)\right] \mathrm{d}x\, \mathrm{d}t = 0,$$
$$\forall\, (w, \beta) \in X_T.$$

In particular, if we choose $\kappa_0 = 1$ and $\varphi(x) = 0$, it then follows from Theorem 2.4.2 that

$$\int_0^T \int_0^\ell \frac{\partial L\left(w^*, \beta^*, t, x\right)}{\partial w} w(t,x)\, \mathrm{d}x\, \mathrm{d}t = \int_0^T \int_0^\ell q(t,x)\phi(t,x)\beta(t)\, \mathrm{d}x\, \mathrm{d}t.$$

Therefore,

$$\int_0^T \left\{\int_0^\ell \left[\frac{\partial L\left(w^*, \beta^*, t, x\right)}{\partial \beta} + q(t,x)\phi(t,x)\right] \mathrm{d}x\right\} \beta(t)\, \mathrm{d}t = 0,$$
$$\forall\, \beta(t) \in L^2(0, T),$$

from which we obtain

$$\int_0^\ell \left[\frac{\partial L\left(w^*, \beta^*, t, x\right)}{\partial \beta} + q(t,x)\phi(t,x)\right] \mathrm{d}x = 0, \quad \forall\, t \in [0, T] \text{ a.e.}$$

Therefore, (2.4.25) still holds.

Finally, if there is a nonzero solution to the adjoint system

$$\begin{cases} \hat{\phi}_{tt}(t,x) + \kappa_1 \hat{\phi}_{xxxx}(t,x) - \kappa_2 \hat{\phi}_{txxxx}(t,x) \\ \quad = \left[g'\left(w_{xx}^*(t,x)\right)\hat{\phi}_{xx}(t,x)\right]_{xx} + \kappa_0 \dfrac{\partial L\left(w^*, \beta^*, t, x\right)}{\partial w}, \\ \hat{\phi}_x(t,0) = \hat{\phi}(t,0) = 0,\ \hat{\phi}_x(t,\ell) = \hat{\phi}(t,\ell) = 0, \\ \kappa_2 \hat{\phi}(T,x) = \mu(x) \end{cases} \tag{2.4.26}$$

such that the following equality holds true

$$\int_0^\ell q(t,x)\hat{\phi}(t,x)\,\mathrm{d}x = 0, \quad \forall\, t \in [0,T] \text{ a.e.},$$

then when we choose $\kappa_0 = 0$, $\mu(x) = \kappa_2\hat{\phi}(T,x)$, which is defined as (2.4.22), (2.4.25) is still valid. Since otherwise, if for any nonzero solution $\hat{\phi}$ of (2.4.26), it has

$$\int_0^\ell q(t,x)\hat{\phi}(t,x)\,\mathrm{d}x \not\equiv 0. \tag{2.4.27}$$

In this case, we refer to the situation as nondegenerate. Consequently, the linearized system (2.4.17) is controllable. In fact, if (2.4.17) is not controllable, then there necessarily exists a $\varphi(\cdot) \in V^*$ such that

$$\int_0^\ell w(T,x)\varphi(x)\,\mathrm{d}x = 0, \quad \varphi(x) \not\equiv 0.$$

Choose $\kappa_0 = 0$, $\hat{\phi}$ to be the solution of (2.4.26). Then it follows from Theorem 2.4.2 that

$$\int_0^T \int_0^\ell q(t,x)\hat{\phi}(t,x)\beta(t)\,\mathrm{d}x\,\mathrm{d}t = 0, \quad \forall\, \beta(t) \in L^2(0,T).$$

Hence,

$$\int_0^\ell q(t,x)\hat{\phi}(t,x)\,\mathrm{d}x = 0, \quad \forall\, t \in [0,T] \text{ a.e.}$$

This is a contradiction. Therefore, under the case of (2.4.26), system (2.4.17) is controllable.

By integrating the preceding findings, we have derived Pontryagin's maximum principle for the optimal control problem (2.4.5), which is subject to system (2.4.1).

Theorem 2.4.3 *Assuming that* (w^*, β^*) *is a solution to the optimal control problem* (2.4.5), *then there exist constants* $\kappa_0 \geq 0$ *and* $\phi(\cdot,\cdot)$, *not identically zero, such that the following maximum principle applies:*

$$\left\{\int_0^\ell \left[\kappa_0 \frac{\partial L\,(w^*,\beta^*,t,x)}{\partial \beta} + q(t,x)\phi(t,x)\right]\mathrm{d}x\right\}\cdot\left[\beta(t)-\beta^*(t)\right] \geq 0,$$
$$\forall\, \beta(t) \in [\beta_0,\beta_1],\ t \in [0,T] \text{ a.e.}, \tag{2.4.28}$$

where the function $\phi(\cdot,\cdot)$ *satisfies the adjoint equation* (2.4.21).

2.5 Optimal Control of Sterilization of Prepackaged Food

In this section, we present the second application, focusing on an optimal control problem with free final horizon and phase constraints. This problem is formulated using a system of PDEs and aims to model the sterilization process of prepackaged solid food. This analysis is conducted in light of critical food industry requirements and technological constraints that govern the sterilization process. By addressing these real-world challenges, we seek to optimize the control strategies employed in food sterilization, ensuring both safety and efficiency.

2.5.1 Background

Food security and hygiene are paramount in people's lives, directly affecting individual well-being and public health. Foods available in markets, particularly prepackaged items, undergo sterilization and inspection before reaching consumers. Food inspection ensures quality and prevents substandard products from entering the market, potentially harming customers. However, even more crucial than inspection is food sterilization. As a vital step in food production, sterilization destroys harmful microorganisms, extending the shelf life of food. Without it, these microorganisms can quickly cause food spoilage, posing a health risk. The ultimate goal of food sterilization is to eliminate harmful bacteria without compromising the food's quality or taste. In food production, heat processing is commonly used for sterilization. Prepackaged food is placed in a large autoclave filled with hot water or steam as the heating medium. By regulating the medium's temperature, workers accomplish the heating sterilization process.

The optimal control problem pertaining to the sterilization of prepackaged food is of critical importance. This topic has garnered significant attention in both industrial and academic circles, with numerous references available [6–8, 11, 24, 25, 49, 133, 134]. Notably, the reference [25] marks a pivotal point in the mathematical exploration of this problem, providing a solid foundation for subsequent research. Building on this foundation, our paper [222] establishes the Pontryagin maximum principle for systems with a fixed final horizon and phase constraints. These findings were derived under two key conditions: the convexity and interior points of the admissible control set, as well as the differentiability of the cost functional with respect to the control variable. During our investigation, we broaden our scope to encompass systems with a free final horizon, thereby relaxing the aforementioned constraints. Our objective is to derive the necessary optimality condition for such optimal control systems, namely, the Pontryagin maximum principle. The sterilization process of prepackaged food is accurately captured by a nonlinear heat equation [25]. This equation serves as the mathematical backbone for our analysis, enabling us to explore the optimal control strategies that maximize sterilization effectiveness while minimizing potential negative impacts on food quality.

The sterilization process of prepackaged food can be described by the following nonlinear heat equation [25]:

$$\begin{cases} \rho c(\theta)\dfrac{\partial \theta(x,t)}{\partial t} - \nabla \cdot (k(\theta)\nabla\theta(x,t)) = 0 & \text{in } \mathcal{D} \times (0,t_1), \\ k(\theta)\dfrac{\partial \theta(x,t)}{\partial n} = \alpha(u(t) - \theta(x,t)) & \text{in } \Gamma \times (0,t_1), \\ \theta(x,0) = \theta_0(x) & \text{in } \mathcal{D}, \end{cases} \tag{2.5.1}$$

where $\theta(x,t)$ is the absolute temperature of the prepackaged food at the point x with the time t. The constant ρ is the density of the product. Other two functions are the heat capacity $c(\theta)$ and the heat conductivity $k(\theta)$ of the product to be heated. The constant α denotes the heat transfer coefficient. The function $u(t)$ denotes the temperature of the heating medium, which is the control. The domain $\mathcal{D}$ with boundary Γ describes the spatial region occupied by the food container, and n denotes the exterior normal to Γ. The process time is $t_1 > 0$. To model (2.5.1), [134] derives its variational formulations as follows.

$$\begin{cases} \left(\dfrac{\partial}{\partial t}\bar{c}(\theta), \vartheta\right) + \left(\nabla\bar{k}(\theta), \nabla\vartheta\right) + \langle\theta, \vartheta\rangle = u(t)\,\langle 1, \vartheta\rangle\,, \quad \forall\, \vartheta \in H^1(\mathcal{D}), \\ \theta(\cdot,0) = \theta_0(\cdot), \end{cases} \tag{2.5.2}$$

in which

$$\langle\vartheta, \omega\rangle = \alpha\int_\Gamma \vartheta(x)\omega(x)\,\mathrm{d}x, \quad (u,\vartheta) = \int_{\mathcal{D}} u(x)\vartheta(x)\,\mathrm{d}x,$$

$$\bar{c}(\theta) = \int_0^\theta \rho c(\xi)\,\mathrm{d}\xi, \quad \bar{k}(\theta) = \int_0^\theta k(\xi)\,\mathrm{d}\xi.$$

One important clarification is that this section specifically focuses on the system pertaining to the sterilization of solid food. This is because the heat equation alone is sufficient for our analysis. However, in the case of liquid food, the heat equation would need to be coupled with the Boussinesq equations to account for natural convection phenomena. Regarding the dimensionality of the system, our investigation is limited to the one-dimensional case. As mentioned in [134], if we consider a cylindrical can that is uniformly heated in an autoclave, due to symmetry, it suffices to analyze one-half of a vertical cross section. This effectively reduces the three-dimensional problem to a two-dimensional one. Furthermore, the one-dimensional version of this system, first introduced by Kleis and Sachs [134], has proven to be valid and exhibits excellent performance in numerical simulations. Its conclusions align with those obtained from the three-dimensional model. Therefore, the state equation under investigation can be transformed from the original system (2.5.1) into a controlled nonlinear heat equation defined on the fixed domain $[a, x_c] \times [0, t_1]$ as follows:

$$\begin{cases} \rho c(\theta)\theta_t(x,t) = k(\theta)\theta_{xx}(x,t) + k'(\theta)\left(\theta_x(x,t)\right)^2, \\ k(\theta(a,t))\theta_x(a,t) = \alpha(\theta(a,t) - u(t)), \\ k(\theta(x_c,t))\theta_x(x_c,t) = 0, \\ \theta(x,0) = \theta_0(x) \end{cases} \tag{2.5.3}$$

with $u \in M \subset \mathbb{R}^+$, M is the admissible control set. Namely, we consider the problem along a straight line from a point a on the boundary of a food container to the center x_c.

In the practical application within the food industry, sterilization procedures must adhere to strict regulations. Therefore, we incorporate additional constraints into the process. Specifically, we introduce an inequality state constraint $F(\theta)(x_c) \geq F_0$ and a terminal state constraint $\theta(x, t_1) = \theta_{end}(x)$ to the system equation (2.5.3). These two constraints represent the industrial sterilization regulations in mathematical form, ensuring that the sterilization process meets the required standards. Here

$$F(\theta)(x) = \int_0^{t_1} 10^{\frac{\theta(x,t)-\theta_r}{z(\theta_r)}}\, \mathrm{d}t = \int_0^{t_1} \mathcal{F}(\theta)(x,t)\, \mathrm{d}t, \quad x \in [a, x_c]$$

is called the F-value at the point x corresponding to the reference temperature θ_r and

$$z(\theta_r) = \frac{R\theta_r^2 \ln 10}{E}$$

is called the z-value. R is the universal gas constant and E is the activation energy.

Take $\theta \in L^2(0, t_1; H^1(a, x_c))$. The control space is $L^\infty(0, t_1)$ and the control constraint is $u \in M \subset \mathbb{R}^+$. It is known that the choice of the function spaces is still an open problem. Here, we assume that for any $u \in L^\infty(0, t_1)$ there exists unique solution $\theta \in L^2(0, t_1; H^1(a, x_c))$ of the state system (2.5.3) with the state constraint $F(\theta)(x_c) \geq F_0$ and the terminal constraint $\theta(x, t_1) = \theta_{end}(x)$.

In the subsequent discussions, whenever we refer to the solution of the state equation (2.5.3), it should be understood as the solution in the variational sense defined by (2.5.2). This variational solution framework allows us to capture the behavior of the system in a more general and flexible manner. For the existence theorem of such solutions, we can refer to [36] for a rigorous mathematical treatment.

The main improvement here compared to [222] lies in the removal of certain restrictions. Firstly, the admissible control set M no longer necessitates convexity or the inclusion of interior points. This relaxation significantly broadens the scope of potential control strategies that can be considered. Secondly, the cost functional $L(\theta, u)$ is no longer required to be differentiable with respect to the control variable u, which allows for more general and possibly non-smooth cost structures. Consequently, the set M can be any set, including the case where it contains only a finite number of points. Moreover, here we investigates a system with a free final time t_1, which differs from the fixed final horizon considered in [222].

2.5.2 Optimal Control Formulation

Consider the following optimal control problem for the sterilization of prepackaged food, which can be expressed mathematically as

$$\text{Problem I:} \quad \text{minimize } J(\theta, u) = \int_0^{t_1} \int_a^{x_c} L(\theta(x,t), u(t)) \, \mathrm{d}x \, \mathrm{d}t \tag{2.5.4}$$

for $\theta(x,t) \in L^2(0, t_1; H^1(a, x_c))$, $u(t) \in L^\infty(0, t_1)$ under the Eqs. (2.5.3) with those two abovementioned state constraints $F(\theta)(x_c) \geq F_0$, $\theta(x, t_1) = \theta_{end}(x)$, where L is a functional defined on $[a, x_c] \times \mathbb{R}^+$ satisfying:

(a) $L(\theta, u)$ is continuous in u;

(b) $\left| \dfrac{\partial L(\theta, u)}{\partial \theta} \right|$ is bounded for every bounded subset of $[a, x_c] \times \mathbb{R}^+$.

For such optimal control problems, [133] establishes the existence of an optimal solution. One can consult it for further details.

Now, we introduce a time transformation $t \to \tau$ that maps the interval $[0, t_1]$ onto $[0, 1]$ and is defined by a certain measurable function $v(\xi) \geq 0$:

$$t(\tau) = \int_0^\tau v(\xi) \, \mathrm{d}\xi, \quad t(1) = t_1. \tag{2.5.5}$$

Let $\theta(x, \tau) = \theta(x, t(\tau))$,

$$u(\tau) = \begin{cases} u(t(\tau)), & \tau \in S_1 = \{\tau \mid \tau \in [0, 1],\ v(\tau) > 0\}, \\ \text{arbitrary}, & \tau \in S_2 = \{\tau \mid \tau \in [0, 1],\ v(\tau) = 0\}. \end{cases} \tag{2.5.6}$$

Then $(\theta(x,\tau), u(\tau))$ satisfies the following equations:

$$\begin{cases} \rho c(\theta)\theta_\tau(x,\tau) = k(\theta)\theta_{xx}(x,\tau)v(\tau) + k'(\theta)\left(\theta_x(x,\tau)\right)^2 v(\tau), \\ k(\theta(a,\tau))\theta_x(a,\tau)v(\tau) = \alpha(\theta(a,\tau) - u(\tau))v(\tau), \\ k(\theta(x_c,\tau))\theta_x(x_c,\tau)v(\tau) = 0, \\ \theta(x,0) = \theta_0(x) \end{cases} \tag{2.5.7}$$

and the constraints

$$\int_0^1 \mathcal{F}(\theta)(x_c,\tau)v(\tau)\,\mathrm{d}\tau \geq F_0, \quad \theta(x,1) = \theta_{end}(x). \tag{2.5.8}$$

The transformation defined in (2.5.5) is one-to-one if $v(\tau) > 0$ for all $0 \leq \tau \leq 1$. However, if $v(\tau)$ vanishes over some closed interval Δ_I, then $t(\tau) = \text{Const}$ within $\tau \in \Delta_I$. This means that a single value of t would correspond to an entire interval on the τ-axis. To ensure that the definition of $\tau(t)$ is one-to-one, we shall assume that

$$\tau(t) = \inf\{\tau \mid t(\tau) = t\}. \tag{2.5.9}$$

Then we formulate a new problem, expressed as

$$\text{Problem II:}\quad \text{minimize } J(\theta,u,v) = \int_0^1 \int_a^{x_c} v(\tau)L(\theta(x,\tau),u(\tau))\,\mathrm{d}x\,\mathrm{d}\tau \tag{2.5.10}$$

for $\theta(x,\tau) \in L^2(0,1;H^1(a,x_c))$, $u(\tau) \in L^\infty(0,1)$, $v(\tau) \in L^\infty(0,1)$ under the constraints (2.5.7), (2.5.8) with $v(\tau) \geq 0$, $u(\tau) \in M$ for almost all $0 \leq \tau \leq 1$.

The relationship between Problem I and Problem II is demonstrated in the following Lemma 2.5.1.

Lemma 2.5.1 *Let $(\theta(x,t), u(t))$ satisfy the Eq.* (2.5.3) *and the constraints $F(\theta)(x_c) \geq F_0$, $\theta(x,t_1) = \theta_{end}(x)$. Then, for any function $v(\tau) \geq 0$ satisfying $\int_0^1 v(\xi)\,\mathrm{d}\xi = t_1$, the functions*

$$\theta(x,\tau) = \theta(x,t(\tau)) \textit{ and } u(\tau) \textit{ in } (2.5.6)$$

satisfy the Eqs. (2.5.7), (2.5.8), *in which $t(\tau)$ is defined by* (2.5.5). *Moreover, $J(\theta,u) = J(\theta,u,v)$.*

Conversely, if $(\theta(x,\tau), u(\tau), v(\tau))$ *satisfies* (2.5.7), (2.5.8), *then the functions defined by* $\theta(x,t) = \theta(x,\tau(t))$, $u(t) = u(\tau(t))$ *with* $t_1 = \int_0^1 v(\tau)\, \mathrm{d}\tau$ *(where* $\tau(t)$ *is given by* (2.5.9)) *satisfy* (2.5.3) *and* $F(\theta)(x_c) \ge F_0$, $\theta(x,t_1) = \theta_{end}(x)$. *Moreover,* $J(\theta,u) = J(\theta,u,v)$.

Remark 2.5.2 If $(\theta^*(x,t), u^*(t))$ is a solution of Problem I, then for any function $v^*(\tau) \ge 0$ satisfying $\int_0^1 v^*(\xi)\, \mathrm{d}\xi = t_1$, the functions $(\theta^*(x,\tau), u^*(\tau), v^*(\tau))$ defined by (2.5.6) furnish a solution to Problem II. Naturally, this also provides a solution to a new Problem III, which is identical to Problem II except that $u^*(\tau)$ is fixed and $J(\theta, u^*, v)$ is determined only over $\theta(x,\tau)$ and $v(\tau)$.

The proof of Lemma 2.5.1 is essentially straightforward. Whenever $v(\tau) > 0$, both sides of Eq. (2.5.7) can be divided by $v(\tau)$. Additionally, one can leverage the fact that

$$\frac{\partial\theta(x,t)}{\partial t} = \frac{\partial\theta(x,t(\tau))}{\partial t} \Big/ \frac{dt(\tau)}{d\tau} = \frac{1}{v(\tau)}\frac{\partial\theta(x,\tau)}{\partial\tau}$$

so that equation in (2.5.7) is transformed into (2.5.3). But if $v(\tau) = 0$, then

$$\frac{\partial\theta(x,\tau)}{\partial\tau} = 0, \quad \theta(x,\tau) = \text{constant (in } \tau) = \theta(x,\tau(t)).$$

Analogous arguments will demonstrate that the functionals $J(\theta,u)$ and $J(\theta,u,v)$ are identical.

Thus, let $\theta^*(x,t)$, $u^*(t)$, t_1 be a solution of the initial problem. Let $v^*(\tau) \ge 0$ be some function satisfying $\int_0^1 v^*(\xi)\, \mathrm{d}\xi = t_1$; the form of the function will be specified later. Define $u^*(\tau)$ in accordance with (2.5.6). Let us solve

$$\text{Problem III:} \quad \text{minimize } J(\theta,u^*,v) = \int_0^1 \int_a^{x_c} v(\tau) L\left(\theta(x,\tau), u^*(\tau)\right)\, \mathrm{d}x\, \mathrm{d}\tau \tag{2.5.11}$$

for $(\theta(x,\tau), v(\tau)) \in X = L^2(0,1; H^1(a,x_c)) \times L^\infty(0,1)$ subject to

$$\begin{cases} \rho c(\theta)\theta_\tau(x,\tau) = k(\theta)\theta_{xx}(x,\tau)v(\tau) + k'(\theta)\left(\theta_x(x,\tau)\right)^2 v(\tau), \\ k(\theta(a,\tau))\theta_x(a,\tau)v(\tau) = \alpha(\theta(a,\tau) - u^*(\tau))v(\tau), \\ k(\theta(x_c,\tau))\theta_x(x_c,\tau)v(\tau) = 0, \\ \theta(x,0) = \theta_0(x), \ \theta(x,1) = \theta_{end}(x), \\ \displaystyle\int_0^1 \mathcal{F}(\theta)(x_c,\tau)v(\tau)\, \mathrm{d}\tau \ge F_0, \ v(\tau) \ge 0, \end{cases} \tag{2.5.12}$$

where $v(\cdot)$ plays the role of the control. Set

$$\Omega_1 = \{(\theta, v) \in X \mid v(\tau) \geq 0,\ \tau \in [0, 1]\},$$

$$\Omega_2 = \left\{(\theta, v) \in X \;\middle|\; \int_0^1 \mathcal{F}(\theta)(x_c, \tau)v(\tau)\,\mathrm{d}\tau \geq F_0\right\},$$

$$\begin{aligned}\Omega_3 = \{(\theta, v) \in X \mid\ & \rho c(\theta)\theta_\tau(x, \tau) = k(\theta)\theta_{xx}(x, \tau)v(\tau) + k'(\theta)\,(\theta_x(x, \tau))^2\, v(\tau),\\ & k(\theta(a, \tau))\theta_x(a, \tau)v(\tau) = \alpha(\theta(a, \tau) - u^*(\tau))v(\tau),\\ & k(\theta(x_c, \tau))\theta_x(x_c, \tau)v(\tau) = 0,\\ & \theta(x, 0) = \theta_0(x),\ \theta(x, 1) = \theta_{end}(x)\}.\end{aligned} \tag{2.5.13}$$

Then Problem III is equivalent to finding $(\theta^*, v^*) \in \Omega = \bigcap_{i=1}^{3} \Omega_i$ such that

$$J(\theta^*, v^*) = \min_{(\theta, v)\in\Omega} J(\theta, v). \tag{2.5.14}$$

It can be observed that problem (2.5.14) is an extremum problem involving inequality constraints $\Omega_1 \cap \Omega_2$ and an equality constraint Ω_3. Consequently, we invoke the Dubovitskiĭ and Milyutin theorem for Problem III, as stated in [89].

Theorem 2.5.3 *Assume that the functional $J(\theta, v)$ attains its minimum at the point (θ^*, v^*) in Ω. Suppose that $J(\theta, v)$ is regularly decreasing at (θ^*, v^*) with the cone of directions of decrease K_0 and the inequality constraints are regular at (θ^*, v^*) with the cones of feasible directions $K_1,\ K_2$; and that the equality constraint is also regular at (θ^*, v^*) with the cone of tangent directions K_3. Then there exist continuous linear functionals $f_0,\ f_1,\ f_2,\ f_3$, not all identically zero, such that $f_i \in K_i^*$, the dual cone of K_i, $i = 0, 1, 2, 3$, which satisfy the condition*

$$f_0 + f_1 + f_2 + f_3 = 0.$$

2.5.3 Determination of the Cones

To apply Theorem 2.5.3, it is necessary to determine all cones $K_i,\ i = 0, 1, 2, 3$. Let us commence by identifying K_0.

2.5.3.1 The Cone of Directions of Decrease K_0

By the assumption, $J(\theta, v)$ is differentiable at any point (θ_0, v_0) in any direction (θ, v) and its directional derivative is

$$\begin{aligned} & J'(\theta_0, u^*, v_0; \theta, v) \\ = & \lim_{\varepsilon\to 0+} \frac{1}{\varepsilon} \left[J\left(\theta_0 + \varepsilon\theta, u^*, v_0 + \varepsilon v\right) - J\left(\theta_0, u^*, v_0\right) \right] \\ = & \lim_{\varepsilon\to 0+} \frac{1}{\varepsilon} \int_0^1 \int_a^{x_c} \left[(v_0 + \varepsilon v) L\left(\theta_0 + \varepsilon\theta, u^*\right) - v_0 L\left(\theta_0, u^*\right) \right] \mathrm{d}x \, \mathrm{d}\tau \\ = & \int_0^1 \int_a^{x_c} \left[v_0 \frac{\partial L\left(\theta_0, u^*\right)}{\partial\theta} \theta + L\left(\theta_0, u^*\right) v \right] \mathrm{d}x \, \mathrm{d}\tau. \end{aligned} \tag{2.5.15}$$

Hence, the cone of directions of decrease of the functional $J(\theta, v)$ at the point (θ^*, v^*) is determined by (Theorem 7.5, [89], p. 48)

$$K_0 = \left\{ (\theta, v) \in X \mid J'(\theta^*, u^*, v^*; \theta, v) < 0 \right\}. \tag{2.5.16}$$

If $K_0 \neq \emptyset$, then for any $f_0 \in K_0^*$, there exists a constant $\lambda_0 \geq 0$ such that (Theorem 10.2, [89], p. 69)

$$f_0(\theta, v) = -\lambda_0 \int_0^1 \int_a^{x_c} \left[v^* \frac{\partial L\left(\theta^*, u^*\right)}{\partial\theta} \theta + L\left(\theta^*, u^*\right) v \right] \mathrm{d}x \, \mathrm{d}\tau. \tag{2.5.17}$$

2.5.3.2 The Cone of Feasible Directions K_1

Since $\Omega_1 = L^2\left(0, 1; H^1(a, x_c)\right) \times \tilde{\Omega}_1$, here $\tilde{\Omega}_1 = \{v \in L^\infty(0, 1) \mid v(\tau) \geq 0\}$ is a closed convex subset of $L^\infty(0, 1)$, the interior $\overset{\circ}{\Omega}_1$ of Ω_1 is not empty and at the point (θ^*, v^*), the cone of feasible directions K_1 of Ω_1 is determined by (Theorem 8.2, [89], p. 59)

$$\begin{aligned} K_1 &= \left\{ \lambda \left(\overset{\circ}{\Omega}_1 - (\theta^*, v^*) \right) \,\middle|\, \lambda > 0 \right\} \\ &= \left\{ \lambda \left(\theta - \theta^*, v - v^* \right) \,\middle|\, (\theta, v) \in \overset{\circ}{\Omega}_1, \ \lambda > 0 \right\}. \end{aligned} \tag{2.5.18}$$

Therefore, for any $f_1 \in K_1^*$, if there exists an $\bar{a}(\tau) \in L(0, 1)$, such that the linear functional defined by

$$f_1(\theta, v) = \int_0^1 \bar{a}(\tau)v(\tau)\, d\tau \qquad (2.5.19)$$

is a support to $\tilde{\Omega}_1$ at the point v^*, then ([89], p. 76–77)

$$\bar{a}(\tau)\left(v(\tau) - v^*(\tau)\right) \geq 0, \quad \forall\, v \in [0, \infty],\; \tau \in [0, 1] \text{ a.e.} \qquad (2.5.20)$$

2.5.3.3 The Cone of Feasible Directions K_2

Note that the second inequality constraint is

$$\Omega_2 = \left\{(\theta, v) \in X \,\middle|\, \int_0^1 \mathcal{F}(\theta)(x_c, \tau)v(\tau)\, d\tau \geq F_0\right\}.$$

Let

$$W(\theta, v) = F_0 - \int_0^1 \mathcal{F}(\theta)(x_c, \tau)v(\tau)\, d\tau. \qquad (2.5.21)$$

Then

$$\Omega_2 = \{(\theta, v) \in X \mid W(\theta, v) \leq 0\}.$$

Here we only consider $W(\theta^*, v^*) = F_0 - \int_0^1 \mathcal{F}(\theta^*)(x_c, \tau)v^*(\tau)\, d\tau = 0$. Since otherwise $W(\theta^*, v^*) < 0$, and (θ^*, v^*) is an inner point of Ω_2. In this case, any direction is feasible and hence the cone of feasible directions K_2 of Ω_2 at (θ^*, v^*) is the whole space, i.e., $K_2 = X$. So $\Omega_2 = \{(\theta, v) \in X \mid W(\theta, v) \leq W(\theta^*, v^*)\}$. By the same argument of [89] at p. 52, we have the following Lemma 2.5.4.

Lemma 2.5.4 *Let $W(\theta, v)$ be defined by (2.5.21). Then $W(\theta, v)$ is differentiable at any point $\left(\hat{\theta}, \hat{v}\right)$ in any direction (θ, v) and its directional derivative is*

$$W'\left(\hat{\theta}, \hat{v}; \theta, v\right) = -\int_0^1 \mathcal{F}\left(\hat{\theta}\right)(x_c, \tau)\left[\frac{\ln 10}{z(\theta_r)}\theta(x_c, \tau)\hat{v}(\tau) + v(\tau)\right] d\tau,$$

and $W(\theta, v)$ satisfies the Lipschitz condition in any ball. Here we assume that $W'(\theta, v; h_1, h_2) \neq 0$ whenever $W(\theta, v) = 0$.

Note that for $W(\theta, v)$ given by (2.5.21), the directional derivative of W at (θ^*, v^*) in the direction $\left(\theta^2 + 1, v^2\right)$ is

$$W'\left(\theta^*, v^*; \theta^2 + 1, v^2\right)$$
$$= -\int_0^1 \mathcal{F}(\theta^*)(x_c, \tau) \left[\frac{\ln 10}{z(\theta_r)} \left(\theta^2(x_c, \tau) + 1\right) v^*(\tau) + v^2(\tau)\right] d\tau < 0.$$

Hence (Theorem 7.3, [89], p. 45)

$$K_2 = \left\{(\theta, v) \in X \mid W'\left(\theta^*, v^*; \theta, v\right) < 0\right\}. \tag{2.5.22}$$

Define the linear operator $A : X \to H^1(a, x_c)$ by

$$A(\theta, v) = \int_0^1 \mathcal{F}(\theta^*)(x_c, \tau) \left[\frac{\ln 10}{z(\theta_r)} \theta(x_c, \tau) v^*(\tau) + v(\tau)\right] d\tau$$

and

$$K = \left\{\xi \in H^1(a, x_c) \,\middle|\, \xi \geq 0\right\}.$$

Then

$$K_2 = \{(\theta, v) \in X \mid A(\theta, v) \in K\}.$$

Given that $A\left(\theta^2 + 1, v^2\right)$ belong to the interior $\overset{\circ}{K}$ of K, one has ([89], p. 72-73)

$$K_2^* = A^* K^*,$$

i.e., for any $f_2 \in K_2^*$, there exists a constant $\delta \geq 0$ such that

$$f_2(\theta, v) = \delta A(\theta, v)$$
$$= \delta \int_0^1 \mathcal{F}(\theta^*)(x_c, \tau) \left[\frac{\ln 10}{z(\theta_r)} \theta(x_c, \tau) v^*(\tau) + v(\tau)\right] d\tau. \tag{2.5.23}$$

2.5.3.4 The Cone of Tangent Directions K_3

Define the operator $G : X \to L^2\left(0, 1; H^1(a, x_c)\right) \times \left(L^2(0, 1)\right)^2 \times \left(H^1(a, x_c)\right)^2$ by

$$G(\theta, v) = \begin{cases} \rho c(\theta)\theta_\tau(x, \tau) - k(\theta)\theta_{xx}(x, \tau)v(\tau) - k'(\theta)\left(\theta_x(x, \tau)\right)^2 v(\tau), \\ k(\theta(a, \tau))\theta_x(a, \tau)v(\tau) - \alpha v(\tau)\left(\theta(a, \tau) - u^*(\tau)\right), \\ k(\theta(x_c, \tau))\theta_x(x_c, \tau)v(\tau), \\ \theta(x, 0) - \theta_0(x), \\ \theta(x, 1) - \theta_{end}(x). \end{cases} \tag{2.5.24}$$

Then

$$\Omega_4 = \{(\theta, v) \in X \mid G(\theta, v) = 0\}. \tag{2.5.25}$$

The Fréchet derivative of the operator $G(\theta, v)$ is

$$\begin{aligned} &G'(\theta, v)\left(\hat{\theta}, \hat{v}\right) \\ &= \begin{cases} \rho c'(\theta)\hat{\theta}(x, \tau)\theta_\tau(x, \tau) - k''(\theta)\hat{\theta}(x, \tau)\left(\theta_x(x, \tau)\right)^2 v(\tau) \\ \quad -k'(\theta)\hat{\theta}(x, \tau)\theta_{xx}(x, \tau)v(\tau) + \rho c(\theta)\hat{\theta}_\tau(x, \tau) - k'(\theta)\left(\theta_x(x, \tau)\right)^2 \hat{v}(\tau) \\ \quad -2k'(\theta)\hat{\theta}_x(x, \tau)\theta_x(x, \tau)v(\tau) - k(\theta)\theta_{xx}(x, \tau)\hat{v}(\tau) - k(\theta)\hat{\theta}_{xx}(x, \tau)v(\tau), \\ k'(\theta(a, \tau))\hat{\theta}(a, \tau)\theta_x(a, \tau)v(\tau) - \alpha\hat{v}(\tau)(\theta(a, \tau) - u^*(\tau)) \\ \quad +k(\theta(a, \tau))\theta_x(a, \tau)\hat{v}(\tau) + k(\theta(a, \tau))\hat{\theta}_x(a, \tau)v(\tau) - \alpha v(\tau)\hat{\theta}(a, \tau), \\ k'(\theta(x_c, \tau))\hat{\theta}(x_c, \tau)\theta_x(x_c, \tau)v(\tau) + k(\theta(x_c, \tau))\theta_x(x_c, \tau)\hat{v}(\tau) \\ \quad +k(\theta(x_c, \tau))\hat{\theta}_x(x_c, \tau)v(\tau), \\ \hat{\theta}(x, 0), \\ \hat{\theta}(x, 1). \end{cases} \end{aligned}$$

Since (θ^*, v^*) is the solution to Problem III, it has $G(\theta^*, v^*) = 0$. Choosing arbitrary

$$(g, g_0, g_1, g_2, g_3) \in L^2\left(0, 1; H^1(a, x_c)\right) \times \left(L^2(0, 1)\right)^2 \times \left(H^1(a, x_c)\right)^2$$

and solving the equation

$$G'\left(\theta^*, v^*\right)\left(\hat{\theta}, \hat{v}\right) = (g(x,\tau), g_0(\tau), g_1(\tau), g_2(x), g_3(x)),$$

we obtain

$$\begin{cases} \rho c'(\theta^*)\hat{\theta}(x,\tau)\theta_\tau^*(x,\tau) - k''(\theta^*)\hat{\theta}(x,\tau)\left(\theta_x^*(x,\tau)\right)^2 v^*(\tau) \\ \quad -k'(\theta^*)\hat{\theta}(x,\tau)\theta_{xx}^*(x,\tau)v^*(\tau) + \rho c(\theta^*)\hat{\theta}_\tau(x,\tau) - k'(\theta^*)\left(\theta_x^*(x,\tau)\right)^2 \hat{v}(\tau) \\ \quad -2k'(\theta^*)\hat{\theta}_x(x,\tau)v^*(\tau)\theta_x^*(x,\tau) - k(\theta^*)\theta_{xx}^*(x,\tau)\hat{v}(\tau) \\ \quad -k(\theta^*)\hat{\theta}_{xx}(x,\tau)v^*(\tau) = g(x,\tau), \\ k'(\theta^*(a,\tau))\hat{\theta}(a,\tau)\theta_x^*(a,\tau)v^*(\tau) - \alpha\hat{v}(\tau)\left(\theta^*(a,\tau) - u^*(\tau)\right) \\ \quad +k(\theta^*(a,\tau))\theta_x^*(a,\tau)\hat{v}(\tau) + k(\theta^*(a,\tau))\hat{\theta}_x(a,\tau)v^*(\tau) - \alpha v^*(\tau)\hat{\theta}(a,\tau) \\ \quad = g_0(\tau), \\ k'(\theta^*(x_c,\tau))\hat{\theta}(x_c,\tau)\theta_x^*(x_c,\tau)v^*(\tau) + k(\theta^*(x_c,\tau))\theta_x^*(x_c,\tau)\hat{v}(\tau) \\ \quad +k(\theta^*(x_c,\tau))\hat{\theta}_x(x_c,\tau)v^*(\tau) = g_1(\tau), \\ \hat{\theta}(x,0) = g_2(x), \\ \hat{\theta}(x,1) = g_3(x). \end{cases} \tag{2.5.26}$$

Next, assume that the linearized system

$$\begin{cases} \rho c(\theta^*)\theta_\tau(x,\tau) - k(\theta^*)v^*(\tau)\theta_{xx}(x,\tau) - 2k'(\theta^*)v^*(\tau)\theta_x^*(x,\tau)\theta_x(x,\tau) \\ \quad +\left[\rho c'(\theta^*)\theta_\tau^*(x,\tau) - k'(\theta^*)v^*(\tau)\theta_{xx}^*(x,\tau) - k''(\theta^*)v^*(\tau)\left(\theta_x^*(x,\tau)\right)^2\right] \\ \qquad \theta(x,\tau) \\ \quad -\left[k(\theta^*)\theta_{xx}^*(x,\tau) + k'(\theta^*)\left(\theta_x^*(x,\tau)\right)^2\right]v(\tau) = 0, \\ k(\theta^*(a,\tau))v^*(\tau)\theta_x(a,\tau) + \left[k'(\theta^*(a,\tau))\theta_x^*(a,\tau) - \alpha\right]v^*(\tau)\theta(a,\tau) \\ \quad +\left[k(\theta^*(a,\tau))\theta_x^*(a,\tau) - \alpha\left(\theta^*(a,\tau) - u^*(\tau)\right)\right]v(\tau) = 0, \\ \left[k(\theta^*(x_c,\tau))\theta_x(x_c,\tau) + k'\left(\theta^*(x_c,\tau)\right)\theta_x^*(x_c,\tau)\theta(x_c,\tau)\right]v^*(\tau) \\ \quad +k(\theta^*(x_c,\tau))\theta_x^*(x_c,\tau)v(\tau) = 0, \\ \theta(x,0) = 0 \end{cases} \tag{2.5.27}$$

is controllable. Then choose $v(\tau) = \hat{v}(\tau) \in L^\infty(0, 1)$ such that $\theta(x, 1) = g_3(x) - \gamma(x, 1)$ and let $\theta(x, \tau)$ be the solution to the linearized system (2.5.27). Choose $\hat{\theta}(x, \tau) = \theta(x, \tau) + \gamma(x, \tau)$, where $\gamma(x, \tau)$ satisfies the following equation:

$$
\begin{cases}
\rho c(\theta^*)\gamma_\tau(x,\tau) - k(\theta^*)v^*(\tau)\gamma_{xx}(x,\tau) - 2k'(\theta^*)v^*(\tau)\theta_x^*(x,\tau)\gamma_x(x,\tau) \\
\quad + \left[\rho c'(\theta^*)\theta_\tau^*(x,\tau) - k'(\theta^*)v^*(\tau)\theta_{xx}^*(x,\tau) - k''(\theta^*)v^*(\tau)\left(\theta_x^*(x,\tau)\right)^2\right] \\
\quad \gamma(x,\tau) = g(x,\tau), \\
k(\theta^*(a,\tau))v^*(\tau)\gamma_x(a,\tau) = \left[\alpha - k'(\theta^*(a,\tau))\theta_x^*(a,\tau)\right] v^*(\tau)\gamma(a,\tau) + g_0(\tau), \\
k(\theta^*(x_c,\tau))v^*(\tau)\gamma_x(x_c,\tau) + k'(\theta^*(x_c,\tau))\theta_x^*(x_c,\tau)v^*(\tau)\gamma(x_c,\tau) = g_1(\tau), \\
\gamma(x,0) = g_2(x).
\end{cases}
$$

In this way, it suffices for $(\hat{\theta}, \hat{v})$ satisfying (2.5.26). Therefore, $G'(\theta^*, v^*)$ maps the space X onto $L^2\left(0, 1; H^1(a, x_c)\right) \times \left(L^2(0, 1)\right)^2 \times \left(H^1(a, x_c)\right)^2$. Moreover, by Theorem 9.1 ([89], p. 61), in X the cone of tangent directions K_3 to the constraint Ω_3 at the point (θ^*, v^*) consists of the kernel of $G'(\theta^*, v^*)$, i.e., (θ, v) satisfies (2.5.27) and

$$\theta(x, 1) = 0. \tag{2.5.28}$$

Define

$$K_{31} = \{(\theta, v) \in X \mid (\theta(x, \tau), v(\tau)) \text{ satisfies (2.5.27)}\},$$

$$K_{32} = \{(\theta, v) \in X \mid (\theta(x, \tau), v(\tau)) \text{ satisfies (2.5.28)}\}.$$

Then the cone of tangent directions $K_3 = K_{31} \bigcap K_{32}$. Hence,

$$K_3^* = K_{31}^* + K_{32}^*,$$

in which K_{3i}^* is the dual cone of K_{3i}, $i = 1, 2$. For any $f_3 \in K_3^*$, decompose $f_3 = f_{31} + f_{32}$, $f_{3i} \in K_{3i}^*$, $i = 1, 2$. Then $f_{31}(\theta, v) = 0$ and for all $\theta(x, \tau) \in L^2\left(0, 1; H^1(a, x_c)\right)$ satisfying $\theta(x, 1) = 0$, there exists a $\varrho(x) \in \mathrm{BMO}(a, x_c)$, the space of functions of bounded mean oscillation [82], such that

$$f_{32}(\theta, v) = \int_a^{x_c} \theta(x, 1)\varrho(x)\,\mathrm{d}x.$$

It then follows from Theorem 2.5.3 that there exist continuous linear functionals, not all identically zero, such that

$$f_0 + f_1 + f_2 + f_{31} + f_{32} = 0.$$

Therefore, when selecting (θ, v) satisfying (2.5.27), $f_{31}(\theta, v) = 0$. Moreover,

$$\begin{aligned} f_1(\theta, v) &= -f_0(\theta, v) - f_2(\theta, v) - f_{32}(\theta, v) \\ &= \lambda_0 \int_0^1 \int_a^{x_c} \left[v^*(\tau) \frac{\partial L(\theta^*, u^*)}{\partial \theta} \theta(x, \tau) + L(\theta^*, u^*) v(\tau) \right] \mathrm{d}x \, \mathrm{d}\tau \\ &\quad - \delta \int_0^1 \mathcal{F}(\theta^*)(x_c, \tau) \left[\frac{\ln 10}{z(\theta_r)} \theta(x_c, \tau) v^*(\tau) + v(\tau) \right] \mathrm{d}\tau \\ &\quad - \int_a^{x_c} \theta(x, 1) \varrho(x) \, \mathrm{d}x. \end{aligned} \tag{2.5.29}$$

2.5.4 Pontryagin's Maximum Principle

Define the adjoint system of (2.5.27) as

$$\begin{cases} \rho c(\theta^*) \phi_\tau(x, \tau) = -k(\theta^*) v^*(\tau) \phi_{xx}(x, \tau) + \lambda_0 v^*(\tau) \dfrac{\partial L(\theta^*, u^*)}{\partial \theta}, \\ k(\theta^*(a, \tau)) v^*(\tau) \phi_x(a, \tau) - \alpha v^*(\tau) \phi(a, \tau) = 0, \\ k(\theta^*(x_c, \tau)) v^*(\tau) \phi_x(x_c, \tau) - \dfrac{\delta \ln 10}{z(\theta_r)} \mathcal{F}(\theta^*)(x_c, \tau) v^*(\tau) = 0, \\ \rho c(\theta^*(x, 1)) \phi(x, 1) = \varrho(x). \end{cases} \tag{2.5.30}$$

Consistent with our previous discussions, when we refer to the solution of (2.5.30), we are referring to the solution in the variational sense.

Lemma 2.5.5 *The solution of the system* (2.5.27) *and that of its adjoint system* (2.5.30) *is interrelated in the following manner:*

$$\begin{aligned} &\lambda_0 \int_0^1 \int_a^{x_c} v^*(\tau) \frac{\partial L(\theta^*, u^*)}{\partial \theta} \theta(x, \tau) \, \mathrm{d}x \, \mathrm{d}\tau \\ &- \frac{\delta \ln 10}{z(\theta_r)} \int_0^1 \mathcal{F}(\theta^*)(x_c, \tau) \theta(x_c, \tau) v^*(\tau) \, \mathrm{d}\tau - \int_a^{x_c} \theta(x, 1) \varrho(x) \, \mathrm{d}x \\ &= - \int_0^1 \int_a^{x_c} \phi(x, \tau) \left[k(\theta^*) \theta^*_{xx}(x, \tau) + k'(\theta^*) \left(\theta^*_x(x, \tau) \right)^2 \right] v(\tau) \, \mathrm{d}x \, \mathrm{d}\tau \end{aligned}$$

$$-\int_0^1 \phi(a,\tau)\left[k(\theta^*(a,\tau))\theta_x^*(a,\tau)-\alpha(\theta^*(a,\tau)-u^*(\tau))\right]v(\tau)\,\mathrm{d}\tau$$

$$+\int_0^1 \phi(x_c,\tau)k\left(\theta^*(x_c,\tau)\right)\theta_x^*(x_c,\tau)v(\tau)\,\mathrm{d}\tau. \tag{2.5.31}$$

Proof Multiplying the first equation in (2.5.30) by $\theta(x,\tau)$ and integrating over $(0,1)\times(a,x_c)$ with respect to τ and x respectively yield

$$\begin{aligned}&\int_0^1\int_a^{x_c} \rho c(\theta^*)\phi_\tau(x,\tau)\theta(x,\tau)\,\mathrm{d}x\,\mathrm{d}\tau\\ =&-\int_0^1\int_a^{x_c} k(\theta^*)v^*(\tau)\phi_{xx}(x,\tau)\theta(x,\tau)\,\mathrm{d}x\,\mathrm{d}\tau\\ &+\lambda_0\int_0^1\int_a^{x_c} v^*(\tau)\frac{\partial L(\theta^*,u^*)}{\partial\theta}\theta(x,\tau)\,\mathrm{d}x\,\mathrm{d}\tau.\end{aligned}$$

To evaluate this integral, we apply the integration by parts technique and shift the derivatives from $\phi(x,\tau)$ to $\theta(x,\tau)$. This completes the proof of the lemma. □

Now, by virtue of (2.5.19) and (2.5.29) and Lemma 2.5.4, there is

$$\begin{aligned}\bar{a}(\tau)=&\int_a^{x_c}\left\{\lambda_0 L(\theta^*,u^*)-\phi(x,\tau)\left[k(\theta^*)\theta_{xx}^*(x,\tau)+k'(\theta^*)\left(\theta_x^*(x,\tau)\right)^2\right]\right\}\mathrm{d}x\\ &-\phi(a,\tau)\left[k(\theta^*(a,\tau))\theta_x^*(a,\tau)-\alpha(\theta^*(a,\tau)-u^*(\tau))\right]\\ &+\phi(x_c,\tau)k(\theta^*(x_c,\tau))\theta_x^*(x_c,\tau)-\delta\mathcal{F}(\theta^*)(x_c,\tau).\end{aligned} \tag{2.5.32}$$

Then (2.5.20) reads

$$\begin{aligned}&\left\{\int_a^{x_c}\left[\lambda_0 L\left(\theta^*,u^*\right)-\phi(x,\tau)\left(k\left(\theta^*\right)\theta_{xx}^*(x,\tau)+k'\left(\theta^*\right)\left(\theta_x^*(x,\tau)\right)^2\right)\right]\mathrm{d}x\right.\\ &-\phi(a,\tau)\left[k(\theta^*(a,\tau))\theta_x^*(a,\tau)-\alpha(\theta^*(a,\tau)-u^*(\tau))\right]\\ &\left.+\phi(x_c,\tau)k(\theta^*(x_c,\tau))\theta_x^*(x_c,\tau)-\delta\mathcal{F}(\theta^*)(x_c,\tau)\right\}\cdot(v(\tau)-v^*(\tau))\geq 0,\\ &\qquad\forall\, v(\tau)\geq 0.\end{aligned} \tag{2.5.33}$$

Therefore,

$$\int_a^{x_c} \left[\lambda_0 L(\theta^*, u^*) - \phi(x,\tau)\left(k(\theta^*)\theta^*_{xx}(x,\tau) + k'(\theta^*)\left(\theta^*_x(x,\tau)\right)^2\right)\right] \mathrm{d}x$$
$$-\phi(a,\tau)\left[k(\theta^*(a,\tau))\theta^*_x(a,\tau) - \alpha(\theta^*(a,\tau) - u^*(\tau))\right]$$
$$+\phi(x_c,\tau)k(\theta^*(x_c,\tau))\theta^*_x(x_c,\tau) - \delta\mathcal{F}(\theta^*)(x_c,\tau) = 0, \quad \forall\, \tau \in S_1, \tag{2.5.34}$$

$$\int_a^{x_c} \left[\lambda_0 L(\theta^*, u^*) - \phi(x,\tau)\left(k(\theta^*)\theta^*_{xx}(x,\tau) + k'(\theta^*)\left(\theta^*_x(x,\tau)\right)^2\right)\right] \mathrm{d}x$$
$$-\phi(a,\tau)\left[k(\theta^*(a,\tau))\theta^*_x(a,\tau) - \alpha(\theta^*(a,\tau) - u^*(\tau))\right]$$
$$+\phi(x_c,\tau)k(\theta^*(x_c,\tau))\theta^*_x(x_c,\tau) - \delta\mathcal{F}(\theta^*)(x_c,\tau) \geq 0, \quad \forall\, \tau \in S_2, \tag{2.5.35}$$

where $S_1 = \{\tau \in [0,1] \mid v^*(\tau) > 0\}$, $S_2 = \{\tau \in [0,1] \mid v^*(\tau) = 0\}$. Note that λ_0, δ and $\phi(x,\tau)$ cannot be zero simultaneously. Since otherwise, there is f_0, f_1, f_2, f_{31}, and f_{32} being zero, which contradicts the choice of these continuous linear functionals.

On the other hand, if K_0 is a null set, we choose $\lambda_0 = 1$, δ and $\varrho(x)$ satisfying

$$\frac{\delta \ln 10}{z(\theta_r)} \int_0^1 \mathcal{F}(\theta^*)(x_c,\tau)\theta(x_c,\tau)v^*(\tau)\,\mathrm{d}\tau + \int_a^{x_c} \theta(x,\tau)\varrho(x)\,\mathrm{d}x = 0. \tag{2.5.36}$$

It then follows from Lemma 2.5.4 that

$$\int_0^1 \int_a^{x_c} \left[v^*(\tau)\frac{\partial L(\theta^*, u^*)}{\partial \theta}\theta(x,\tau) + L(\theta^*, u^*)\,v(\tau)\right] \mathrm{d}x\,\mathrm{d}\tau$$
$$-\delta \int_0^1 \mathcal{F}(\theta^*)(x_c,\tau)v(\tau)\,\mathrm{d}\tau$$
$$= \int_0^1 \Bigg\{ \int_a^{x_c} \left[L(\theta^*, u^*) - \phi(x,\tau)\left(k(\theta^*)\theta^*_{xx}(x,\tau) + k'(\theta^*)\left(\theta^*_x(x,\tau)\right)^2\right)\right] \mathrm{d}x$$
$$-\phi(a,\tau)\left[k(\theta^*(a,\tau))\theta^*_x(a,\tau) - \alpha\left(\theta^*(a,\tau) - u^*(\tau)\right)\right]$$
$$+\phi(x_c,\tau)k(\theta^*(x_c,\tau))\theta^*_x(x_c,\tau) - \delta\mathcal{F}(\theta^*)(x_c,\tau)\Bigg\} v(\tau)\,\mathrm{d}\tau.$$

Therefore, (2.5.34), (2.5.35) are still valid.

Finally, if there is a nonzero solution to the adjoint system (2.5.30) such that

$$\begin{aligned}&\int_a^{x_c}\left[L\left(\theta^*,u^*\right)-\phi(x,\tau)\left(k(\theta^*)\theta^*_{xx}(x,\tau)+k'(\theta^*)\left(\theta^*_x(x,\tau)\right)^2\right)\right]\mathrm{d}x\\&\quad-\phi(a,\tau)\left[k(\theta^*(a,\tau))\theta^*_x(a,\tau)-\alpha(\theta^*(a,\tau)-u^*(\tau))\right]\\&\quad+\phi(x_c,\tau)k(\theta^*(x_c,\tau))\theta^*_x(x_c,\tau)-\delta\mathcal{F}(\theta^*)(x_c,\tau)=0,\end{aligned}\tag{2.5.37}$$

then take $\lambda_0=1$, and (2.5.33) is still valid. On the other hand, for any nonzero solution of (2.5.30),

$$\begin{aligned}&\int_a^{x_c}\left[L\left(\theta^*,u^*\right)-\phi(x,\tau)\left(k(\theta^*)\theta^*_{xx}(x,\tau)+k'(\theta^*)\left(\theta^*_x(x,\tau)\right)^2\right)\right]\mathrm{d}x\\&\quad-\phi(a,\tau)\left[k(\theta^*(a,\tau))\theta^*_x(a,\tau)-\alpha(\theta^*(a,\tau)-u^*(\tau))\right]\\&\quad+\phi(x_c,\tau)k(\theta^*(x_c,\tau))\theta^*_x(x_c,\tau)-\delta\mathcal{F}(\theta^*)(x_c,\tau)\neq 0.\end{aligned}\tag{2.5.38}$$

This situation is referred to as the nondegenerate case, as the linearized system is undoubtedly controllable in this instance. Alternatively, if not, there exists a $\varrho(x)\in \mathrm{BMO}(a,x_c)$ such that $\int_a^{x_c}\theta(x,1)\varrho(x)\,\mathrm{d}x=0$, $\varrho(x)\neq 0$. Taking $\lambda_0=1$, δ and $\varrho(x)$ satisfying the expression (2.5.36), we have a contradiction to (2.5.38). That is to say, (2.5.34) and (2.5.35) are always valid.

Define $\phi(x,t)=\phi(x,\tau(t))$ so (2.5.34) can be written as

$$\begin{aligned}&\int_a^{x_c}\left[\lambda_0 L\left(\theta^*,u^*\right)-\phi(x,t)\left(k(\theta^*)\theta^*_{xx}(x,t)+k'(\theta^*)\left(\theta^*_x(x,t)\right)^2\right)\right]\mathrm{d}x\\&\quad-\phi(a,t)\left[k(\theta^*(a,t))\theta^*_x(a,t)-\alpha(\theta^*(a,t)-u^*(t))\right]\\&\quad+\phi(x_c,t)k(\theta^*(x_c,t))\theta^*_x(x_c,t)-\delta\mathcal{F}(\theta^*)(x_c,t)=0,\quad \forall\, t\in[0,t_1]\text{ a.e.}\end{aligned}\tag{2.5.39}$$

Choose S_1 to be a perfect nowhere dense subset of [0, 1] (see [89]) and define

$$v^*(\tau)=\begin{cases}t_1/\mu(S_1), & \tau\in S_1,\\ 0, & \tau\in S_2=[0,1]\setminus S_1,\end{cases}\tag{2.5.40}$$

where $\mu(S)$ denotes the measure of the set S. By analyzing (2.5.35) in a similar manner to [89], we can define $u^*(\tau)$ on S_2 and get

$$\begin{aligned}
&\int_a^{x_c} \left[\lambda_0 L\left(\theta^*, u\right) - \phi(x,t)\left(k(\theta^*)\theta^*_{xx}(x,t) + k'(\theta^*)\left(\theta^*_x(x,t)\right)^2\right)\right] \mathrm{d}x \\
&\quad -\phi(a,t)\left[k\left(\theta^*(a,t)\right)\theta^*_x(a,t) - \alpha(\theta^*(a,t) - u(t))\right] \\
&\quad +\phi(x_c,t)k(\theta^*(x_c,t))\theta^*_x(x_c,t) - \delta\mathcal{F}(\theta^*)(x_c,t) \geq 0, \quad \forall\, u \in M,\ t \in [0,t_1].
\end{aligned} \tag{2.5.41}$$

Consequently, we arrive at Pontryagin's maximum principle for Problem I, outlined as follows.

Theorem 2.5.6 *Suppose that (θ^*, u^*, t_1) is a solution to Problem* I. *Then there exist $\lambda_0 \geq 0$, $\delta \geq 0$ and $\phi(\cdot,\cdot)$, not identically zero, such that*

$$\begin{aligned}
&\int_a^{x_c} \left[\lambda_0 L\left(\theta^*(x,t), u^*(t)\right) - \phi(x,t)\left(k(\theta^*)\theta^*_{xx}(x,t) + k'(\theta^*)\left(\theta^*_x(x,t)\right)^2\right)\right] \mathrm{d}x \\
&\quad -\phi(a,t)\left[k(\theta^*(a,t))\theta^*_x(a,t) - \alpha(\theta^*(a,t) - u^*(t))\right] \\
&\quad +\phi(x_c,t)k(\theta^*(x_c,t))\theta^*_x(x_c,t) - \delta\mathcal{F}(\theta^*)(x_c,t) = 0, \quad \forall\, t \in [0,t_1] \text{ a.e.},
\end{aligned} \tag{2.5.42}$$

$$\begin{aligned}
&\int_a^{x_c} \left[\lambda_0 L\left(\theta^*(x,t), u^*(t)\right) - \phi(x,t)\left(k(\theta^*)\theta^*_{xx}(x,t) + k'(\theta^*)\left(\theta^*_x(x,t)\right)^2\right)\right] \mathrm{d}x \\
&\quad -\phi(a,t)\left[k(\theta^*(a,t))\theta^*_x(a,t) - \alpha(\theta^*(a,t) - u)\right] \\
&\quad +\phi(x_c,t)k(\theta^*(x_c,t))\theta^*_x(x_c,t) - \delta\mathcal{F}(\theta^*)(x_c,t) \geq 0, \quad \forall\, u \in M, \\
&\quad t \in [0,t_1] \text{ a.e.},
\end{aligned} \tag{2.5.43}$$

where the function $\phi(\cdot,\cdot)$ satisfies

$$\begin{cases}
\rho c(\theta^*)\phi_t(x,t) = -k(\theta^*)\phi_{xx}(x,t) + \lambda_0 \dfrac{\partial L(\theta^*, u^*)}{\partial \theta}, \\
k(\theta^*(a,t))\phi_x(a,t) - \alpha\phi(a,t) = 0, \\
k(\theta^*(x_c,t))\phi_x(x_c,t) - \dfrac{\delta \ln 10}{z(\theta_r)}\mathcal{F}(\theta^*)(x_c,t) = 0, \\
\rho c(\theta^*(x,t_1))\phi(x,t_1) = \varrho(x).
\end{cases} \tag{2.5.44}$$

Chapter 3
Spectral Method

As seen, optimal control theory finds widespread application across various scientific, technological, and engineering domains. Its applications range from technical to biological dynamic systems, highlighting its versatility. In robotics, optimal control theory ensures precise arm movements, mitigating unwanted vibrations that could compromise operational accuracy. Similarly, in the aerospace industry, vibration analysis of solar cells is crucial for the seamless operation of communication satellites, while missile vibration analysis prevents accidents during space exploration. Furthermore, fiscal and monetary policies employ optimal control theory to minimize unemployment, stabilizing national economies. In the medical field, optimal control theory answers crucial questions such as the optimal timing for HIV therapy initiation and derives effective treatment schedules. These are just a few examples from a vast array of applications detailed in various studies [57, 153]. Notably, optimal control problems are inherently nonlinear, often lacking analytical solutions. Consequently, numerical methods play a pivotal role in addressing these challenges. The numerical solution has indeed become a crucial component in the digitization trend of modern science and technology [223].

This chapter is dedicated to the spectral method for numerically solving optimal control problems. Using this introduced spectral method, we delve into two PDE-constrained optimal control problems. The first problem concerns the optimal control of a fourth-order PDE with an H^1-norm state constraint. The second problem focuses on the optimal control of the Stokes equation, with an L^2-norm constraint on the control variable. Firstly, we derive the necessary optimality conditions for both problems. Subsequently, we establish rigorous a priori and a posteriori error estimates for the spectral approximation of these problems. These error estimates provide insights into the accuracy and convergence properties of the spectral method, enabling us to assess its performance and reliability in solving these complex optimal control problems.

B. Sun et al., *Maximum Principle and Dynamic Programming Viscosity Solution Approach*, Systems & Control: Foundations & Applications,
https://doi.org/10.1007/978-981-96-5739-1_3

3.1 The Spectral Approximation

There is a rich tapestry of numerical methods available for solving optimal control problems, ranging from the finite element method, finite difference methods, to spectral approximation techniques. Spectral methods occupy a unique position in this pantheon of techniques, offering a powerful tool for solving differential equations, including PDEs and ODEs, as well as optimization problems.

Spectral methods trace their origins to series solutions of PDEs, a testament to their deep-rooted mathematical foundation. The essence of these methods lies in employing global polynomials as trial functions for discretizing PDEs. This approach yields highly accurate approximations with a comparably small number of unknowns, assuming the solutions possess sufficient smoothness.

The core idea behind spectral methods is to approximate solutions of PDEs using finite series of orthogonal functions, such as complex exponentials, Chebyshev polynomials, or Legendre polynomials. The resulting expansion serves as an approximate spectral expansion of the solution. By combining this expansion with the original equation, one arrives at a system of equations for the expansion coefficients. In essence, the spectral method generalizes the standard variable separation technique, with accuracy directly tied to the number of terms in the series expansion.

The versatility and precision of spectral methods have made them a valuable tool in various fields, from physics to engineering, where they have been successfully employed to solve complex problems. Their ability to capture global behavior while maintaining high accuracy offers a unique advantage over other numerical methods.

In the computational realm, the spectral method has undergone robust evolution since the early 1970s, particularly in the domain of computational fluid dynamics (see [40, 94] and the citations within). Readers can trace the spectral method's initial application to PDEs in [209]. Shen et al. [208] delve deeply into various spectral method algorithms and their applications to PDEs.

For the volume-conserved phase-field lipid vesicle model, [41] constructs an effective fully discrete numerical scheme that boasts second-order accuracy in time, spectral accuracy in space, and unconditional energy stability. In [248], an alternating direction implicit Legendre-Laguerre spectral scheme is introduced for the two-dimensional time distributed-order diffusion wave equation on a semi-infinite domain. The numerical approximation's stability and convergence analysis are thoroughly examined. Additional insights can be found in [94, 99, 206].

In general, developing a spectral method involves three critical decisions: selecting the appropriate expansion functions, determining the format of the approximation, and ultimately specifying the procedure to determine the unknown solution components. Notably, the selection of trial functions is a distinguishing factor between spectral methods and finite-element or finite-difference methods. Spectral methods employ globally defined trial functions that are infinitely differentiable. Conversely, in finite-element methods, the trial functions are local, and the same applies to finite-difference trial functions. This attribute contributes to the excellent error characteristics of spectral methods, as they exhibit exponential convergence, the fastest possible when dealing with smooth solutions.

Furthermore, spectral methods can often be computationally more efficient and simpler to implement compared to finite element methods. They excel in scenarios where high accuracy is crucial in straightforward domains with smooth solutions. However, for larger problems or those involving non-smooth solutions, finite element methods generally offer superior performance [95].

Moreover, within the realm of spectral methods, three spectral schemes stand out as the most commonly employed: the Galerkin, collocation, and tau versions. These schemes offer distinct approaches to solving various problems. As for the trial functions, trigonometric polynomials, Chebyshev polynomials, and Legendre polynomials are the most frequently utilized options, each with their unique properties and applications.

The tau approach, as a modification of the Galerkin method, is particularly useful for problems with non-periodic boundary conditions. It can be seen as a specialized instance of the Petrov-Galerkin method, providing additional flexibility and adaptability to a wider range of problems [40].

For those interested in delving deeper into the spectral methods, [40, 99, 138, 206] are excellent resources that offer comprehensive coverage of the theory, implementation, and applications of these methods. These references provide not only the mathematical foundation but also practical insights into the usage and advantages of spectral methods in various fields.

Now, through a specific example, we demonstrate the primary steps involved in applying spectral methods to solve the PDE [239]. Let's consider the Poisson equation, which can be physically interpreted as a heat conduction problem or a potential theory problem:

$$\left(\frac{\partial^2}{\partial x^2} + \frac{\partial^2}{\partial y^2}\right) f(x, y) = g(x, y) \quad \text{for all } x, y,$$

in which $g(x, y)$ is a known, complex-valued function of two real variables and

$$g(x, y) = g(x + 2\pi, y) = g(x, y + 2\pi).$$

Namely, the problem exhibits periodicity in both x and y. Our objective is to determine a function $f(x, y)$ that satisfies the given equation. To achieve this, we postulate that f possesses a continuous second derivative. Subsequently, we rewrite both f and g in Fourier series representations, as outlined below.

$$f := \sum a_{j,k} e^{i(jx+ky)},$$
$$g := \sum b_{j,k} e^{i(jx+ky)}.$$

Afterward, substitute the equations mentioned above into the Poisson equation to obtain

$$-\sum a_{j,k}\left(j^2+k^2\right)e^{i(jx+ky)}=\sum b_{j,k}e^{i(jx+ky)},$$

where we have interchanged partial differentiation with an infinite sum. By invoking the uniqueness theorem for Fourier expansions, we can equate the Fourier coefficients term by term, thereby obtaining an explicit formula for the coefficients

$$a_{j,k}=-\frac{b_{j,k}}{j^2+k^2}. \tag{3.1.1}$$

Moreover, under periodic boundary conditions, the Poisson equation possesses a solution only if $b_{0,0}=0$. In such a case, we are free to choose an $a_{0,0}$, which will be equivalent to the mean of the solution. Consequently, an algorithm can be developed to solve this problem. Firstly, we compute the Fourier transform $(b_{j,k})$ of the function g. Secondly, using Eq. (3.1.1), we compute the Fourier transform $(a_{j,k})$ of f. Finally, we reconstruct f by taking an inverse Fourier transform of $(a_{j,k})$.

Then, consider a time-dependent PDE

$$\frac{\partial u}{\partial t}+\mathcal{N}[u]=0,\quad t>0,\ x\in(b_1,b_2),$$

with initial condition $u|_{t=0}=u_0$ and appropriate boundary conditions in the interval $[0,2\pi]$ [170]. Here the differential operator $\mathcal{N}$ is possibly nonlinear. Given an orthonormal basis $\{\phi_j\}_{j=1}^r$, a Galerkin method proceeds by discretizing the solution as

$$u^r(t,x)=\sum_{j=1}^r a_j(t)\phi_j(x),$$

and imposing the constraints

$$\left\langle\phi_l,u_t^r+\mathcal{N}[u^r]\right\rangle=0,\quad 1\le l\le r,$$

where $\langle\cdot,\cdot\rangle$ is the L^2 inner product on the spatial domain.

In this manner, we obtain the following ODEs:

$$a_l'(t)=-\left\langle\phi_l,\mathcal{N}\left[\sum_{j=1}^r a_j(t)\phi_j\right]\right\rangle, \tag{3.1.2}$$

complemented by the initial condition

$$a_l(0) = \langle \phi_l, u_0 \rangle$$

for $1 \leq l \leq r$.

Furthermore, if the basis functions are inherently periodic, the boundary conditions are automatically satisfied. This approach is employed to construct the ground-truth solutions, u_G, for periodic problems that serve as the basis for training and error computation. In nonlinear scenarios, the quadratic terms are evaluated in modal space, whereas the required triple product integrals are precomputed. Equations (3.1.2) are numerically integrated over time using suitable adaptive schemes. Subsequently, the relative errors in the numerical solution are computed by

$$E(t) = \frac{\|u^r(t, \cdot) - u_G(t, \cdot)\|}{\|u_G(t, \cdot)\|}.$$

In the subsequent sections, we shall employ the spectral method to investigate three distinct optimal control problems. These problems encompass the optimal control of a fourth-order PDE subject to a state constraint, a flow optimal control problem constrained by control limitations, and an optimal control problem governed by a fourth-order parabolic PDE, addressing three distinct scenarios of control constraints.

Firstly, let us delve into the first of these problems.

3.2 Optimal Control of a Fourth-Order PDE with State Constraint

Among the optimal control problems we are studying, a particular class holds fundamental importance in practical applications. These problems may seem ordinary, equipped with standard cost functions and dynamics, but they involve intricate joint or mixed constraints on both the state and control variables. These are known as optimal control problems with mixed constraints. Typically, dealing with such problems, especially those involving state or mixed constraints, is more challenging than those solely constrained by control variables.

The existence of mixed constraints has long been recognized as a hurdle in deriving appropriate necessary conditions of the maximum principle type. Scholars worldwide have systematically studied problems with mixed constraints ([45, 57, 86, 87, 111, 158] and the references cited therein).

In this section, our focus is on a state-constrained optimal control problem involving a fourth-order PDE [226]. In fluid mechanics and solid mechanics, certain mathematical models can be reduced to a fourth-order PDE, specifically, the biharmonic equation. This equation plays a pivotal role in describing various physical phenomena in these fields. By studying this specific control problem, we aim to gain insights into the optimal strategies for manipulating the system while adhering to state constraints.

Let $\Omega \subset \mathbb{R}^d$ ($d = 1$ or 2) be a bounded domain with the Lipschitz continuous boundary $\partial\Omega$. Consider the following optimal control problem governed by the biharmonic equation:

$$\min_{y\in K} J(y,u) = \frac{1}{2}\|y - y_d\|^2_{L^2(\Omega)} + \frac{\alpha}{2}\|u\|^2_{L^2(\Omega)}, \tag{3.2.1}$$

$$s.t. \begin{cases} \Delta^2 y = f + u, & \text{in } \Omega, \\ y|_{\partial\Omega} = \left.\dfrac{\partial y}{\partial n}\right|_{\partial\Omega} = 0, \end{cases} \tag{3.2.2}$$

where u is chosen as the control and y is the state. Set α to be a positive constant and n the outward normal unit vector of Ω. We take $y_d \in L^2(\Omega)$ the objective state and the function $f \in L^2(\Omega)$. Let γ be a given positive constant. The set

$$K = \left\{ v \in H^1(\Omega) \mid \|v\|_{H^1(\Omega)} \le \gamma \right\}$$

is defined as a constraint set on the state variable.

This class of control problems involving H^1-norm state constraints holds significant importance. In practical scenarios, such constraints often arise in research on potential flow, where the flow velocity (the gradient of a potential function) must be kept within certain bounds. Additionally, these constraints can also describe steering problems, where rapid steering is undesirable for safety reasons, necessitating a limit on the speed (the derivative of displacement). Notably, problems with H^1-norm state constraints are generally considered more challenging to investigate than those involving integral state constraints. Obtaining high-accuracy approximations for these problems can be particularly difficult. Furthermore, the optimality conditions for optimal control problems with H^1-norm state constraints become more intricate, posing additional challenges in their analysis and solution.

Currently, there is a wealth of research on numerical approximations for optimal control problems. These include various methods such as finite element approximation [46, 47, 64, 156, 158, 160, 175], spectral approximation [55, 100, 119, 205, 206], and others. Spectral approximation, which utilizes global polynomials as trial functions, has been successfully applied to various optimal control problems. For instance, it has been used to solve second-order elliptic optimal control problems with control constraints such as L^2-norm constraints [152] and integral constraints [55], as well as state constraints including H^1-norm constraints [52] and integral constraints [252]. It has also been applied to problems with both state and control constraints [119, 250].

Furthermore, spectral approximation has been particularly effective in addressing flow optimal control problems with H^1-norm state constraints. For instance, [53] derives both a priori and a posteriori error estimates for the spectral approximation of such problems. Additionally, the Legendre-Galerkin spectral method is employed for an optimal control problem governed by the Stokes equations with an integral constraint on control [56]. For time-dependent control problems with integral constraints on control, [122] presents a priori error estimates for their spectral approximation.

We employ the standard notation $W^{m,q}(\Omega)$ to denote the Sobolev spaces on Ω, endowed with the norm $\|\cdot\|_{W^{m,q}(\Omega)}$. In particular, when $q = 2$, we abbreviate $W^{m,2}(\Omega)$ as $H^m(\Omega)$. For brevity, we assume that C represents a positive constant that is independent of N (which refers to the degree of polynomials in the discretization systems introduced later). It is worth noting that the value of C may vary in different contexts.

3.2.1 Galerkin Spectral Approximation and Optimality Conditions

In this subsection, our focus is on exploring the Galerkin spectral approximation of the optimal control problem (3.2.1)–(3.2.2). Additionally, we aim to establish the optimality conditions for both the original control problem (3.2.1)–(3.2.2) and its corresponding spectral discrete system.

3.2.1.1 Preliminary

Firstly, it is imperative to discuss the existence and uniqueness of the solution to the control problem (3.2.1)–(3.2.2). To facilitate this discussion, we define the control space as $U = L^2(\Omega)$ and the state space

$$Y = H_0^2(\Omega) := \left\{ w \in H^2(\Omega) \;\middle|\; w|_{\partial\Omega} = \left.\frac{\partial w}{\partial n}\right|_{\partial\Omega} = 0 \right\}.$$

In fact, we can rewrite the control problem (3.2.1)–(3.2.2) as a control-constrained problem on a subset $U_{ad} \subset U$ of the control space U by employing certain transformations. Here, the control constraint set U_{ad} is defined as the subset of U that satisfies

$$U_{ad} = \left\{ v \in U \;\middle|\; \left\| \left(\Delta^2\right)^{-1} v \right\|_{H^1(\Omega)} \le \gamma \right\}.$$

Then the control problem (3.2.1)–(3.2.2) is equivalent to the following control-constrained problem:

Find $(u, y) \in U \times Y$ such that

$$\begin{cases} \min\limits_{u \in U_{ad}} J(y, u) = \frac{1}{2} \|y - y_d\|^2_{L^2(\Omega)} + \frac{\alpha}{2} \|u - f\|^2_{L^2(\Omega)}, \\ \Delta^2 y = u, \quad \text{in } \Omega, \quad y|_{\partial\Omega} = \left.\frac{\partial y}{\partial n}\right|_{\partial\Omega} = 0. \end{cases} \tag{3.2.3}$$

The set K is a closed convex subset of Y. By the continuity property of the operator $\left(\Delta^2\right)^{-1} : U \mapsto Y$, we know that U_{ad} is a closed convex subset of U.

Based on the results from [153], we are assured that the control problem (3.2.3) possesses a unique solution. Similarly, the control problem defined by (3.2.1)–(3.2.2) also enjoys uniqueness of its solution. While this scheme appears theoretically robust, in practical applications, the control problem (3.2.3) cannot be directly utilized due to the complexities involved in calculating and determining elements within the admissible control set U_{ad}.

To proceed with our analysis of the spectral approximation system, we must establish the weak formulation of the system described by (3.2.2). For this purpose, we utilize the L^2-inner product:

$$(u, v) = \int_\Omega uv, \quad u, v \in U,$$

and

$$a(y, w) = \int_\Omega \Delta y \cdot \Delta w, \quad y, w \in Y.$$

By the Poincaré inequality (Theorem 1.1.10), there exist two constants $C_0,\ c_0 > 0$ such that

$$|a(y, w)| \le C_0 \|y\|_Y \|w\|_Y, \quad \forall\, y, w \in Y, \tag{3.2.4}$$

$$a(y, y) \ge c_0 \|y\|^2_Y, \quad \forall\, y \in Y. \tag{3.2.5}$$

Therefore, the standard weak formulation of the state equation (3.2.2) reads as follows.

Find $y \in Y$ such that

$$a\,(y(u), w) = (f + u, w), \quad \forall\, w \in Y. \tag{3.2.6}$$

Then, we present the weak formulation of the control problem (3.2.1)–(3.2.2) as follows. Find $(u, y) \in U \times Y$ such that

$$(\mathscr{P}) \quad \begin{cases} \min\limits_{y \in K} J(y, u) = \frac{1}{2}\|y - y_d\|^2_{L^2(\Omega)} + \frac{\alpha}{2}\|u\|^2_{L^2(\Omega)}, \\ a(y(u), w) = (f + u, w), \quad \forall\, w \in Y. \end{cases}$$

Moreover, we know that the control problem $(\mathscr{P})$ has a unique solution.

3.2.1.2 Galerkin Spectral Approximation

With the necessary preparations in place, we now proceed to investigate the Galerkin spectral approximation of the control problem $(\mathscr{P})$. Let Q_N denote the space of all algebraic polynomials with a degree less than or equal to N in each individual variable x_i $(i = 1, 2)$. We consider the Q_N-spectral discretization of the control problem $(\mathscr{P})$ and introduce finite-dimensional spaces to facilitate this approximation. Denote the finite-dimensional spaces

$$Q_N^0 := \left\{ w_N \in Q_N \;\middle|\; w_N|_{\partial\Omega} = \left.\frac{\partial w_N}{\partial n}\right|_{\partial\Omega} = 0 \right\}$$

Denote the approximation of the control space U and the state space Y as U_N and Y_N, where $U_N = Q_N$ and $Y_N = Q_N^0$, and $K_N = K \cap Y_N$ as the approximation of the state constraint set K. Then, the spectral discrete formulation of the weak formulation (3.2.6) can be written as

$$a\,(y_N, w_N) = (f + u_N, w_N), \quad \forall\, w_N \in Y_N.$$

Subsequently, we can write the Galerkin spectral approximation of the control problem $(\mathscr{P})$ as follows.

Find $(u_N, y_N) \in U_N \times Y_N$ such that

$$(\mathscr{P}_N) \quad \begin{cases} \min\limits_{y_N \in K_N} J(y_N, u_N) = \frac{1}{2}\|y_N - y_d\|^2_{L^2(\Omega)} + \frac{\alpha}{2}\|u_N\|^2_{L^2(\Omega)}, \\ a(y_N, w_N) = (f + u_N, w_N), \quad \forall\, w_N \in Y_N. \end{cases} \tag{3.2.7}$$

By employing similar techniques used in the preceding subsections, we reformulate the spectral approximation problem $(\mathscr{P}_N)$ as a control-constrained problem defined on a subset $U_{ad}^N \subset U_N$, in which

$$U_{ad}^N = \left\{ v_N \in U_N \;\middle|\; \left\| \left(\Delta_N^2\right)^{-1} v_N \right\|_{H^1(\Omega)} \le \gamma \right\},$$

and Δ_N^2 is a discretized bi-harmonic operator satisfying

$$\left(\Delta_N^2 v_N, w_N\right) = (\Delta v_N, \Delta w_N), \quad v_N, w_N \in Y_N.$$

Thus, the spectral approximation problem $(\mathscr{P}_N)$ is equivalent to the following optimal control problem with the control constraint.

Find $(u_N, y_N) \in U_N \times Y_N$ such that

$$\begin{cases} \min\limits_{u_N \in U_{ad}^N} J(y_N, u_N) = \dfrac{1}{2}\|y_N - y_d\|^2_{L^2(\Omega)} + \dfrac{\alpha}{2}\|u_N - f_N\|^2_{L^2(\Omega)}, \\ a(y_N, w_N) = (u_N, w_N), \quad \forall\, w_N \in Y_N, \end{cases}$$

where $f_N = \mathbb{P}_N f$, and $\mathbb{P}_N$ is the L^2-orthogonal projection operator defined in Lemma 3.2.2. It is evident that this problem possesses a unique solution, as detailed in [153]. Consequently, the spectral approximation problem $(\mathscr{P}_N)$ also has a unique solution.

3.2.1.3 Optimality Conditions

In this subsection, we derive the optimality conditions for both the control problem $(\mathscr{P})$ and its spectral approximation problem $(\mathscr{P}_N)$.

Theorem 3.2.1 *The pair $(u, y) \in U \times Y$ is the solution to the optimal control problem $(\mathscr{P})$, if and only if there exists a pair $(p, \lambda) \in Y \times [0, +\infty)$ such that (u, y, p, λ) satisfies the optimality conditions:*

$$(\mathscr{Q}) \begin{cases} \text{(a) } a(y, w) = (f + u, w), \quad \forall\, w \in Y, \\ \text{(b) } a(q, p) = (y - y_d, q) + \lambda(y - \Delta y, q), \quad \forall\, q \in Y, \\ \text{(c) } \lambda(y - \Delta y, v - y) \le 0, \quad \forall\, v \in K, \\ \text{(d) } p + \alpha u = 0, \end{cases} \tag{3.2.8}$$

where

$$\lambda = \begin{cases} \text{constant} \ge 0, & \|y\|_{H^1(\Omega)} = \gamma, \\ 0, & \|y\|_{H^1(\Omega)} < \gamma. \end{cases} \tag{3.2.9}$$

Furthermore, the solution to the optimality conditions $(\mathscr{Q})$ is unique.

Proof In order to obtain the optimality conditions ($\mathscr{Q}$), a Lagrange functional $\mathcal{L}(u,t) : U \times \mathbb{R} \to \mathbb{R}$ associated with the control problem ($\mathscr{P}$) is introduced as the following:

$$\mathcal{L}(u,t) = J(y(u),u) + tF(u), \quad F(u) = \frac{1}{2}\|y(u)\|_{H^1(\Omega)}^2 - \frac{1}{2}\gamma^2.$$

It can be verified that both $J(y(u),u)$ and $F(u)$ are Gâteaux differentiable. By the Kuhn-Tucker theorem (see [23, 48] for details), if u is the solution, then there exists a real number $\tilde{t}$ such that

$$\begin{aligned} &(\mathrm{a}')\ \tilde{t} \geq 0, \quad \tilde{t}F(u) = 0, \\ &(\mathrm{b}')\ \frac{\partial \mathcal{L}}{\partial u}\left(u, \tilde{t}\right) v = 0. \end{aligned} \tag{3.2.10}$$

From (3.2.10)(a′), when $\|y\|_{H^1(\Omega)} < \gamma$, we obtain $\tilde{t} = 0$. Set $\lambda = \tilde{t}$. Clearly, it satisfies (3.2.9), and $\lambda(y - \Delta y, v - y) = 0$ holds. When $\|y\|_{H^1(\Omega)} = \gamma$, we have $\lambda \geq 0$, which also satisfies (3.2.9). Because for any $v \in K$,

$$\begin{aligned} \lambda\,(y - \Delta y, v - y) =& \lambda\left((y - \Delta y, v) - \|y\|_{H^1(\Omega)}^2\right) = \lambda\left((y - \Delta y, v) - \gamma^2\right) \\ \leq& \lambda\left(\|y\|_{H^1(\Omega)}\|v\|_{H^1(\Omega)} - \gamma^2\right) \leq 0 \end{aligned}$$

holds, (3.2.8)(c) is true.

From (3.2.10)(b′), for any $v \in U$, it has

$$\begin{aligned} 0 = \frac{\partial \mathcal{L}}{\partial u}\left(u, \tilde{t}\right) v &= \left(J'\left(y(u),u\right) + \tilde{t}F'(u)\right) v \\ &= (y - y_d, D(y(u))v) + \alpha(u,v) + \tilde{t}\,(y - \Delta y, D(y(u))v) \\ &= \left(D(y(u))^*\,(y - y_d + \lambda\,(y - \Delta y)) + \alpha u, v\right), \end{aligned}$$

where $D(y(u))$ represents the Gâteaux derivative of y at u and $D(y(u))^*$ is its adjoint. Define a costate $p \in Y$ that satisfies

$$p = D(y(u))^*\,(y - y_d + \lambda\,(y - \Delta y)).$$

It yields

$$a(q,p) = (y - y_d, q) + \lambda\,(y - \Delta y, q), \quad \forall\, q \in Y \text{ and } p + \alpha u = 0.$$

Thus (3.2.8)(b) and (d) hold.

The subsequent analysis focuses on proving the uniqueness of the solution to the optimality conditions ($\mathscr{Q}$). Let's assume that there exist two solutions $(u_1, y_1, p_1, \lambda_1)$ and $(u_2, y_2, p_2, \lambda_2)$ to (3.2.8). Then we have

$$a(y_1 - y_2, w) = (u_1 - u_2, w), \quad \forall\, w \in Y,$$
$$a(q, p_1 - p_2) = (y_1 - y_2, q) + \lambda_1 (y_1 - \Delta y_1, q) - \lambda_2 (y_2 - \Delta y_2, q), \quad \forall\, q \in Y. \tag{3.2.11}$$

In the two equations mentioned above, take $w = p_1 - p_2 \in Y$ and $q = y_1 - y_2 \in Y$. It has

$$\begin{aligned} a(y_1 - y_2, p_1 - p_2) &= (u_1 - u_2, p_1 - p_2) = -\alpha(u_1 - u_2, u_1 - u_2), \\ a(y_1 - y_2, p_1 - p_2) &= (y_1 - y_2, y_1 - y_2) + \lambda_1 (y_1 - \Delta y_1, y_1 - y_2) \\ &\quad -\lambda_2 (y_2 - \Delta y_2, y_1 - y_2), \end{aligned}$$

which yields

$$\begin{aligned} \alpha(u_1 - u_2, u_1 - u_2) + (y_1 - y_2, y_1 - y_2) + \lambda_1 (y_1 - \Delta y_1, y_1 - y_2) \\ -\lambda_2 (y_2 - \Delta y_2, y_1 - y_2) = 0. \end{aligned}$$

Given that $\lambda_1 (y_1 - \Delta y_1, y_1 - y_2) \geq 0$ and $\lambda_2 (y_2 - \Delta y_2, y_2 - y_1) \geq 0$, we have

$$\alpha(u_1 - u_2, u_1 - u_2) + (y_1 - y_2, y_1 - y_2) \leq 0,$$

which implies that $u_1 = u_2$ and $y_1 = y_2$. Thus $p_1 = p_2$ from (3.2.8)(d). By the second equation in (3.2.11), we have $\lambda_1 = \lambda_2$. This completes the proof of the theorem. □

Using similar techniques, it can be shown that the spectral approximation problem ($\mathscr{P}_N$) possesses a unique solution denoted as $(u_N, y_N) \in U_N \times Y_N$. Furthermore, (u_N, y_N) constitutes the solution of the spectral approximation problem ($\mathscr{P}_N$) if and only if there is a distinct pair $(p_N, \lambda_N) \in Y_N \times [0, +\infty)$ such that the quartet $(u_N, y_N, p_N, \lambda_N)$ satisfies the following discrete optimality conditions:

$$(\mathscr{Q}_N) \begin{cases} (\tilde{\text{a}})\ a(y_N, w_N) = (f + u_N, w_N), \quad \forall\, w_N \in Y_N, \\ (\tilde{\text{b}})\ a(q_N, p_N) = (y_N - y_d, q_N) + \lambda_N (y_N - \Delta y_N, q_N), \quad \forall\, q_N \in Y_N, \\ (\tilde{\text{c}})\ \lambda_N (y_N - \Delta y_N, v_N - y_N) \leq 0, \quad \forall\, v_N \in K_N, \\ (\tilde{\text{d}})\ p_N + \alpha u_N = 0, \end{cases} \tag{3.2.12}$$

where

$$\lambda_N = \begin{cases} \text{constant} \geq 0, & \|y_N\|_{H^1(\Omega)} = \gamma, \\ 0, & \|y_N\|_{H^1(\Omega)} < \gamma. \end{cases} \tag{3.2.13}$$

The optimality conditions $(\mathscr{Q}_N)$ above give an easy approach to approximate the spectral approximation problem $(\mathscr{P}_N)$, which is helpful for us to analyze the property of the solution (u_N, y_N) to the spectral approximation problem $(\mathscr{P}_N)$.

3.2.2 A Priori Error Estimates

In this subsection, we investigate a priori error estimates for the spectral approximation problem $(\mathscr{P}_N)$ including the L^2-error estimation for the control and H^2-error estimations for the state and the costate. For this purpose, we firstly recall two important facts listed as Lemmas 3.2.2 and 3.2.3.

Lemma 3.2.2 (see [40]) *Let $\mathbb{P}_N : L^2(\Omega) \to Q_N$ be the L^2-orthogonal projection operator, which satisfies for any $r \in L^2(\Omega)$ that*

$$(r - \mathbb{P}_N r, w_N) = 0, \quad \forall\, w_N \in Q_N.$$

Then, we have

$$\|r - \mathbb{P}_N r\|_{L^2(\Omega)} \leq C N^{-m} \|r\|_{H^m(\Omega)}$$

provided that $r \in H^m(\Omega)$ with $m \geq 0$.

Lemma 3.2.3 (see [26, 27]) *Define the projection operator $\mathbb{P}^0_{2,N} : H^2_0(\Omega) \to Q^0_N$, which satisfies for any $v \in H^2_0(\Omega)$ that*

$$a\left(v - \mathbb{P}^0_{2,N} v, w_N\right) = 0, \quad \forall\, w_N \in Q^0_N.$$

Then for any $v \in H^m(\Omega) \cap H^2_0(\Omega)$ and $m \geq 2$, we have

$$\left\|v - \mathbb{P}^0_{2,N} v\right\|_{H^k(\Omega)} \leq C N^{k-m} \|v\|_{H^m(\Omega)}, \quad k = 0, 1, 2.$$

Subsequently, we list Lemmas 3.2.4–3.2.7. Moreover, a priori error estimates can be derived quickly as soon as these facts are proved.

Lemma 3.2.4 *Let $(u_N, y_N, p_N, \lambda_N)$ satisfy the optimality conditions $(\mathscr{Q}_N)$. Then there exists a constant C independent of N such that*

$$\max\left\{\|u_N\|_{L^2(\Omega)}, \|y_N\|_{H^2(\Omega)}, \|p_N\|_{H^2(\Omega)}, |\lambda_N|\right\} \le C. \tag{3.2.14}$$

Proof Due to the fact that the pair (u_N, y_N) constitutes the solution of the spectral problem $(\mathscr{P}_N)$,

$$\begin{aligned} J(y_N, u_N) &= \frac{1}{2}\|y_N - y_d\|^2_{L^2(\Omega)} + \frac{\alpha}{2}\|u_N\|^2_{L^2(\Omega)} \\ &\le J(y_N(0), 0) = \frac{1}{2}\|y_N(0) - y_d\|^2_{L^2(\Omega)} \le C. \end{aligned}$$

By (3.2.5), (3.2.12)($\tilde{a}$) and ($\tilde{d}$), it yields

$$\|u_N\|_{L^2(\Omega)} + \|p_N\|_{L^2(\Omega)} + \|y_N\|_{H^2(\Omega)} \le C, \tag{3.2.15}$$

in which C is a constant independent of N.

Next, we proceed to estimate the norm $\|p_N\|_{H^2(\Omega)}$. The discussions will be divided into two separate cases. One case involves $\|y_N\|_{H^1(\Omega)} < \gamma$, and the other is $\|y_N\|_{H^1(\Omega)} = \gamma$.

If $\|y_N\|_{H^1(\Omega)} < \gamma$, we have $\lambda_N = 0$ by (3.2.13). Let $q_N = p_N$ in (3.2.12)($\tilde{b}$). By (3.2.5) and (3.2.15), we have

$$\|p_N\|_{H^2(\Omega)} \le c_0^{-1}\|y_N - y_d\|_{L^2(\Omega)} \le c_0^{-1}\left(\|y_N\|_{L^2(\Omega)} + \|y_d\|_{L^2(\Omega)}\right) \le C. \tag{3.2.16}$$

If $\|y_N\|_{H^1(\Omega)} = \gamma$, let $q_N = p_N - \frac{1}{\gamma^2}(y_N - \Delta y_N, p_N)\, y_N$ in (3.2.12)($\tilde{b}$). It holds

$$\begin{aligned} & a\left(p_N - \frac{1}{\gamma^2}(y_N - \Delta y_N, p_N)y_N, p_N\right) \\ &= \left(y_N - y_d + \lambda_N(y_N - \Delta y_N), p_N - \frac{1}{\gamma^2}(y_N - \Delta y_N, p_N)y_N\right) \\ &= \left(y_N - y_d, p_N - \frac{1}{\gamma^2}(y_N - \Delta y_N, p_N)y_N\right). \end{aligned}$$

Let $\varepsilon > 0$ be a constant to be determined. Using (3.2.15), the equation stated above implies that

$$\begin{aligned} a(p_N, p_N) &= a\left(\frac{1}{\gamma^2}(y_N - \Delta y_N, p_N)\, y_N, p_N\right) \\ &\quad + \left(y_N - y_d, p_N - \frac{1}{\gamma^2}(y_N - \Delta y_N, p_N) y_N\right) \\ &= \frac{1}{\gamma^2}(y_N - \Delta y_N, p_N) a(y_N, p_N) + (y_N - y_d, p_N) \\ &\quad - \frac{1}{\gamma^2}(y_N - \Delta y_N, p_N)(y_N - y_d, y_N) \\ &\le \varepsilon \|p_N\|^2_{H^2(\Omega)} + \frac{C^2}{4\varepsilon}. \end{aligned} \tag{3.2.17}$$

Choose $\varepsilon = \frac{c_0}{2}$. By (3.2.5), (3.2.16), and (3.2.17), we obtain

$$\|p_N\|_{H^2(\Omega)} \le C. \tag{3.2.18}$$

Furthermore, if $\|y_N\|_{H^1(\Omega)} = \gamma$, let $q_N = y_N$ in (3.2.12)($\tilde{\text{b}}$). We get

$$\begin{aligned} \lambda_N \|y_N\|^2_{H^1(\Omega)} &= \lambda_N\,(y_N - \Delta y_N, y_N) = a(y_N, p_N) - (y_N - y_d, y_N) \\ &\le C_0 \|y_N\|_{H^2(\Omega)} \|p_N\|_{H^2(\Omega)} + \|y_N - y_d\|_{L^2(\Omega)} \|y_N\|_{L^2(\Omega)} \le C, \end{aligned}$$

which leads to

$$|\lambda_N| \le C. \tag{3.2.19}$$

Thus, taking into account (3.2.15), (3.2.18), and (3.2.19), we have completed the proof of the lemma. □

Next, we introduce a highly useful auxiliary system.
Find $(y_N(u), p_N(u)) \in Y_N \times Y_N$ such that

$$a(y_N(u), w_N) = (f + u, w_N), \quad \forall\, w_N \in Y_N, \tag{3.2.20}$$

$$a(q_N, p_N(u)) = (y_N(u) - y_d, q_N) + \lambda(y_N(u) - \Delta y_N(u), q_N), \quad \forall\, q_N \in Y_N. \tag{3.2.21}$$

Firstly, we estimate $\|y_N - y_N(u)\|_{H^2(\Omega)}$ and $\|p_N - p_N(u)\|_{H^2(\Omega)}$ in the following lemma.

Lemma 3.2.5 *Let* (u, y, p, λ) *and* $(u_N, y_N, p_N, \lambda_N)$ *satisfy the optimality conditions* $(\mathscr{Q})$ *and* $(\mathscr{Q}_N)$*, respectively. Then the following estimates hold:*

$$\|y_N - y_N(u)\|_{H^2(\Omega)} \le C\|u - u_N\|_{L^2(\Omega)}, \tag{3.2.22}$$

$$\|p_N - p_N(u)\|_{H^2(\Omega)} \le C\|y_N - y_N(u)\|_{H^2(\Omega)} + C|\lambda - \lambda_N|. \tag{3.2.23}$$

Proof Subtract (3.2.12)(ã)-($\tilde{\text{b}}$) from (3.2.20) and (3.2.21), respectively, to obtain that

$$a(y_N(u) - y_N, w_N) = (u - u_N, w_N), \quad \forall\, w_N \in Y_N, \tag{3.2.24}$$

and

$$\begin{aligned} &a(q_N, p_N(u) - p_N) \\ &= (y_N(u) - y_N, w_N) + \lambda(y_N(u) - \Delta y_N(u), q_N) \\ &\quad -\lambda_N(y_N - \Delta y_N, q_N), \quad \forall\, q_N \in Y_N. \end{aligned} \tag{3.2.25}$$

Set $w_N = y_N(u) - y_N$ in (3.2.24). By (3.2.5), we get

$$\begin{aligned} c_0\|y_N(u) - y_N\|^2_{H^2(\Omega)} &\le a(y_N(u) - y_N, y_N(u) - y_N) = (u - u_N, y_N(u) - y_N) \\ &\le \|u - u_N\|_{L^2(\Omega)}\|y_N(u) - y_N\|_{L^2(\Omega)} \\ &\le \|u - u_N\|_{L^2(\Omega)}\|y_N(u) - y_N\|_{H^2(\Omega)}, \end{aligned}$$

which implies that

$$\|y_N(u) - y_N\|_{H^2(\Omega)} \le C\|u - u_N\|_{L^2(\Omega)}.$$

Similarly, substituting $q_N = p_N(u) - p_N$ into (3.2.25) and invoking (3.2.5) along with the triangular inequality, we obtain

$$\begin{aligned} &c_0\|p_N(u) - p_N\|^2_{H^2(\Omega)} \\ &\le a(p_N(u) - p_N, p_N(u) - p_N) \\ &= (y_N(u) - y_N, p_N(u) - p_N) + \lambda\,(y_N(u) - \Delta y_N(u) - y_N + \Delta y_N, p_N(u) - p_N) \\ &\quad +(\lambda - \lambda_N)(y_N - \Delta y_N, p_N(u) - p_N) \\ &\le \|y_N(u) - y_N\|_{L^2(\Omega)}\|p_N(u) - p_N\|_{L^2(\Omega)} \\ &\quad +\lambda\|y_N(u) - y_N\|_{H^1(\Omega)}\|p_N(u) - p_N\|_{H^1(\Omega)} \\ &\quad +|\lambda - \lambda_N| \cdot \|y_N\|_{H^1(\Omega)}\|p_N(u) - p_N\|_{H^1(\Omega)} \\ &\le C\left(\|y_N(u) - y_N\|_{H^2(\Omega)} + |\lambda - \lambda_N|\right)\|p_N(u) - p_N\|_{H^2(\Omega)}. \end{aligned}$$

Therefore,

$$\|p_N(u) - p_N\|_{H^2(\Omega)} \leq C\left(\|y_N(u) - y_N\|_{H^2(\Omega)} + |\lambda - \lambda_N|\right).$$

This complete the proof of the lemma. □

Lemma 3.2.6 *Let (u, y, p, λ) and $(u_N, y_N, p_N, \lambda_N)$ satisfy the optimality conditions $(\mathscr{Q})$ and $(\mathscr{Q}_N)$, respectively. Then we have the following estimate:*

$$\begin{aligned}|\lambda - \lambda_N| \leq C\Big(&\|p - p_N(u)\|_{H^2(\Omega)} + \|u - u_N\|_{L^2(\Omega)}\\ &+\|y - y_N(u)\|_{H^1(\Omega)} + \left\|\mathbb{P}^0_{2,N}y - y\right\|_{H^1(\Omega)}\Big).\end{aligned} \tag{3.2.26}$$

Proof By (3.2.8)(b) and (3.2.12)($\tilde{\text{b}}$), for any $q_N \in Y_N$, it has

$$\begin{aligned}&a(q_N, p - p_N)\\ &= (y - y_N, q_N) + \lambda(y - \Delta y, q_N) - \lambda_N(y_N - \Delta y_N, q_N)\\ &= (y - y_N, q_N) + (\lambda - \lambda_N)(y - \Delta y, q_N) + \lambda_N(y - \Delta y - y_N + \Delta y_N, q_N).\end{aligned} \tag{3.2.27}$$

Substituting $q_N = \mathbb{P}^0_{2,N}y$ into (3.2.27), and employing both (3.2.8)(a) and Lemma 3.2.3, we derive

$$\begin{aligned}&(\lambda - \lambda_N)\|y\|^2_{H^1(\Omega)}\\ &= (\lambda - \lambda_N)(y - \Delta y, y)\\ &= a\left(\mathbb{P}^0_{2,N}y, p - p_N\right) - \left(y - y_N, \mathbb{P}^0_{2,N}y\right)\\ &\quad +(\lambda - \lambda_N)\left(y - \Delta y, y - \mathbb{P}^0_{2,N}y\right) - \lambda_N\left(y - \Delta y - y_N + \Delta y_N, \mathbb{P}^0_{2,N}y\right)\\ &= a\left(\mathbb{P}^0_{2,N}y - y, p - p_N(u)\right) + a(y, p - p_N) - \left(y - y_N, \mathbb{P}^0_{2,N}y\right)\\ &\quad +(\lambda - \lambda_N)\left(y - \Delta y, y - \mathbb{P}^0_{2,N}y\right) - \lambda_N\left(y - \Delta y - y_N + \Delta y_N, \mathbb{P}^0_{2,N}y\right)\\ &= a\left(\mathbb{P}^0_{2,N}y - y, p - p_N(u)\right) + (u + f, p - p_N) - \left(y - y_N, \mathbb{P}^0_{2,N}y\right)\\ &\quad +(\lambda - \lambda_N)\left(y - \Delta y, y - \mathbb{P}^0_{2,N}y\right) - \lambda_N\left(y - y_N - \Delta y + \Delta y_N, \mathbb{P}^0_{2,N}y\right).\end{aligned} \tag{3.2.28}$$

If $\|y\|_{H^1(\Omega)} = \gamma$, then considering (3.2.4), (3.2.8)(d), (3.2.12)($\tilde{\text{d}}$), (3.2.14), (3.2.22), and (3.2.28) collectively, and applying Lemma 3.2.5 along with the triangular inequality, we deduce that

$$\begin{aligned} |\lambda - \lambda_N| &\le C \left\| \mathbb{P}^0_{2,N} y - y \right\|_{H^2(\Omega)} \|p - p_N(u)\|_{H^2(\Omega)} + C\|p - p_N\|_{L^2(\Omega)} \\ &\quad + C\|y - y_N\|_{L^2(\Omega)} + C \left\| \mathbb{P}^0_{2,N} y - y \right\|_{H^1(\Omega)} + C\|y - y_N\|_{H^1(\Omega)} \\ &\le C\|p - p_N(u)\|_{H^2(\Omega)} + C\|u - u_N\|_{L^2(\Omega)} + C\|y - y_N(u)\|_{H^1(\Omega)} \\ &\quad + C \left\| \mathbb{P}^0_{2,N} y - y \right\|_{H^1(\Omega)}. \end{aligned} \tag{3.2.29}$$

If $\|y\|_{H^1(\Omega)} < \gamma$ and $\|y_N\|_{H^1(\Omega)} < \gamma$, by (3.2.9) and (3.2.13), it has

$$\lambda = \lambda_N = 0, \tag{3.2.30}$$

which satisfies (3.2.26).

If $\|y\|_{H^1(\Omega)} < \gamma$ and $\|y_N\|_{H^1(\Omega)} = \gamma$, it follows from (3.2.9) and (3.2.25),

$$a\left(y_N, p_N(u) - p_N\right) = \left(y_N(u) - y_N, y_N\right) - \lambda_N \left(y_N - \Delta y_N, y_N\right).$$

Furthermore, by (3.2.12)($\tilde{\text{a}}$), we have

$$\begin{aligned} \lambda_N \|y_N\|^2_{H^1(\Omega)} &= \lambda_N \left(y_N - \Delta y_N, y_N\right) \\ &= (y_N(u) - y_N, y_N) - a(y_N, p_N(u) - p_N) \\ &= (y_N(u) - y_N, y_N) - (u_N + f, p_N(u) - p_N). \end{aligned} \tag{3.2.31}$$

Therefore, combining (3.2.8)(d), (3.2.12)($\tilde{\text{d}}$), (3.2.14), (3.2.22), and (3.2.31), we obtain

$$\begin{aligned} |\lambda - \lambda_N| = |\lambda_N| &\le C\|y_N(u) - y_N\|_{L^2(\Omega)} + C\|p_N(u) - p_N\|_{L^2(\Omega)} \\ &\le C\|u - u_N\|_{L^2(\Omega)} + C\|p - p_N\|_{L^2(\Omega)} + C\|p - p_N(u)\|_{L^2(\Omega)} \\ &\le C\|u - u_N\|_{L^2(\Omega)} + C\|p - p_N(u)\|_{L^2(\Omega)}. \end{aligned} \tag{3.2.32}$$

The lemma directly follows from (3.2.29), (3.2.30), and (3.2.32). □

Lemma 3.2.7 *Let (u, y, p, λ) and $(u_N, y_N, p_N, \lambda_N)$ satisfy the optimality conditions $(\mathscr{Q})$ and $(\mathscr{Q}_N)$, respectively. Then the following estimate holds:*

$$\begin{aligned}
&\|u - u_N\|_{L^2(\Omega)} \\
&\le C\Big(\|p - p_N(u)\|_{H^2(\Omega)} + \|y - y_N(u)\|_{H^1(\Omega)} + \left\|\mathbb{P}^0_{2,N} y - y\right\|_{H^1(\Omega)} \\
&+ \|\mathbb{P}_N u - u\|_{L^2(\Omega)}\Big).
\end{aligned}$$

Proof Set $w_N = p_N(u) - p_N$ in (3.2.24) and $q_N = y_N(u) - y_N$ in (3.2.25). We obtain

$$\begin{aligned}
&(y_N(u) - y_N, y_N(u) - y_N) \\
&= (u - u_N, p_N(u) - p_N) - \lambda\,(y_N(u) - \Delta y_N(u), y_N(u) - y_N) \\
&\quad +\lambda_N\,(y_N - \Delta y_N, y_N(u) - y_N) \\
&= (u - u_N, p_N(u) - p_N) - \lambda\,(y_N(u) - \Delta y_N(u), y_N(u) - y) \\
&\quad -\lambda\,(y_N(u) - \Delta y_N(u), y - y_N) \\
&\quad +\lambda_N\,(y_N - \Delta y_N, y_N(u) - y) + \lambda_N\,(y_N - \Delta y_N, y - y_N)
\end{aligned} \tag{3.2.33}$$

By employing (3.2.13), it becomes effortless to prove that

$$\lambda_N\,(y_N - \Delta y_N, y - y_N) \le 0. \tag{3.2.34}$$

Employing both (3.2.8)(c) and (3.2.34) in the context of equation (3.2.33), we derive that

$$\begin{aligned}
&(y_N(u) - y_N, y_N(u) - y_N) \\
&\le (u - u_N, p_N(u) - p_N) - \lambda\,(y_N(u) - \Delta y_N(u), y_N(u) - y) \\
&\quad -\lambda\,(y_N(u) - \Delta y_N(u), y - y_N) \\
&\quad +\lambda_N\,(y_N - \Delta y_N, y_N(u) - y) + \lambda\,(y - \Delta y, y - y_N) \\
&\le (u - u_N, p_N(u) - p_N) + \lambda\,(y_N - \Delta y_N - y_N(u) + \Delta y_N(u), y_N(u) - y) \\
&\quad +\lambda\,(y - \Delta y - y_N(u) + \Delta y_N(u), y - y_N) + (\lambda_N - \lambda)\,(y_N - \Delta y_N, y_N(u) - y).
\end{aligned} \tag{3.2.35}$$

Since

$$\begin{aligned}
&\alpha\,\|\mathbb{P}_N u-u_N\|^2_{L^2(\Omega)}+(u-u_N,p_N(u)-p_N)\\
&=\alpha\,(\mathbb{P}_N u-u_N,u-u_N)+(u-u_N,p_N(u)-p_N)\\
&=\alpha\,(\mathbb{P}_N u-u,u-\mathbb{P}_N u)+(p_N(u)-p,u-u_N)\\
&=-\alpha\,\|\mathbb{P}_N u-u\|^2_{L^2(\Omega)}+(p_N(u)-p,u-u_N),
\end{aligned} \tag{3.2.36}$$

by (3.2.35) and (3.2.36), we have

$$\begin{aligned}
&\alpha\,\|\mathbb{P}_N u-u_N\|^2_{L^2(\Omega)}+\|y_N(u)-y_N\|^2_{L^2(\Omega)}\\
&\le-\alpha\,\|\mathbb{P}_N u-u\|^2_{L^2(\Omega)}+(p_N(u)-p,u-u_N)\\
&\quad+\lambda\,(y_N-\Delta y_N-y_N(u)+\Delta y_N(u),y_N(u)-y)\\
&\quad+\lambda\,(y-\Delta y-y_N(u)+\Delta y_N(u),y-y_N)+(\lambda_N-\lambda)\,(y_N-\Delta y_N,y_N(u)-y)\\
&\le-\alpha\,\|\mathbb{P}_N u-u\|^2_{L^2(\Omega)}+\|p_N(u)-p\|_{L^2(\Omega)}\|u-u_N\|_{L^2(\Omega)}\\
&\quad+C\,\|y_N-y_N(u)\|_{H^1(\Omega)}\,\|y_N(u)-y\|_{H^1(\Omega)}\\
&\quad+C\,\|y-y_N(u)\|_{H^1(\Omega)}\,\|y-y_N\|_{H^1(\Omega)}+C\,|\lambda_N-\lambda|\cdot\|y-y_N(u)\|_{L^2(\Omega)}\,.
\end{aligned} \tag{3.2.37}$$

Taking into account (3.2.37) in conjunction with Lemmas 3.2.4 through 3.2.6, we can infer that

$$\begin{aligned}
&\frac{\alpha}{2}\,\|u-u_N\|^2_{L^2(\Omega)}+\|y_N(u)-y_N\|^2_{L^2(\Omega)}\\
&\le\alpha\,\|\mathbb{P}_N u-u\|^2_{L^2(\Omega)}+\alpha\,\|\mathbb{P}_N u-u_N\|^2_{L^2(\Omega)}+\|y_N(u)-y_N\|^2_{L^2(\Omega)}\\
&\le\alpha\,\|\mathbb{P}_N u-u\|^2_{L^2(\Omega)}+\|p_N(u)-p\|_{L^2(\Omega)}\,\|u-u_N\|_{L^2(\Omega)}\\
&\quad+C\,\|y_N-y_N(u)\|_{H^1(\Omega)}\,\|y_N(u)-y\|_{H^1(\Omega)}\\
&\quad+C\,\|y-y_N(u)\|_{H^1(\Omega)}\,\|y-y_N\|_{H^1(\Omega)}+C\,|\lambda_N-\lambda|\cdot\|y-y_N(u)\|_{L^2(\Omega)}\\
&\le\alpha\,\|\mathbb{P}_N u-u\|^2_{L^2(\Omega)}+\|p_N(u)-p\|_{L^2(\Omega)}\,\|u-u_N\|_{L^2(\Omega)}\\
&\quad+C\,\|u-u_N\|_{L^2(\Omega)}\,\|y_N(u)-y\|_{H^1(\Omega)}\\
&\quad+C\,\|y-y_N(u)\|_{L^2(\Omega)}\Big(\|p-p_N(u)\|_{H^2(\Omega)}+\|u-u_N\|_{L^2(\Omega)}\\
&\quad+\|y-y_N(u)\|_{H^1(\Omega)}+\left\|\mathbb{P}^0_{2,N}y-y\right\|_{H^1(\Omega)}\Big)+C\|y-y_N(u)\|^2_{H^1(\Omega)}\\
&\le\varepsilon\,\|u-u_N\|^2_{L^2(\Omega)}+C\varepsilon^{-1}\left(\|p_N(u)-p\|^2_{L^2(\Omega)}+\|y_N(u)-y\|^2_{H^1(\Omega)}\right)
\end{aligned}$$

$$+ C\Big(\|p - p_N(u)\|^2_{H^2(\Omega)} + \|y - y_N(u)\|^2_{H^1(\Omega)} + \left\|\mathbb{P}^0_{2,N} y - y\right\|^2_{H^1(\Omega)}$$
$$+ \|\mathbb{P}_N u - u\|^2_{L^2(\Omega)}\Big).$$

Choosing $\varepsilon = \frac{\alpha}{4}$, we obtain

$$\begin{aligned}
&\|u - u_N\|_{L^2(\Omega)} \\
&\leq C\Big(\|p - p_N(u)\|_{H^2(\Omega)} + \|y - y_N(u)\|_{H^1(\Omega)} + \left\|\mathbb{P}^0_{2,N} y - y\right\|_{H^1(\Omega)} \\
&\quad + \|\mathbb{P}_N u - u\|_{L^2(\Omega)}\Big).
\end{aligned} \tag{3.2.38}$$

This completes the proof of the lemma. □

Now, we proceed to present the a priori error estimates for the spectral approximation problem $(\mathscr{P}_N)$ as follows.

Theorem 3.2.8 *Let (u, y, p, λ) and $(u_N, y_N, p_N, \lambda_N)$ be the solutions to the optimality conditions $(\mathscr{Q})$ and $(\mathscr{Q}_N)$, respectively. Assume that $(u, y, p) \in H^{m-2}(\Omega) \times \left(H^m(\Omega) \cap H^2_0(\Omega)\right)^2$ where $m \geq 2$, then the following estimate holds:*

$$\begin{aligned}
&\|u - u_N\|_{L^2(\Omega)} + \|y - y_N\|_{H^2(\Omega)} + \|p - p_N\|_{H^2(\Omega)} + |\lambda - \lambda_N| \\
&\leq CN^{2-m}\left(\|u\|_{H^{m-2}(\Omega)} + \|y\|_{H^m(\Omega)} + \|p\|_{H^m(\Omega)}\right).
\end{aligned}$$

Proof Using Lemmas 3.2.6–3.2.7 in conjunction with the triangular inequality, we obtain

$$\begin{aligned}
&\|u - u_N\|_{L^2(\Omega)} + \|y - y_N\|_{H^2(\Omega)} + \|p - p_N\|_{H^2(\Omega)} + |\lambda - \lambda_N| \\
&\leq \|u - u_N\|_{L^2(\Omega)} + \left(\|y - y_N(u)\|_{H^2(\Omega)} + \|y_N(u) - y_N\|_{H^2(\Omega)}\right) \\
&\quad + \left(\|p - p_N(u)\|_{H^2(\Omega)} + \|p_N(u) - p_N\|_{H^2(\Omega)}\right) + |\lambda - \lambda_N| \\
&\leq C\,\|y - y_N(u)\|_{H^2(\Omega)} + C\,\|p - p_N(u)\|_{H^2(\Omega)} \\
&\quad + C\,\|\mathbb{P}_N u - u\|_{L^2(\Omega)} + C\left\|\mathbb{P}^0_{2,N} y - y\right\|_{H^1(\Omega)}.
\end{aligned} \tag{3.2.39}$$

Next, we just need to estimate $\|y - y_N(u)\|_{H^2(\Omega)}$ and $\|p - p_N(u)\|_{H^2(\Omega)}$. Subtract (3.2.20)–(3.2.21) from (3.2.8)(a)–(b), respectively, to obtain

$$a(y - y_N(u), w_N) = 0, \quad \forall\, w_N \in Y_N, \tag{3.2.40}$$

and

$$a(q_N, p-p_N(u)) = (y-y_N(u), q_N)+\lambda(y-\Delta y-y_N(u)+\Delta y_N(u), q_N), \quad \forall\, q_N \in Y_N. \tag{3.2.41}$$

Using (3.2.4), (3.2.5), and (3.2.40) together, we find that

$$\begin{aligned} & c_0 \|y - y_N(u)\|^2_{H^2(\Omega)} \\ & \le a\left(y - y_N(u), y - y_N(u)\right) = a\left(y - y_N(u), y - \mathbb{P}^0_{2,N} y\right) \\ & \le C_0 \|y - y_N(u)\|_{H^2(\Omega)} \left\| y - \mathbb{P}^0_{2,N} y \right\|_{H^2(\Omega)}. \end{aligned}$$

Thus,

$$\|y - y_N(u)\|_{H^2(\Omega)} \le C \left\| y - \mathbb{P}^0_{2,N} y \right\|_{H^2(\Omega)}. \tag{3.2.42}$$

Next estimate $\left\|\mathbb{P}^0_{2,N} p - p_N(u)\right\|_{H^2(\Omega)}$. By (3.2.4), (3.2.5), (3.2.41), and (3.2.42), we have

$$\begin{aligned} & c_0 \left\|\mathbb{P}^0_{2,N} p - p_N(u)\right\|^2_{H^2(\Omega)} \le a\left(\mathbb{P}^0_{2,N} p - p_N(u), \mathbb{P}^0_{2,N} p - p_N(u)\right) \\ & \le a\left(\mathbb{P}^0_{2,N} p - p_N(u), \mathbb{P}^0_{2,N} p - p\right) + \left(y - y_N(u), \mathbb{P}^0_{2,N} p - p_N(u)\right) \\ & \quad + \lambda\left(y - \Delta y - y_N(u) + \Delta y_N(u), \mathbb{P}^0_{2,N} p - p_N(u)\right) \\ & \le C \left\|\mathbb{P}^0_{2,N} p - p_N(u)\right\|_{H^2(\Omega)} \left(\|p - \mathbb{P}^0_{2,N} p\|_{H^2(\Omega)} + \|y - y_N(u)\|_{H^2(\Omega)}\right) \\ & \le C \left\|\mathbb{P}^0_{2,N} p - p_N(u)\right\|_{H^2(\Omega)} \left(\|p - \mathbb{P}^0_{2,N} p\|_{H^2(\Omega)} + \|y - \mathbb{P}^0_{2,N} y\|_{H^2(\Omega)}\right), \end{aligned}$$

which implies that

$$\left\|\mathbb{P}^0_{2,N} p - p_N(u)\right\|_{H^2(\Omega)} \le C \left\|p - \mathbb{P}^0_{2,N} p\right\|_{H^2(\Omega)} + C \left\|y - \mathbb{P}^0_{2,N} y\right\|_{H^2(\Omega)}.$$

Therefore,

$$\begin{aligned} \|p - p_N(u)\|_{H^2(\Omega)} & \le \left\|p - \mathbb{P}^0_{2,N} p\right\|_{H^2(\Omega)} + \left\|\mathbb{P}^0_{2,N} p - p_N(u)\right\|_{H^2(\Omega)} \\ & \le C \left\|p - \mathbb{P}^0_{2,N} p\right\|_{H^2(\Omega)} + C \left\|y - \mathbb{P}^0_{2,N} y\right\|_{H^2(\Omega)}. \end{aligned} \tag{3.2.43}$$

Finally, combining (3.2.39), (3.2.42), and (3.2.43) with Lemmas 3.2.2 and 3.2.3, we arrive at

$$\begin{aligned}
&\|u-u_N\|_{L^2(\Omega)}+\|y-y_N\|_{H^2(\Omega)}+\|p-p_N\|_{H^2(\Omega)}+|\lambda-\lambda_N|\\
&\le C\left\|p-\mathbb{P}^0_{2,N}p\right\|_{H^2(\Omega)}+C\left\|y-\mathbb{P}^0_{2,N}y\right\|_{H^2(\Omega)}+C\,\|\mathbb{P}_Nu-u\|_{L^2(\Omega)}\\
&\le CN^{2-m}\left(\|u\|_{H^{m-2}(\Omega)}+\|y\|_{H^m(\Omega)}+\|p\|_{H^m(\Omega)}\right).
\end{aligned}$$

This completes the proof of the theorem. □

3.2.3 A Posteriori Error Estimates

It is widely recognized that the initial step in implementing an adaptive strategy involves the establishment of effective error indicators, which are founded upon a posteriori error estimates. In this subsection, our objective is to formulate a posteriori error estimates specifically tailored for the spectral approximation problem $(\mathscr{P}_N)$. Additionally, we aim to provide an upper bound for the spectral approximation error. To elaborate, we will construct H^2-error estimates for both the state and costate, as well as an L^2- error estimate for the control.

In order to construct a posteriori error estimates for the spectral approximation system $(\mathscr{P}_N)$, we introduce a crucial auxiliary system.

Find $(y(u_N), p(u_N)) \in Y \times Y$ such that

$$a(y(u_N), w) = (f + u_N, w), \quad \forall\, w \in Y, \tag{3.2.44}$$

and

$$a(q, p(u_N)) = (y_N - y_d, q) + \lambda_N(y_N - \Delta y_N, q), \quad \forall\, q \in Y. \tag{3.2.45}$$

Firstly, let us present several lemmas.

Lemma 3.2.9 *Let (u, y, p, λ) and $(u_N, y_N, p_N, \lambda_N)$ satisfy the optimality conditions $(\mathscr{Q})$ and $(\mathscr{Q}_N)$, respectively. Then the following estimates hold:*

$$\|y(u_N)-y_N\|_{H^2(\Omega)} \le CN^{-2}\left\|f+u_N-\Delta^2 y_N\right\|_{L^2(\Omega)},$$

$$\|p(u_N)-p_N\|_{H^2(\Omega)} \le CN^{-2}\left\|y_N-y_d+\lambda_N\left(y_N-\Delta y_N\right)-\Delta^2 p_N\right\|_{L^2(\Omega)}.$$

Proof From (3.2.12)(ã)-($\tilde{\text{b}}$), (3.2.44), and (3.2.45), it has

$$a\left(y(u_N)-y_N, w_N\right)=0, \quad \forall\, w_N \in Y_N, \tag{3.2.46}$$

$$a\left(q_N, p(u_N)-p_N\right)=0, \quad \forall\, q_N \in Y_N. \tag{3.2.47}$$

By applying (3.2.5), (3.2.44), and (3.2.46) and invoking Lemma 3.2.3, we obtain

$$\begin{aligned}
& c_0\left\|y(u_N)-y_N\right\|_{H^2(\Omega)}^2 \\
\leq\; & a\left(y(u_N)-y_N, y(u_N)-y_N\right) \\
=\; & a\Big(y(u_N)-y_N, y(u_N)-y_N-\mathbb{P}_{2,N}^0\left(y(u_N)-y_N\right)\Big) \\
=\; & a\Big(y(u_N), y(u_N)-y_N-\mathbb{P}_{2,N}^0(y(u_N)-y_N)\Big) \\
& -a\Big(y_N, y(u_N)-y_N-\mathbb{P}_{2,N}^0(y(u_N)-y_N)\Big) \\
=\; & \Big(f+u_N-\Delta^2 y_N, y(u_N)-y_N-\mathbb{P}_{2,N}^0\left(y(u_N)-y_N\right)\Big) \\
\leq\; & \left\|f+u_N-\Delta^2 y_N\right\|_{L^2(\Omega)}\left\|y(u_N)-y_N-\mathbb{P}_{2,N}^0(y(u_N)-y_N)\right\|_{L^2(\Omega)} \\
\leq\; & C N^{-2}\left\|f+u_N-\Delta^2 y_N\right\|_{L^2(\Omega)}\left\|y(u_N)-y_N\right\|_{H^2(\Omega)}.
\end{aligned}$$

Hence,

$$\left\|y(u_N)-y_N\right\|_{H^2(\Omega)} \leq C N^{-2}\left\|f+u_N-\Delta^2 y_N\right\|_{L^2(\Omega)}. \tag{3.2.48}$$

Likewise, by utilizing (3.2.5), (3.2.45), and (3.2.47) and employing Lemma 3.2.3, we arrive at

$$\begin{aligned}
& c_0\left\|p(u_N)-p_N\right\|_{H^2(\Omega)}^2 \\
\leq\; & a\left(p(u_N)-p_N, p(u_N)-p_N\right) \\
=\; & a\Big(p(u_N)-p_N, p(u_N)-p_N-\mathbb{P}_{2,N}^0(p(u_N)-p_N)\Big) \\
=\; & a\Big(p(u_N), p(u_N)-p_N-\mathbb{P}_{2,N}^0(p(u_N)-p_N)\Big) \\
& -a\Big(p_N, p(u_N)-p_N-\mathbb{P}_{2,N}^0(p(u_N)-p_N)\Big) \\
=\; & \Big(y_N-y_d+\lambda_N(y_N-\Delta y_N)-\Delta^2 p_N, p(u_N)-p_N-\mathbb{P}_{2,N}^0(p(u_N)-p_N)\Big) \\
\leq\; & \left\|y_N-y_d+\lambda_N(y_N-\Delta y_N)-\Delta^2 p_N\right\|_{L^2(\Omega)}
\end{aligned}$$

$$\cdot \left\| p(u_N) - p_N - \mathbb{P}^0_{2,N}(p(u_N) - p_N) \right\|_{L^2(\Omega)}$$

$$\le CN^{-2} \left\| y_N - y_d + \lambda_N(y_N - \Delta y_N) - \Delta^2 p_N \right\|_{L^2(\Omega)} \| p(u_N) - p_N \|_{H^2(\Omega)} .$$

Furthermore,

$$\| p(u_N) - p_N \|_{H^2(\Omega)} \le CN^{-2} \left\| y_N - y_d + \lambda_N (y_N - \Delta y_N) - \Delta^2 p_N \right\|_{L^2(\Omega)} . \tag{3.2.49}$$

By combining (3.2.48) and (3.2.49), we finalize the proof. □

Lemma 3.2.10 *Let* (u, y, p, λ) *and* $(u_N, y_N, p_N, \lambda_N)$ *satisfy the optimality conditions* $(\mathcal{Q})$ *and* $(\mathcal{Q}_N)$*, respectively. Then we have*

$$\| y - y(u_N) \|_{H^2(\Omega)} \le C \| u - u_N \|_{L^2(\Omega)} , \tag{3.2.50}$$

$$\| p - p(u_N) \|_{H^2(\Omega)} \le C \left(\| y(u_N) - y_N \|_{H^1(\Omega)} + \| u - u_N \|_{L^2(\Omega)} + |\lambda - \lambda_N| \right) , \tag{3.2.51}$$

and

$$|\lambda - \lambda_N| \le C \| y_N - y(u_N) \|_{H^1(\Omega)} + C \| p_N - p(u_N) \|_{L^2(\Omega)} + C \| u - u_N \|_{L^2(\Omega)} . \tag{3.2.52}$$

Proof It follows from (3.2.8)(a)–(b), (3.2.44), and (3.2.45) that

$$a(y - y(u_N), w) = (u - u_N, w), \quad \forall\, w \in Y, \tag{3.2.53}$$

$$a(q, p - p(u_N)) = (y - y_N, q) + \lambda(y - \Delta y, q) - \lambda_N(y_N - \Delta y_N, q), \quad \forall\, q \in Y. \tag{3.2.54}$$

Set $w = y - y(u_N)$ in (3.2.53). By (3.2.5), we get

$$\begin{aligned} c_0 \| y - y(u_N) \|^2_{H^2(\Omega)} &\le a(y - y(u_N), y - y(u_N)) = (u - u_N, y - y(u_N)) \\ &\le \| y - y(u_N) \|_{L^2(\Omega)} \| u - u_N \|_{L^2(\Omega)} \\ &\le \| y - y(u_N) \|_{H^2(\Omega)} \| u - u_N \|_{L^2(\Omega)} , \end{aligned}$$

which implies that (3.2.50) holds.

Analogously, let $q = p - p(u_N)$ in (3.2.54). Utilizing (3.2.5) and (3.2.50), we obtain

$$\begin{aligned}
&c_0 \|p - p(u_N)\|^2_{H^2(\Omega)} \\
&\le a\,(p - p(u_N), p - p(u_N)) \\
&= \lambda\,(y - \Delta y, p - p(u_N)) - \lambda_N\,(y_N - \Delta y_N, p - p(u_N)) \\
&= \lambda\,(y - y_N, p - p(u_N)) - \lambda\,(\Delta(y - y_N), p - p(u_N)) \\
&\quad + (\lambda - \lambda_N)\,(y_N - \Delta y_N, p - p(u_N)) \\
&\le C\,\|p - p(u_N)\|_{H^1(\Omega)} \left(\|y - y_N\|_{H^1(\Omega)} + |\lambda - \lambda_N|\right) \\
&\le C\,\|p - p(u_N)\|_{H^2(\Omega)} \left(\|y(u_N) - y_N\|_{H^1(\Omega)} + \|u - u_N\|_{L^2(\Omega)} + |\lambda - \lambda_N|\right).
\end{aligned}$$

This implies that (3.2.51) is valid.

Next, we proceed to estimate the absolute difference $|\lambda - \lambda_N|$.

Given that $\|y\|_{H^1(\Omega)} = \gamma$, in this scenario, we choose $q = y$ in (3.2.54). Consequently, we have

$$a\,(y, p - p(u_N)) = (y - y_N, y) + \lambda\,(y - \Delta y, y) - \lambda_N\,(y_N - \Delta y_N, y)\,. \tag{3.2.55}$$

From (3.2.8)(a), we know

$$a\,(y, p - p(u_N)) = (u + f, p - p(u_N))\,. \tag{3.2.56}$$

By (3.2.55) and (3.2.56), it has

$$\begin{aligned}
&(\lambda - \lambda_N)\,(y - \Delta y, y) \\
&\quad = (u + f, p - p(u_N)) - (y - y_N, y) - \lambda_N\,(y - \Delta y - y_N + \Delta y_N, y)\,.
\end{aligned}$$

By combining (3.2.8)(d), (3.2.12)(d), and (3.2.50) and invoking Lemma 3.2.4, we obtain

$$\begin{aligned}
|\lambda - \lambda_N| &\le C\,\|p - p(u_N)\|_{L^2(\Omega)} + C\,\|y - y_N\|_{L^2(\Omega)} + C\,\|y - y_N\|_{H^1(\Omega)} \\
&\le C\,\|p - p_N\|_{L^2(\Omega)} + C\,\|p(u_N) - p_N\|_{L^2(\Omega)} \\
&\quad + C\,\|y - y(u_N)\|_{H^1(\Omega)} + C\,\|y(u_N) - y_N\|_{H^1(\Omega)} \\
&\le C\,\|p(u_N) - p_N\|_{L^2(\Omega)} + C\,\|y(u_N) - y_N\|_{H^1(\Omega)} + C\,\|u - u_N\|_{L^2(\Omega)}\,.
\end{aligned} \tag{3.2.57}$$

The second scenario occurs when both $\|y\|_{H^1(\Omega)} < \gamma$ and $\|y_N\|_{H^1(\Omega)} < \gamma$. In this situation, employing (3.2.9) and (3.2.13), we arrive at

$$|\lambda - \lambda_N| = 0. \tag{3.2.58}$$

Next, we address the final case. When $\|y\|_{H^1(\Omega)} < \gamma$ and $\|y_N\|_{H^1(\Omega)} = \gamma$, it follows that $\lambda = 0$. To deduce the desired outcome, we set $q = y(u_N)$ in (3.2.54) and obtain

$$a\left(y(u_N), p - p(u_N)\right) = \left(y - y_N, y(u_N)\right) - \lambda_N\left(y_N - \Delta y_N, y(u_N)\right). \tag{3.2.59}$$

From (3.2.44) and (3.2.59), it has

$$\left(y - y_N, y(u_N)\right) - \lambda_N\left(y_N - \Delta y_N, y(u_N)\right) = \left(f + u_N, p - p(u_N)\right),$$

which leads to

$$\begin{aligned}\lambda_N\left(y_N - \Delta y_N, y_N\right) = {} & \left(y - y_N, y(u_N)\right) + \lambda_N\left(y_N - \Delta y_N, y_N - y(u_N)\right) \\ & - \left(f + u_N, p - p(u_N)\right).\end{aligned} \tag{3.2.60}$$

By amalgamating (3.2.8)(d), (3.2.12)($\tilde{\text{d}}$), (3.2.50), and (3.2.60) and incorporating Lemma 3.3, we arrive at

$$\begin{aligned}|\lambda - \lambda_N| &= |\lambda_N| \\ &\leq C\,\|y - y_N\|_{L^2(\Omega)} + C\,\|y_N - y(u_N)\|_{L^2(\Omega)} + C\,\|p - p(u_N)\|_{L^2(\Omega)} \\ &\leq C\,\|y_N - y(u_N)\|_{L^2(\Omega)} + C\,\|p_N - p(u_N)\|_{L^2(\Omega)} + C\,\|u - u_N\|_{L^2(\Omega)}.\end{aligned} \tag{3.2.61}$$

Subsequently, (3.2.52) is derived from (3.2.57), (3.2.58), and (3.2.61). This concludes the proof. □

Next, we proceed to derive the error estimate $\|u - u_N\|_{L^2(\Omega)}$, which holds significant importance in constructing a posteriori error estimates.

Lemma 3.2.11 *Let (u, y, p, λ) and $(u_N, y_N, p_N, \lambda_N)$ satisfy the optimality conditions $(\mathcal{Q})$ and $(\mathcal{Q}_N)$, respectively. Then,*

$$\|u - u_N\|_{L^2(\Omega)} \leq C\left(\|p(u_N) - p_N\|_{L^2(\Omega)} + \|y(u_N) - y_N\|_{H^1(\Omega)}\right).$$

Proof By (3.2.53) and (3.2.54),

$$
\begin{aligned}
&(y - y(u_N), y - y(u_N)) \\
&= (u - u_N, p - p(u_N)) - (y(u_N) - y_N, y - y(u_N)) \\
&\quad -\lambda (y - \Delta y, y - y(u_N)) + \lambda_N (y_N - \Delta y_N, y - y(u_N)) \\
&= (u - u_N, p - p(u_N)) - (y(u_N) - y_N, y - y(u_N)) \\
&\quad -\lambda (y - \Delta y, y - y_N) - \lambda (y - \Delta y, y_N - y(u_N)) \\
&\quad +\lambda_N (y_N - \Delta y_N, y - y_N) + \lambda_N (y_N - \Delta y_N, y_N - y(u_N)) .
\end{aligned}
\tag{3.2.62}
$$

By (3.2.8)(c) and (3.2.34), for (3.2.62) above,

$$
\begin{aligned}
&(y - y(u_N), y - y(u_N)) \\
&\leq (u - u_N, p - p(u_N)) - (y(u_N) - y_N, y - y(u_N)) \\
&\quad -\lambda (y - \Delta y, y_N - y(u_N)) + \lambda_N (y_N - \Delta y_N, y_N - y(u_N)) \\
&= (u - u_N, p - p(u_N)) - (y(u_N) - y_N, y - y(u_N)) \\
&\quad - (\lambda - \lambda_N) (y_N - \Delta y_N, y_N - y(u_N)) \\
&\quad -\lambda (y - \Delta y - y_N + \Delta y_N, y_N - y(u_N)) .
\end{aligned}
\tag{3.2.63}
$$

Furthermore, since

$$
\begin{aligned}
&\alpha \|u - u_N\|^2_{L^2(\Omega)} + (u - u_N, p - p(u_N)) \\
&= \alpha (u - u_N, u - u_N) + (u - u_N, p - p(u_N)) = (u - u_N, p_N - p(u_N)),
\end{aligned}
$$

by (3.2.63) and Lemma 3.2.10,

$$
\begin{aligned}
&\alpha \|u - u_N\|^2_{L^2(\Omega)} + \|y - y(u_N)\|^2_{L^2(\Omega)} \\
&\leq \|u - u_N\|_{L^2(\Omega)} \|p(u_N) - p_N\|_{L^2(\Omega)} + \|y(u_N) - y_N\|_{L^2(\Omega)} \|y - y(u_N)\|_{L^2(\Omega)} \\
&\quad +C \|y - y_N\|_{H^1(\Omega)} \|y(u_N) - y_N\|_{H^1(\Omega)} + C |\lambda - \lambda_N| \cdot \|y(u_N) - y_N\|_{L^2(\Omega)} \\
&\leq \varepsilon \|u - u_N\|^2_{L^2(\Omega)} + C\varepsilon^{-1} \left(\|p(u_N) - p_N\|^2_{L^2(\Omega)} + \|y(u_N) - y_N\|^2_{H^1(\Omega)}\right) \\
&\quad +C \left(\|p(u_N) - p_N\|^2_{L^2(\Omega)} + \|y(u_N) - y_N\|^2_{H^1(\Omega)}\right) .
\end{aligned}
$$

Choose $\varepsilon = \frac{\alpha}{2}$ to yield

$$\|u - u_N\|_{L^2(\Omega)} \leq C \left(\|p(u_N) - p_N\|_{L^2(\Omega)} + \|y(u_N) - y_N\|_{H^1(\Omega)}\right).$$

This completes the proof of the lemma. □

Finally, we introduce a posteriori error estimates pertaining to L^2-error estimation for the control, as well as H^2-error estimates for both the state and costate, as outlined below.

Theorem 3.2.12 *Let (u, y, p, λ) and $(u_N, y_N, p_N, \lambda_N)$ satisfy the optimality conditions $(\mathscr{Q})$ and $(\mathscr{Q}_N)$, respectively. The following estimate*

$$\|u - u_N\|_{L^2(\Omega)} + \|y - y_N\|_{H^2(\Omega)} + \|p - p_N\|_{H^2(\Omega)} + |\lambda - \lambda_N| \leq CN^{-2}\eta \tag{3.2.64}$$

holds, where

$$\eta = \left\|f + u_N - \Delta^2 y_N\right\|_{L^2(\Omega)} + \left\|y_N - y_d + \lambda_N(y_N - \Delta y_N) - \Delta^2 p_N\right\|_{L^2(\Omega)}.$$

Proof Using the triangular inequality,

$$\begin{aligned}
&\|u - u_N\|_{L^2(\Omega)} + \|y - y_N\|_{H^2(\Omega)} + \|p - p_N\|_{H^2(\Omega)} + |\lambda - \lambda_N| \\
\leq{}& \|u - u_N\|_{L^2(\Omega)} + \left(\|y - y(u_N)\|_{H^2(\Omega)} + \|y(u_N) - y_N\|_{H^2(\Omega)}\right) \\
&+ \left(\|p - p(u_N)\|_{H^2(\Omega)} + \|p(u_N) - p_N\|_{H^2(\Omega)}\right) + |\lambda - \lambda_N|.
\end{aligned}$$

Then, (3.2.64) is derived from Lemmas 3.2.9–3.2.11. □

3.2.4 Numerical Examples

In this subsection, we present two numerical examples to validate the theoretical results obtained regarding the error estimate for the spectral discrete system $(\mathscr{P}_N)$ discussed in Sects. 3.2.2 and 3.2.3. During this process, we utilize a gradient projection algorithm to solve these control problems. The iteration parameter is selected as $\rho < 1$. Additionally, it is straightforward to demonstrate its convergence using the methods outlined in [253]. The primary steps of the gradient projection algorithm are outlined below.

Algorithm 1 Gradient projection algorithm

1: Given the initial values $\tilde{u}_N^0 \in Y_N$.
 1). Seek $\tilde{y}_N^0 \in Y_N$ such that

$$a\left(\tilde{y}_N^0, w_N\right) = \left(f + \tilde{u}_N^0, w_N\right), \quad \forall\, w_N \in Y_N.$$

 2). Seek $g_N^0 \in Y_N$ such that

$$a\left(q_N, g_N^0\right) = \left(\tilde{y}_N^0 - \Delta \tilde{y}_N^0, q_N\right), \quad \forall\, q_N \in Y_N.$$

 3). Set $\kappa_N^0 = \dfrac{\max\left\{\gamma, \|\tilde{y}_N^0\|_{H^1(\Omega)}\right\} - \gamma}{\rho \left\|g_N^0\right\|_{L^2(\Omega)}^2} \dfrac{\|\tilde{y}_N^0\|_{H^1(\Omega)}^2}{\max\left\{\gamma, \|\tilde{y}_N^0\|_{H^1(\Omega)}\right\}}$ and $u_N^0 = \tilde{u}_N^0 - \rho\kappa_N^0 g_N^0$.
 4). Seek $y_N^0 \in Y_N$ such that

$$a\left(y_N^0, w_N\right) = \left(f + u_N^0, w_N\right), \quad \forall\, w_N \in Y_N.$$

2: For $n = 0, 1, 2, \ldots$, seek $\tilde{p}_N^{n+1} \in Y_N$ such that

$$a\left(q_N, \tilde{p}_N^{n+1}\right) = \left(y_N^n - y_d, q_N\right), \quad \forall\, q_N \in Y_N.$$

3: Set $\tilde{u}_N^{n+1} = u_N^n - \rho\left(\alpha u_N^n + \tilde{p}_N^{n+1}\right)$ and seek $\tilde{y}_N^{n+1} \in Y_N$ such that

$$a\left(\tilde{y}_N^{n+1}, w_N\right) = \left(f + \tilde{u}_N^{n+1}, w_N\right), \quad \forall\, w_N \in Y_N.$$

4: Seek $g_N^{n+1} \in Y_N$ such that

$$a\left(q_N, g_N^{n+1}\right) = \left(\tilde{y}_N^{n+1} - \Delta \tilde{y}_N^{n+1}, q_N\right), \quad \forall\, q_N \in Y_N.$$

5: Set $\kappa_N^{n+1} = \dfrac{\max\left\{\gamma, \|\tilde{y}_N^{n+1}\|_{H^1(\Omega)}\right\} - \gamma}{\rho \left\|g_N^{n+1}\right\|_{L^2(\Omega)}^2} \dfrac{\left\|\tilde{y}_N^{n+1}\right\|_{H^1(\Omega)}^2}{\max\left\{\gamma, \left\|\tilde{y}_N^{n+1}\right\|_{H^1(\Omega)}\right\}}$ and

$$u_N^{n+1} = \tilde{u}_N^{n+1} - \rho\kappa_N^{n+1} g_N^{n+1}.$$

6: Seek $y_N^{n+1} \in Y_N$ such that

$$a\left(y_N^{n+1}, w_N\right) = \left(f + u_N^{n+1}, w_N\right), \quad \forall\, w_N \in Y_N.$$

7: Stop if the stopping criterion $\left\| u_N^{n+1} - u_N^n \right\|_{L^2(\Omega)} \leq$ Tol (the tolerance) is satisfied. Otherwise, set $n = n + 1$ and go to Step 2.

Furthermore, the costate p_N can be determined using (3.2.12)($\tilde{\text{d}}$), while the value of λ_N can be calculated via the least square method in conjunction with (3.2.12) ($\tilde{\text{b}}$).

Additionally, across all numerical examples, we consistently set the state constraint parameter $\gamma = 1$.

3.2.4.1 One-Dimensional Case

Let $L_n(x)$ be the n-th degree Legendre polynomial and denote (see [204, 253])

$$c_k = \frac{1}{\sqrt{2(2k+3)^2(2k+5)}}, \quad k = 0, 1, ..., N-4,$$

$$\phi_k(x) = c_k\left(L_k(x) - \frac{2(2k+5)}{2k+7} L_{k+2}(x) + \frac{2k+3}{2k+7} L_{k+4}(x)\right), \quad k = 0, 1, \ldots, N-4.$$

Both of them satisfy $\phi(\pm 1) = \frac{\partial \phi}{\partial n}(\pm 1) = 0$. Subsequently, we proceed to construct the function space as follows:

$$U_N = \text{span}\{L_0(x), L_1(x), \ldots, L_N(x)\},$$

$$Y_N = \text{span}\{\phi_0(x), \phi_1(x), \ldots, \phi_{N-4}(x)\}, \quad K_N = K \cap Y_N.$$

The functions y_N, p_N, and u_N are presented as

$$y_N = \sum_{i=0}^{N-4} y_i \phi_i(x), \quad p_N = \sum_{i=0}^{N-4} p_i \phi_i(x), \quad u_N = \sum_{i=0}^{N} u_i L_i(x).$$

By substituting the aforementioned three terms into (3.2.12)($\tilde{\text{a}}$) and ($\tilde{\text{b}}$), we derive two $(N-3)$-dimensional systems of linear equations. For a detailed understanding of this transformation process, readers can refer to [204]. These two $(N-3)$-dimensional systems of linear equations are instrumental in executing Algorithm 1.

Example 3.2.13 Set $\alpha = 1$ and $\Omega = (-1, 1)$. The exact solution is given by

$$\begin{cases} y = \dfrac{2}{\sqrt{4\pi^2+3}} \sin^2(\pi x), \\ u = \dfrac{4}{\sqrt{4\pi^2+3}} \sin^2(\pi x), \\ \lambda = 0.4, \\ p = -\alpha u = -\dfrac{4}{\sqrt{4\pi^2+3}} \sin^2(\pi x). \end{cases}$$

The data for the optimal control problem are provided as follows:

$$f = y^{(4)} - u, \quad y_d = y + \lambda(y - \Delta y) - p^{(4)}.$$

Table 3.1 not only showcases the numerical results pertaining to $\|u - u_N\|_{L^2(\Omega)}$, $\|y-y_N\|_{H^2(\Omega)}$, $\|p-p_N\|_{H^2(\Omega)}$, and $|\lambda-\lambda_N|$ but also includes data on $\|y-y_N\|_{L^2(\Omega)}$ and $\|p - p_N\|_{L^2(\Omega)}$ (despite the fact that we did not theoretically analyze the L^2-error estimates for the state and costate). The error results are visually represented in Fig. 3.1, where the curves exhibit a rapid decline. This rapid decrease indicates that the errors are diminishing quickly. Moreover, the convergence orders are clearly illustrated by

$$\log_{\frac{N_{i+1}}{N_i}} \frac{\Sigma_i}{\Sigma_{i+1}},$$

where

$$\Sigma_i = \|u - u_N\|_{L^2(\Omega)} + \|y - y_N\|_{H^2(\Omega)} + \|p - p_N\|_{H^2(\Omega)}, \quad i = 1, 2, 3, 4, 5,$$

Table 3.1 Error results between the numerical solutions and exact solutions in Example 3.2.13

N	8	10	12	14	16
$\|u - u_N\|_{L^2(\Omega)}$	1.5638e-02	4.8263e-03	2.1084e-04	3.6586e-06	6.4550e-07
$\|y - y_N\|_{L^2(\Omega)}$	7.7335e-03	6.7837e-04	4.0115e-05	1.7541e-06	5.9992e-08
$\|p - p_N\|_{L^2(\Omega)}$	1.5638e-02	4.8263e-03	2.1084e-04	3.6586e-06	6.4550e-07
$\|y - y_N\|_{H^2(\Omega)}$	8.4155e-02	1.1307e-01	9.7148e-03	5.8293e-04	2.5893e-05
$\|p - p_N\|_{H^2(\Omega)}$	1.6843	2.3621e-01	1.9641e-02	1.1662e-03	5.5254e-05
$\|\lambda - \lambda_N\|$	4.0000e-01	4.0000e-01	1.6856e-02	1.7794e-04	1.1661e-04
Order	/	8.8324	13.6185	18.3290	22.9518

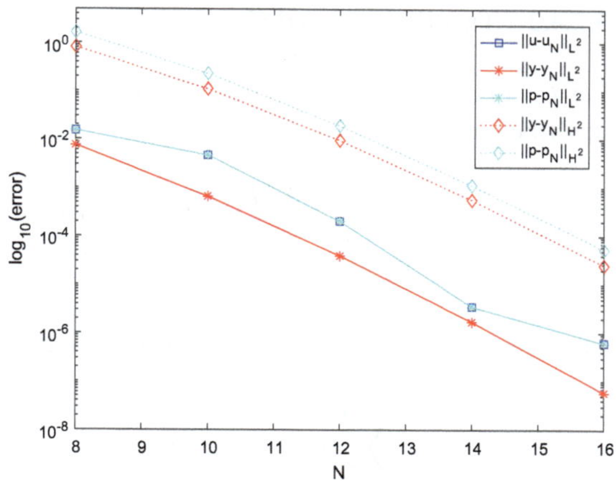

Fig. 3.1 Error estimates in Example 3.2.13

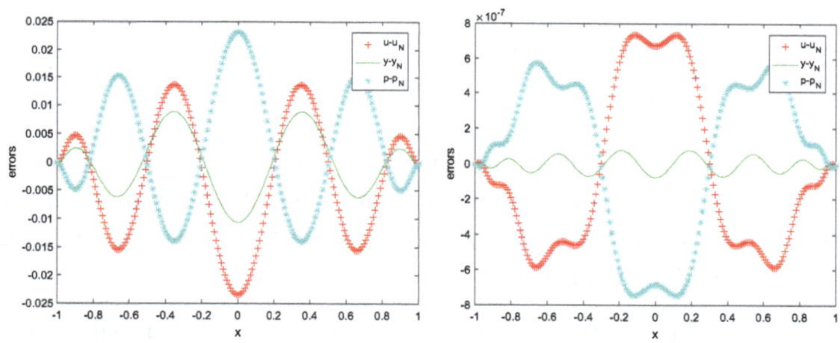

Fig. 3.2 The point-wise error estimates of $u - u_N$, $y - y_N$, and $p - p_N$ in Example 3.2.13. (**a**) $N = 8$. (**b**) $N = 16$

with subscript representing the corresponding information for the i-th step. Upon analyzing the convergence order data presented in Table 3.1, it becomes evident that spectral accuracy has been attained.

Moreover, Fig. 3.2 depicts the point-wise error values for u, y, and p with $N = 8$ and $N = 16$. It reveals that the magnitude of point-wise errors for u decreases significantly, from 10^{-7} to 10^{-2}. These numerical findings demonstrate that the Galerkin spectral method described in this section can yield a highly accurate approximation for the H^1-norm state-constrained optimal control problem governed by a fourth-order PDE.

Next, we delve into the examination of a two-dimensional instance of the H^1-norm state-constrained optimal control problem, which is governed by a fourth-order partial differential equation.

3.2.4.2 Two-Dimensional Case

Let $L_n(x)$ denote the n-th degree Legendre polynomial, and let $\phi_i(x)$ be defined as described in Sect. 3.2.4.1. Define

$$U_N = \text{span}\left\{L_i(x_1)L_j(x_2)\right\}_{i,j=0}^{N},$$

$$Y_N = \text{span}\left\{\phi_i(x_1)\phi_j(x_2)\right\}_{i,j=0}^{N-4}, \quad K_N = K \cap Y_N.$$

In this manner, y_N, p_N, and u_N are, respectively, expressed as follows:

$$y_N = \sum_{i,j=0}^{N-4} y_{ij}\phi_i(x_1)\phi_j(x_2), \quad p_N = \sum_{i,j=0}^{N-4} p_{ij}\phi_i(x_1)\phi_j(x_2),$$

$$u_N = \sum_{i,j=0}^{N} u_{ij}L_i(x_1)L_j(x_2).$$

As we did previously, substitute the three terms mentioned above into Eq. (3.2.12)(ã) to obtain a system of linear matrix equations. Employing a similar technique, we can solve Eq. (3.2.12)($\tilde{\text{b}}$). Subsequently, a specific example in two dimensions is provided as follows.

Example 3.2.14 The optimal control problem is considered within the square domain $\Omega = (-1, 1)^2$. Setting $\alpha = 1$, the exact solution is provided as follows:

$$\begin{cases} y = \dfrac{4}{\sqrt{24\pi^2+9}}\sin^2(\pi x_1)\sin^2(\pi x_2), \\ u = \dfrac{2}{\sqrt{24\pi^2+9}}\sin^2(\pi x_1)\sin^2(\pi x_2), \\ \lambda = 0.4, \\ p = -\alpha u = -\dfrac{2}{\sqrt{24\pi^2+9}}\sin^2(\pi x_1)\sin^2(\pi x_2), \end{cases}$$

The data in the optimal control problem are given by

$$f = \Delta^2 y - u, \quad y_d = y + \lambda(y - \Delta y) - \Delta^2 p.$$

We present the error results for u, y, and p in Table 3.2, and these results are visually represented in Fig. 3.3. Additionally, the corresponding convergence order data are provided in Table 3.2. These numerical results demonstrate spectral accuracy for the two-dimensional fourth-order optimal control problem with an H^1-norm state constraint. Furthermore, the spectral solutions align with the theoretical results obtained.

Table 3.2 Error results between the numerical solutions and exact solutions in Example 3.2.14

N	8	10	12	14	16
$\|u - u_N\|_{L^2(\Omega)}$	3.8781e-03	2.0618e-03	3.9953e-04	1.3221e-06	6.6432e-08
$\|y - y_N\|_{L^2(\Omega)}$	7.3043e-03	6.4197e-04	3.9116e-05	1.8401e-06	7.2719e-08
$\|p - p_N\|_{L^2(\Omega)}$	3.8781e-03	2.0618e-03	3.9953e-04	1.3221e-06	6.6432e-08
$\|y - y_N\|_{H^2(\Omega)}$	9.0999e-01	1.1954e-01	1.0143e-02	6.0429e-04	2.6784e-05
$\|p - p_N\|_{H^2(\Omega)}$	4.5853e-01	8.1932e-02	1.2034e-02	3.0346e-04	1.3575e-05
$\|\lambda - \lambda_N\|$	4.0000e-01	4.0000e-01	7.8035e-02	2.0234e-04	9.7857e-06
Order	/	8.5528	12.0606	20.8383	23.3126

Fig. 3.3 Error estimates in Example 3.2.14

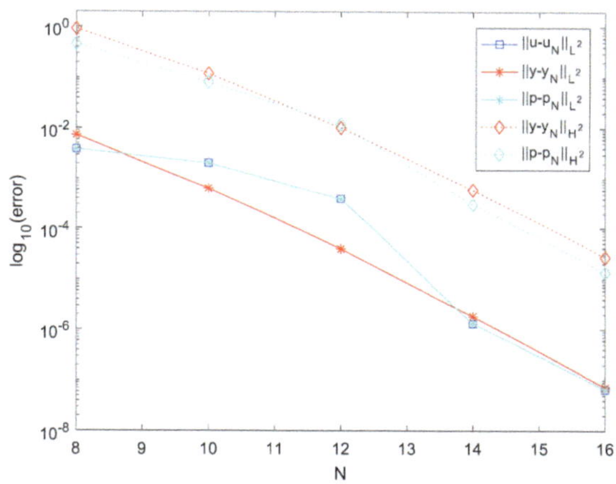

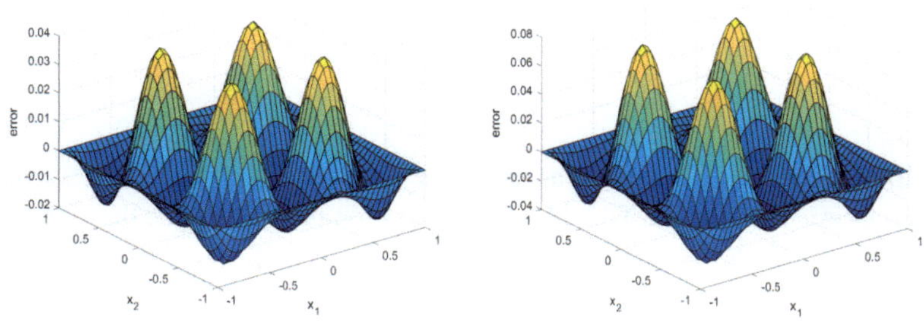

(a) (b)

Fig. 3.4 The point-wise error estimates when $N = 8$ in Example 3.2.14. **(a)** $u - u_N$. **(b)** $y - y_N$

Furthermore, the point-wise errors for u and y in the cases of $N = 8$ and $N = 16$ are plotted in Figs. 3.4 and 3.5, respectively. In both cases, the magnitude of the errors decreases significantly, reaching down to 10^{-7} from an initial level of 10^{-2}. Therefore, by increasing the value of N from 8 to 16, we can achieve a more precise spectral approximation of the exact solution.

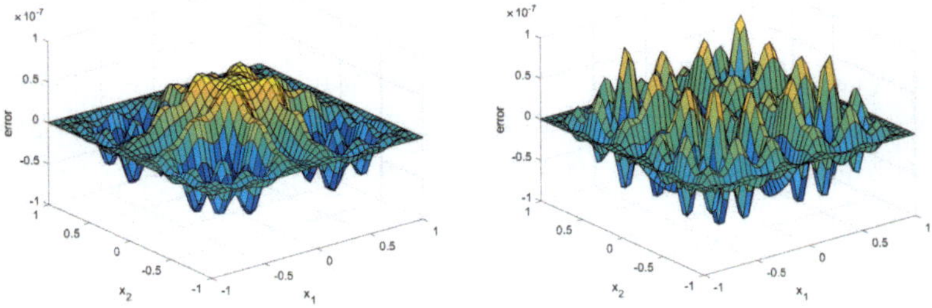

(a) (b)

Fig. 3.5 The point-wise error estimates when $N = 16$ in Example 3.2.14. (**a**) $u - u_N$. (**b**) $y - y_N$

3.3 Flow Optimal Control Problem with Control Constraint

3.3.1 Introduction

Let $\Omega \subset \mathbb{R}^2$ be a rectangular domain. We consider the following optimal control problem for the Stokes equation with an L^2-norm constraint on the control variable: [229].

$$\min_{\boldsymbol{u}\in \boldsymbol{K}} J(\boldsymbol{y}, \boldsymbol{u}) = \frac{1}{2}\left\|\boldsymbol{y} - \boldsymbol{y}_d\right\|_{0,\Omega}^2 + \frac{\alpha}{2}\left\|\boldsymbol{u}\right\|_{0,\Omega}^2, \tag{3.3.1}$$

$$s.t. \begin{cases} -\Delta \boldsymbol{y} + \nabla r = \boldsymbol{f} + \boldsymbol{u}, & \text{in } \Omega, \\ \nabla \cdot \boldsymbol{y} = 0, & \text{in } \Omega, \\ \boldsymbol{y} = 0, & \text{on } \partial\Omega, \end{cases} \tag{3.3.2}$$

where the control $\boldsymbol{u} = (u_1, u_2)$, the state $\boldsymbol{y} = (y_1, y_2)$ denotes the flow velocity, r is the pressure, $\alpha > 0$ is the Tikhonov regularization parameter, the desired velocity field $\boldsymbol{y}_d = (y_{d1}, y_{d2}) \in L^2(\Omega)^2$, and the forcing term $\boldsymbol{f} = (f_1, f_2) \in L^2(\Omega)^2$. Denote the norm of the space $L^2(\Omega)$ or $L^2(\Omega)^2$ by $\|\cdot\|_{0,\Omega}$, and

$$\|\boldsymbol{w}\|_{0,\Omega} = \left[\int_\Omega \left(|w_1|^2 + |w_2|^2\right)\,\mathrm{d}x\right]^{\frac{1}{2}},$$

for $\boldsymbol{w} = (w_1, w_2)$. The control constraint set

$$\boldsymbol{K} = \left\{\boldsymbol{w} \in L^2(\Omega)^2 \;\middle|\; \|\boldsymbol{w}\|_{0,\Omega} \le \gamma\right\},$$

where $\gamma > 0$ is a given constant. Let Δ and ∇, respectively, denote the Laplace operator and the gradient operator with respect to the variable $\boldsymbol{x} = (x_1, x_2)$.

Indeed, the analysis of the validity of numerical methods for optimal control problems has been a long-standing challenge, as documented in [79, 87, 165]. Surrounding this topic, numerous compelling papers have reported impressive results, including those employing the finite difference method [223], the finite element method [46, 47, 64, 160, 175], and spectral methods [52, 53]. Among these approaches, the finite element approximation stands out as the most widely used, while the finite difference method lacks sufficient universal validity.

Furthermore, several novel algorithms have emerged by integrating solution techniques based on these three primary approaches. For instance, a new finite element approximation technique is introduced in [159], which circumvents the first-order optimality conditions. Kohls et al. [137] consider finite element discretizations with both discretized and non-discretized controls, where a fundamental error equivalence significantly simplifies a posteriori error analysis for optimal control problems. The paper [115] presents a novel discretization concept for optimal control problems with control constraints. Rather than discretizing the space of admissible controls, this approach implicitly utilizes the first-order optimality conditions and the discretization of the state and costate equations for the discretization of the control. Remarkably, this new concept is numerically feasible with only a minor increase in program complexity.

Additionally, several other effective algorithms are available, including the augmented Lagrangian method [22], a semi-smooth Newton method [65], and many more.

There have been several interesting developments regarding the spectral method for constrained optimal control problems. Chen et al. have made significant contributions to the spectral approximation of second-order elliptic optimal control problems with control or state constraints. Their work includes the analysis of a priori and a posteriori error estimates, as detailed in [52] and related references.

Zhou et al. [251, 253] explore optimal control problems governed by the first biharmonic equation, incorporating an integral constraint on the state. They achieve a priori error estimates exhibiting high-order spectral accuracy.

For parabolic control problems featuring integral control constraints, [122] examines a priori error estimates. Importantly, all the aforementioned papers present numerical examples that consistently validate their theoretical analysis results.

When it comes to the numerical approximation of optimal control for the Stokes equation, numerous works have been dedicated to this area, with the majority of them emphasizing the finite element method. Niu et al. have established a priori error estimates and introduced equivalent residual-type a posteriori error estimators for the finite element approximation of the Stokes optimal control problem with an L^2-norm state constraint, as reported in [176, 177].

Liu and Yan, on the other hand, explore a posteriori error estimation and demonstrate global superconvergence for the Stokes optimal control problem in their studies [156, 157].

Furthermore, it's worth noting that in certain algorithms, solving the discrete optimality system in saddle point form can often lead to an ill-conditioned system. To address this computational challenge, Rees and Wathen have devised two highly effective preconditioners and successfully implemented them for the Stokes optimal control problem, as documented in [189]. These preconditioners significantly enhance the numerical stability and efficiency of the solution process.

Despite significant advancements in numerical methods for optimal control problems, the spectral approximation of flow optimal control problems remains relatively unexplored. The development and design of effective approximation techniques for these problems continue to pose challenges.

In the context of the Stokes equations with an H^1-norm state constraint, [53] employs the spectral method approximation and derives both a priori and a posteriori error estimates. This work represents an important contribution to the understanding of spectral approximation techniques for such problems.

Furthermore, [56] discusses the Legendre-Galerkin spectral approximation for distributed optimal control problems governed by the Stokes equations. They first derive a priori error estimates in both H^1 and L^2-norms for the approximation of unconstrained control problems. Subsequently, they obtain both a priori and a posteriori error estimates for control problems with constraints of an integral type, leveraging the higher regularity of the optimal control.

However, it's worth noting that, to the best of our knowledge, there has not been a similar discussion on the spectral approximation of optimal control problems governed by the Stokes equations with an H^2-norm control constraint. This area remains an open and challenging topic for future investigation.

In this section, we undertake a comprehensive study of the Galerkin spectral approximation for an optimal control problem governed by the Stokes equation, specifically incorporating an L^2-norm constraint on the control variable. We delve into the a priori error estimate, providing detailed insights into the theoretical bounds of the approximation error. Furthermore, we construct an a posteriori error estimator that proves instrumental in determining an adaptive strategy within the spectral-element method.

It's worth highlighting that our a posteriori error estimate not only encompasses the H^1-norm error for state and costate variables but also extends to the L^2-norm error for control, state, and costate. This comprehensive approach ensures a thorough evaluation of the approximation's accuracy.

To validate the effectiveness and performance of the spectral method, we present a numerical example that clearly demonstrates its validity. This example offers tangible evidence of the spectral approximation's high accuracy and the efficacy of the error estimators in practical applications.

Moreover, we extend our investigation to the optimal control problem governed by the Navier-Stokes equation, which differs from the Stokes equation solely by the addition of a nonlinear term $(\boldsymbol{y} \cdot \nabla)\, \boldsymbol{y}$. Based on our findings for the Stokes equation, we conjecture that similar conclusions regarding spectral approximation hold true for the Navier-Stokes case. This conjecture is subsequently supported by another numerical example, providing further confirmation of our approach's versatility and applicability.

Through this investigation, we establish a solid foundation for future research on the spectral approximation of optimal control for the Navier-Stokes equation.

Remark 3.3.12, presented later, offers additional details on this intriguing prospect, inviting further exploration in this exciting field.

Throughout this section, for any integers $m \geq 0$, $q \geq 0$, we adopt the standard notations $W^{m,q}(\Omega)$ for the Sobolev spaces on domain Ω with norm $\|\cdot\|_{m,q,\Omega}$. Denote

$$W_0^{m,q}(\Omega) = \left\{w \in W^{m,q}(\Omega) \middle| \, w\big|_{\partial\Omega} = 0\right\}.$$

When $q = 2$, we denote $W^{m,q}(\Omega)$ $\left(W_0^{m,q}(\Omega)\right)$ as $H^m(\Omega)$ $\left(H_0^m(\Omega)\right)$ and the corresponding norm as $\|\cdot\|_{m,\Omega}$. Note that $H^m(\Omega)$ is the Lebesgue space $L^2(\Omega)$ with $m = 0$. Moreover, we define the space $H^m(\Omega)^2$ equipped with the corresponding norm

$$\|\boldsymbol{v}\|_{m,\Omega} = \left(\|v_1\|_{m,\Omega}^2 + \|v_2\|_{m,\Omega}^2\right)^{\frac{1}{2}},$$

for $\boldsymbol{v} = (v_1, v_2) \in H^m(\Omega)^2$. Furthermore, for convenience, we assume that $C > 0$ denotes a general constant that is independent of any functions and N (the order of spectral approximation). It is important to note that while its value may vary depending on the context, it remains constant within each specific instance or calculation. This notation allows us to simplify expressions and facilitate the analysis of our spectral approximation methods.

3.3.2 *Galerkin Spectral Approximation and Optimality Conditions*

In this subsection, we focus on the control problem given by Eqs. (3.3.1) and (3.3.2). Firstly, we explore the Galerkin spectral approximation of this problem. Subsequently, we derive the optimality conditions for both the original system and its spectral discrete counterpart.

Denote the velocity space $\boldsymbol{Y} = H_0^1(\Omega)^2$, the control space $\boldsymbol{U} = L^2(\Omega)^2$, and the pressure space

$$M = L_0^2(\Omega) := \left\{q \in L^2(\Omega) \middle| \int_\Omega q \,\mathrm{d}x = 0\right\}.$$

3.3.2.1 Galerkin Spectral Approximation

Firstly, we give a weak formula for the state equation (3.3.2). Let

$$a(\boldsymbol{y}, \boldsymbol{w}) = \int_\Omega \nabla \boldsymbol{y} \cdot \nabla \boldsymbol{w} \,\mathrm{d}x = \sum_{i,j=1}^{2} \int_\Omega \frac{\partial y_i}{\partial x_j} \frac{\partial w_i}{\partial x_j} \,\mathrm{d}x, \quad \boldsymbol{y}, \boldsymbol{w} \in \boldsymbol{Y},$$

$$b(\boldsymbol{w}, r) = \int_\Omega r \nabla \cdot \boldsymbol{w} \, \mathrm{d}x, \quad \boldsymbol{w} \in \boldsymbol{Y}, \ r \in M,$$

$$(\boldsymbol{u}, \boldsymbol{v}) = \int_\Omega \boldsymbol{u} \cdot \boldsymbol{v} \, \mathrm{d}x, \quad \boldsymbol{u}, \boldsymbol{v} \in \boldsymbol{U},$$

where $(\cdot,\cdot)$ is the inner product in $L^2(\Omega)$ or $L^2(\Omega)^2$.

Next, we introduce some important inequalities that will be frequently used in the derivation of error estimates. There are two constants $\sigma > 0$, $\delta > 0$, such that

$$|a(\boldsymbol{y}, \boldsymbol{w})| \le \sigma \|\boldsymbol{y}\|_{\boldsymbol{Y}} \|\boldsymbol{w}\|_{\boldsymbol{Y}}, \quad \forall\, \boldsymbol{y}, \boldsymbol{w} \in \boldsymbol{Y}, \tag{3.3.3}$$

$$|b(\boldsymbol{w}, r)| \le \delta \|\boldsymbol{w}\|_{\boldsymbol{Y}} \|r\|_M, \quad \forall\, \boldsymbol{w} \in \boldsymbol{Y}, \ r \in M. \tag{3.3.4}$$

It is observed that $\|\cdot\|_{\boldsymbol{Y}} := \|\cdot\|_{1,\Omega}$ is the norm of the space $\boldsymbol{Y}$, and $\|\cdot\|_M := \|\cdot\|_{0,\Omega}$ is that of the space M.

Moreover, by the Poincaré inequality (Theorem 1.1.10) [2], there exists a constant $\theta > 0$ such that

$$a(\boldsymbol{y}, \boldsymbol{y}) \ge \theta \|\boldsymbol{y}\|_{\boldsymbol{Y}}^2, \quad \forall\, \boldsymbol{y} \in \boldsymbol{Y}. \tag{3.3.5}$$

By [88], there exists a constant $\beta > 0$ such that

$$\sup_{\boldsymbol{w} \in \boldsymbol{Y}} \frac{b(\boldsymbol{w}, r)}{\|\boldsymbol{w}\|_{\boldsymbol{Y}}} \ge \beta \|r\|_M, \quad \forall\, r \in M. \tag{3.3.6}$$

Then, the standard weak formulation for the state equation (3.3.2) can be stated as follows.

Find $(\boldsymbol{y}(\boldsymbol{u}), r(\boldsymbol{u})) \in \boldsymbol{Y} \times M$ such that

$$\begin{aligned} a(\boldsymbol{y}(\boldsymbol{u}), \boldsymbol{w}) - b(\boldsymbol{w}, r(\boldsymbol{u})) &= (\boldsymbol{f} + \boldsymbol{u}, \boldsymbol{w}), \quad \forall\, \boldsymbol{w} \in \boldsymbol{Y}, \\ b(\boldsymbol{y}(\boldsymbol{u}), \phi) &= 0, \quad \forall\, \phi \in M. \end{aligned} \tag{3.3.7}$$

By (3.3.5), (3.3.6), and the Babuška-Brezzi theorem [88], we know that the state equation (3.3.7) is well-posed. Furthermore, the weak formulation of the control problem (3.3.1)–(3.3.2) can be presented as the problem $(\mathscr{P})$ below.

Find $(\boldsymbol{y}, r, \boldsymbol{u}) \in \boldsymbol{Y} \times M \times \boldsymbol{U}$ such that

$$\begin{gathered} \min_{\boldsymbol{u} \in \boldsymbol{K}} J(\boldsymbol{y}, \boldsymbol{u}) = \frac{1}{2} \|\boldsymbol{y} - \boldsymbol{y}_d\|_{0,\Omega}^2 + \frac{\alpha}{2} \|\boldsymbol{u}\|_{0,\Omega}^2, \\ a(\boldsymbol{y}, \boldsymbol{w}) - b(\boldsymbol{w}, r) = (\boldsymbol{f} + \boldsymbol{u}, \boldsymbol{w}), \quad \forall\, \boldsymbol{w} \in \boldsymbol{Y}, \\ b(\boldsymbol{y}, \phi) = 0, \quad \forall\, \phi \in M. \end{gathered}$$

Furthermore, it is evident that the control constraint set $\boldsymbol{K}$ constitutes a nonempty, closed, and convex subset within $\boldsymbol{U}$. Referring to [153], we can affirm that the control problem $(\mathscr{P})$ possesses a unique solution.

Now, we proceed to investigate the Galerkin spectral approximation of the control problem $(\mathscr{P})$. Let Q_N represent the space comprising all algebraic polynomials whose degree is less than or equal to the integer $N > 0$ with respect to each individual variable x_i, $i = 1, 2$. The first step involves spectral discretization. We define

$$Q_N^0 = \{w_N \in Q_N \mid w_N|_{\partial\Omega} = 0\}.$$

And let the finite-dimensional spaces

$$\boldsymbol{Y}_N = \left(Q_N^0\right)^2, \quad M_{N-2} = Q_{N-2} \cap L_0^2(\Omega), \quad \boldsymbol{U}_N = \left(Q_N \cap L^2(\Omega)\right)^2,$$

denote the corresponding approximation spaces of $\boldsymbol{Y}$, M, and $\boldsymbol{U}$, respectively (see Chapter 7 of [206]).

Then, rewrite the $Q_N - Q_{N-2}$ spectral approximation of the control problem $(\mathscr{P})$ as the following problem $(\mathscr{P}_N)$.

Find $\left(\boldsymbol{y}_N, r_N, \boldsymbol{u}_N\right) \in \boldsymbol{Y}_N \times M_{N-2} \times \boldsymbol{U}_N$ such that

$$\begin{aligned}
\min_{\boldsymbol{u}_N \in \boldsymbol{K}_N} J\left(\boldsymbol{y}_N, \boldsymbol{u}_N\right) &= \frac{1}{2}\left\|\boldsymbol{y}_N - \boldsymbol{y}_d\right\|_{0,\Omega}^2 + \frac{\alpha}{2}\left\|\boldsymbol{u}_N\right\|_{0,\Omega}^2, \\
a\left(\boldsymbol{y}_N, \boldsymbol{w}_N\right) - b\left(\boldsymbol{w}_N, r_N\right) &= \left(\boldsymbol{f} + \boldsymbol{u}_N, \boldsymbol{w}_N\right), \quad \forall\, \boldsymbol{w}_N \in \boldsymbol{Y}_N, \\
b\left(\boldsymbol{y}_N, \phi_N\right) &= 0, \quad \forall\, \phi_N \in M_{N-2},
\end{aligned}$$

where $\boldsymbol{K}_N = \boldsymbol{K} \cap \boldsymbol{U}_N$.

By Chapter 11 of [40] and Chapter 7 of [206], we know that there exists a constant

$$\beta_N = \mathcal{O}\left(N^{-\frac{1}{2}}\right),$$

such that

$$\sup_{\boldsymbol{w}_N \in \boldsymbol{Y}_N} \frac{b\left(\boldsymbol{w}_N, r_N\right)}{\left\|\boldsymbol{w}_N\right\|_{\boldsymbol{Y}}} \geq \beta_N \left\|r_N\right\|_M, \quad \forall\, r_N \in M_{N-2}. \tag{3.3.8}$$

Furthermore, according to the Babuška-Brezzi theorem, the state equation of the spectral approximation problem $(\mathscr{P}_N)$ is well-posed, as stated in [88]. Consequently, it follows from [14, 58] that the problem $(\mathscr{P}_N)$ possesses a unique solution.

3.3.2.2 Optimality Conditions

Next, we turn our attention to the optimality conditions of the control $(\mathscr{P})$.

Theorem 3.3.1 *The triplet* $(\boldsymbol{y}, r, \boldsymbol{u}) \in \boldsymbol{Y} \times M \times \boldsymbol{U}$ *is the solution to the control problem* $(\mathscr{P})$*, if and only if there exists* $(\boldsymbol{y}^*, r^*, \lambda) \in \boldsymbol{Y} \times M \times [0, +\infty)$ *such that* $(\boldsymbol{y}, r, \boldsymbol{u}, \boldsymbol{y}^*, r^*, \lambda)$ *satisfies the optimality conditions* $(\mathscr{Q})$ *as follows:*

$$\begin{aligned}
&\text{(a) } a(\boldsymbol{y}, \boldsymbol{w}) - b(\boldsymbol{w}, r) = (\boldsymbol{f} + \boldsymbol{u}, \boldsymbol{w}), \quad \forall\, \boldsymbol{w} \in \boldsymbol{Y}, \\
&\text{(b) } b(\boldsymbol{y}, \phi) = 0, \quad \forall\, \phi \in M, \\
&\text{(c) } a\left(\boldsymbol{q}, \boldsymbol{y}^*\right) + b\left(\boldsymbol{q}, r^*\right) = \left(\boldsymbol{y} - \boldsymbol{y}_d, \boldsymbol{q}\right), \quad \forall\, \boldsymbol{q} \in \boldsymbol{Y}, \\
&\text{(d) } b\left(\boldsymbol{y}^*, \psi\right) = 0, \quad \forall\, \psi \in M, \\
&\text{(e) } \lambda\,(\boldsymbol{u}, \boldsymbol{v} - \boldsymbol{u}) \le 0, \quad \forall\, \boldsymbol{v} \in \boldsymbol{K}, \\
&\text{(f) } \boldsymbol{y}^* + (\lambda + \alpha)\,\boldsymbol{u} = 0,
\end{aligned} \tag{3.3.9}$$

where

$$\lambda = \begin{cases} \text{constant} \ge 0, & \|\boldsymbol{u}\|_{0,\Omega} = \gamma, \\ 0, & \|\boldsymbol{u}\|_{0,\Omega} < \gamma. \end{cases}$$

Moreover, the solution to the optimality conditions $(\mathscr{Q})$ *is unique.*

Proof To obtain the optimality conditions $(\mathscr{Q})$, we introduce the following Lagrange functional $\mathcal{L}(\boldsymbol{u}, \lambda) : \boldsymbol{U} \times \mathbb{R} \to \mathbb{R}$ associated with the problem (3.3.1)–(3.3.2) as

$$\mathcal{L}(\boldsymbol{u}, \lambda) = J(\boldsymbol{y}(\boldsymbol{u}), \boldsymbol{u}) + \lambda F(\boldsymbol{u}), \quad F(\boldsymbol{u}) = \frac{1}{2}\|\boldsymbol{u}\|_{0,\Omega}^2 - \frac{1}{2}\gamma^2.$$

It can be verified that $J(\boldsymbol{y}(\boldsymbol{u}), \boldsymbol{u})$ and $F(\boldsymbol{u})$ are the Fréchet differentiable with respect to $\boldsymbol{u}$, and the Slater type condition $F(\mathbf{0}) \le 0$ holds. Referring to [14, 58], if u represents the solution to the control problem $(\mathscr{P})$, there must exist a corresponding $\lambda \in \mathbb{R}$ such that

$$\begin{aligned}
&\text{(a) } \lambda \ge 0, \quad \lambda F(\boldsymbol{u}) = 0, \\
&\text{(b) } \frac{\partial \mathcal{L}}{\partial \boldsymbol{u}}(\boldsymbol{u}, \lambda) = 0.
\end{aligned} \tag{3.3.10}$$

By (3.3.10)(a), when $\|\boldsymbol{u}\|_{0,\Omega} < \gamma$, we obtain $\lambda = 0$ and $\lambda(\boldsymbol{u}, \boldsymbol{w} - \boldsymbol{u}) = 0$. While $\|\boldsymbol{u}\|_{0,\Omega} = \gamma$, we have $\lambda \geq 0$, and

$$\lambda(\boldsymbol{u}, \boldsymbol{v} - \boldsymbol{u}) = \lambda\left((\boldsymbol{u}, \boldsymbol{v}) - \gamma^2\right) \leq \lambda\left(\|\boldsymbol{u}\|_{0,\Omega}\|\boldsymbol{v}\|_{0,\Omega} - \gamma^2\right) \leq 0,$$

for all $\boldsymbol{v} \in \boldsymbol{K}$. This is (3.3.9)(e).

Differentiate the state equation (3.3.7) at $\boldsymbol{u}$ in the direction $\boldsymbol{v}$ to get

$$\begin{aligned} a\left(\boldsymbol{y}'(\boldsymbol{u})\boldsymbol{v}, \boldsymbol{w}\right) - b\left(\boldsymbol{w}, r'(\boldsymbol{u})\boldsymbol{v}\right) &= (\boldsymbol{v}, \boldsymbol{w}), \quad \forall\, \boldsymbol{w} \in \boldsymbol{Y}, \\ b\left(\boldsymbol{y}'(\boldsymbol{u})\boldsymbol{v}, \phi\right) &= 0, \quad \forall\, \phi \in M. \end{aligned} \tag{3.3.11}$$

By (3.3.10)(b), it has, for all $\boldsymbol{v} \in \boldsymbol{U}$,

$$\begin{aligned} 0 = \frac{\partial \mathcal{L}}{\partial \boldsymbol{u}}(\boldsymbol{u}, \lambda)\boldsymbol{v} &= \left(J'(\boldsymbol{y}(\boldsymbol{u}), \boldsymbol{u}) + \lambda F'(\boldsymbol{u})\right)\boldsymbol{v} \\ &= \left(\boldsymbol{y} - \boldsymbol{y}_d, \boldsymbol{y}'(\boldsymbol{u})\boldsymbol{v}\right) + \alpha(\boldsymbol{u}, \boldsymbol{v}) + \lambda(\boldsymbol{u}, \boldsymbol{v}) \\ &= \left(\boldsymbol{y} - \boldsymbol{y}_d, \boldsymbol{y}'(\boldsymbol{u})\boldsymbol{v}\right) + (\alpha + \lambda)(\boldsymbol{u}, \boldsymbol{v}). \end{aligned} \tag{3.3.12}$$

Given that $\boldsymbol{y} - \boldsymbol{y}_d \in L^2(\Omega)^2$, we introduce the costate equation as follows:

$$\begin{aligned} a\left(\boldsymbol{q}, \boldsymbol{y}^*\right) + b\left(\boldsymbol{q}, r^*\right) &= \left(\boldsymbol{y} - \boldsymbol{y}_d, \boldsymbol{q}\right), \quad \forall\, \boldsymbol{q} \in \boldsymbol{Y}, \\ b\left(\boldsymbol{y}^*, \psi\right) &= 0, \quad \forall\, \psi \in M. \end{aligned} \tag{3.3.13}$$

Set $\boldsymbol{w} = \boldsymbol{y}^*$ in (3.3.11) and $\boldsymbol{q} = \boldsymbol{y}'(\boldsymbol{u})\boldsymbol{v}$ in (3.3.13). Therefore,

$$\left(\boldsymbol{y} - \boldsymbol{y}_d, \boldsymbol{y}'(\boldsymbol{u})\boldsymbol{v}\right) = \left(\boldsymbol{v}, \boldsymbol{y}^*\right), \quad \forall\, \boldsymbol{v} \in \boldsymbol{U}. \tag{3.3.14}$$

By (3.3.12) and (3.3.14),

$$\left(\boldsymbol{y}^* + (\alpha + \lambda)\boldsymbol{u}, \boldsymbol{v}\right) = 0, \quad \forall\, \boldsymbol{v} \in \boldsymbol{U}.$$

Thus, Eq. (3.3.9)(f) holds.

Next, we proceed to prove uniqueness. Suppose there exist two solutions $(\boldsymbol{y}_1, r_1, \boldsymbol{u}_1, \boldsymbol{y}_1^*, r_1^*, \lambda_1)$ and $\left(\boldsymbol{y}_2, r_2, \boldsymbol{u}_2, \boldsymbol{y}_2^*, r_2^*, \lambda_2\right)$ satisfying (3.3.9). Then,

$$\begin{aligned} a\left(\boldsymbol{y}_1 - \boldsymbol{y}_2, \boldsymbol{w}\right) - b\left(\boldsymbol{w}, r_1 - r_2\right) &= \left(\boldsymbol{u}_1 - \boldsymbol{u}_2, \boldsymbol{w}\right), \quad \forall\, \boldsymbol{w} \in \boldsymbol{Y}, \\ b\left(\boldsymbol{y}_1 - \boldsymbol{y}_2, \phi\right) &= 0, \quad \forall\, \phi \in M, \end{aligned} \tag{3.3.15}$$

and

$$
\begin{aligned}
a\left(\boldsymbol{q}, \boldsymbol{y}_1^* - \boldsymbol{y}_2^*\right) + b\left(\boldsymbol{q}, r_1^* - r_2^*\right) &= \left(\boldsymbol{y}_1 - \boldsymbol{y}_2, \boldsymbol{q}\right), \quad \forall\, \boldsymbol{q} \in \boldsymbol{Y}, \\
b\left(\boldsymbol{y}_1^* - \boldsymbol{y}_2^*, \psi\right) &= 0, \quad \forall\, \psi \in M.
\end{aligned} \tag{3.3.16}
$$

Taking $\boldsymbol{w} = \boldsymbol{y}_1^* - \boldsymbol{y}_2^*$ in (3.3.15) and $\boldsymbol{q} = \boldsymbol{y}_1 - \boldsymbol{y}_2$ in (3.3.16), respectively, we obtain

$$
\left(\boldsymbol{u}_1 - \boldsymbol{u}_2, \boldsymbol{y}_1^* - \boldsymbol{y}_2^*\right) = \left(\boldsymbol{y}_1 - \boldsymbol{y}_2, \boldsymbol{y}_1 - \boldsymbol{y}_2\right). \tag{3.3.17}
$$

By (3.3.9)(f), there is

$$
\begin{aligned}
\left(\boldsymbol{u}_1 - \boldsymbol{u}_2, \boldsymbol{y}_1^* - \boldsymbol{y}_2^*\right) &= -\left(\boldsymbol{u}_1 - \boldsymbol{u}_2, (\alpha + \lambda_1)\, \boldsymbol{u}_1 - (\alpha + \lambda_2)\, \boldsymbol{u}_2\right) \\
&= -\alpha\left(\boldsymbol{u}_1 - \boldsymbol{u}_2, \boldsymbol{u}_1 - \boldsymbol{u}_2\right) - \left(\boldsymbol{u}_1 - \boldsymbol{u}_2, \lambda_1 \boldsymbol{u}_1\right) + \left(\boldsymbol{u}_1 - \boldsymbol{u}_2, \lambda_2 \boldsymbol{u}_2\right).
\end{aligned}
$$

Therefore, referring to (3.3.17), we derive the following conclusion:

$$
\begin{aligned}
&\alpha\left(\boldsymbol{u}_1 - \boldsymbol{u}_2, \boldsymbol{u}_1 - \boldsymbol{u}_2\right) + \left(\boldsymbol{y}_1 - \boldsymbol{y}_2, \boldsymbol{y}_1 - \boldsymbol{y}_2\right) \\
&= \left(\boldsymbol{u}_2 - \boldsymbol{u}_1, \lambda_1 \boldsymbol{u}_1\right) + \left(\boldsymbol{u}_1 - \boldsymbol{u}_2, \lambda_2 \boldsymbol{u}_2\right).
\end{aligned} \tag{3.3.18}
$$

Then, according to (3.3.9)(e), we deduce that

$$
\left(\boldsymbol{u}_2 - \boldsymbol{u}_1, \lambda_1 \boldsymbol{u}_1\right) \le 0, \quad \left(\boldsymbol{u}_1 - \boldsymbol{u}_2, \lambda_2 \boldsymbol{u}_2\right) \le 0.
$$

So, referring to (3.3.18), we obtain the following conclusion:

$$
\alpha\left(\boldsymbol{u}_1 - \boldsymbol{u}_2, \boldsymbol{u}_1 - \boldsymbol{u}_2\right) + \left(\boldsymbol{y}_1 - \boldsymbol{y}_2, \boldsymbol{y}_1 - \boldsymbol{y}_2\right) \le 0,
$$

which implies that $\boldsymbol{y}_1 = \boldsymbol{y}_2$ and $\boldsymbol{u}_1 = \boldsymbol{u}_2$.

Moreover, by (3.3.15), we derive that $r_1 = r_2$. Additionally, due to (3.3.16), it can be proven that $\boldsymbol{y}_1^* = \boldsymbol{y}_2^*$ and $r_1^* = r_2^*$. Subsequently, $\lambda_1 = \lambda_2$ follows from (3.3.9)(f). This completes the proof of the theorem. □

Remark 3.3.2 When $\|\boldsymbol{u}\|_{0,\Omega} = \gamma$, it is easy to get

$$
\lambda = \gamma^{-1} \left\| \boldsymbol{y}^* + \alpha \boldsymbol{u} \right\|_{0,\Omega},
$$

by (3.3.9)(f).

By employing similar arguments, we deduce that the spectral approximation problem $(\mathscr{P}_N)$ possesses a unique solution denoted by $\left(\boldsymbol{y}_N, r_N, \boldsymbol{u}_N\right) \in \boldsymbol{Y}_N \times M_{N-2} \times \boldsymbol{U}_N$. Furthermore, $\left(\boldsymbol{y}_N, r_N, \boldsymbol{u}_N\right)$ constitutes the solution to the spectral approximation problem $(\mathscr{P}_N)$ if and only if there exists a corresponding triplet $(\boldsymbol{y}_N^*, r_N^*, \lambda_N) \in \boldsymbol{Y}_N \times M_{N-2} \times [0, +\infty)$ such that the aggregated vector $(\boldsymbol{y}_N, r_N, \boldsymbol{u}_N, \boldsymbol{y}_N^*, r_N^*, \lambda_N)$ satisfies the optimality conditions $(\mathscr{Q}_N)$ described below.

$$\begin{aligned}
&(a)\ a(\boldsymbol{y}_N, \boldsymbol{w}_N) - b(\boldsymbol{w}_N, r_N) = (\boldsymbol{f} + \boldsymbol{u}_N, \boldsymbol{w}_N), \quad \forall\, \boldsymbol{w}_N \in \boldsymbol{Y}_N,\\
&(b)\ b(\boldsymbol{y}_N, \phi_N) = 0, \quad \forall\, \phi_N \in M_{N-2},\\
&(c)\ a(\boldsymbol{q}_N, \boldsymbol{y}_N^*) + b(\boldsymbol{q}_N, r_N^*) = (\boldsymbol{y}_N - \boldsymbol{y}_d, \boldsymbol{q}_N), \quad \forall\, \boldsymbol{q}_N \in \boldsymbol{Y}_N,\\
&(d)\ b(\boldsymbol{y}_N^*, \psi_N) = 0, \quad \forall\, \psi_N \in M_{N-2},\\
&(e)\ \lambda_N(\boldsymbol{u}_N, \boldsymbol{v}_N - \boldsymbol{u}_N) \le 0, \quad \forall\, \boldsymbol{v}_N \in \boldsymbol{K}_N,\\
&(f)\ \boldsymbol{y}_N^* + (\lambda_N + \alpha)\,\boldsymbol{u}_N = 0,
\end{aligned} \tag{3.3.19}$$

where

$$\lambda_N = \begin{cases} \text{constant} \ge 0, & \|\boldsymbol{u}_N\|_{0,\Omega} = \gamma, \\ 0, & \|\boldsymbol{u}_N\|_{0,\Omega} < \gamma. \end{cases}$$

In addition, when $\|\boldsymbol{u}_N\|_{0,\Omega} = \gamma$, we find

$$\lambda_N = \gamma^{-1} \left\| \boldsymbol{y}_N^* + \alpha \boldsymbol{u}_N \right\|_{0,\Omega},$$

by (3.3.19)(f).

3.3.3 A Priori Error Estimates

The primary objective of this subsection is to derive a priori error estimates for the spectral approximation of the control problem $(\mathscr{P})$. Additionally, we aim to establish L^2-error estimations for the control and H^1-error estimations for both the state and costate.

To begin, let us introduce some useful results.

Let $\mathbb{P}_N : L^2(\Omega) \to Q_N$ be the L^2-orthogonal projection operator (see [40, Chapter 5]), which satisfies, for any $r \in L^2(\Omega)$,

$$(r - \mathbb{P}_N r, w_N) = 0, \quad \forall\, w_N \in Q_N. \tag{3.3.20}$$

Then,

$$\|r - \mathbb{P}_N r\|_{0,\Omega} \le C N^{-m} \|r\|_{m,\Omega},$$

provided that $r \in H^m(\Omega)$ with $m \geq 0$. Moreover, this result can be extended to vector functions.

Subsequently, define the projection operator

$$\mathbb{P}^0_{1,N} : H^1_0(\Omega)^2 \to \left(Q^0_N\right)^2,$$

(see [40, Chapter 9]), which satisfies, for any $\boldsymbol{v} \in H^1_0(\Omega)^2$, that

$$a\left(\boldsymbol{v} - \mathbb{P}^0_{1,N}\boldsymbol{v}, \boldsymbol{w}_N\right) = 0, \quad \forall\, \boldsymbol{w}_N \in \left(Q^0_N\right)^2.$$

Then we have

$$\left\|\boldsymbol{v} - \mathbb{P}^0_{1,N}\boldsymbol{v}\right\|_{k,\Omega} \leq CN^{k-m} \|\boldsymbol{v}\|_{m,\Omega}, \quad k = 0, 1, \tag{3.3.21}$$

provided $\boldsymbol{v} \in \left(H^m(\Omega) \cap H^1_0(\Omega)\right)^2$ with $m \geq 1$.

Next, we highlight a crucial fact concerning the spectral approximation of the flow problem.

Lemma 3.3.3 (see [40, 206]) *Let $\Omega \subset \mathbb{R}^2$ be a rectangular domain. And $(\boldsymbol{\epsilon}, \rho) \in \boldsymbol{Y} \times M$ is the solution to the following equations:*

$$\begin{aligned} a(\boldsymbol{\epsilon}, \boldsymbol{w}) \pm b(\boldsymbol{w}, \rho) &= (\Phi, \boldsymbol{w}), \quad \forall\, \boldsymbol{w} \in \boldsymbol{Y}, \\ b(\boldsymbol{\epsilon}, \phi) &= 0, \quad \forall\, \phi \in M. \end{aligned} \tag{3.3.22}$$

Then, it holds that

$$\|\boldsymbol{\epsilon}\|_{2,\Omega} + \|\rho\|_{1,\Omega} \leq C\, \|\Phi\|_{0,\Omega}.$$

Moreover, if

$$(\boldsymbol{\epsilon}, \rho) \in \left(H^1_0(\Omega) \cap H^m(\Omega)\right)^2 \times \left(L^2_0(\Omega) \cap H^{m-1}(\Omega)\right),$$

and $m \geq 1$, there are

$$\begin{aligned} \|\boldsymbol{\epsilon} - \boldsymbol{\epsilon}_N\|_{1,\Omega} &\leq CN^{1-m}\left(\|\boldsymbol{\epsilon}\|_{m,\Omega} + \|\rho\|_{m-1,\Omega}\right), \\ \|\rho - \rho_N\|_{0,\Omega} &\leq CN^{\frac{3}{2}-m}\left(\|\boldsymbol{\epsilon}\|_{m,\Omega} + \|\rho\|_{m-1,\Omega}\right), \\ \|\boldsymbol{\epsilon} - \boldsymbol{\epsilon}_N\|_{0,\Omega} &\leq CN^{-m}\left(\|\boldsymbol{\epsilon}\|_{m,\Omega} + \|\rho\|_{m-1,\Omega}\right), \end{aligned}$$

where $\boldsymbol{\epsilon}_N$ and ρ_N are the Q_N-Q_{N-2} approximations of $\boldsymbol{\epsilon}$ and ρ, respectively.

Subsequently, we introduce a highly useful auxiliary system as follows.

Find $\left(y_N(u), r_N(u), y_N^*(u), r_N^*(u)\right)$ such that

$$a\left(y_N(u), w_N\right) - b\left(w_N, r_N(u)\right) = \left(f + u, w_N\right), \quad \forall\, w_N \in Y_N, \tag{3.3.23}$$

$$b\left(y_N(u), \phi_N\right) = 0, \quad \forall\, \phi_N \in M_{N-2}, \tag{3.3.24}$$

$$a\left(q_N, y_N^*(u)\right) + b\left(q_N, r_N^*(u)\right) = \left(y - y_d, q_N\right), \quad \forall\, q_N \in Y_N, \tag{3.3.25}$$

$$b\left(y_N^*(u), \psi_N\right) = 0, \quad \forall\, \psi_N \in M_{N-2}, \tag{3.3.26}$$

in which $\left(y_N(u), r_N(u), y_N^*(u), r_N^*(u)\right)$ is the spectral approximation of (y, r, y^*, r^*). By the Babuška-Brezzi theorem [88], the auxiliary system (3.3.23)–(3.3.26) is well-posed for a given $(y, u) \in Y \times U$.

By Lemma 3.3.3 mentioned above, if

$$(y, r),\ (y^*, r^*) \in \left(H_0^1(\Omega) \cap H^m(\Omega)\right)^2 \times \left(L_0^2(\Omega) \cap H^{m-1}(\Omega)\right),$$

and $m \geq 1$, there are

$$\begin{aligned} N \left\| y - y_N(u) \right\|_{0,\Omega} + \left\| y - y_N(u) \right\|_{1,\Omega} + N^{-\frac{1}{2}} \left\| r - r_N(u) \right\|_{0,\Omega} \\ \leq C N^{1-m} \left(\|y\|_{m,\Omega} + \|r\|_{m-1,\Omega} \right), \end{aligned} \tag{3.3.27}$$

and

$$\begin{aligned} N \left\| y^* - y_N^*(u) \right\|_{0,\Omega} + \left\| y^* - y_N^*(u) \right\|_{1,\Omega} + N^{-\frac{1}{2}} \left\| r^* - r_N^*(u) \right\|_{0,\Omega} \\ \leq C N^{1-m} \left(\left\| y^* \right\|_{m,\Omega} + \left\| r^* \right\|_{m-1,\Omega} \right). \end{aligned} \tag{3.3.28}$$

Subsequently, we present some auxiliary results in Lemmas 3.3.4–3.3.5. These lemmas will facilitate the derivation of a priori error estimates.

Lemma 3.3.4 *Let* $(y, r, u, y^*, r^*, \lambda)$ *and* $\left(y_N, r_N, u_N, y_N^*, r_N^*, \lambda_N\right)$ *satisfy the optimality conditions* ($\mathscr{Q}$) *and* ($\mathscr{Q}_N$)*, respectively. We have the following estimates:*

$$\left\| y_N(u) - y_N \right\|_{1,\Omega} + N^{-\frac{1}{2}} \left\| r_N(u) - r_N \right\|_{0,\Omega} \leq C \left\| u - u_N \right\|_{0,\Omega}, \tag{3.3.29}$$

$$\left\| y_N^*(u) - y_N^* \right\|_{1,\Omega} + N^{-\frac{1}{2}} \left\| r_N^*(u) - r_N^* \right\|_{0,\Omega} \leq C \left\| y - y_N \right\|_{0,\Omega}. \tag{3.3.30}$$

Proof Subtracting (3.3.19)(a)–(b) by (3.3.23)–(3.3.24), respectively, we obtain

$$\begin{aligned} a\left(y_N(u) - y_N, w_N\right) - b\left(w_N, r_N(u) - r_N\right) = \left(u - u_N, w_N\right), \\ \forall\, w_N \in Y_N, \\ b\left(y_N(u) - y_N, \phi_N\right) = 0, \quad \forall\, \phi_N \in M_{N-2}. \end{aligned} \tag{3.3.31}$$

Let $\boldsymbol{w}_N = \boldsymbol{y}_N(\boldsymbol{u}) - \boldsymbol{y}_N$ and $\phi_N = r_N(\boldsymbol{u}) - r_N$ in (3.3.31). By (3.3.5), we have

$$\begin{aligned}
&\theta \left\| \boldsymbol{y}_N(\boldsymbol{u}) - \boldsymbol{y}_N \right\|_{1,\Omega}^2 \\
&\leq a\left(\boldsymbol{y}_N(\boldsymbol{u}) - \boldsymbol{y}_N, \boldsymbol{y}_N(\boldsymbol{u}) - \boldsymbol{y}_N\right) = \left(\boldsymbol{u} - \boldsymbol{u}_N, \boldsymbol{y}_N(\boldsymbol{u}) - \boldsymbol{y}_N\right) \\
&\leq \|\boldsymbol{u} - \boldsymbol{u}_N\|_{0,\Omega} \left\| \boldsymbol{y}_N(\boldsymbol{u}) - \boldsymbol{y}_N \right\|_{0,\Omega} \leq \|\boldsymbol{u} - \boldsymbol{u}_N\|_{0,\Omega} \left\| \boldsymbol{y}_N(\boldsymbol{u}) - \boldsymbol{y}_N \right\|_{1,\Omega},
\end{aligned}$$

which yields

$$\left\| \boldsymbol{y}_N(\boldsymbol{u}) - \boldsymbol{y}_N \right\|_{1,\Omega} \leq C \|\boldsymbol{u} - \boldsymbol{u}_N\|_{0,\Omega}. \tag{3.3.32}$$

Moreover, according to (3.3.3), (3.3.8), (3.3.31), and (3.3.32), we obtain

$$\begin{aligned}
\beta_N \|r_N(\boldsymbol{u}) - r_N\|_{0,\Omega} &\leq \sup_{\boldsymbol{w}_N \in \boldsymbol{Y}_N} \frac{b\left(\boldsymbol{w}_N, r_N(\boldsymbol{u}) - r_N\right)}{\|\boldsymbol{w}_N\|_{1,\Omega}} \\
&= \sup_{\boldsymbol{w}_N \in \boldsymbol{Y}_N} \frac{a\left(\boldsymbol{y}_N(\boldsymbol{u}) - \boldsymbol{y}_N, \boldsymbol{w}_N\right) - \left(\boldsymbol{u} - \boldsymbol{u}_N, \boldsymbol{w}_N\right)}{\|\boldsymbol{w}_N\|_{1,\Omega}} \\
&\leq \sigma \left\| \boldsymbol{y}_N(\boldsymbol{u}) - \boldsymbol{y}_N \right\|_{1,\Omega} + \|\boldsymbol{u} - \boldsymbol{u}_N\|_{0,\Omega} \\
&\leq C \|\boldsymbol{u} - \boldsymbol{u}_N\|_{0,\Omega}.
\end{aligned} \tag{3.3.33}$$

The estimate (3.3.29) then follows from (3.3.32) and (3.3.33).

Next, we estimate $\left\| \boldsymbol{y}_N^*(\boldsymbol{u}) - \boldsymbol{y}_N^* \right\|_{1,\Omega}$ and $\left\| r_N^*(\boldsymbol{u}) - r_N^* \right\|_{0,\Omega}$. Subtract (3.3.19)(c)–(d) by (3.3.25)–(3.3.26), respectively, to obtain

$$\begin{aligned}
a\left(\boldsymbol{q}_N, \boldsymbol{y}_N^*(\boldsymbol{u}) - \boldsymbol{y}_N^*\right) + b\left(\boldsymbol{q}_N, r_N^*(\boldsymbol{u}) - r_N^*\right) &= \left(\boldsymbol{y} - \boldsymbol{y}_N, \boldsymbol{q}_N\right), \\
&\quad \forall\, \boldsymbol{q}_N \in \boldsymbol{Y}_N, \\
b\left(\boldsymbol{y}_N^*(\boldsymbol{u}) - \boldsymbol{y}_N^*, \psi_N\right) = 0, \quad \forall\, \psi_N \in M_{N-2}.&
\end{aligned} \tag{3.3.34}$$

Choose $\boldsymbol{q}_N = \boldsymbol{y}_N^*(\boldsymbol{u}) - \boldsymbol{y}_N^*$ and $\psi_N = r_N^*(\boldsymbol{u}) - r_N^*$ in (3.3.34). Then,

$$\begin{aligned}
&\theta \left\| \boldsymbol{y}_N^*(\boldsymbol{u}) - \boldsymbol{y}_N^* \right\|_{1,\Omega}^2 \\
&\leq a\left(\boldsymbol{y}_N^*(\boldsymbol{u}) - \boldsymbol{y}_N^*, \boldsymbol{y}_N^*(\boldsymbol{u}) - \boldsymbol{y}_N^*\right) = \left(\boldsymbol{y} - \boldsymbol{y}_N, \boldsymbol{y}_N^*(\boldsymbol{u}) - \boldsymbol{y}_N^*\right) \\
&\leq \left\| \boldsymbol{y} - \boldsymbol{y}_N \right\|_{0,\Omega} \left\| \boldsymbol{y}_N^*(\boldsymbol{u}) - \boldsymbol{y}_N^* \right\|_{0,\Omega} \\
&\leq \left\| \boldsymbol{y} - \boldsymbol{y}_N \right\|_{0,\Omega} \left\| \boldsymbol{y}_N^*(\boldsymbol{u}) - \boldsymbol{y}_N^* \right\|_{1,\Omega},
\end{aligned}$$

which leads to

$$\left\| \boldsymbol{y}_N^*(\boldsymbol{u}) - \boldsymbol{y}_N^* \right\|_{1,\Omega} \leq C \left\| \boldsymbol{y} - \boldsymbol{y}_N \right\|_{0,\Omega}. \tag{3.3.35}$$

Furthermore, by (3.3.3), (3.3.8), (3.3.34), and (3.3.35), we have

$$\begin{aligned}\beta_N \left\| r_N^* (\boldsymbol{u}) - r_N^* \right\|_{0,\Omega} &\leq \sup_{\boldsymbol{q}_N \in \boldsymbol{Y}_N} \frac{b\left(\boldsymbol{q}_N, r_N^* (\boldsymbol{u}) - r_N^*\right)}{\left\|\boldsymbol{q}_N\right\|_{1,\Omega}} \\ = \sup_{\boldsymbol{q}_N \in \boldsymbol{Y}_N} &\frac{-a\left(\boldsymbol{y}_N^* (\boldsymbol{u}) - \boldsymbol{y}_N^*, \boldsymbol{q}_N\right) + \left(\boldsymbol{y} - \boldsymbol{y}_N, \boldsymbol{q}_N\right)}{\left\|\boldsymbol{q}_N\right\|_{1,\Omega}} \\ \leq \sigma &\left\| \boldsymbol{y}_N^* (\boldsymbol{u}) - \boldsymbol{y}_N^* \right\|_{1,\Omega} + \left\| \boldsymbol{y} - \boldsymbol{y}_N \right\|_{0,\Omega} \\ \leq C &\left\| \boldsymbol{y} - \boldsymbol{y}_N \right\|_{0,\Omega}.\end{aligned} \tag{3.3.36}$$

The estimate stated in (3.3.30) follows directly from considering both (3.3.35) and (3.3.36). The proof is now complete. □

Lemma 3.3.5 *Let* $(\boldsymbol{y}, r, \boldsymbol{u}, \boldsymbol{y}^*, r^*, \lambda)$ *and* $\left(\boldsymbol{y}_N, r_N, \boldsymbol{u}_N, \boldsymbol{y}_N^*, r_N^*, \lambda_N\right)$ *satisfy the optimality conditions* $(\mathscr{Q})$ *and* $(\mathscr{Q}_N)$*, respectively. The following estimate holds:*

$$\left\|\boldsymbol{u} - \boldsymbol{u}_N\right\|_{0,\Omega} \leq C \left(\left\| \boldsymbol{y} - \boldsymbol{y}_N (\boldsymbol{u}) \right\|_{0,\Omega} + \left\| \boldsymbol{y}^* - \boldsymbol{y}_N^* (\boldsymbol{u}) \right\|_{0,\Omega} \right).$$

Proof In (3.3.31), let

$$\boldsymbol{w}_N = \boldsymbol{y}_N^* (\boldsymbol{u}) - \boldsymbol{y}_N^*, \quad \phi_N = r_N^* (\boldsymbol{u}) - r_N^*.$$

Meanwhile, defining

$$\boldsymbol{q}_N = \boldsymbol{y}_N (\boldsymbol{u}) - \boldsymbol{y}_N, \quad \psi_N = r_N (\boldsymbol{u}) - r_N,$$

in Eq. (3.3.34), we arrive at the following conclusion:

$$\left(\boldsymbol{u} - \boldsymbol{u}_N, \boldsymbol{y}_N^* (\boldsymbol{u}) - \boldsymbol{y}_N^*\right) = \left(\boldsymbol{y} - \boldsymbol{y}_N, \boldsymbol{y}_N (\boldsymbol{u}) - \boldsymbol{y}_N\right),$$

which results in

$$\begin{aligned}&\left(\boldsymbol{y}_N (\boldsymbol{u}) - \boldsymbol{y}_N, \boldsymbol{y}_N (\boldsymbol{u}) - \boldsymbol{y}_N\right) \\ &= \left(\boldsymbol{u} - \boldsymbol{u}_N, \boldsymbol{y}_N^* (\boldsymbol{u}) - \boldsymbol{y}_N^*\right) - \left(\boldsymbol{y} - \boldsymbol{y}_N (\boldsymbol{u}), \boldsymbol{y}_N (\boldsymbol{u}) - \boldsymbol{y}_N\right).\end{aligned} \tag{3.3.37}$$

By referring to (3.3.20), it becomes evident that the projection operator $\mathbb{P}_N$ preserves feasibility, i.e., $\boldsymbol{v} \in \boldsymbol{K} \Rightarrow \mathbb{P}_N \boldsymbol{v} \in \boldsymbol{K}$, because the projection is non-expansive. Additionally, due to $\mathbb{P}_N \boldsymbol{u} \in \boldsymbol{U}_N$, $\mathbb{P}_N \boldsymbol{u} \in \boldsymbol{K}_N$. Then, by combining (3.3.9)(e)–(f), (3.3.19)(e)–(f) and (3.3.20), we obtain

$$\begin{aligned}\left(\boldsymbol{u} - \boldsymbol{u}_N, \alpha \boldsymbol{u} + \boldsymbol{y}^*\right) \leq 0, \quad &\left(\boldsymbol{u} - \mathbb{P}_N \boldsymbol{u}, \alpha \boldsymbol{u}_N + \boldsymbol{y}_N^*\right) = 0, \\ \left(\mathbb{P}_N \boldsymbol{u} - \boldsymbol{u}_N, \alpha \boldsymbol{u}_N + \boldsymbol{y}_N^*\right) &\geq 0.\end{aligned} \tag{3.3.38}$$

By (3.3.37) and (3.3.38), it has

$$\begin{aligned}
&\alpha \|\boldsymbol{u}-\boldsymbol{u}_N\|_{0,\Omega}^2+\left\|y_N(\boldsymbol{u})-y_N\right\|_{0,\Omega}^2\\
&=\alpha\left(\boldsymbol{u}-\boldsymbol{u}_N,\boldsymbol{u}-\boldsymbol{u}_N\right)+\left(\boldsymbol{u}-\boldsymbol{u}_N,y_N^*(\boldsymbol{u})-y_N^*\right)-\left(y-y_N(\boldsymbol{u}),y_N(\boldsymbol{u})-y_N\right)\\
&=\left(\boldsymbol{u}-\boldsymbol{u}_N,\alpha\boldsymbol{u}+y^*\right)-\left(\boldsymbol{u}-\mathbb{P}_N\boldsymbol{u},\alpha\boldsymbol{u}_N+y_N^*\right)-\left(\mathbb{P}_N\boldsymbol{u}-\boldsymbol{u}_N,\alpha\boldsymbol{u}_N+y_N^*\right)\\
&\quad+\left(\boldsymbol{u}-\boldsymbol{u}_N,y_N^*(\boldsymbol{u})-y^*\right)-\left(y-y_N(\boldsymbol{u}),y_N(\boldsymbol{u})-y_N\right)\\
&\le\left(\boldsymbol{u}-\boldsymbol{u}_N,y_N^*(\boldsymbol{u})-y^*\right)-\left(y-y_N(\boldsymbol{u}),y_N(\boldsymbol{u})-y_N\right)\\
&\le\|\boldsymbol{u}-\boldsymbol{u}_N\|_{0,\Omega}\left\|y_N^*(\boldsymbol{u})-y^*\right\|_{0,\Omega}+\left\|y-y_N(\boldsymbol{u})\right\|_{0,\Omega}\left\|y_N(\boldsymbol{u})-y_N\right\|_{0,\Omega}\\
&\le\frac{\alpha}{2}\|\boldsymbol{u}-\boldsymbol{u}_N\|_{0,\Omega}^2+\frac{1}{2\alpha}\left\|y_N^*(\boldsymbol{u})-y^*\right\|_{0,\Omega}^2+\frac{1}{2}\left\|y-y_N(\boldsymbol{u})\right\|_{0,\Omega}^2\\
&\quad+\frac{1}{2}\left\|y_N(\boldsymbol{u})-y_N\right\|_{0,\Omega}^2.
\end{aligned}$$

Thus,

$$\|\boldsymbol{u}-\boldsymbol{u}_N\|_{0,\Omega}+\left\|y_N(\boldsymbol{u})-y_N\right\|_{0,\Omega}\le C\left(\left\|y_N^*(\boldsymbol{u})-y^*\right\|_{0,\Omega}+\left\|y-y_N(\boldsymbol{u})\right\|_{0,\Omega}\right),$$

which completes the proof. □

Theorem 3.3.6 *Let* $(y,r,\boldsymbol{u},y^*,r^*,\lambda)$ *and* $\left(y_N,r_N,\boldsymbol{u}_N,y_N^*,r_N^*,\lambda_N\right)$ *be the solutions to the optimality conditions* ($\mathscr{Q}$) *and* ($\mathscr{Q}_N$)*, respectively. If*

$$(y,r),\ \left(y^*,r^*\right)\in\left(H_0^1(\Omega)\cap H^m(\Omega)\right)^2\times\left(L_0^2(\Omega)\cap H^{m-1}(\Omega)\right),$$

and $m\ge 1$*, the following estimate holds:*

$$\begin{aligned}
\|\boldsymbol{u}-\boldsymbol{u}_N\|_{0,\Omega}+\left\|y-y_N\right\|_{1,\Omega}+N^{-\frac{1}{2}}\|r-r_N\|_{0,\Omega}+\left\|y-y_N^*\right\|_{1,\Omega}\\
+N^{-\frac{1}{2}}\left\|r-r_N^*\right\|_{0,\Omega}\le CN^{1-m}\xi,
\end{aligned}$$

where

$$\xi=\|y\|_{m,\Omega}+\left\|y^*\right\|_{m,\Omega}+\|r\|_{m-1,\Omega}+\left\|r^*\right\|_{m-1,\Omega}.$$

Proof Employing the triangle inequality alongside Lemmas 3.3.4–3.3.5, we deduce that

$$\begin{aligned}
&\|\boldsymbol{u}-\boldsymbol{u}_N\|_{0,\Omega}+\left\|y-y_N\right\|_{1,\Omega}+N^{-\frac{1}{2}}\|r-r_N\|_{0,\Omega}+\left\|y^*-y_N^*\right\|_{1,\Omega}\\
&+N^{-\frac{1}{2}}\left\|r^*-r_N^*\right\|_{0,\Omega}
\end{aligned}$$

$$
\begin{aligned}
&\leq \|u - u_N\|_{0,\Omega} + \left\|y - y_N\right\|_{1,\Omega} + N^{-\frac{1}{2}}\|r - r_N\|_{0,\Omega} + \left\|y^* - y_N^*(u)\right\|_{1,\Omega} \\
&\quad + N^{-\frac{1}{2}}\left\|r^* - r_N^*(u)\right\|_{0,\Omega} + \left\|y_N^*(u) - y_N^*\right\|_{1,\Omega} + N^{-\frac{1}{2}}\left\|r_N^*(u) - r_N^*\right\|_{0,\Omega} \\
&\leq \|u - u_N\|_{0,\Omega} + C\left\|y - y_N\right\|_{1,\Omega} + N^{-\frac{1}{2}}\|r - r_N\|_{0,\Omega} + \left\|y^* - y_N^*(u)\right\|_{1,\Omega} \\
&\quad + N^{-\frac{1}{2}}\left\|r^* - r_N^*(u)\right\|_{0,\Omega} \\
&\leq \|u - u_N\|_{0,\Omega} + C\left\|y - y_N(u)\right\|_{1,\Omega} + N^{-\frac{1}{2}}\|r - r_N(u)\|_{0,\Omega} \\
&\quad + C\left\|y_N(u) - y_N\right\|_{1,\Omega} + N^{-\frac{1}{2}}\|r_N(u) - r_N\|_{0,\Omega} + \left\|y^* - y_N^*(u)\right\|_{1,\Omega} \\
&\quad + N^{-\frac{1}{2}}\left\|r^* - r_N^*(u)\right\|_{0,\Omega} \\
&\leq C\|u - u_N\|_{0,\Omega} + C\left\|y - y_N(u)\right\|_{1,\Omega} + N^{-\frac{1}{2}}\|r - r_N(u)\|_{0,\Omega} \\
&\quad + \left\|y^* - y_N^*(u)\right\|_{1,\Omega} + N^{-\frac{1}{2}}\left\|r^* - r_N^*(u)\right\|_{0,\Omega} \\
&\leq C\Big(\left\|y - y_N(u)\right\|_{1,\Omega} + N^{-\frac{1}{2}}\|r - r_N(u)\|_{0,\Omega} + \left\|y^* - y_N^*(u)\right\|_{1,\Omega} \\
&\quad + N^{-\frac{1}{2}}\left\|r^* - r_N^*(u)\right\|_{0,\Omega}\Big).
\end{aligned}
$$

Moreover, by leveraging (3.3.27) and (3.3.28), the inequality stated above leads to the following conclusion:

$$
\begin{aligned}
&\|u - u_N\|_{0,\Omega} + \left\|y - y_N\right\|_{1,\Omega} + N^{-\frac{1}{2}}\|r - r_N\|_{0,\Omega} + \left\|y^* - y_N^*\right\|_{1,\Omega} \\
&\quad + N^{-\frac{1}{2}}\left\|r^* - r_N^*\right\|_{0,\Omega} \\
&\leq CN^{1-m}\left(\|y\|_{m,\Omega} + \left\|y^*\right\|_{m,\Omega} + \|r\|_{m-1,\Omega} + \left\|r^*\right\|_{m-1,\Omega}\right).
\end{aligned}
$$

□

Theorem 3.3.7 *Let* $(y, r, u, y^*, r^*, \lambda)$ *and* $\left(y_N, r_N, u_N, y_N^*, r_N^*, \lambda_N\right)$ *be the solutions to the optimality conditions* $(\mathscr{Q})$ *and* $(\mathscr{Q}_N)$*, respectively. Under the conditions of Theorem* 3.3.6*, the following estimate holds:*

$$
\|u - u_N\|_{0,\Omega} + \left\|y - y_N\right\|_{0,\Omega} + \left\|y - y_N^*\right\|_{0,\Omega} \leq CN^{-m}\xi,
$$

where ξ *is defined as in Theorem* 3.3.6.

Proof Utilizing the triangle inequality in conjunction with Lemmas 3.3.4–3.3.5, we arrive at the following deduction:

$$\begin{aligned}
&\|\boldsymbol{u}-\boldsymbol{u}_N\|_{0,\Omega}+\left\|\boldsymbol{y}-\boldsymbol{y}_N\right\|_{0,\Omega}+\left\|\boldsymbol{y}^*-\boldsymbol{y}_N^*\right\|_{0,\Omega}\\
\leq\,&\|\boldsymbol{u}-\boldsymbol{u}_N\|_{0,\Omega}+\left\|\boldsymbol{y}-\boldsymbol{y}_N\right\|_{0,\Omega}+\left\|\boldsymbol{y}^*-\boldsymbol{y}_N^*(\boldsymbol{u})\right\|_{0,\Omega}+\left\|\boldsymbol{y}_N^*(\boldsymbol{u})-\boldsymbol{y}_N^*\right\|_{0,\Omega}\\
\leq\,&\|\boldsymbol{u}-\boldsymbol{u}_N\|_{0,\Omega}+C\left\|\boldsymbol{y}-\boldsymbol{y}_N\right\|_{0,\Omega}+\left\|\boldsymbol{y}^*-\boldsymbol{y}_N^*(\boldsymbol{u})\right\|_{0,\Omega}\\
\leq\,&\|\boldsymbol{u}-\boldsymbol{u}_N\|_{0,\Omega}+C\left\|\boldsymbol{y}-\boldsymbol{y}_N(\boldsymbol{u})\right\|_{0,\Omega}+C\left\|\boldsymbol{y}_N(\boldsymbol{u})-\boldsymbol{y}_N\right\|_{0,\Omega}+\left\|\boldsymbol{y}^*-\boldsymbol{y}_N^*(\boldsymbol{u})\right\|_{0,\Omega}\\
\leq\,&C\|\boldsymbol{u}-\boldsymbol{u}_N\|_{0,\Omega}+C\left\|\boldsymbol{y}-\boldsymbol{y}_N(\boldsymbol{u})\right\|_{0,\Omega}+\left\|\boldsymbol{y}^*-\boldsymbol{y}_N^*(\boldsymbol{u})\right\|_{0,\Omega}\\
\leq\,&C\left(\left\|\boldsymbol{y}-\boldsymbol{y}_N(\boldsymbol{u})\right\|_{0,\Omega}+\left\|\boldsymbol{y}^*-\boldsymbol{y}_N^*(\boldsymbol{u})\right\|_{0,\Omega}\right).
\end{aligned}$$

In light of (3.3.27) and (3.3.28), we obtain the following result:

$$\begin{aligned}
&\|\boldsymbol{u}-\boldsymbol{u}_N\|_{0,\Omega}+\left\|\boldsymbol{y}-\boldsymbol{y}_N\right\|_{0,\Omega}+\left\|\boldsymbol{y}^*-\boldsymbol{y}_N^*\right\|_{0,\Omega}\\
\leq\,&CN^{-m}\left(\|\boldsymbol{y}\|_{m,\Omega}+\left\|\boldsymbol{y}^*\right\|_{m,\Omega}+\|r\|_{m-1,\Omega}+\left\|r^*\right\|_{m-1,\Omega}\right),
\end{aligned}$$

which completes the proof. □

3.3.4 A Posteriori Error Estimates

The primary goal of this subsection is to derive a posteriori error estimates for the spectral approximation of the control problem ($\mathscr{P}$). These estimates are highly beneficial for constructing reliable hp-spectral element approximations and adaptive algorithms tailored for control problems.

Furthermore, we aim to establish error bounds in terms of the H^1-norm and the L^2-norm for both the state and costate variables.

To achieve these objectives, we begin by introducing an essential auxiliary system as follows.

Find $(\boldsymbol{y}(\boldsymbol{u}_N), r(\boldsymbol{u}_N), \boldsymbol{y}^*(\boldsymbol{u}_N), r^*(\boldsymbol{u}_N))$ such that

$$a(\boldsymbol{y}(\boldsymbol{u}_N), \boldsymbol{w}) - b(\boldsymbol{w}, r(\boldsymbol{u}_N)) = (\boldsymbol{f}+\boldsymbol{u}_N, \boldsymbol{w}), \quad \forall\, \boldsymbol{w}\in \boldsymbol{Y}, \tag{3.3.39}$$

$$b(\boldsymbol{y}(\boldsymbol{u}_N), \phi) = 0, \quad \forall\, \phi \in M, \tag{3.3.40}$$

$$a\left(\boldsymbol{q}, \boldsymbol{y}^*(\boldsymbol{u}_N)\right) + b\left(\boldsymbol{q}, r^*(\boldsymbol{u}_N)\right) = \left(\boldsymbol{y}_N - \boldsymbol{y}_d, \boldsymbol{q}\right), \quad \forall\, \boldsymbol{q}\in \boldsymbol{Y}, \tag{3.3.41}$$

$$b\left(\boldsymbol{y}^*(\boldsymbol{u}_N), \psi\right) = 0, \quad \forall\, \psi \in M. \tag{3.3.42}$$

Then, we proceed to prove Lemma 3.3.8.

Lemma 3.3.8 *Let* $(\boldsymbol{y}, r, \boldsymbol{u}, \boldsymbol{y}^*, r^*, \lambda)$ *and* $\left(\boldsymbol{y}_N, r_N, \boldsymbol{u}_N, \boldsymbol{y}_N^*, r_N^*, \lambda_N\right)$ *satisfy the optimality conditions* $(\mathscr{Q})$ *and* $(\mathscr{Q}_N)$*, respectively. The following results hold:*

$$\begin{aligned} \|\boldsymbol{y} - \boldsymbol{y}(\boldsymbol{u}_N)\|_{1,\Omega} + \|r - r(\boldsymbol{u}_N)\|_{0,\Omega} &\le C\,\|\boldsymbol{u} - \boldsymbol{u}_N\|_{0,\Omega}, \\ \left\|\boldsymbol{y}^* - \boldsymbol{y}^*(\boldsymbol{u}_N)\right\|_{1,\Omega} + \left\|r^* - r^*(\boldsymbol{u}_N)\right\|_{0,\Omega} &\le C\,\left\|\boldsymbol{y} - \boldsymbol{y}_N\right\|_{0,\Omega}. \end{aligned} \tag{3.3.43}$$

Proof Subtracting (3.3.39)–(3.3.40) by (3.3.9)(a)–(b), we have

$$\begin{aligned} a(\boldsymbol{y} - \boldsymbol{y}(\boldsymbol{u}_N), \boldsymbol{w}) - b(\boldsymbol{w}, r - r(\boldsymbol{u}_N)) &= (\boldsymbol{u} - \boldsymbol{u}_N, \boldsymbol{w}), \quad \forall\, \boldsymbol{w} \in \boldsymbol{Y}, \\ b(\boldsymbol{y} - \boldsymbol{y}(\boldsymbol{u}_N), \phi) &= 0, \quad \forall\, \phi \in M. \end{aligned} \tag{3.3.44}$$

Set $\boldsymbol{w} = \boldsymbol{y} - \boldsymbol{y}(\boldsymbol{u}_N)$ and $\phi = r - r(\boldsymbol{u}_N)$ in (3.3.44). By (3.3.5), we have

$$\begin{aligned} \theta\,\|\boldsymbol{y} - \boldsymbol{y}(\boldsymbol{u}_N)\|_{1,\Omega}^2 &\le a(\boldsymbol{y} - \boldsymbol{y}(\boldsymbol{u}_N), \boldsymbol{y} - \boldsymbol{y}(\boldsymbol{u}_N)) = (\boldsymbol{u} - \boldsymbol{u}_N, \boldsymbol{y} - \boldsymbol{y}(\boldsymbol{u}_N)) \\ &\le \|\boldsymbol{u} - \boldsymbol{u}_N\|_{0,\Omega}\,\|\boldsymbol{y} - \boldsymbol{y}(\boldsymbol{u}_N)\|_{0,\Omega}. \end{aligned}$$

This yields

$$\|\boldsymbol{y} - \boldsymbol{y}(\boldsymbol{u}_N)\|_{1,\Omega} \le C\,\|\boldsymbol{u} - \boldsymbol{u}_N\|_{0,\Omega}. \tag{3.3.45}$$

Utilizing (3.3.3), (3.3.6), (3.3.44), and (3.3.45), we have

$$\begin{aligned} \beta\,\|r - r(\boldsymbol{u}_N)\|_{0,\Omega} &\le \sup_{\boldsymbol{w}\in\boldsymbol{Y}} \frac{b(\boldsymbol{w}, r - r(\boldsymbol{u}_N))}{\|\boldsymbol{w}\|_{1,\Omega}} \\ &= \sup_{\boldsymbol{w}\in\boldsymbol{Y}} \frac{a(\boldsymbol{y} - \boldsymbol{y}(\boldsymbol{u}_N), \boldsymbol{w}) - (\boldsymbol{u} - \boldsymbol{u}_N, \boldsymbol{w})}{\|\boldsymbol{w}\|_{1,\Omega}} \\ &\le \sigma\,\|\boldsymbol{y} - \boldsymbol{y}(\boldsymbol{u}_N)\|_{1,\Omega} + \|\boldsymbol{u} - \boldsymbol{u}_N\|_{0,\Omega} \\ &\le C\,\|\boldsymbol{u} - \boldsymbol{u}_N\|_{0,\Omega}. \end{aligned}$$

Thus,

$$\|r - r(\boldsymbol{u}_N)\|_{0,\Omega} \le C\,\|\boldsymbol{u} - \boldsymbol{u}_N\|_{0,\Omega}. \tag{3.3.46}$$

Similarly, by subtracting (3.3.9)(c)–(d) from (3.3.41)–(3.3.42), we obtain

$$\begin{aligned} a\left(\boldsymbol{q}, \boldsymbol{y}^* - \boldsymbol{y}^*(\boldsymbol{u}_N)\right) + b\left(\boldsymbol{q}, r^* - r^*(\boldsymbol{u}_N)\right) &= \left(\boldsymbol{y} - \boldsymbol{y}_N, \boldsymbol{q}\right), \quad \forall\, \boldsymbol{q} \in \boldsymbol{Y}, \\ b\left(\boldsymbol{y}^* - \boldsymbol{y}^*(\boldsymbol{u}_N), \psi\right) &= 0, \quad \forall\, \psi \in M. \end{aligned} \tag{3.3.47}$$

Choose $\boldsymbol{q} = \boldsymbol{y}^* - \boldsymbol{y}^*(\boldsymbol{u}_N)$ and $\psi = r^* - r^*(\boldsymbol{u}_N)$ in (3.3.47). By (3.3.5), we have

$$\theta \left\|\boldsymbol{y}^* - \boldsymbol{y}^*(\boldsymbol{u}_N)\right\|_{1,\Omega}^2 \le a\left(\boldsymbol{y}^* - \boldsymbol{y}^*(\boldsymbol{u}_N), \boldsymbol{y}^* - \boldsymbol{y}^*(\boldsymbol{u}_N)\right) = \left(\boldsymbol{y} - \boldsymbol{y}_N, \boldsymbol{y}^* - \boldsymbol{y}^*(\boldsymbol{u}_N)\right)$$
$$\le \left\|\boldsymbol{y} - \boldsymbol{y}_N\right\|_{0,\Omega} \|\boldsymbol{y}^* - \boldsymbol{y}^*(\boldsymbol{u}_N)\|_{0,\Omega},$$

so

$$\left\|\boldsymbol{y}^* - \boldsymbol{y}^*(\boldsymbol{u}_N)\right\|_{1,\Omega} \le C\left\|\boldsymbol{y} - \boldsymbol{y}_N\right\|_{0,\Omega}. \tag{3.3.48}$$

It follows from (3.3.3), (3.3.6), (3.3.47), and (3.3.48) that

$$\begin{aligned}
\beta\left\|r^* - r^*(\boldsymbol{u}_N)\right\|_{0,\Omega} &\le \sup_{\boldsymbol{q}\in Y} \frac{b\left(\boldsymbol{w}, r^* - r^*(\boldsymbol{u}_N)\right)}{\|\boldsymbol{q}\|_{1,\Omega}} \\
&= \sup_{\boldsymbol{q}\in Y} \frac{-a\left(\boldsymbol{q}, \boldsymbol{y}^* - \boldsymbol{y}^*(\boldsymbol{u}_N)\right) + \left(\boldsymbol{y} - \boldsymbol{y}_N, \boldsymbol{q}\right)}{\|\boldsymbol{q}\|_{1,\Omega}} \\
&\le \sigma\left\|\boldsymbol{y}^* - \boldsymbol{y}^*(\boldsymbol{u}_N)\right\|_{1,\Omega} + \left\|\boldsymbol{y} - \boldsymbol{y}_N\right\|_{0,\Omega} \\
&\le C\left\|\boldsymbol{y} - \boldsymbol{y}_N\right\|_{0,\Omega}.
\end{aligned}$$

Therefore,

$$\left\|r^* - r^*(\boldsymbol{u}_N)\right\|_{0,\Omega} \le C\left\|\boldsymbol{y} - \boldsymbol{y}_N\right\|_{0,\Omega}. \tag{3.3.49}$$

Then, (3.3.43) follows from (3.3.45), (3.3.46), (3.3.48), and (3.3.49). □

Lemma 3.3.9 *Let* $(\boldsymbol{y}, r, \boldsymbol{u}, \boldsymbol{y}^*, r^*, \lambda)$ *and* $\left(\boldsymbol{y}_N, r_N, \boldsymbol{u}_N, \boldsymbol{y}_N^*, r_N^*, \lambda_N\right)$ *satisfy the optimality conditions* ($\mathcal{Q}$) *and* ($\mathcal{Q}_N$), *respectively. It holds that*

$$\|\boldsymbol{u} - \boldsymbol{u}_N\|_{0,\Omega} \le C\left(\left\|\boldsymbol{y}(\boldsymbol{u}_N) - \boldsymbol{y}_N\right\|_{0,\Omega} + \left\|\boldsymbol{y}^*(\boldsymbol{u}_N) - \boldsymbol{y}_N^*\right\|_{0,\Omega}\right).$$

Proof Set $\boldsymbol{w} = \boldsymbol{y}^* - \boldsymbol{y}^*(\boldsymbol{u}_N)$ in (3.3.44) and $\boldsymbol{q} = \boldsymbol{y} - \boldsymbol{y}(\boldsymbol{u}_N)$ in (3.3.47). We get

$$\left(\boldsymbol{y} - \boldsymbol{y}_N, \boldsymbol{y} - \boldsymbol{y}(\boldsymbol{u}_N)\right) = \left(\boldsymbol{u} - \boldsymbol{u}_N, \boldsymbol{y}^* - \boldsymbol{y}^*(\boldsymbol{u}_N)\right).$$

Hence,

$$\begin{aligned}
&\left(\boldsymbol{y} - \boldsymbol{y}(\boldsymbol{u}_N), \boldsymbol{y} - \boldsymbol{y}(\boldsymbol{u}_N)\right) \\
&= \left(\boldsymbol{u} - \boldsymbol{u}_N, \boldsymbol{y}^* - \boldsymbol{y}^*(\boldsymbol{u}_N)\right) - \left(\boldsymbol{y}(\boldsymbol{u}_N) - \boldsymbol{y}_N, \boldsymbol{y} - \boldsymbol{y}(\boldsymbol{u}_N)\right).
\end{aligned} \tag{3.3.50}$$

Drawing upon (3.3.9)(e)–(f), (3.3.19)(e)–(f), (3.3.20), and (3.3.50), we deduce that

$$\begin{aligned}
&\alpha \|\boldsymbol{u}-\boldsymbol{u}_N\|_{0,\Omega}^2+\|\boldsymbol{y}-\boldsymbol{y}(\boldsymbol{u}_N)\|_{0,\Omega}^2\\
&=\alpha(\boldsymbol{u}-\boldsymbol{u}_N,\boldsymbol{u}-\boldsymbol{u}_N)+\left(\boldsymbol{u}-\boldsymbol{u}_N,\boldsymbol{y}^*-\boldsymbol{y}^*(\boldsymbol{u}_N)\right)-\left(\boldsymbol{y}(\boldsymbol{u}_N)-\boldsymbol{y}_N,\boldsymbol{y}-\boldsymbol{y}(\boldsymbol{u}_N)\right)\\
&=\left(\boldsymbol{u}-\boldsymbol{u}_N,\alpha\boldsymbol{u}+\boldsymbol{y}^*\right)+\left(\boldsymbol{u}-\boldsymbol{u}_N,-\alpha\boldsymbol{u}_N-\boldsymbol{y}^*(\boldsymbol{u}_N)\right)-\left(\boldsymbol{y}(\boldsymbol{u}_N)-\boldsymbol{y}_N,\boldsymbol{y}-\boldsymbol{y}(\boldsymbol{u}_N)\right)\\
&=\left(\boldsymbol{u}-\boldsymbol{u}_N,\alpha\boldsymbol{u}+\boldsymbol{y}^*\right)+\left(\boldsymbol{u}-\mathbb{P}_N\boldsymbol{u},-\alpha\boldsymbol{u}_N-\boldsymbol{y}_N^*\right)+\left(\boldsymbol{u}-\boldsymbol{u}_N,\boldsymbol{y}_N^*-\boldsymbol{y}^*(\boldsymbol{u}_N)\right)\\
&\quad+\left(\mathbb{P}_N\boldsymbol{u}-\boldsymbol{u}_N,-\alpha\boldsymbol{u}_N-\boldsymbol{y}_N^*\right)-\left(\boldsymbol{y}(\boldsymbol{u}_N)-\boldsymbol{y}_N,\boldsymbol{y}-\boldsymbol{y}(\boldsymbol{u}_N)\right)\\
&\le\left(\boldsymbol{u}-\boldsymbol{u}_N,\boldsymbol{y}_N^*-\boldsymbol{y}^*(\boldsymbol{u}_N)\right)-\left(\boldsymbol{y}(\boldsymbol{u}_N)-\boldsymbol{y}_N,\boldsymbol{y}-\boldsymbol{y}(\boldsymbol{u}_N)\right)\\
&\le\|\boldsymbol{u}-\boldsymbol{u}_N\|_{0,\Omega}\left\|\boldsymbol{y}_N^*-\boldsymbol{y}^*(\boldsymbol{u}_N)\right\|_{0,\Omega}+\left\|\boldsymbol{y}(\boldsymbol{u}_N)-\boldsymbol{y}_N\right\|_{0,\Omega}\|\boldsymbol{y}-\boldsymbol{y}(\boldsymbol{u}_N)\|_{0,\Omega}\\
&\le\frac{\alpha}{2}\|\boldsymbol{u}-\boldsymbol{u}_N\|_{0,\Omega}^2+\frac{1}{2\alpha}\left\|\boldsymbol{y}_N^*-\boldsymbol{y}^*(\boldsymbol{u}_N)\right\|_{0,\Omega}^2+\frac{1}{2}\left\|\boldsymbol{y}(\boldsymbol{u}_N)-\boldsymbol{y}_N\right\|_{0,\Omega}^2\\
&\quad+\frac{1}{2}\|\boldsymbol{y}-\boldsymbol{y}(\boldsymbol{u}_N)\|_{0,\Omega}^2.
\end{aligned}$$

In this manner, we obtain

$$\|\boldsymbol{u}-\boldsymbol{u}_N\|_{0,\Omega}\le C\left\|\boldsymbol{y}_N^*-\boldsymbol{y}^*(\boldsymbol{u}_N)\right\|_{0,\Omega}+C\left\|\boldsymbol{y}(\boldsymbol{u}_N)-\boldsymbol{y}_N\right\|_{0,\Omega},$$

which completes the proof. □

We are now ready to proceed with deriving the a posteriori error estimates.

Theorem 3.3.10 *Let* $(\boldsymbol{y}, r, \boldsymbol{u}, \boldsymbol{y}^*, r^*, \lambda)$ *and* $\left(\boldsymbol{y}_N, r_N, \boldsymbol{u}_N, \boldsymbol{y}_N^*, r_N^*, \lambda_N\right)$ *satisfy the optimality conditions* $(\mathscr{Q})$ *and* $(\mathscr{Q}_N)$*, respectively. Then,*

$$\begin{aligned}
&\|\boldsymbol{u}-\boldsymbol{u}_N\|_{0,\Omega}+\left\|\boldsymbol{y}-\boldsymbol{y}_N\right\|_{1,\Omega}+\left\|\boldsymbol{y}^*-\boldsymbol{y}_N^*\right\|_{1,\Omega}\\
&\quad+\|r-r_N\|_{0,\Omega}+\left\|r^*-r_N^*\right\|_{0,\Omega}\le C\eta,
\end{aligned}$$

where the estimator

$$\begin{aligned}
\eta=&N^{-1}\left\|\Delta\boldsymbol{y}_N-\nabla r_N+\boldsymbol{u}_N+\boldsymbol{f}\right\|_{0,\Omega}+\left\|\nabla\cdot\boldsymbol{y}_N\right\|_{0,\Omega}\\
&+N^{-1}\left\|\Delta\boldsymbol{y}_N^*+\nabla r_N^*+\boldsymbol{y}_N-\boldsymbol{y}_d\right\|_{0,\Omega}+\left\|\nabla\cdot\boldsymbol{y}_N^*\right\|_{0,\Omega}.
\end{aligned}$$

Proof Using Lemmas 3.3.8–3.3.9 and applying the triangle inequality, we obtain

$$\begin{aligned}
&\|\boldsymbol{u}-\boldsymbol{u}_N\|_{0,\Omega}+\left\|\boldsymbol{y}-\boldsymbol{y}_N\right\|_{1,\Omega}+\left\|\boldsymbol{y}^*-\boldsymbol{y}_N^*\right\|_{1,\Omega}+\|r-r_N\|_{0,\Omega}+\left\|r^*-r_N^*\right\|_{0,\Omega}\\
&\le \|\boldsymbol{u}-\boldsymbol{u}_N\|_{0,\Omega}+\|\boldsymbol{y}-\boldsymbol{y}(\boldsymbol{u}_N)\|_{1,\Omega}+\left\|\boldsymbol{y}(\boldsymbol{u}_N)-\boldsymbol{y}_N\right\|_{1,\Omega}+\left\|\boldsymbol{y}^*-\boldsymbol{y}(\boldsymbol{u}_N)\right\|_{1,\Omega}\\
&\quad+\left\|\boldsymbol{y}(\boldsymbol{u}_N)-\boldsymbol{y}_N^*\right\|_{1,\Omega}+\|r-r(\boldsymbol{u}_N)\|_{0,\Omega}+\|r(\boldsymbol{u}_N)-r_N\|_{0,\Omega}\\
&\quad+\left\|r^*-r^*(\boldsymbol{u}_N)\right\|_{0,\Omega}+\left\|r^*(\boldsymbol{u}_N)-r_N^*\right\|_{0,\Omega}\\
&\le C\Big(\left\|\boldsymbol{y}(\boldsymbol{u}_N)-\boldsymbol{y}_N\right\|_{1,\Omega}+\left\|\boldsymbol{y}^*(\boldsymbol{u}_N)-\boldsymbol{y}_N^*\right\|_{1,\Omega}+\|r(\boldsymbol{u}_N)-r_N\|_{0,\Omega}\\
&\quad+\left\|r^*(\boldsymbol{u}_N)-r_N^*\right\|_{0,\Omega}\Big).
\end{aligned}$$

Now, we only need to estimate four terms, specifically: $\left\|\boldsymbol{y}(\boldsymbol{u}_N)-\boldsymbol{y}_N\right\|_{1,\Omega}$, $\|r(\boldsymbol{u}_N)- r_N\|_{0,\Omega}$, $\left\|\boldsymbol{y}^*(\boldsymbol{u}_N)-\boldsymbol{y}_N^*\right\|_{1,\Omega}$, and $\left\|r^*(\boldsymbol{u}_N)-r_N^*\right\|_{0,\Omega}$.

First and foremost, let us estimate the first two terms. Considering (3.3.19)(a)–(b), (3.3.39), and (3.3.40), we deduce that

$$\begin{aligned}
a\left(\boldsymbol{y}(\boldsymbol{u}_N)-\boldsymbol{y}_N,\boldsymbol{w}_N\right)-b\left(\boldsymbol{w}_N,r(\boldsymbol{u}_N)-r_N\right)&=0,\quad\forall\,\boldsymbol{w}_N\in\boldsymbol{Y}_N,\\
b\left(\boldsymbol{y}(\boldsymbol{u}_N)-\boldsymbol{y}_N,\phi_N\right)&=0,\quad\forall\,\phi_N\in M_{N-2}.
\end{aligned}\tag{3.3.51}$$

Set $\boldsymbol{w}_N=\mathbb{P}_{1,N}^0\left(\boldsymbol{y}(\boldsymbol{u}_N)-\boldsymbol{y}_N\right)$ in (3.3.51). By (3.3.5), (3.3.39), and (3.3.40) to obtain

$$\begin{aligned}
\theta&\left\|\boldsymbol{y}(\boldsymbol{u}_N)-\boldsymbol{y}_N\right\|_{1,\Omega}^2\le a\left(\boldsymbol{y}(\boldsymbol{u}_N)-\boldsymbol{y}_N,\boldsymbol{y}(\boldsymbol{u}_N)-\boldsymbol{y}_N\right)\\
&=a\Big(\boldsymbol{y}(\boldsymbol{u}_N)-\boldsymbol{y}_N,\boldsymbol{y}(\boldsymbol{u}_N)-\boldsymbol{y}_N-\mathbb{P}_{1,N}^0\left(\boldsymbol{y}(\boldsymbol{u}_N)-\boldsymbol{y}_N\right)\Big)\\
&\quad-b\Big(\boldsymbol{y}(\boldsymbol{u}_N)-\boldsymbol{y}_N-\mathbb{P}_{1,N}^0\left(\boldsymbol{y}(\boldsymbol{u}_N)-\boldsymbol{y}_N\right),r(\boldsymbol{u}_N)-r_N\Big)\\
&\quad+b\left(\boldsymbol{y}(\boldsymbol{u}_N)-\boldsymbol{y}_N,r(\boldsymbol{u}_N)-r_N\right)\\
&=\Big(\Delta\boldsymbol{y}_N-\nabla r_N+\boldsymbol{u}_N+\boldsymbol{f},\boldsymbol{y}(\boldsymbol{u}_N)-\boldsymbol{y}_N-\mathbb{P}_{1,N}^0\left(\boldsymbol{y}(\boldsymbol{u}_N)-\boldsymbol{y}_N\right)\Big)\\
&\quad+b\left(\boldsymbol{y}(\boldsymbol{u}_N)-\boldsymbol{y}_N,r(\boldsymbol{u}_N)-r_N\right)\\
&=\Big(\Delta\boldsymbol{y}_N-\nabla r_N+\boldsymbol{u}_N+\boldsymbol{f},\boldsymbol{y}(\boldsymbol{u}_N)-\boldsymbol{y}_N-\mathbb{P}_{1,N}^0\left(\boldsymbol{y}(\boldsymbol{u}_N)-\boldsymbol{y}_N\right)\Big)\\
&\quad-b\left(\boldsymbol{y}_N,r(\boldsymbol{u}_N)-r_N\right).
\end{aligned}$$

Using (3.3.21), we obtain the following result:

$$
\begin{aligned}
&\theta \left\| \boldsymbol{y}\left(\boldsymbol{u}_N\right)-\boldsymbol{y}_N \right\|_{1,\Omega}^2 \\
\leq\ & CN^{-1} \left\| \Delta \boldsymbol{y}_N - \nabla r_N + \boldsymbol{u}_N + \boldsymbol{f} \right\|_{0,\Omega} \left\| \boldsymbol{y}\left(\boldsymbol{u}_N\right)-\boldsymbol{y}_N \right\|_{1,\Omega} + \left\| \nabla \cdot \boldsymbol{y}_N \right\|_{0,\Omega} \\
&\cdot \left\| r\left(\boldsymbol{u}_N\right) - r_N \right\|_{0,\Omega} \\
\leq\ & \frac{C}{4\varepsilon}\left(N^{-2} \left\| \Delta \boldsymbol{y}_N - \nabla r_N + \boldsymbol{u}_N + \boldsymbol{f} \right\|_{0,\Omega}^2 + \left\| \nabla \cdot \boldsymbol{y}_N \right\|_{0,\Omega}^2 \right) \\
&+\varepsilon \left(\left\| \boldsymbol{y}\left(\boldsymbol{u}_N\right)-\boldsymbol{y}_N \right\|_{1,\Omega}^2 + \left\| r\left(\boldsymbol{u}_N\right) - r_N \right\|_{0,\Omega}^2 \right),
\end{aligned}
\tag{3.3.52}
$$

where $\varepsilon > 0$ is an undetermined constant. Subsequently, by employing (3.3.39), (3.3.40), and (3.3.51), we derive that

$$
\begin{aligned}
& b\left(\boldsymbol{w}, r\left(\boldsymbol{u}_N\right)-r_N\right) \\
=\ & b\left(\mathbb{P}_{1,N}^0 \boldsymbol{w}, r\left(\boldsymbol{u}_N\right)-r_N\right) + b\left(\boldsymbol{w} - \mathbb{P}_{1,N}^0 \boldsymbol{w}, r\left(\boldsymbol{u}_N\right)-r_N\right) \\
=\ & a\left(\boldsymbol{y}\left(\boldsymbol{u}_N\right)-\boldsymbol{y}_N, \mathbb{P}_{1,N}^0 \boldsymbol{w}\right) + b\left(\boldsymbol{w} - \mathbb{P}_{1,N}^0 \boldsymbol{w}, r\left(\boldsymbol{u}_N\right)-r_N\right) \\
=\ & a\left(\boldsymbol{y}\left(\boldsymbol{u}_N\right)-\boldsymbol{y}_N, \mathbb{P}_{1,N}^0 \boldsymbol{w} - \boldsymbol{w}\right) + a\left(\boldsymbol{y}\left(\boldsymbol{u}_N\right)-\boldsymbol{y}_N, \boldsymbol{w}\right) \\
& + b\left(\boldsymbol{w} - \mathbb{P}_{1,N}^0 \boldsymbol{w}, r\left(\boldsymbol{u}_N\right)-r_N\right) \\
=\ & -\left(\Delta \boldsymbol{y}_N - \nabla r_N + \boldsymbol{u}_N + \boldsymbol{f}, \boldsymbol{w} - \mathbb{P}_{1,N}^0 \boldsymbol{w}\right) + a\left(\boldsymbol{y}\left(\boldsymbol{u}_N\right)-\boldsymbol{y}_N, \boldsymbol{w}\right).
\end{aligned}
$$

Therefore, by invoking (3.3.3) and (3.3.6), we obtain the following derivation:

$$
\begin{aligned}
& \left\| r\left(\boldsymbol{u}_N\right) - r_N \right\|_{0,\Omega} \leq \beta^{-1} \sup_{\boldsymbol{w} \in \boldsymbol{Y}} \frac{b\left(\boldsymbol{w}, r\left(\boldsymbol{u}_N\right)-r_N\right)}{\|\boldsymbol{w}\|_{1,\Omega}} \\
=\ & \beta^{-1} \sup_{\boldsymbol{w} \in \boldsymbol{Y}} \frac{-\left(\Delta \boldsymbol{y}_N - \nabla r_N + \boldsymbol{u}_N + \boldsymbol{f}, \boldsymbol{w} - \mathbb{P}_{1,N}^0 \boldsymbol{w}\right) + a\left(\boldsymbol{y}\left(\boldsymbol{u}_N\right)-\boldsymbol{y}_N, \boldsymbol{w}\right)}{\|\boldsymbol{w}\|_{1,\Omega}} \\
\leq\ & CN^{-1} \left\| \Delta \boldsymbol{y}_N - \nabla r_N + \boldsymbol{u}_N + \boldsymbol{f} \right\|_{0,\Omega} + \frac{\sigma}{\beta} \left\| \boldsymbol{y}\left(\boldsymbol{u}_N\right)-\boldsymbol{y}_N \right\|_{1,\Omega}.
\end{aligned}
\tag{3.3.53}
$$

Substituting (3.3.53) into (3.3.52) and choosing

$$
\varepsilon = \frac{\theta \beta^2}{2\left(\beta^2 + 2\sigma^2\right)},
$$

we obtain

$$\begin{aligned}&\left\|\boldsymbol{y}\left(\boldsymbol{u}_N\right)-\boldsymbol{y}_N\right\|_{1,\Omega}\\&\leq C\left(N^{-1}\left\|\Delta\boldsymbol{y}_N-\nabla r_N+\boldsymbol{u}_N+\boldsymbol{f}\right\|_{0,\Omega}+\left\|\nabla\cdot\boldsymbol{y}_N\right\|_{0,\Omega}\right).\end{aligned}\tag{3.3.54}$$

It then follows from (3.3.53) and (3.3.54) that

$$\left\|r\left(\boldsymbol{u}_N\right)-r_N\right\|_{0,\Omega}\leq C\left(N^{-1}\left\|\Delta\boldsymbol{y}_N-\nabla r_N+\boldsymbol{u}_N+\boldsymbol{f}\right\|_{0,\Omega}+\left\|\nabla\cdot\boldsymbol{y}_N\right\|_{0,\Omega}\right).\tag{3.3.55}$$

Next, we estimate $\left\|\boldsymbol{y}^*\left(\boldsymbol{u}_N\right)-\boldsymbol{y}_N^*\right\|_{1,\Omega}$ and $\left\|r^*\left(\boldsymbol{u}_N\right)-r_N^*\right\|_{0,\Omega}$. Subtract (3.3.19)(c)–(d) from (3.3.41)–(3.3.42), respectively, to obtain

$$\begin{aligned}a\left(\boldsymbol{q}_N,\boldsymbol{y}^*\left(\boldsymbol{u}_N\right)-\boldsymbol{y}_N^*\right)+b\left(\boldsymbol{q}_N,r^*\left(\boldsymbol{u}_N\right)-r_N^*\right)=0,\quad\forall\,\boldsymbol{q}_N\in\boldsymbol{Y}_N,\\b\left(\boldsymbol{y}^*\left(\boldsymbol{u}_N\right)-\boldsymbol{y}_N^*,\psi_N\right)=0,\quad\forall\,\psi_N\in M_{N-2}.\end{aligned}$$

Using the similar technique, by (3.3.3), (3.3.5), (3.3.6), (3.3.19)(c)–(d), (3.3.21), (3.3.41), and (3.3.42), we obtain

$$\begin{aligned}\theta\left\|\boldsymbol{y}^*\left(\boldsymbol{u}_N\right)-\boldsymbol{y}_N^*\right\|_{1,\Omega}^2\leq{}&\frac{C}{4\varepsilon}\left(N^{-2}\left\|\Delta\boldsymbol{y}_N^*+\nabla r_N^*+\boldsymbol{y}_N-\boldsymbol{y}_d\right\|_{0,\Omega}^2+\left\|\nabla\cdot\boldsymbol{y}_N^*\right\|_{0,\Omega}^2\right)\\&+\varepsilon\left(\left\|\boldsymbol{y}^*\left(\boldsymbol{u}_N\right)-\boldsymbol{y}_N^*\right\|_{1,\Omega}^2+\left\|r^*\left(\boldsymbol{u}_N\right)-r_N^*\right\|_{0,\Omega}^2\right),\end{aligned}$$

and

$$\left\|r^*\left(\boldsymbol{u}_N\right)-r_N^*\right\|_{0,\Omega}\leq CN^{-1}\left\|\Delta\boldsymbol{y}_N^*+\nabla r_N^*+\boldsymbol{y}_N-\boldsymbol{y}_d\right\|_{0,\Omega}+\frac{\sigma}{\beta}\left\|\boldsymbol{y}^*\left(\boldsymbol{u}_N\right)-\boldsymbol{y}_N^*\right\|_{1,\Omega}.$$

Once again, choosing

$$\varepsilon=\frac{\theta\beta^2}{2\left(\beta^2+2\sigma^2\right)},$$

we derive that

$$\begin{aligned}&\left\|\boldsymbol{y}^*\left(\boldsymbol{u}_N\right)-\boldsymbol{y}_N^*\right\|_{1,\Omega}\\&\leq C\left(N^{-1}\left\|\Delta\boldsymbol{y}_N^*+\nabla r_N^*+\boldsymbol{y}_N-\boldsymbol{y}_d\right\|_{0,\Omega}+\left\|\nabla\cdot\boldsymbol{y}_N^*\right\|_{0,\Omega}\right),\end{aligned}\tag{3.3.56}$$

and

$$\left\|r^*(\boldsymbol{u}_N)-r_N^*\right\|_{0,\Omega}\leq C\left(N^{-1}\left\|\Delta\boldsymbol{y}_N^*+\nabla r_N^*+\boldsymbol{y}_N-\boldsymbol{y}_d\right\|_{0,\Omega}+\left\|\nabla\cdot\boldsymbol{y}_N^*\right\|_{0,\Omega}\right).\tag{3.3.57}$$

It immediately completes the proof by combining (3.3.54), (3.3.55), (3.3.56), and (3.3.57). □

In Theorem 3.3.10 presented earlier, we established an H^1-norm error bound for both the state $\boldsymbol{y}$ and the costate $\boldsymbol{y}^*$. Moving forward, our objective is to derive the corresponding L^2-norm error for these quantities.

Theorem 3.3.11 *Let* $(\boldsymbol{y}, r, \boldsymbol{u}, \boldsymbol{y}^*, r^*, \lambda)$ *and* $\left(\boldsymbol{y}_N, r_N, \boldsymbol{u}_N, \boldsymbol{y}_N^*, r_N^*, \lambda_N\right)$ *satisfy the optimality conditions* $(\mathscr{Q})$ *and* $(\mathscr{Q}_N)$*, respectively. Then the following estimate holds:*

$$\|\boldsymbol{u}-\boldsymbol{u}_N\|_{0,\Omega}+\left\|\boldsymbol{y}-\boldsymbol{y}_N\right\|_{0,\Omega}+\left\|\boldsymbol{y}^*-\boldsymbol{y}_N^*\right\|_{0,\Omega} \leq C\zeta,$$

where the estimator

$$\begin{aligned}\zeta = & N^{-2}\left\|\Delta \boldsymbol{y}_N-\nabla r_N+\boldsymbol{u}_N+\boldsymbol{f}\right\|_{0,\Omega}+N^{-\frac{1}{2}}\left\|\nabla\cdot \boldsymbol{y}_N\right\|_{0,\Omega} \\ & +N^{-2}\left\|\Delta \boldsymbol{y}_N^*+\nabla r_N^*+\boldsymbol{y}_N-\boldsymbol{y}_d\right\|_{0,\Omega}+N^{-\frac{1}{2}}\left\|\nabla\cdot \boldsymbol{y}_N^*\right\|_{0,\Omega}.\end{aligned}$$

Proof By Lemmas 3.3.8–3.3.9 and applying the triangle inequality, we can derive the following:

$$\begin{aligned}&\|\boldsymbol{u}-\boldsymbol{u}_N\|_{0,\Omega}+\left\|\boldsymbol{y}-\boldsymbol{y}_N\right\|_{0,\Omega}+\left\|\boldsymbol{y}^*-\boldsymbol{y}_N^*\right\|_{0,\Omega} \\ \leq\ & \|\boldsymbol{u}-\boldsymbol{u}_N\|_{0,\Omega}+\|\boldsymbol{y}-\boldsymbol{y}(\boldsymbol{u}_N)\|_{0,\Omega}+\left\|\boldsymbol{y}(\boldsymbol{u}_N)-\boldsymbol{y}_N\right\|_{0,\Omega}+\left\|\boldsymbol{y}^*-\boldsymbol{y}(\boldsymbol{u}_N)\right\|_{0,\Omega} \\ & +\left\|\boldsymbol{y}(\boldsymbol{u}_N)-\boldsymbol{y}_N^*\right\|_{0,\Omega} \\ \leq\ & C\left(\left\|\boldsymbol{y}(\boldsymbol{u}_N)-\boldsymbol{y}_N\right\|_{0,\Omega}+\left\|\boldsymbol{y}^*(\boldsymbol{u}_N)-\boldsymbol{y}_N^*\right\|_{0,\Omega}\right).\end{aligned}$$

Then, we derive the error estimates $\left\|\boldsymbol{y}(\boldsymbol{u}_N)-\boldsymbol{y}_N\right\|_{0,\Omega}$ and $\left\|\boldsymbol{y}^*(\boldsymbol{u}_N)-\boldsymbol{y}_N^*\right\|_{0,\Omega}$. Consider first $\left\|\boldsymbol{y}(\boldsymbol{u}_N)-\boldsymbol{y}_N\right\|_{0,\Omega}$. Set $\Phi=\boldsymbol{y}(\boldsymbol{u}_N)-\boldsymbol{y}_N$ in Lemma 3.3.3, and $(\boldsymbol{\epsilon}, \rho)$ is the solution to Eqs. (3.3.22). By (3.3.19)(b), (3.3.39), (3.3.40), and (3.3.51), we have

$$\begin{aligned}&\left\|\boldsymbol{y}(\boldsymbol{u}_N)-\boldsymbol{y}_N\right\|_{0,\Omega}^2=\left(\Phi, \boldsymbol{y}(\boldsymbol{u}_N)-\boldsymbol{y}_N\right) \\ =\ & a\left(\boldsymbol{\epsilon}, \boldsymbol{y}(\boldsymbol{u}_N)-\boldsymbol{y}_N\right)-b\left(\boldsymbol{y}(\boldsymbol{u}_N)-\boldsymbol{y}_N, \rho\right) \\ =\ & a\left(\boldsymbol{\epsilon}-\mathbb{P}_{1,N}^0\boldsymbol{\epsilon}, \boldsymbol{y}(\boldsymbol{u}_N)-\boldsymbol{y}_N\right)-b\left(-\mathbb{P}_{1,N}^0\boldsymbol{\epsilon}, r(\boldsymbol{u}_N)-r_N\right)-b\left(\boldsymbol{y}(\boldsymbol{u}_N)-\boldsymbol{y}_N, \rho\right) \\ =\ & a\left(\boldsymbol{\epsilon}-\mathbb{P}_{1,N}^0\boldsymbol{\epsilon}, \boldsymbol{y}(\boldsymbol{u}_N)-\boldsymbol{y}_N\right)-b\left(\boldsymbol{\epsilon}-\mathbb{P}_{1,N}^0\boldsymbol{\epsilon}, r(\boldsymbol{u}_N)-r_N\right)+b\left(\boldsymbol{y}_N, \rho-\rho_N\right) \\ =\ & \left(\boldsymbol{f}+\boldsymbol{u}_N+\Delta \boldsymbol{y}_N-\nabla r_N, \boldsymbol{\epsilon}-\mathbb{P}_{1,N}^0\boldsymbol{\epsilon}\right)+b\left(\boldsymbol{y}_N, \rho-\rho_N\right) \\ \leq\ & \left\|\boldsymbol{f}+\boldsymbol{u}_N+\Delta \boldsymbol{y}_N-\nabla r_N\right\|_{0,\Omega}\left\|\boldsymbol{\epsilon}-\mathbb{P}_{1,N}^0\boldsymbol{\epsilon}\right\|_{0,\Omega}+\left\|\nabla\cdot \boldsymbol{y}_N\right\|_{0,\Omega}\|\rho-\rho_N\|_{0,\Omega}.\end{aligned}$$

Using (3.3.21), the inequality above leads to the conclusion that

$$\begin{aligned}
&\left\| \boldsymbol{y}\left(\boldsymbol{u}_N\right) - \boldsymbol{y}_N \right\|_{0,\Omega}^2 \\
&\leq CN^{-2} \left\| \boldsymbol{f} + \boldsymbol{u}_N + \Delta \boldsymbol{y}_N - \nabla r_N \right\|_{0,\Omega} \|\boldsymbol{\epsilon}\|_{2,\Omega} + CN^{-\frac{1}{2}} \left\| \nabla \cdot \boldsymbol{y}_N \right\|_{0,\Omega} \left(\|\boldsymbol{\epsilon}\|_{2,\Omega} \right. \\
&\quad \left. + \|\rho\|_{1,\Omega} \right) \\
&\leq C \left(N^{-2} \left\| \boldsymbol{f} + \boldsymbol{u}_N + \Delta \boldsymbol{y}_N - \nabla r_N \right\|_{0,\Omega} + N^{-\frac{1}{2}} \left\| \nabla \cdot \boldsymbol{y}_N \right\|_{0,\Omega} \right) \left(\|\boldsymbol{\epsilon}\|_{2,\Omega} + \|\rho\|_{1,\Omega} \right) \\
&\leq C \left(N^{-2} \left\| \boldsymbol{f} + \boldsymbol{u}_N + \Delta \boldsymbol{y}_N - \nabla r_N \right\|_{0,\Omega} + N^{-\frac{1}{2}} \left\| \nabla \cdot \boldsymbol{y}_N \right\|_{0,\Omega} \right) \|\Phi\|_{0,\Omega} \\
&\leq C \left(N^{-2} \left\| \boldsymbol{f} + \boldsymbol{u}_N + \Delta \boldsymbol{y}_N - \nabla r_N \right\|_{0,\Omega} + N^{-\frac{1}{2}} \left\| \nabla \cdot \boldsymbol{y}_N \right\|_{0,\Omega} \right) \left\| \boldsymbol{y}\left(\boldsymbol{u}_N\right) - \boldsymbol{y}_N \right\|_{0,\Omega} .
\end{aligned}$$

Thus,

$$\begin{aligned}
&\left\| \boldsymbol{y}\left(\boldsymbol{u}_N\right) - \boldsymbol{y}_N \right\|_{0,\Omega} \\
&\quad \leq C \left(N^{-2} \left\| \boldsymbol{f} + \boldsymbol{u}_N + \Delta \boldsymbol{y}_N - \nabla r_N \right\|_{0,\Omega} + N^{-\frac{1}{2}} \left\| \nabla \cdot \boldsymbol{y}_N \right\|_{0,\Omega} \right) .
\end{aligned} \tag{3.3.58}$$

Similarly, let $(\boldsymbol{\epsilon}, \rho)$ be the solution to Eqs. (3.3.22) for $\Phi = \boldsymbol{y}^*\left(\boldsymbol{u}_N\right) - \boldsymbol{y}_N^*$. By invoking (3.3.19)(d), (3.3.41), (3.3.42), and (3.3.54) and Lemma 3.3.3, we obtain

$$\begin{aligned}
&\left\| \boldsymbol{y}^*\left(\boldsymbol{u}_N\right) - \boldsymbol{y}_N^* \right\|_{0,\Omega}^2 = \left(\Phi, \boldsymbol{y}^*\left(\boldsymbol{u}_N\right) - \boldsymbol{y}_N^* \right) \\
&= a\left(\boldsymbol{\epsilon}, \boldsymbol{y}^*\left(\boldsymbol{u}_N\right) - \boldsymbol{y}_N^* \right) + b\left(\boldsymbol{y}^*\left(\boldsymbol{u}_N\right) - \boldsymbol{y}_N^*, \rho \right) \\
&= a\left(\boldsymbol{\epsilon} - \mathbb{P}_{1,N}^0 \boldsymbol{\epsilon}, \boldsymbol{y}^*\left(\boldsymbol{u}_N\right) - \boldsymbol{y}_N^* \right) + b\left(-\mathbb{P}_{1,N}^0 \boldsymbol{\epsilon}, r^*\left(\boldsymbol{u}_N\right) - r_N^* \right) \\
&+ b\left(\boldsymbol{y}^*\left(\boldsymbol{u}_N\right) - \boldsymbol{y}_N^*, \rho \right) \\
&= a\left(\boldsymbol{\epsilon} - \mathbb{P}_{1,N}^0 \boldsymbol{\epsilon}, \boldsymbol{y}^*\left(\boldsymbol{u}_N\right) - \boldsymbol{y}_N^* \right) + b\left(\boldsymbol{\epsilon} - \mathbb{P}_{1,N}^0 \boldsymbol{\epsilon}, r^*\left(\boldsymbol{u}_N\right) - r_N^* \right) \\
&- b\left(\boldsymbol{y}_N^*, \rho - \rho_N \right) \\
&= \left(\Delta \boldsymbol{y}_N^* + \nabla r_N^* + \boldsymbol{y}_N - \boldsymbol{y}_d, \boldsymbol{\epsilon} - \mathbb{P}_{1,N}^0 \boldsymbol{\epsilon} \right) - b\left(\boldsymbol{y}_N^*, \rho - \rho_N \right) \\
&\leq \left\| \Delta \boldsymbol{y}_N^* + \nabla r_N^* + \boldsymbol{y}_N - \boldsymbol{y}_d \right\|_{0,\Omega} \left\| \boldsymbol{\epsilon} - \mathbb{P}_{1,N}^0 \boldsymbol{\epsilon} \right\|_{0,\Omega} + \left\| \nabla \cdot \boldsymbol{y}_N^* \right\|_{0,\Omega} \|\rho - \rho_N\|_{0,\Omega} .
\end{aligned}$$

By combining (3.3.21) and Lemma 3.3.3, we arrive at the following conclusion:

$$\begin{aligned}
&\left\| y^* (u_N) - y_N^* \right\|_{0,\Omega}^2 \\
&\le CN^{-2} \left\| \Delta y_N^* + \nabla r_N^* + y_N - y_d \right\|_{0,\Omega} \|\epsilon\|_{2,\Omega} \\
&\quad + CN^{-\frac{1}{2}} \left\| \nabla \cdot y_N^* \right\|_{0,\Omega} \left(\|\epsilon\|_{2,\Omega} + \|\rho\|_{1,\Omega} \right) \\
&\le C \left(N^{-2} \left\| \Delta y_N^* + \nabla r_N^* + y_N - y_d \right\|_{0,\Omega} + N^{-\frac{1}{2}} \left\| \nabla \cdot y_N \right\|_{0,\Omega} \right) \|\Phi\|_{0,\Omega} \\
&\le C \left(N^{-2} \left\| \Delta y_N^* + \nabla r_N^* + y_N - y_d \right\|_{0,\Omega} + N^{-\frac{1}{2}} \left\| \nabla \cdot y_N \right\|_{0,\Omega} \right) \\
&\quad \left\| y^* (u_N) - y_N^* \right\|_{0,\Omega} .
\end{aligned}$$

Hence,

$$\begin{aligned}
&\left\| y^* (u_N) - y_N^* \right\|_{0,\Omega} \\
&\le C \left(N^{-2} \left\| \Delta y_N^* + \nabla r_N^* + y_N - y_d \right\|_{0,\Omega} + N^{-\frac{1}{2}} \left\| \nabla \cdot y_N^* \right\|_{0,\Omega} \right) .
\end{aligned} \tag{3.3.59}$$

It completes the proof by invoking (3.3.58) and (3.3.59). □

Next we focus on the Stokes equation as the state equation for the control problem defined by (3.3.1)–(3.3.2). However, we aim to extend the spectral method to address more complex flow control scenarios, such as those governed by the Navier-Stokes equations. To this end, we provide a corresponding discussion outlining our approach.

Remark 3.3.12 We consider an optimal control problem that is governed by the Navier-Stokes equation, subject to an L^2-norm control constraint. The detailed formulation of this problem is presented below.

Find $(y, r, u) \in Y \times M \times U$ such that

$$\min_{u \in K} J(y, u) = \frac{1}{2} \left\| y - y_d \right\|_{0,\Omega}^2 + \frac{\alpha}{2} \|u\|_{0,\Omega}^2 , \tag{3.3.60}$$

$$s.t. \begin{cases} -\Delta y + Re\,(y \cdot \nabla)\, y + \nabla r = f + u, & \text{in } \Omega, \\ \nabla \cdot y = 0, \quad \text{in } \Omega, \\ y = 0, \quad \text{on } \partial\Omega, \end{cases} \tag{3.3.61}$$

where Re denotes the Reynolds number and all other variables and constants retain their original meanings as defined in (3.3.1)–(3.3.2).

As mentioned previously, the Navier-Stokes equation bears strong similarities to the Stokes equation, with the primary difference lying in the nonlinear term $Re\,(y \cdot \nabla)\, y$. We postulate that a posteriori error estimates for the system defined by

(3.3.60)–(3.3.61) can be analogously derived, contingent upon certain assumptions. Our rationale for this contention is partly informed by the work presented in [156]. In the said paper, a posteriori error estimates are established for the Navier-Stokes optimal control problem in the regime of small Reynolds numbers Re. While the complete and rigorous proof details remain in progress, the findings therein support our hypothesis.

In fact, by applying the techniques outlined in Theorem 3.3.1, we are able to reach a comparable conclusion regarding optimality conditions, as is evidenced in the literature by [156].

The triplet $(\boldsymbol{y}, r, \boldsymbol{u}) \in \boldsymbol{Y} \times M \times \boldsymbol{U}$ is the solution to the control problem (3.3.60)–(3.3.61) if and only if there exists $(\boldsymbol{y}^*, r^*, \lambda) \in \boldsymbol{Y} \times M \times [0, +\infty)$ such that $(\boldsymbol{y}, r, \boldsymbol{u}, \boldsymbol{y}^*, r^*, \lambda)$ satisfies the optimality conditions $(\mathscr{Q}_{N-S})$ that followed.

$$(a)\ a(\boldsymbol{y}, \boldsymbol{w}) + Re\,((\boldsymbol{y} \cdot \nabla)\,\boldsymbol{y}, \boldsymbol{w}) - b(\boldsymbol{w}, r) = (\boldsymbol{f} + \boldsymbol{u}, \boldsymbol{w}), \quad \forall\, \boldsymbol{w} \in \boldsymbol{Y},$$

$$(b)\ b(\boldsymbol{y}, \phi) = 0, \quad \forall\, \phi \in M,$$

$$(c)\ a\left(\boldsymbol{q}, \boldsymbol{y}^*\right) - Re\left((\boldsymbol{y} \cdot \nabla)\,\boldsymbol{y}^*, \boldsymbol{q}\right) + Re\left((\nabla \boldsymbol{y})^\top \boldsymbol{y}^*, \boldsymbol{q}\right) + b\left(\boldsymbol{q}, r^*\right) = (\boldsymbol{y} - \boldsymbol{y}_d, \boldsymbol{q}),$$
$$\forall\, \boldsymbol{q} \in \boldsymbol{Y},$$

$$(d)\ b\left(\boldsymbol{y}^*, \psi\right) = 0, \quad \forall\, \psi \in M,$$

$$(e)\ \lambda\,(\boldsymbol{u}, \boldsymbol{v} - \boldsymbol{u}) \le 0, \quad \forall \boldsymbol{v} \in \boldsymbol{K},$$

$$(f)\ \boldsymbol{y}^* + (\lambda + \alpha)\,\boldsymbol{u} = 0,$$

where

$$\lambda = \begin{cases} \text{constant} \ge 0, & \|\boldsymbol{u}\|_{0,\Omega} = \gamma, \\ 0, & \|\boldsymbol{u}\|_{0,\Omega} < \gamma. \end{cases}$$

To clarify any potential ambiguity, the notation $A^\top$ in this context signifies the transpose of the matrix A.

We acknowledge that the analysis involving the nonlinear case is significantly more intricate than that of the linear-quadratic scenario. Consequently, a rigorous proof of a posteriori error estimates for the control problem defined by (3.3.60)–(3.3.61) remains elusive. Nonetheless, when the Reynolds number Re is very small, and the well-posedness of the investigated system is assured, we can approach the spectral approximation of the Navier-Stokes optimal control problem analogously to our treatment of the Stokes equation, albeit with certain assumptions. By employing similar techniques, we postulate that the subsequent conclusion pertaining to a posteriori error estimates should be applicable to the Navier-Stokes optimal control problem.

For example, let $(\boldsymbol{y}, r, \boldsymbol{u}, \boldsymbol{y}^*, r^*, \lambda)$ satisfy the optimality conditions $(\mathscr{Q}_{N-S})$. And $(\boldsymbol{y}_N, r_N, \boldsymbol{u}_N, \boldsymbol{y}_N^*, r_N^*, \lambda_N)$ is its spectral approximation. Then it holds

$$\|\boldsymbol{u}-\boldsymbol{u}_N\|_{0,\Omega}+\left\|\boldsymbol{y}-\boldsymbol{y}_N\right\|_{1,\Omega}+\left\|\boldsymbol{y}^*-\boldsymbol{y}_N^*\right\|_{1,\Omega}+\|r-r_N\|_{0,\Omega}+\left\|r^*-r_N^*\right\|_{0,\Omega} \leq C\eta_1,$$

where the estimator

$$\begin{aligned}\eta_1 &= N^{-1}\left\|\Delta \boldsymbol{y}_N - Re\left(\boldsymbol{y}_N\cdot\nabla\right)\boldsymbol{y}_N - \nabla r_N + \boldsymbol{u}_N + \boldsymbol{f}\right\|_{0,\Omega} + \left\|\nabla\cdot\boldsymbol{y}_N\right\|_{0,\Omega}\\ &\quad + N^{-1}\left\|\Delta \boldsymbol{y}_N^* + Re\left(\boldsymbol{y}_N\cdot\nabla\right)\boldsymbol{y}_N^* - Re\left(\nabla\boldsymbol{y}_N\right)^{\top}\boldsymbol{y}_N^* + \nabla r_N^* + \boldsymbol{y}_N - \boldsymbol{y}_d\right\|_{0,\Omega}\\ &\quad + \left\|\nabla\cdot\boldsymbol{y}_N^*\right\|_{0,\Omega}.\end{aligned}$$

In the upcoming subsection, we will validate our hypothesis through a numerical example.

3.3.5 Numerical Simulation

In this subsection, we present numerical simulations for two distinct cases: one involving the Stokes equation and the other pertaining to the Navier-Stokes equation. For the Stokes equation, we conduct a numerical example primarily to illustrate the method's effectiveness and the indicators' high performance. Subsequently, we validate our conjecture regarding the Navier-Stokes equation by means of a second numerical example.

Let $L_k(x)$ be the k-th degree Legendre polynomial [205] and

$$\begin{gathered}\phi_k(x) = \frac{1}{\sqrt{4k+6}}\left(L_k(x) - L_{k+2}(x)\right), \quad k = 0, 1, \cdots, N-2,\\ H_N = \operatorname{span}\left\{\phi_i(x_1)\phi_j(x_2)\right\}_{i,j=0}^{N-2}, \quad Q_{N-2} = \operatorname{span}\left\{L_i(x_1)L_j(x_2)\right\}_{i,j=0}^{N-2},\\ \boldsymbol{Y}_N = (H_N)^2, \quad M_{N-2} = Q_{N-2}\cap L_0^2(\Omega).\end{gathered}$$

Subsequently, several approximations are provided as follows:

$$\begin{gathered}\boldsymbol{y}_N = (y_{N1}, y_{N2}), \quad \boldsymbol{u}_N = (u_{N1}, u_{N2}), \quad \boldsymbol{y}_N^* = \left(y_{N1}^*, y_{N2}^*\right),\\ y_{Nm} = \sum_{i,j=0}^{N-2} y_{ijm}\phi_i(x_1)\phi_j(x_2), \quad u_{Nm} = \sum_{i,j=0}^{N} u_{ijm}L_i(x_1)L_j(x_2),\end{gathered}$$

$$y_{Nm}^* = \sum_{i,j=0}^{N-2} y_{ijm}^* \phi_i(x_1)\phi_j(x_2), \quad m = 1, 2,$$

$$r_N = \sum_{i,j=0}^{N-2} r_{ij} L_i(x_1) L_j(x_2), \quad r_N^* = \sum_{i,j=0}^{N-2} r_{ij}^* L_i(x_1) L_j(x_2), \quad r_{00} = 0, \quad r_{00}^* = 0.$$

We utilize a gradient projection algorithm to address the control problem. During the computation, we set the iteration parameter $\rho < 1$. The simulation is carried out using MATLAB software on a desktop computer equipped with a core processor (i5-6600M) and 8GB of random access memory.

Algorithm 2 (Gradient projection algorithm) The primary steps are outlined below.

1: Give the initial values $\boldsymbol{u}_N^{(0)} \in \boldsymbol{K}_N$.

2: Let $n = 0, 1, 2, \ldots$, and get $\left(\boldsymbol{y}_N^{(n)}, r_N^{(n)}\right) \in \boldsymbol{Y}_N \times M_{N-2}$ such that

$$a\left(\boldsymbol{y}_N^{(n)}, \boldsymbol{w}_N\right) - b\left(\boldsymbol{w}_N, r_N^{(n)}\right) = \left(\boldsymbol{f} + \boldsymbol{u}_N^{(n)}, \boldsymbol{w}_N\right), \quad \forall\, \boldsymbol{w}_N \in \boldsymbol{Y}_N,$$
$$b\left(\boldsymbol{y}_N^{(n)}, \phi_N\right) = 0, \quad \forall\, \phi_N \in M_{N-2}.$$

3: Get $\left(\boldsymbol{y}_N^{*(n)}, r_N^{*(n)}\right) \in \boldsymbol{Y}_N \times M_{N-2}$ such that

$$a\left(\boldsymbol{q}_N, \boldsymbol{y}_N^{*(n)}\right) + b\left(\boldsymbol{q}_N, r_N^{*(n)}\right) = \left(\boldsymbol{y}_N^{(n)} - \boldsymbol{y}_d, \boldsymbol{q}_N\right), \quad \forall\, \boldsymbol{q}_N \in \boldsymbol{Y}_N,$$
$$b\left(\boldsymbol{y}_N^{*(n)}, \psi_N\right) = 0, \quad \forall\, \psi_N \in M_{N-2}.$$

4: Set $\boldsymbol{u}_N^{\left(n+\frac{1}{2}\right)} = \boldsymbol{u}_N^{(n)} - \rho\left(\alpha \boldsymbol{u}_N^{(n)} + \boldsymbol{y}_N^{*(n)}\right)$, and

$$\boldsymbol{u}_N^{(n+1)} = \frac{\gamma \boldsymbol{u}_N^{\left(n+\frac{1}{2}\right)}}{\max\left\{\left\|\boldsymbol{u}_N^{\left(n+\frac{1}{2}\right)}\right\|_{0,\Omega}, \gamma\right\}}.$$

5: Stop if the stopping criterion $\left\|\boldsymbol{u}_N^{(n+1)} - \boldsymbol{u}_N^{(n)}\right\|_{0,\Omega} \leq$ Tol (the tolerance) is satisfied. Otherwise set $n = n + 1$ and go to Step 2.

By the definition of $\boldsymbol{u}_N^{(n+1)}$ in Step 4, we have

$$\left\|\boldsymbol{u}_N^{(n+1)}\right\|_{0,\Omega} = \frac{\gamma \left\|\boldsymbol{u}_N^{\left(n+\frac{1}{2}\right)}\right\|_{0,\Omega}}{\max\left\{\left\|\boldsymbol{u}_N^{\left(n+\frac{1}{2}\right)}\right\|_{0,\Omega}, \gamma\right\}} \leq \gamma,$$

which implies that $\boldsymbol{u}_N^{(n+1)} \in \boldsymbol{K}_N$. Therefore, this algorithm converges based on the techniques described in [160]. Furthermore, λ_N can be determined through (3.3.19)(f) by employing the least squares method.

In the subsequent analysis, we define the domain as $\Omega = (-1, 1)^2$ and $\alpha = \frac{3}{4}$. Additionally, we adopt the control constraint set $\boldsymbol{K}$ as previously defined, with $\gamma = 1$.

3.3.5.1 The Stokes Case

In this subsection, we focus on the optimal control problem governed by the Stokes equation.

Find $(\boldsymbol{y}, r, \boldsymbol{u}) \in \boldsymbol{Y} \times M \times \boldsymbol{U}$ such that

$$\min_{\boldsymbol{u}\in\boldsymbol{K}} J(\boldsymbol{y}, \boldsymbol{u}) = \frac{1}{2}\|\boldsymbol{y} - \boldsymbol{y}_d\|_{0,\Omega}^2 + \frac{\alpha}{2}\|\boldsymbol{u}\|_{0,\Omega}^2,$$

$$-\Delta \boldsymbol{y} + \nabla r = \boldsymbol{f} + \boldsymbol{u},$$

$$\nabla \cdot \boldsymbol{y} = 0 \text{ in } \Omega, \quad \boldsymbol{y} = 0 \text{ on } \partial\Omega.$$

Furthermore, let

$$\boldsymbol{f} = \left(1 - \frac{1}{\sqrt{6}}\right)\boldsymbol{\omega} + 2\boldsymbol{h}, \quad \boldsymbol{y}_d = \left(\frac{1}{\pi^2} + \frac{\pi^2}{\sqrt{6}}\right)\boldsymbol{\omega} + \frac{2\pi^2}{\sqrt{6}}\boldsymbol{h},$$

where

$$\boldsymbol{\omega} = \left((1 + \cos(\pi x_1))\sin(\pi x_2), -\sin(\pi x_1)(1 + \cos(\pi x_2))\right),$$

$$\boldsymbol{h} = (\cos(\pi x_1)\sin(\pi x_2), 0).$$

For this problem, its exact solutions are

$$\boldsymbol{y} = \frac{1}{\pi^2}\boldsymbol{\omega}, \quad \boldsymbol{u} = \frac{1}{\sqrt{6}}\boldsymbol{\omega}, \quad \boldsymbol{y}^* = -\frac{1}{\sqrt{6}}\boldsymbol{\omega}, \quad \lambda = \frac{1}{4},$$

$$r = \frac{1}{\pi}\sin(\pi x_1)\sin(\pi x_2), \quad r^* = \frac{\pi}{\sqrt{6}}\sin(\pi x_1)\sin(\pi x_2).$$

Table 3.3 Error results and indicators η and ζ for the Stokes problem

N	4	8	12	16	20
$\|\boldsymbol{u}-\boldsymbol{u}_N\|_{0,\Omega}$	2.1205e-1	1.0128e-3	5.5297e-7	9.7982e-11	6.9339e-15
$\left\|\boldsymbol{y}-\boldsymbol{y}_N\right\|_{1,\Omega}$	3.4839e-1	2.9315e-3	2.4041e-6	4.1206e-10	2.8449e-14
$\|r-r_N\|_{0,\Omega}$	2.5714e-1	2.7753e-3	2.3353e-6	3.9898e-10	3.0040e-14
$\left\|\boldsymbol{y}^*-\boldsymbol{y}_N^*\right\|_{1,\Omega}$	1.3120e+0	1.1464e-2	9.5451e-6	1.6466e-9	1.1391e-13
$\left\|r^*-r_N^*\right\|_{0,\Omega}$	1.0238e+0	1.1186e-2	9.4062e-6	1.6070e-9	1.2138e-13
$\left\|\boldsymbol{y}-\boldsymbol{y}_N\right\|_{0,\Omega}$	5.6836e-2	2.5945e-4	1.3929e-7	2.4527e-11	1.2809e-15
$\left\|\boldsymbol{y}^*-\boldsymbol{y}_N^*\right\|_{0,\Omega}$	2.1187e-1	1.0128e-3	5.5297e-7	9.7981e-11	4.8888e-15
Σ	3.1534e+0	2.9370e-2	2.4244e-5	4.1626e-9	3.0071e-13
Σ_1	4.8075e-1	2.2850e-3	1.2452e-6	2.2049e-10	1.3104e-14
η	3.7276e+0	3.0111e-2	2.2922e-5	3.6442e-9	2.4850e-13
ζ	1.2136e+0	6.7677e-3	4.2722e-6	6.0057e-10	3.8516e-14
Order($\boldsymbol{u}$)	\	13.1803	26.1153	38.7117	52.4135
Order($\boldsymbol{y}$)	\	13.2918	26.1739	38.7398	54.0802

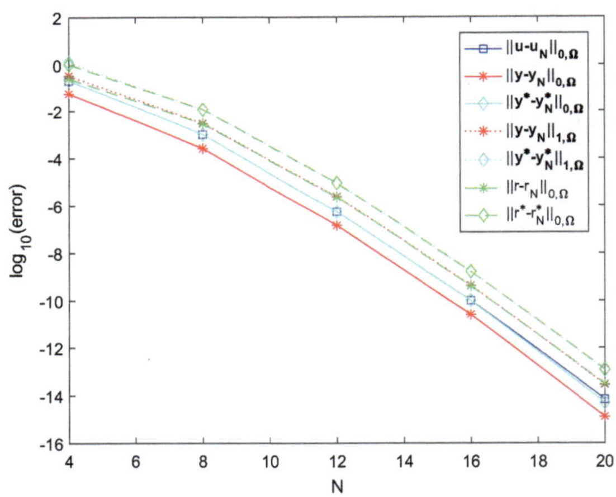

Fig. 3.6 Error estimates for the Stokes problem

Subsequently, in Table 3.3 and Fig. 3.6, we exhibit the error results pertaining to $\|\boldsymbol{u}-\boldsymbol{u}_N\|_{0,\Omega}$, $\left\|\boldsymbol{y}-\boldsymbol{y}_N\right\|_{1,\Omega}$, $\|r-r_N\|_{0,\Omega}$, $\left\|\boldsymbol{y}^*-\boldsymbol{y}_N^*\right\|_{1,\Omega}$, $\left\|r^*-r_N^*\right\|_{0,\Omega}$, $\left\|\boldsymbol{y}-\boldsymbol{y}_N\right\|_{0,\Omega}$, and $\left\|\boldsymbol{y}^*-\boldsymbol{y}_N^*\right\|_{0,\Omega}$. These errors exhibit an exponential decrease with respect to N. Additionally, Figs. 3.7 and 3.8 illustrate the point-wise error estimates for the velocity, control, and pressure when $N = 20$. These figures reveal that the final point-wise error estimates for $\boldsymbol{y}_N$, $\boldsymbol{u}_N$, and r_N are of the orders 10^{-15}, 10^{-14}, and 10^{-13}, respectively. All these numerical findings, presented in tables and figures, demonstrate a rapid decrease in errors and the achievement of spectral accuracy. They confirm that the spectral method provides a highly accurate approximation for the Stokes optimal control problem with an L^2-norm control constraint.

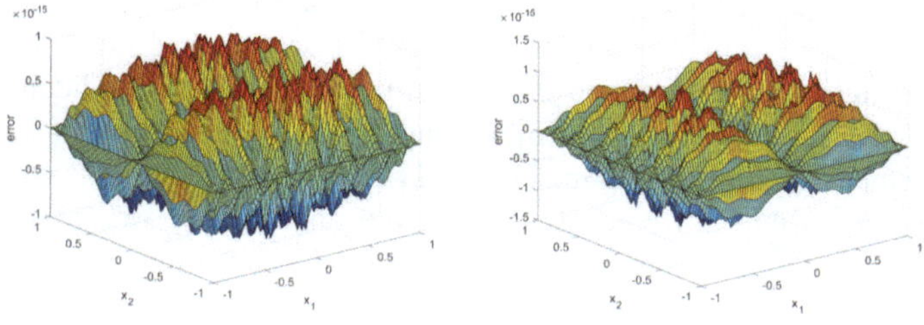

(a) (b)

Fig. 3.7 The point-wise error estimates of velocity when $N = 20$ for the Stokes problem. (**a**) $y_1 - y_{N1}$. (**b**) $y_2 - y_{N2}$

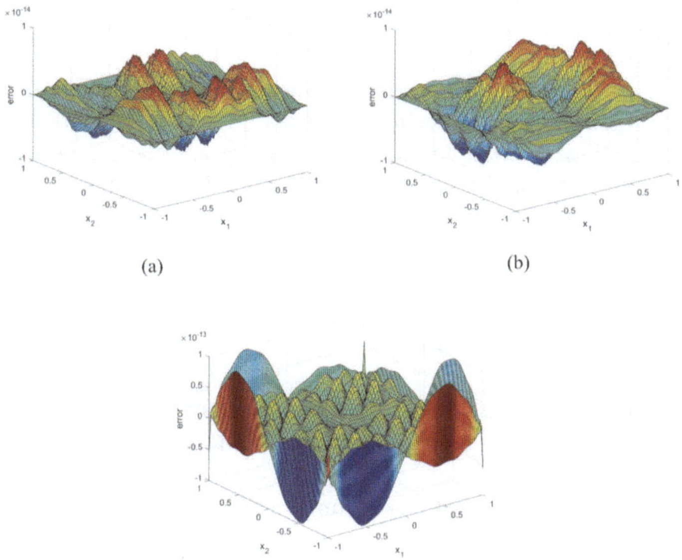

(a) (b)

(c)

Fig. 3.8 The point-wise error estimates of control and pressure when $N = 20$ for the Stokes problem. (**a**) $u_1 - u_{N1}$. (**b**) $u_2 - u_{N2}$. (**c**) $r - r_N$

Furthermore, in Table 3.3, the total error Σ and the error Σ_1 are defined as follows:

$$\Sigma = \|\boldsymbol{u} - \boldsymbol{u}_N\|_{0,\Omega} + \left\|\boldsymbol{y} - \boldsymbol{y}_N\right\|_{1,\Omega} + \left\|\boldsymbol{y}^* - \boldsymbol{y}_N^*\right\|_{1,\Omega} + \|r - r_N\|_{0,\Omega} + \left\|r^* - r_N^*\right\|_{0,\Omega},$$

$$\Sigma_1 = \|\boldsymbol{u} - \boldsymbol{u}_N\|_{0,\Omega} + \left\|\boldsymbol{y} - \boldsymbol{y}_N\right\|_{0,\Omega} + \left\|\boldsymbol{y}^* - \boldsymbol{y}_N^*\right\|_{0,\Omega}.$$

From Table 3.3, it is evident that the error estimator η provides a good indication of the total error Σ. Similarly, the error estimator ζ accurately reflects the error Σ_1. These observations align with the theoretical results obtained in Theorems 3.3.10 and 3.3.11, further validating their importance in the development of hp-adaptive spectral element methods for optimal control problems.

Furthermore, Table 3.3 presents data on the convergence rate of the spectral method. We calculate the convergence order using the following formula:

$$\log_{\frac{N_{i+1}}{N_i}} \frac{\mathrm{err}_i}{\mathrm{err}_{i+1}},$$

where $\mathrm{err}_i = \|\boldsymbol{u} - \boldsymbol{u}_N\|_{0,\Omega}$ or $\|\boldsymbol{y} - \boldsymbol{y}_N\|_{0,\Omega}$, $i = 1, 2, \cdots, 5$, The subscripts denote information from the corresponding i-th and $(i + 1)$-th steps. Order($\boldsymbol{u}$) and Order($\boldsymbol{y}$) indicate the convergence orders of L^2-norm errors for control $\boldsymbol{u}$ and velocity $\boldsymbol{y}$, respectively.

When comparing the spectral method to the finite element method, it's notable that the convergence orders Order($\boldsymbol{u}$) and Order($\boldsymbol{y}$) for the finite element method do not exceed 4 (as referenced in [251]). This is significantly lower than the convergence orders achieved by the spectral method. Consequently, the spectral method demonstrates superior performance in terms of convergence when compared to the finite element method.

3.3.5.2 The Navier-Stokes Case

To demonstrate the feasibility of the spectral method for the Navier-Stokes optimal control problem and justify the assertions made in Remark 3.3.12, we present an illustrative example below.

Let's consider the Navier-Stokes optimal control problem with a Reynolds number of $Re = 0.001$.

Find $(\boldsymbol{y}, r, \boldsymbol{u}) \in \boldsymbol{Y} \times M \times \boldsymbol{U}$ such that

$$\begin{aligned}
\min_{\boldsymbol{u}\in \boldsymbol{K}} J(\boldsymbol{y}, \boldsymbol{u}) &= \frac{1}{2}\|\boldsymbol{y} - \boldsymbol{y}_d\|_{0,\Omega}^2 + \frac{\alpha}{2}\|\boldsymbol{u}\|_{0,\Omega}^2, \\
-\Delta \boldsymbol{y} + Re(\boldsymbol{y}\cdot\nabla)\boldsymbol{y} + \nabla r &= \boldsymbol{f} + \boldsymbol{u}, \\
\nabla\cdot\boldsymbol{y} = 0 \text{ in } \Omega, &\quad \boldsymbol{y} = 0 \text{ on } \partial\Omega.
\end{aligned}$$

Correspondingly, the data of this example are listed as follows.

$$\boldsymbol{f} = \left(1 - \frac{1}{\sqrt{6}}\right)\boldsymbol{\omega} + 2\boldsymbol{h} - Re\pi^{-3}(\widetilde{\boldsymbol{\omega}} + \overline{\boldsymbol{\omega}}),$$

$$\boldsymbol{y}_d = \left(\frac{1}{\pi^2} + \frac{\pi^2}{\sqrt{6}}\right)\boldsymbol{\omega} + \frac{2\pi^2}{\sqrt{6}}\boldsymbol{h} + \frac{2Re}{\sqrt{6}\pi}\widetilde{\boldsymbol{\omega}} + \frac{Re}{\sqrt{6}\pi}(\overline{\boldsymbol{\omega}} - \widehat{\boldsymbol{\omega}}),$$

where

$$\boldsymbol{\omega} = \left(\left(1+\cos\left(\pi x_1\right)\right)\sin\left(\pi x_2\right), -\sin\left(\pi x_1\right)\left(1+\cos\left(\pi x_2\right)\right)\right),$$

$$\boldsymbol{h} = \left(\cos\left(\pi x_1\right)\sin\left(\pi x_2\right), 0\right),$$

$$\widetilde{\boldsymbol{\omega}} = \left(\sin\left(\pi x_1\right)\cos\left(\pi x_1\right)\left(1+\cos\left(\pi x_2\right)\right), \left(1+\cos\left(\pi x_1\right)\right)\sin\left(\pi x_2\right)\cos\left(\pi x_2\right)\right),$$

$$\overline{\boldsymbol{\omega}} = \left(\sin\left(\pi x_1\right)\left(1+\cos\left(\pi x_2\right)\right), \left(1+\cos\left(\pi x_1\right)\right)\sin\left(\pi x_2\right)\right),$$

$$\widehat{\boldsymbol{\omega}} = \Big(\sin\left(\pi x_1\right)\left(1+2\cos\left(\pi x_1\right)\right)\left(\sin\left(\pi x_2\right)\right)^2,$$
$$\left(\sin\left(\pi x_1\right)\right)^2\left(1+2\cos\left(\pi x_2\right)\right)\sin\left(\pi x_2\right)\Big).$$

Its exact solutions are given by

$$\boldsymbol{y} = \frac{1}{\pi^2}\boldsymbol{\omega}, \quad \boldsymbol{u} = \frac{1}{\sqrt{6}}\boldsymbol{\omega}, \quad \boldsymbol{y}^* = -\frac{1}{\sqrt{6}}\boldsymbol{\omega}, \quad \lambda = \frac{1}{4},$$
$$r = \frac{1}{\pi}\sin\left(\pi x_1\right)\sin\left(\pi x_2\right), \quad r^* = \frac{\pi}{\sqrt{6}}\sin\left(\pi x_1\right)\sin\left(\pi x_2\right).$$

Then, to solve the problem described above, we employ the gradient projection algorithm with modifications to Steps 2 and 3 of Algorithm 2. The updated steps are denoted as Steps 2′ and 3′, respectively.

Step 2′ : Get $\left(\boldsymbol{y}_N^{(n)}, r_N^{(n)}\right) \in \boldsymbol{Y}_N \times M_{N-2}$ such that

$$a\left(\boldsymbol{y}_N^{(n)}, \boldsymbol{w}_N\right) + Re\left(\left(\boldsymbol{y}_N^{(n)}\cdot\nabla\right)\boldsymbol{y}_N^{(n)}, \boldsymbol{w}_N\right) - b\left(\boldsymbol{w}_N, r_N^{(n)}\right) = \left(\boldsymbol{f} + \boldsymbol{u}_N^{(n)}, \boldsymbol{w}_N\right),$$
$$\forall\, \boldsymbol{w}_N \in \boldsymbol{Y}_N,$$
$$b\left(\boldsymbol{y}_N^{(n)}, \phi_N\right) = 0, \quad \forall\, \phi_N \in M_{N-2}.$$

Step 3′ : Get $\left(\boldsymbol{y}_N^{*(n)}, r_N^{*(n)}\right) \in \boldsymbol{Y}_N \times M_{N-2}$ such that

$$a\left(\boldsymbol{q}_N, \boldsymbol{y}_N^{*(n)}\right) - Re\left(\left(\boldsymbol{y}_N^{(n)}\cdot\nabla\right)\boldsymbol{y}_N^{*(n)}, \boldsymbol{q}_N\right) + Re\left(\left(\nabla\boldsymbol{y}_N^{(n)}\right)^\top\boldsymbol{y}_N^{*(n)}, \boldsymbol{q}_N\right)$$
$$+b\left(\boldsymbol{q}_N, r_N^{*(n)}\right) = \left(\boldsymbol{y}_N^{(n)} - \boldsymbol{y}_d, \boldsymbol{q}_N\right), \quad \forall\, \boldsymbol{q}_N \in \boldsymbol{Y}_N,$$
$$b\left(\boldsymbol{y}_N^{*(n)}, \psi_N\right) = 0, \quad \forall\, \psi_N \in M_{N-2}.$$

Moreover, we adopt the Stokes iteration to deal with $Re\left((\boldsymbol{y}_N^{(n)} \cdot \nabla)\boldsymbol{y}_N^{(n)},\ \boldsymbol{w}_N\right)$ in solving the state equation [114]. The same techniques are utilized in

$$Re\left(\left(\boldsymbol{y}_N^{(n)} \cdot \nabla\right) \boldsymbol{y}_N^{*(n)}, \boldsymbol{q}_N\right), \quad Re\left(\left(\nabla \boldsymbol{y}_N^{(n)}\right)^\top \boldsymbol{y}_N^{*(n)}, \boldsymbol{q}_N\right),$$

of the costate equation.

The numerical results, presented in Table 3.4 and Fig. 3.9, clearly demonstrate that the error estimates decrease rapidly. Furthermore, Figs. 3.10 and 3.11 show that, for $N = 12$, the magnitudes of point-wise error estimates for $\boldsymbol{y}_N$, $\boldsymbol{u}_N$ and r_N are all at the level of 10^{-6}. These empirical findings strongly support the validity of the spectral method for the Navier-Stokes optimal control problem.

Additionally, as can be seen from Table 3.4, the error estimator η_1 provides an excellent indication of the total error Σ. This, to some extent, confirms our hypothesis stated in Remark 3.3.12.

Table 3.4 Error results and indicators η_1 for the Navier-Stokes problem

N	4	6	8	10	12
$\lVert \boldsymbol{u} - \boldsymbol{u}_N \rVert_{0,\Omega}$	2.1205e-1	2.1146e-2	1.0127e-3	2.9748e-5	4.6690e-6
$\lVert \boldsymbol{y} - \boldsymbol{y}_N \rVert_{1,\Omega}$	3.4839e-1	4.7157e-2	2.9317e-3	1.0468e-4	8.8421e-6
$\lVert r - r_N \rVert_{0,\Omega}$	2.5715e-1	4.4363e-2	2.7753e-3	1.0680e-4	3.6780e-5
$\lVert \boldsymbol{y}^* - \boldsymbol{y}_N^* \rVert_{1,\Omega}$	3.1197e+0	1.8114e-1	1.1461e-2	4.1248e-4	3.4989e-5
$\lVert r^* - r_N^* \rVert_{0,\Omega}$	1.0238e+0	1.7913e-1	1.1186e-2	4.1640e-4	1.0156e-4
$\lVert \boldsymbol{y} - \boldsymbol{y}_N \rVert_{0,\Omega}$	5.6837e-2	5.5325e-3	2.5949e-4	7.6005e-6	1.4564e-6
$\lVert \boldsymbol{y}^* - \boldsymbol{y}_N^* \rVert_{0,\Omega}$	2.1187e-1	2.1145e-2	1.0127e-3	2.9990e-5	6.0261e-6
Σ	3.1534e+0	4.7293e-1	2.9367e-2	1.0701e-3	1.8684e-4
η_1	3.7276e+0	5.1293e-1	3.0100e-2	1.0323e-3	6.4454e-5

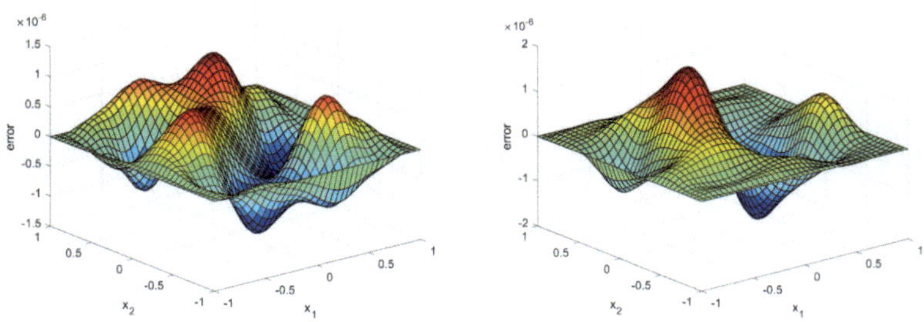

Fig. 3.9 Error estimates for the Navier-Stokes problem

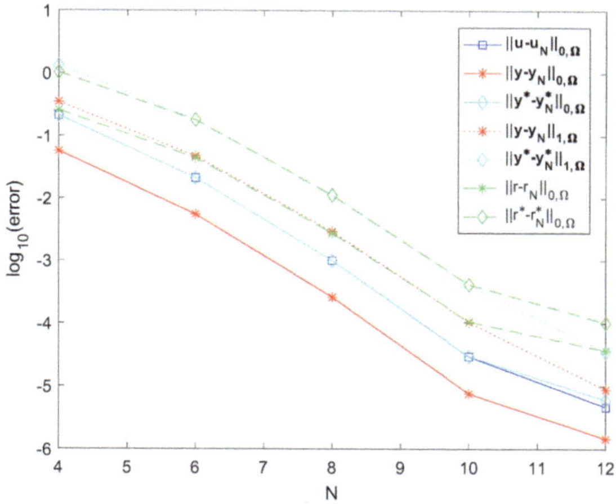

Fig. 3.10 The point-wise error estimates of velocity when $N = 12$ for the Navier-Stokes problem. (**a**) $y_1 - y_{N1}$. (**b**) $y_2 - y_{N2}$

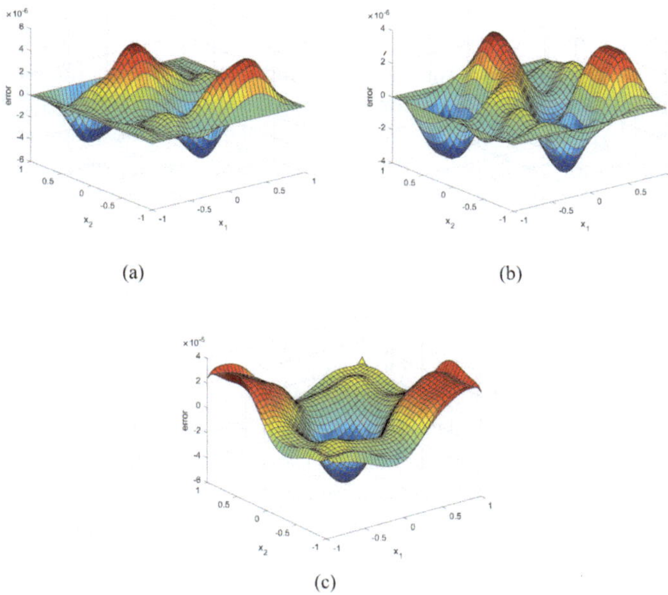

Fig. 3.11 The point-wise error estimates of control and pressure when $N = 12$ for the Navier-Stokes problem. (**a**) $u_1 - u_{N1}$. (**b**) $u_2 - u_{N2}$. (**c**) $r - r_N$

3.4 A Fourth-Order Parabolic Optimal Control Problem

In this section, we focus on an optimal control problem governed by fourth-order parabolic PDEs under three distinct control constraint scenarios: no constraint, integral mean constraint on control, and energy constraint on control, as referenced in [228].

Let $T > 0$. The spatial domain Ω represents an open interval in $\mathbb{R}$ or an open rectangular region in $\mathbb{R}^2$. The optimal control problem model that we consider involves the state variable y and the control variable u and is formulated as follows:

$$\min_{u \in U_{ad}} J(y, u) = \frac{1}{2} \int_0^T \|y(\cdot, t) - y_d(\cdot, t)\|^2_{L^2(\Omega)} \, dt + \frac{\alpha}{2} \int_0^T \|u(\cdot, t)\|^2_{L^2(\Omega)} \, dt, \tag{3.4.1}$$

subject to

$$\begin{cases} \partial_t y + \Delta (a_1(x) \Delta y) = f + u, \quad x \in \Omega, \ t \in (0, T], \\ y|_{\partial\Omega} = \left.\dfrac{\partial y}{\partial \mathbf{n}}\right|_{\partial\Omega} = 0, \quad t \in [0, T], \\ y(x, 0) = y_0(x), \quad x \in \Omega. \end{cases} \tag{3.4.2}$$

In this context, f represents a known force term, y_d denotes the desired objective state, and $\alpha > 0$ is a preset regularization parameter. The coefficient $a_1(\cdot)$ is a given function that solely depends on the spatial variable x within the domain Ω. Moreover, $\mathbf{n}$ signifies the outward unit normal vector on the boundary $\partial\Omega$. The symbol Δ stands for the Laplace operator with respect to the spatial variable $x \in \Omega$. The control constraint set U_{ad} is a nonempty, convex, and closed subset within the control set U. It's worth mentioning that specific assumptions regarding these known functions will be introduced in the subsequent subsection.

As a matter of fact, many important physical processes, including phase transitions [71], thin-film theory [155], and lubrication theory [240], can be accurately modeled using fourth-order parabolic equations. For example, when the state variable y represents the height from the surface of a film, these equations can depict the progression of epitaxial growth in nanoscale thin films [113]. Evidently, among the vast array of optimal control problems encountered in engineering and sciences, studying optimal control for fourth-order parabolic equations holds fundamental importance. Furthermore, given the analytical complexity inherent in practical engineering problems, deriving analytical solutions for optimal control problems is often not feasible, except in a few highly specialized cases. Therefore, there's a need to devise effective algorithms to obtain numerical solutions.

In general, the methods chosen to discretize the space and time of parabolic optimization problems can be similar but are not necessarily the same. For instance, researchers have discretized parabolic optimization problems using various approaches such as the FE-FD method (combining the finite element method in space with the finite difference method in time) [92, 169], the space-time finite element method [168, 174], the SP-FD method (integrating the spectral method in space with the finite difference method in time) [54], and the finite difference method alone [223]. Moreover, for parabolic optimization problems, studies such as [92, 168, 169, 174] explore a priori error analysis for these discretization systems.

Meanwhile, [54] derives a posteriori error estimates. The paper [161] introduces a hp-spectral element approximation scheme tailored for a parabolic optimal control problem and provided a posteriori error estimates for this approach.

Up to now, a priori error analysis for the spectral discretization of optimal control problems has been extensively discussed in various publications. For instance, Chen et al. [52, 55, 119, 175] explore this analysis in the context of second-order elliptic optimization problems, while Zhou et al. [251, 253] and Tao et al. [226] focus on fourth-order elliptic optimization problems. Additionally, [53] addresses optimal control problems governed by the Stokes equations.

However, when it comes to parabolic optimization problems, a priori error analysis for their space-time spectral discretization remains scarce. The paper [122] stands as the sole paper that delves into this discussion. It effectively tackles the challenge posed by the accuracy mismatch inherent in the SP-FD method, specifically targeting a second-order parabolic PDE within the control system.

Importantly, no prior research has established a priori error estimates for the space-time spectral discretization of a fourth-order parabolic optimal control system. This gap in the literature underscores the novelty and significance of the present study.

In this section, we focus on the space-time spectral discretization of an optimal control problem governed by fourth-order parabolic PDEs with control constraints. Furthermore, we provide detailed constructions of a priori error estimates. Significantly, our investigations encompass not only the scenario without control constraints but also two other types of control constraints for the parabolic control system represented by Eqs. (3.4.1)–(3.4.2). This comprehensive approach allows for a more thorough understanding and analysis of the problem at hand.

Next, we introduce some notations to facilitate our discussion. Let k be a non-negative integer. The standard notation $H^k(I)$ represents the usual Sobolev space on the domain I equipped with the norm $\|\cdot\|_{H^k(I)}$ and the inner product $\langle\cdot,\cdot\rangle_{H^k(I)}$. It's worth noting that the Lebesgue space $L^2(I)$ is a special case of $H^k(I)$ when $k = 0$. For the domain Ω, we adopt similar notations:

$$H_0^2(\Omega) := \left\{ v \in H^2(\Omega) \;\middle|\; v|_{\partial\Omega} = \left.\frac{\partial v}{\partial \mathbf{n}}\right|_{\partial\Omega} = 0 \right\}.$$

Define the time interval $I_t := (a, b)$. Let ω be a positive weighted function in I_t. For a Sobolev space H, define the function space

$$H^m(I_t; H) = \left\{ w \;\middle|\; \int_a^b \left\| \partial_t^s w(\cdot, t) \right\|_H^2 \, \mathrm{d}t < \infty, \;\; 0 \le s \le m \right\},$$

$$\|w\|_{H^m(I_t;H)} = \left[\sum_{s=0}^{m} \int_a^b \left\| \partial_t^s w(\cdot, t) \right\|_H^2 \, \mathrm{d}t \right]^{\frac{1}{2}}, \quad m \ge 0,$$

and the weighted function space

$$L^2_\omega(I_t; H) = \left\{ w \,\middle|\, \int_a^b \omega \, \|w(\cdot, t)\|_H^2 \, \mathrm{d}t < \infty \right\},$$

$$\|w\|_{L^2_\omega(I_t;H)} = \left[\int_a^b \omega \, \|w(\cdot, t)\|_H^2 \, \mathrm{d}t \right]^{\frac{1}{2}},$$

where $\partial_t^s w = \frac{\partial^s w}{\partial t^s}$. Without any confusion arising from this, ω (if $\omega \equiv 1$) can be omitted from the notations. Furthermore, it is straightforward to verify that $H^m(I_t; H)$ is $L^2(I_t; H)$ when $m = 0$.

In addition, within this section, the letters C_i and c_i ($i = 0, 1, 2, \cdots$) represent some general positive constants. These constants are independent of the functions y_d and f, as well as the spectral orders N and M (which refer to the degrees of polynomials in the spectral discrete system that will be introduced later).

3.4.1 The Spectral Discretization

In this subsection, we present the optimal control problem, subject to control constraints, which is governed by a fourth-order parabolic PDE. We also introduce some assumptions that underlie our analysis. Our objective is to discretize the parabolic optimization problem using the space-time spectral method and establish the optimality conditions for both the original control system and its spectral discrete counterpart.

3.4.1.1 Preliminary

We assume without loss of generality that $y_0(\cdot) \equiv 0$. If not, the transformation $\hat{y} = y - y_0$ can be applied to ensure this condition. Furthermore, to simplify the representation of Eqs. (3.4.1)–(3.4.2), we set $\Omega := (-1, 1)^d$, $d = 1$ or 2 and remap the time interval $(0, T)$ to $I_t := (-1, 1)$. Denote $X = L^2\left(I_t; L^2(\Omega)\right)$, $Y = H_0^2(\Omega)$, and $V = L^2(I_t; Y) \cap H^1\left(I_t; L^2(\Omega)\right)$. Take the control space U, the state space V_l, and the adjoint state space V_r as

$$U = X, \quad V_l = \{w \in V \mid w(x, -1) = 0, \ \forall\, x \in \Omega\},$$
$$V_r = \{w \in V \mid w(x, 1) = 0, \ \forall\, x \in \Omega\}.$$

The standard inner product in X is denoted as $\langle w, v \rangle_X = \int_{-1}^1 \int_\Omega wv \, \mathrm{d}x \, \mathrm{d}t$ for $w, v \in X$.

Then, we consider the parabolic optimal control problem of the subsequent form. Find $(y, u) \in V_l \times U$ such that

$$\min_{u \in U_{ad}} J(y, u) = \frac{1}{2} \int_{-1}^{1} \|y(\cdot, t) - y_d(\cdot, t)\|_{L^2(\Omega)}^2 \,\mathrm{d}t + \frac{\alpha}{2} \int_{-1}^{1} \|u(\cdot, t)\|_{L^2(\Omega)}^2 \,\mathrm{d}t, \tag{3.4.3}$$

$$\begin{cases} \partial_t y + \Delta(a_1(x)\Delta y) = f + u, \quad x \in \Omega,\ t \in (-1, 1), \\ y|_{\partial\Omega} = \left.\dfrac{\partial y}{\partial \mathbf{n}}\right|_{\partial\Omega} = 0, \quad t \in [-1, 1], \\ y(x, -1) = 0, \quad x \in \Omega. \end{cases} \tag{3.4.4}$$

Furthermore, to guarantee the existence and uniqueness of the solution to the aforementioned parabolic optimization problem, we impose certain assumptions on the known functions f, y_d, and the coefficient function a_1 appearing in Eqs. (3.4.3)–(3.4.4). These assumptions are enumerated below.

Assumption 3.4.1 *Suppose that* $f \in L^2\left(I_t; L^2(\Omega)\right)$ *and* $y_d \in L^2\left(I_t; L^2(\Omega)\right)$. *Set* $a_1(\cdot) \in L^\infty(\Omega)$ *and*

$$c_1 = \|a_1\|_{L^\infty(\Omega)} := \operatorname*{ess\,sup}_{x \in \Omega} |a_1(x)| < +\infty.$$

In addition, here exists a constant $c_0 > 0$ *such that* $a_1(x) \geq c_0$ *almost everywhere in* Ω.

In the subsequent subsections, we firstly introduce the standard weak formulation of the state equation given by (3.4.4). Specifically, the objective is to find $y \in V_l$ satisfying the following conditions:

$$\langle \partial_t y, w \rangle_X + \langle a_1(x)\Delta y, \Delta w \rangle_X = \langle f + u, w \rangle_X, \quad \forall\, w \in V_r. \tag{3.4.5}$$

Next, we endeavor to derive the sufficient conditions that guarantee the existence and uniqueness of solutions to Eq. (3.4.5). Leveraging Assumption 3.4.1 and Hölder's inequality [2], for $y, w \in V$,

$$|\langle a_1(x)\Delta y, \Delta w \rangle_X| \leq \|a_1\|_{L^\infty(\Omega)} \int_{-1}^{1} \int_{\Omega} |\Delta y \cdot \Delta w| \,\mathrm{d}x \,\mathrm{d}t \leq c_1 \|\Delta y\|_X \|\Delta w\|_X,$$

which leads to

$$|\langle a_1(x)\Delta y, \Delta w \rangle_X| \leq c_1 \|\Delta y\|_X \|\Delta w\|_X \leq c_1 \|y\|_V \|w\|_V. \tag{3.4.6}$$

By the Poincaré inequality (Theorem 1.1.10) [2], there exist two constants c_Ω and C_Ω (dependent on Ω only) such that

$$\|v\|_{L^2(\Omega)} \leq c_\Omega \|\Delta v\|_{L^2(\Omega)}, \quad \|v\|_{H^2(\Omega)} \leq C_\Omega \|\Delta v\|_{L^2(\Omega)}, \quad \forall\, v \in H_0^2(\Omega).$$

This result along with Assumption 3.4.1 implies that, for $y \in V$,

$$|\langle a_1(x)\Delta y, \Delta y\rangle_X| \geq c_0 \|\Delta y\|_X^2 \geq c_0 C_\Omega^{-2} \|y\|_V^2 . \tag{3.4.7}$$

Based on the derived results (3.4.6) and (3.4.7), we conclude that the state equation (3.4.5) possesses a unique solution for any given $u \in U$, as stated by Theorem 1.2 in Chapter 3 of [153].

Consequently, we arrive at the following weak formulation for the parabolic optimization problem described by (3.4.3)–(3.4.4).

Find $(y, u) \in V_l \times U$ such that

$$(\mathscr{P}) \quad \begin{cases} \min\limits_{u \in U_{ad}} J(y, u) = \dfrac{1}{2}\displaystyle\int_{-1}^{1} \|y(\cdot, t) - y_d(\cdot, t)\|_{L^2(\Omega)}^2 \mathrm{d}t \\ +\dfrac{\alpha}{2}\displaystyle\int_{-1}^{1} \|u(\cdot, t)\|_{L^2(\Omega)}^2 \,\mathrm{d}t, \\ \langle \partial_t y, w\rangle_X + \langle a_1(x)\Delta y, \Delta w\rangle_X = \langle f + u, w\rangle_X, \quad \forall\, w \in V_r. \end{cases}$$

Moreover, given that U_{ad} is nonempty, convex, and closed, the solution to the parabolic optimization problem $(\mathscr{P})$ exists and is unique (refer to Chapter 3 in [153] for details).

Theorem 3.4.1 *The control-constrained parabolic optimization problem $(\mathscr{P})$ has a unique solution $(y, u) \in V_l \times U$. Furthermore, alongside the adjoint state $p \in V$, the triplet $(u, y, p) \in U \times V_l \times V_r$ is uniquely characterized by satisfying the first-order optimality conditions, denoted as $(\mathscr{Q})$ and stated as follows:*

$$\begin{aligned} &\text{(a) The state equation } \langle \partial_t y, w\rangle_X + \langle a_1(x)\Delta y, \Delta w\rangle_X = \langle f + u, w\rangle_X, \ \forall\, w \in V_r, \\ &\text{(b) The adjoint equation } -\langle q, \partial_t p\rangle_X + \langle a_1(x)\Delta q, \Delta p\rangle_X = \langle y - y_d, q\rangle_X, \\ &\quad \forall\, q \in V_l, \\ &\text{(c) The gradient equation } u = \Pi(p), \end{aligned} \tag{3.4.8}$$

where the continuous projection operator $\Pi : U \to U_{ad}$ represents the best approximation of $\frac{1}{\alpha}p$ in U_{ad} for a given p. Namely, $\Pi(p)$ is uniquely determined by

$$\langle \alpha\Pi(p) + p, \Pi(p) - v\rangle_X \leq 0, \quad \forall\, v \in U_{ad}. \tag{3.4.9}$$

Remark 3.4.2 The conclusions stated above also hold true for constrained optimal control problems that can be reformulated as control-constrained problems with a convex, closed, and nonempty control constraint set U_{ad}. Moreover, it can be easily deduced that (3.4.8)(c) and (3.4.9) are equivalent to

$$\langle \alpha u + p, u - v \rangle_X \le 0, \quad \forall\, v \in U_{ad}. \tag{3.4.10}$$

Now, we enumerate three specific cases of the control constraint set that we mentioned at the beginning of this subsection. It should be noted that, in this subsection, our focus is solely on these three cases.

Case 1 (the case with no constraint). Let $U_{ad} = U$, representing the simplest scenario. In this case, it is straightforward to observe that the projection operator Π takes the following form:

$$\Pi(p) = -\frac{1}{\alpha} p.$$

Case 2 (integral mean constraint on control). The control constraint set is given by

$$U_{ad} = \left\{ v \in U \,\middle|\, \int_{-1}^{1} \int_{\Omega} v(x,t)\, \mathrm{d}x\, \mathrm{d}t \ge 0 \right\} \ne \emptyset.$$

At this point, one can verify that the projection operator Π satisfies the following properties:

$$\Pi(p) = \begin{cases} -\dfrac{1}{\alpha} p, & \displaystyle\int_{-1}^{1} \int_{\Omega} p\, \mathrm{d}x\, \mathrm{d}t < 0, \\ -\dfrac{1}{\alpha} \left(p - \overline{p} \right), & \text{otherwise}, \end{cases} \tag{3.4.11}$$

where

$$\overline{p} = \frac{\displaystyle\int_{-1}^{1} \int_{\Omega} p\, \mathrm{d}x\, \mathrm{d}t}{\displaystyle\int_{-1}^{1} \int_{\Omega} 1\, \mathrm{d}x\, \mathrm{d}t}.$$

Moreover, (3.4.11) can be simplified to

$$\Pi(p) = \frac{1}{\alpha} \left(\max\{0, \overline{p}\} - p \right).$$

Case 3 (energy constraint on control). The control constraint set is

$$U_{ad} = \{ v \in U \mid \|v\|_X \le \gamma \} \ne \emptyset,$$

where γ is a positive constant. The projection operator Π satisfies

$$\Pi(p) = \begin{cases} -\dfrac{1}{\alpha}p, & \dfrac{1}{\alpha}\|p\|_X \le \gamma, \\ -\dfrac{\gamma}{\|p\|_X}p, & \text{otherwise}, \end{cases}$$

which can be simplified to

$$\Pi(p) = -\frac{\gamma}{\max\left\{\|p\|_X, \alpha\gamma\right\}}p.$$

3.4.1.2 The Discretization

In this subsection, we employ the space-time spectral method to discretize the parabolic optimization problem ($\mathscr{P}$).

Let Q_N (Q_M) be the space of all algebraic polynomials of degree less than or equal to the spectral orders N (M) with respect to each single variable x_i, $i = 1, 2$, (t). Take the finite-dimensional space

$$Y_N = Q_N^0 := \left\{ w \in Q_N \;\middle|\; w|_{\partial\Omega} = \left.\frac{\partial w}{\partial \mathbf{n}}\right|_{\partial\Omega} = 0 \right\}$$

as the approximation space to Y. Then, denote $\otimes$ a tensor product and

$$A \otimes B = \text{span}\{w_1 \cdot w_2 \mid w_1 \in A,\; w_2 \in B\}$$

for the linear spaces A and B. We introduce the finite-dimensional spaces U_S, V_{Sl}, and V_{Sr} as approximations to spaces U, V_l, and V_r, respectively. These finite-dimensional spaces are defined as follows:

$$U_S = Q_N \otimes Q_M, \quad V_{Sl} = Y_N \otimes Q_{Ml}, \quad V_{Sr} = Y_N \otimes Q_{Mr},$$

where

$$Q_{Ml} = \{w \in Q_M \mid w(-1) = 0\}, \quad Q_{Mr} = \{w \in Q_M \mid w(1) = 0\}.$$

Let $U_{ad}^S = U_S \cap U_{ad}$. The space-time spectral approximation of the parabolic optimization problem is then formulated as follows.

Find $(y_S, u_S) \in V_{Sl} \times U_S$ such that

$$(\mathscr{P}_S)\quad \begin{cases} \min\limits_{u_S \in U_{ad}^S} J(y_S, u_S) = \dfrac{1}{2}\displaystyle\int_{-1}^{1} \|y_S(\cdot,t) - y_d(\cdot,t)\|^2_{L^2(\Omega)}\,\mathrm{d}t \\ \qquad\qquad + \dfrac{\alpha}{2}\displaystyle\int_{-1}^{1} \|u_S(\cdot,t)\|^2_{L^2(\Omega)}\,\mathrm{d}t, \\ \langle \partial_t y_S, w_S\rangle_X + \langle a_1(x)\Delta y_S, \Delta w_S\rangle_X = \langle f + u_S, w_S\rangle_X\,, \quad \forall\, w_S \in V_{Sr}. \end{cases}$$

Here, we introduce the discrete projection operator $\Pi_S : U_S \to U_{ad}^S$, which fulfills the following condition:

$$\langle \alpha \Pi_S(p_S) + p_S, \Pi_S(p_S) - v_S\rangle_X \le 0, \quad \forall\, v_S \in U_{ad}^S. \tag{3.4.12}$$

Clearly, the discrete projection operator Π_S possesses a form analogous to that of the operator Π (refer to Cases 1–3 in Sect. 3.4.1.1), although we will not elaborate on the specifics.

By employing similar reasoning as in Sect. 3.4.1.1, we can prove the existence and uniqueness of a spectral discrete solution $(y_S, u_S) \in V_{Sl} \times U_S$ for the parabolic optimization problem $(\mathscr{P}_S)$. Together with the discrete adjoint state $p_S \in V_{Sr}$, the triplet $(u_S, y_S, p_S) \in U_S \times V_{Sl} \times V_{Sr}$ fulfills the following discrete optimality conditions $(\mathscr{Q}_S)$:

$$\begin{aligned} &\text{(a) } \textit{The state equation } \langle \partial_t y_S, w_S\rangle_X + \langle a_1(x)\Delta y_S, \Delta w_S\rangle_X = \langle f + u_S, w_S\rangle_X\,, \\ &\qquad\qquad \forall\, w_S \in V_{Sr}, \\ &\text{(b) } \textit{The adjoint equation } -\langle q_S, \partial_t p_S\rangle_X + \langle a_1(x)\Delta q_S, \Delta p_S\rangle_X = \langle y_S - y_d, q_S\rangle_X\,, \\ &\qquad\qquad \forall\, q_S \in V_{Sl}, \\ &\text{(c) } \textit{The gradient equation } u_S = \Pi_S(p_S). \end{aligned} \tag{3.4.13}$$

Then, it is straightforward to deduce that (3.4.12) and (3.4.13)(c) are equivalent to the following inequality:

$$\langle \alpha u_S + p_S, u_S - v_S\rangle_X \le 0, \quad \forall\, v_S \in U_{ad}^S. \tag{3.4.14}$$

3.4.2 A Priori Error Estimates

In this subsection, our primary objective is to establish a priori error estimates for the spectral approximation of the parabolic optimization problem $(\mathscr{P})$. To achieve this, we first recall three crucial properties of projection operators, which are listed below:

Assumption 3.4.2 *Let* $\mathbb{P}_Z : L^2(I) \to Q_Z$ *be an* L^2*-orthogonal projection operator, which satisfies, for any* $r \in L^2(I)$,

$$\langle r - \mathbb{P}_Z r, w \rangle_{L^2(I)} = 0, \quad \forall\, w \in Q_Z.$$

Then (see, e.g., [40], Chapter 9, §7)

$$\|r - \mathbb{P}_Z r\|_{L^2(I)} \leq C_0 Z^{-m} \|r\|_{H^m(I)},$$

provided $r \in H^m(I)$ *with* $m \geq 0$.

Assumption 3.4.3 *Define the projection operator* $\mathbb{P}^0_{2,N} : H^2_0(\Omega) \to Q^0_N$, *which satisfies, for any* $v \in H^2_0(\Omega)$, *that*

$$\left\langle \Delta \left(v - \mathbb{P}^0_{2,N} v \right), \Delta w \right\rangle_{L^2(\Omega)} = 0, \quad \forall\, w \in Q^0_N.$$

Then there is (see [40], Chapter 9, §7),

$$\left\| v - \mathbb{P}^0_{2,N} v \right\|_{H^k(\Omega)} \leq C_1 N^{k-m} \|v\|_{H^m(\Omega)}, \quad k = 0, 1, 2,$$

for any $v \in H^m(\Omega) \cap H^2_0(\Omega)$ *and* $m \geq 2$.

In addition, we introduce the Jacobi weighted function, denoted as follows:

$$\omega^{\alpha,\beta}(t) = (1-t)^\alpha (1+t)^\beta,$$

where $\alpha \geq -1$, $\beta \geq -1$. We denote the special cases $\omega^{0,-1}$ as ω_1 and $\omega^{-1,0}$ as ω_2, respectively, which will be frequently used in the subsequent analysis. The notation $L^2_\omega(I_t)$ refers to the weighted Lebesgue space equipped with the inner product defined as

$$\langle w, v \rangle_{L^2_\omega(I_t)} = \int_{I_t} \omega w v \, \mathrm{d}t$$

and the norm

$$\|w\|_{L^2_\omega(I_t)} = \int_{I_t} \omega w^2 \, \mathrm{d}t.$$

We now introduce two projection operators, defined as follows:

Assumption 3.4.4 *The projection operator* $\mathbb{P}_{\omega_1}^M : L^2_{\omega_1}(I_t) \to Q_{Ml}$ *is defined by, for any* $v \in L^2_{\omega_1}(I_t)$,

$$\left\langle v - \mathbb{P}_{\omega_1}^M v, w \right\rangle_{L^2_{\omega_1}(I_t)} = 0, \quad \forall\, w \in Q_{Ml}.$$

Then, if $v \in L^2_{\omega_1}(I_t)$ *and* $\partial_t^k v \in L^2_{\omega^{k,k-1}}(I_t)$ *for* $1 \le k \le l$, *there is* (see [207])

$$\left\| v - \mathbb{P}_{\omega_1}^M v \right\|_{L^2_{\omega_1}(I_t)} \le C_2 M^{-l} \left\| \partial_t^l v \right\|_{L^2_{\omega^{l,l-1}}(I_t)}.$$

For another, define the projection operator $\mathbb{P}_{\omega_2}^M : L^2_{\omega_2}(I_t) \to Q_{Mr}$, *which satisfies, for any* $v \in L^2_{\omega_2}(I_t)$,

$$\left\langle v - \mathbb{P}_{\omega_2}^M v, w \right\rangle_{L^2_{\omega_2}(I_t)} = 0, \quad \forall\, w \in Q_{Mr}.$$

Then, if $v \in L^2_{\omega_2}(I_t)$ *and* $\partial_t^k v \in L^2_{\omega^{k-1,k}}(I_t)$ *for* $1 \le k \le l$, *the following estimate* (see [28])

$$\left\| v - \mathbb{P}_{\omega_2}^M v \right\|_{L^2_{\omega_2}(I_t)} \le C_3 M^{-l} \left\| \partial_t^l v \right\|_{L^2_{\omega^{l-1,l}}(I_t)}$$

holds.

Furthermore, to simplify the notation, we define the following quantities:

$$W = L^2\left(I_t; H^\theta(\Omega)\right) \cap H^1\left(I_t; L^2(\Omega)\right).$$

The primary outcome of this subsection can be summarized in Theorem 3.4.3.

Theorem 3.4.3 (A priori Error Estimates). *Let* (u, y, p) *satisfy the optimality conditions* $(\mathscr{Q})$, *and let* (u_S, y_S, p_S) *satisfy the discrete optimality conditions* $(\mathscr{Q}_S)$, *respectively. If*

$$u \in W, \quad y \in W \cap H^\sigma\left(I_t; H^2(\Omega)\right) \cap L^2_{\omega_1}\left(I_t; H^\theta(\Omega)\right),$$

$$p \in W \cap H^\sigma\left(I_t; H^2(\Omega)\right) \cap L^2_{\omega_2}\left(I_t; H^\theta(\Omega)\right),$$

where θ *and* σ *are integers satisfying* $\theta \ge 2$ *and* $\sigma \ge 1$, *the following estimate holds:*

$$\|u - u_S\|_X + \|y - y_S\|_X + \|p - p_S\|_X \le \widehat{C}_1 N^{-\theta}\eta_1 + \widehat{C}_2 N^{2-\theta}\eta_2 + \widehat{C}_3 M^{-\sigma}\eta_3$$

with

$$\widehat{C}_1 = 2\sqrt{2}C_1 \max\left\{1 + 2c_\Omega c_0^{-\frac{1}{2}} C_4, C_4\right\},$$

$$\widehat{C}_2 = \sqrt{2}c_0^{-\frac{1}{2}} c_1 C_1 \max\left\{1 + 2c_\Omega c_0^{-\frac{1}{2}} C_4, \sqrt{2}C_4\right\},$$

$$\widehat{C}_3 = 4\max\{C_2, C_3\}\max\left\{2 + c_0^{-\frac{1}{2}} c_1\left(1 + 2c_\Omega c_0^{-\frac{1}{2}} C_4\right), \left(2 + \sqrt{2}c_1 c_0^{-\frac{1}{2}}\right) C_4\right\},$$

$$C_4 = 1 + \alpha^{-1}\left(1 + c_0^{-\frac{1}{2}} c_\Omega + c_0^{-1} c_\Omega^2\right),$$

and

$$\eta_1 = \|y\|_{L^2_{\omega_1}(I_t; H^\theta(\Omega))} + \|p\|_{L^2_{\omega_2}(I_t; H^\theta(\Omega))},$$

$$\eta_2 = \|y\|_{L^2(I_t; H^\theta(\Omega))} + \|p\|_{L^2(I_t; H^\theta(\Omega))},$$

$$\eta_3 = \|y\|_{H^\sigma(I_t; H^2(\Omega))} + \|p\|_{H^\sigma(I_t; H^2(\Omega))}.$$

Subsequently, to derive a priori error estimates for the problem $(\mathscr{P})$, we introduce a highly useful auxiliary system as follows.

Find $(y_S(u), p_S(u)) \in V_{Sl} \times V_{Sr}$ such that

$$\begin{aligned}
&\text{(a)}\ \langle \partial_t y_S(u), w_S\rangle_X + \langle a_1(x)\Delta y_S(u), \Delta w_S\rangle_X = \langle f + u, w_S\rangle_X, \ \ \forall\, w_S \in V_{Sr},\\
&\text{(b)}\ -\langle q_S, \partial_t p_S(u)\rangle_X + \langle a_1(x)\Delta q_S, \Delta p_S(u)\rangle_X = \langle y_S(u) - y_d, q_S\rangle_X, \ \ \forall\, q_S \in V_{Sl}.
\end{aligned} \tag{3.4.15}$$

We are now prepared to prove the theorem stated above, which we will do in four parts. The first three parts, comprising Lemmas 3.4.4, 3.4.5, and 3.4.6, constitute significant preparatory work. Specifically, Lemma 3.4.4 estimates the errors $\|y - y_S(u)\|_X$ and $\|p - p_S(u)\|_X$. Lemma 3.4.5 addresses the estimation of the errors $\|y_S(u) - y_S\|_X$ and $\|p_S(u) - p_S\|_X$. Lemma 3.4.6 focuses on estimating the error $\|u - u_S\|_X$. With the aid of these lemmas, we are left with only one critical section, namely, the ***Proof of Theorem* 3.4.3**.

Firstly, we estimate the errors $\|y - y_S(u)\|_X$ and $\|p - p_S(u)\|_X$.

Lemma 3.4.4 *There holds estimates:*

$$\|y - y_S(u)\|_X \le 2\sqrt{2}\,\|y - \widehat{y}\|_{L^2_{\omega_1}(I_t; L^2(\Omega))} + \sqrt{2}c_0^{-\frac{1}{2}} c_1\,\|\Delta(\widehat{y} - y)\|_X$$

and

$$\|p - p_S(u)\|_X \le 2\sqrt{2}\,\|\widehat{p} - p\|_{L^2_{\omega_2}(I_t;L^2(\Omega))} + 2c_1 c_0^{-\frac{1}{2}}\,\|\Delta(\widehat{p} - p)\|_X$$
$$+2c_\Omega c_0^{-\frac{1}{2}}\,\|y - y_S(u)\|_X\,, \tag{3.4.16}$$

where $\widehat{y} = \mathbb{P}^0_{2,N}\mathbb{P}^M_{\omega_1} y$ *and* $\widehat{p} = \mathbb{P}^0_{2,N}\mathbb{P}^M_{\omega_2} p$.

Proof We proceed in two steps to estimate the errors.

Step 1. By employing the continuous state equation (3.4.8)(a) and the auxiliary equation (3.4.15)(a), we obtain

$$\langle \partial_t(y - y_S(u)), w_S\rangle_X + \langle a_1(x)\Delta\,(y - y_S(u))\,, \Delta w_S\rangle_X = 0, \quad \forall\ w_S \in V_{Sr}.$$

Therefore, for any $w_S \in V_{Sr}$,

$$\begin{aligned}&\langle \partial_t(\widehat{y} - y_S(u)), w_S\rangle_X + \langle a_1(x)\Delta\,(\widehat{y} - y_S(u))\,, \Delta w_S\rangle_X\\ &\quad = \langle \partial_t(\widehat{y} - y), w_S\rangle_X + \langle a_1(x)\Delta\,(\widehat{y} - y)\,, \Delta w_S\rangle_X\,.\end{aligned} \tag{3.4.17}$$

In (3.4.17), set $w_S = \int_t^1 \omega_1(s)\,(\widehat{y}(\cdot,s) - y_S(u)(\cdot,s))\,\mathrm{d}s$, namely, the integral of $\omega_1\,(\widehat{y} - y_S(u))$ from t to 1 with respect to its time variable. To avoid any confusion caused by this, we briefly record that $w_S = \int_t^1 \omega_1\,(\widehat{y} - y_S(u))\,\mathrm{d}s$. Then, the first term on the left-hand side of (3.4.17) can be expressed as

$$\begin{aligned}\langle \partial_t\,(\widehat{y} - y_S(u))\,, w_S\rangle_X &= \int_{-1}^1\int_\Omega \partial_t\,(\widehat{y} - y_S(u))\int_t^1 \omega_1\,(\widehat{y} - y_S(u))\,\mathrm{d}s\,\mathrm{d}x\,\mathrm{d}t\\ &= \int_\Omega\left[\int_{-1}^1 \partial_t\,(\widehat{y} - y_S(u))\int_t^1 \omega_1\,(\widehat{y} - y_S(u))\,\mathrm{d}s\,\mathrm{d}t\right]\mathrm{d}x\\ &= \int_\Omega\int_{-1}^1 \omega_1\,(\widehat{y} - y_S(u))^2\,\mathrm{d}t\,\mathrm{d}x\\ &= \|\widehat{y} - y_S(u)\|^2_{L^2_{\omega_1}(I_t;L^2(\Omega))}\,,\end{aligned}$$

where we applied the technique of interchanging integral orders in the second equality.

For the second term on the left-hand side of (3.4.17), using the assumption of the lower bound on $a_1(\cdot)$ from Assumption 3.4.1 and applying integration by parts, we obtain

$$\begin{aligned}&\langle a_1(x)\Delta\,(\widehat{y} - y_S(u))\,, \Delta w_S\rangle_X\\ &= \int_{-1}^1\int_\Omega a_1(x)\Delta\,(\widehat{y} - y_S(u))\int_t^1 \omega_1\Delta\,(\widehat{y} - y_S(u))\,\mathrm{d}s\,\mathrm{d}x\,\mathrm{d}t\end{aligned}$$

$$
\begin{aligned}
&= -\frac{1}{2}\int_{\Omega} a_1(x)\left[\int_{-1}^{1} \omega_1^{-1}\,\mathrm{d}\left|\int_t^1 \omega_1 \Delta\left(\widehat{y}-y_S(u)\right)\,\mathrm{d}s\right|^2\right]\mathrm{d}x\\
&= \frac{1}{2}\int_{\Omega}\int_{-1}^{1} a_1(x)\left|\int_t^1 \omega_1 \Delta\left(\widehat{y}-y_S(u)\right)\,\mathrm{d}s\right|^2 \mathrm{d}t\,\mathrm{d}x\\
&= \frac{1}{2}\int_{-1}^{1}\int_{\Omega} a_1(x)\left|\int_t^1 \omega_1 \Delta\left(\widehat{y}-y_S(u)\right)\,\mathrm{d}s\right|^2 \mathrm{d}x\,\mathrm{d}t\\
&\geq \frac{1}{2}c_0\left\|\int_t^1 \omega_1 \Delta\left(\widehat{y}-y_S(u)\right)\,\mathrm{d}s\right\|_X^2 .
\end{aligned}
$$

Next, considering the two terms on the right-hand side of (3.4.17), we employ the technique of integration by parts along with the Cauchy-Schwarz inequality [2] to obtain the following estimates:

$$
\begin{aligned}
\langle \partial_t(\widehat{y}-y), w_S\rangle_X &= \int_{-1}^{1}\int_{\Omega} \partial_t\left(\widehat{y}-y\right)\int_t^1 \omega_1\left(\widehat{y}-y_S(u)\right)\,\mathrm{d}s\,\mathrm{d}x\,\mathrm{d}t\\
&= \int_{\Omega}\left[\int_{-1}^{1}\int_t^1 \omega_1\left(\widehat{y}-y_S(u)\right)\,\mathrm{d}s\,\mathrm{d}\left(\widehat{y}-y\right)\right]\mathrm{d}x\\
&= \int_{\Omega}\int_{-1}^{1} \omega_1\left(\widehat{y}-y_S(u)\right)\left(\widehat{y}-y\right)\,\mathrm{d}t\,\mathrm{d}x\\
&= \int_{-1}^{1}\int_{\Omega} \omega_1\left(\widehat{y}-y_S(u)\right)\left(\widehat{y}-y\right)\,\mathrm{d}x\,\mathrm{d}t\\
&\leq \frac{1}{2}\|\widehat{y}-y_S(u)\|^2_{L^2_{\omega_1}(I_t;L^2(\Omega))} + \frac{1}{2}\|\widehat{y}-y\|^2_{L^2_{\omega_1}(I_t;L^2(\Omega))}
\end{aligned}
$$

and

$$
\begin{aligned}
\langle a_1(x)\Delta\left(\widehat{y}-y\right), \Delta w_S\rangle_X &= \int_{-1}^{1}\int_{\Omega} a_1(x)\Delta\left(\widehat{y}-y\right)\int_t^1 \omega_1\Delta\left(\widehat{y}-y_S(u)\right)\,\mathrm{d}s\,\mathrm{d}x\,\mathrm{d}t\\
&\leq c_1\left\|\Delta\left(\widehat{y}-y\right)\right\|_X \left\|\int_t^1 \omega_1 \Delta\left(\widehat{y}-y_S(u)\right)\,\mathrm{d}s\right\|_X\\
&\leq \frac{1}{2}c_0\left\|\int_t^1 \omega_1 \Delta\left(\widehat{y}-y_S(u)\right)\,\mathrm{d}s\right\|_X^2 + \frac{c_1^2}{2c_0}\left\|\Delta\left(\widehat{y}-y\right)\right\|_X^2 .
\end{aligned}
$$

Combining the results obtained above, we arrive at the following conclusion:

$$\begin{aligned}
&\|\widehat{y} - y_S(u)\|^2_{L^2_{\omega_1}(I_t;L^2(\Omega))} + \frac{1}{2}c_0 \left\| \int_t^1 \omega_1 \Delta\left(\widehat{y} - y_S(u)\right)\, ds \right\|_X^2 \\
&\le \frac{1}{2}\|\widehat{y} - y_S(u)\|^2_{L^2_{\omega_1}(I_t;L^2(\Omega))} + \frac{1}{2}\|\widehat{y} - y\|^2_{L^2_{\omega_1}(I_t;L^2(\Omega))} \\
&+\frac{1}{2}c_0 \left\| \int_t^1 \omega_1 \Delta\left(\widehat{y} - y_S(u)\right)\, ds \right\|_X^2 + \frac{c_1^2}{2c_0}\|\Delta\left(\widehat{y} - y\right)\|_X^2,
\end{aligned}$$

which leads to

$$\|\widehat{y} - y_S(u)\|^2_{L^2_{\omega_1}(I_t;L^2(\Omega))} \le \|\widehat{y} - y\|^2_{L^2_{\omega_1}(I_t;L^2(\Omega))} + \frac{c_1^2}{c_0}\|\Delta\left(\widehat{y} - y\right)\|_X^2.$$

Thus,

$$\|\widehat{y} - y_S(u)\|_{L^2_{\omega_1}(I_t;L^2(\Omega))} \le \|\widehat{y} - y\|_{L^2_{\omega_1}(I_t;L^2(\Omega))} + c_0^{-\frac{1}{2}}c_1\|\Delta\left(\widehat{y} - y\right)\|_X.$$

By applying the triangle inequality, this immediately yields

$$\begin{aligned}
\| y - y_S(u)\|_{L^2_{\omega_1}(I_t;L^2(\Omega))} &\le \|\widehat{y} - y\|_{L^2_{\omega_1}(I_t;L^2(\Omega))} + \|\widehat{y} - y_S(u)\|_{L^2_{\omega_1}(I_t;L^2(\Omega))} \\
&\le 2\|\widehat{y} - y\|_{L^2_{\omega_1}(I_t;L^2(\Omega))} + c_0^{-\frac{1}{2}}c_1\|\Delta\left(\widehat{y} - y\right)\|_X.
\end{aligned}$$

For $w \in L^2_{\omega_2}(I_t; L^2(\Omega))$, one has

$$\begin{aligned}
\|w\|_X^2 = \int_{-1}^1\int_\Omega \omega_1^{-1}\omega_1 w^2\, dx\, dt &\le \max_{t\in[-1,1]}\left|\omega_1^{-1}\right|\int_{-1}^1\int_\Omega \omega_1 w^2\, dx\, dt \\
&= 2\|w\|^2_{L^2_{\omega_1}(I_t;L^2(\Omega))}.
\end{aligned}$$

Namely,

$$\|\cdot\|_X \le \sqrt{2}\|\cdot\|_{L^2_{\omega_1}(I_t;L^2(\Omega))}.$$

Similarly, we can obtain

$$\|\cdot\|_X \le \sqrt{2}\|\cdot\|_{L^2_{\omega_2}(I_t;L^2(\Omega))}.$$

Consequently, by combining the inequalities derived above, we arrive at

$$\|y - y_S(u)\|_X \leq 2\sqrt{2}\,\|\widehat{y} - y\|_{L^2_{\omega_1}(I_t;L^2(\Omega))} + \sqrt{2}c_0^{-\frac{1}{2}}c_1\,\|\Delta\,(\widehat{y} - y)\|_X\,. \tag{3.4.18}$$

Step 2. In the following, we proceed to estimate (3.4.16). Using the continuous state equation (3.4.8)(b) and the auxiliary equation (3.4.15)(b), we obtain, for any $q_S \in V_{SI}$, that

$$-\langle q_S, \partial_t(p - p_S(u))\rangle_X + \langle a_1(x)\Delta q_S, \Delta\,(p - p_S(u))\rangle_X = \langle y - y_S(u), q_S\rangle_X\,.$$

As a result,

$$\begin{aligned}
&-\langle q_S, \partial_t\,(\widehat{p} - p_S(u))\rangle_X + \langle a_1(x)\Delta q_S, \Delta\,(\widehat{p} - p_S(u))\rangle_X \\
&= -\langle q_S, \partial_t\,(\widehat{p} - p)\rangle_X + \langle a_1(x)\Delta q_S, \Delta\,(\widehat{p} - p)\rangle_X + \langle y - y_S(u), q_S\rangle_X\,.
\end{aligned} \tag{3.4.19}$$

Set

$$q_S = \int_{-1}^{t} \omega_2\,(\widehat{p} - p_S(u))\,\mathrm{d}s.$$

Using similar arguments as before, we derive that the left-hand side of (3.4.19) can be expressed as:

$$\begin{aligned}
&-\langle q_S, \partial_t\,(\widehat{p} - p_S(u))\rangle_X + \langle a_1(x)\Delta q_S, \Delta\,(\widehat{p} - p_S(u))\rangle_X \\
&\geq \|\widehat{p} - p_S(u)\|^2_{L^2_{\omega_2}(I_t;L^2(\Omega))} + \frac{1}{2}c_0\left\|\int_{-1}^{t}\omega_2\Delta\,(\widehat{p} - p_S(u))\,\mathrm{d}s\right\|_X^2.
\end{aligned}$$

And for the right-hand side of (3.4.19), we obtain

$$\begin{aligned}
&-\langle q_S, \partial_t(\widehat{p} - p)\rangle_X + \langle a_1(x)\Delta q_S, \Delta\,(\widehat{p} - p)\rangle_X \\
&\leq \frac{1}{2}\,\|\widehat{p} - p_S(u)\|^2_{L^2_{\omega_2}(I_t;L^2(\Omega))} + \frac{1}{2}\,\|\widehat{p} - p\|^2_{L^2_{\omega_2}(I_t;L^2(\Omega))} \\
&\quad + \frac{1}{4}c_0\left\|\int_{-1}^{t}\omega_2\Delta\,(\widehat{p} - p_S(u))\,\mathrm{d}s\right\|_X^2 + \frac{c_1^2}{c_0}\,\|\Delta\,(\widehat{p} - p)\|_X^2\,.
\end{aligned}$$

Next, applying the Schwarz inequality (see [2]), we deduce that

$$\begin{aligned}\langle y - y_S(u), q_S\rangle_X &\leq \|y - y_S(u)\|_X \left\| \int_{-1}^{t} \omega_2(\widehat{p} - p_S(u))\, \mathrm{d}s \right\|_X \\ &\leq c_\Omega \|y - y_S(u)\|_X \left\| \int_{-1}^{t} \omega_2 \Delta (\widehat{p} - p_S(u))\, \mathrm{d}s \right\|_X \\ &\leq \frac{c_\Omega^2}{c_0} \|y - y_S(u)\|_X^2 + \frac{1}{4} c_0 \left\| \int_{-1}^{t} \omega_2 \Delta (\widehat{p} - p_S(u))\, \mathrm{d}s \right\|_X^2 .\end{aligned}$$

To sum up, we conclude that

$$\begin{aligned}&\|\widehat{p} - p_S(u)\|^2_{L^2_{\omega_2}(I_t;L^2(\Omega))} \\ &\leq \|\widehat{p} - p\|^2_{L^2_{\omega_2}(I_t;L^2(\Omega))} + 2\frac{c_1^2}{c_0} \|\Delta (\widehat{p} - p)\|_X^2 + \frac{2c_\Omega^2}{c_0} \|y - y_S(u)\|_X^2 .\end{aligned}$$

Furthermore,

$$\begin{aligned}&\|\widehat{p} - p_S(u)\|_{L^2_{\omega_2}(I_t;L^2(\Omega))} \\ &\leq |\widehat{p} - p\|_{L^2_{\omega_2}(I_t;L^2(\Omega))} + \sqrt{2} c_1 c_0^{-\frac{1}{2}} \|\Delta (\widehat{p} - p)\|_X + \sqrt{2} c_\Omega c_0^{-\frac{1}{2}} \|y - y_S(u)\|_X .\end{aligned}$$

Subsequently, by leveraging the relationship $\|\cdot\|_X \leq \sqrt{2}\|\cdot\|_{L^2_{\omega_2}(I_t;L^2(\Omega))}$ and invoking the triangle inequality, we arrive at the following inequality:

$$\begin{aligned}&\|p - p_S(u)\|_X \\ &\leq \sqrt{2}\left(\|p - \widehat{p}\|_{L^2_{\omega_2}(I_t;L^2(\Omega))} + \|\widehat{p} - p_S(u)\|_{L^2_{\omega_2}(I_t;L^2(\Omega))}\right) \\ &\leq 2\sqrt{2}\,\|\widehat{p} - p\|_{L^2_{\omega_2}(I_t;L^2(\Omega))} + 2c_1 c_0^{-\frac{1}{2}} \|\Delta (\widehat{p} - p)\|_X + 2c_\Omega c_0^{-\frac{1}{2}} \|y - y_S(u)\|_X .\end{aligned} \tag{3.4.20}$$

Therefore, by integrating (3.4.18) and (3.4.20), we have successfully concluded the proof. □

Next, we proceed to estimate $\|y_S(u) - y_S\|_X$ and $\|p_S(u) - p_S\|_X$ in Lemma 3.4.5 as follows.

Lemma 3.4.5 *There holds*

$$\|y_S(u) - y_S\|_X \leq c_0^{-\frac{1}{2}} c_\Omega \|u - u_S\|_X , \tag{3.4.21}$$

and

$$\|p_S(u) - p_S\|_X \leq c_0^{-\frac{1}{2}} c_\Omega \|y_S(u) - y_S\|_X .$$

Proof First, upon consideration of the discrete system given by (3.4.13)(a)–(b) and the auxiliary system described by (3.4.15)(a)–(b), we are able to ascertain the following facts:

$$\begin{aligned}&\langle \partial_t (y_S(u) - y_S), w_S\rangle_X + \langle a_1(x)\Delta (y_S(u) - y_S), \Delta w_S\rangle_X \\ &= \langle u - u_S, w_S\rangle_X, \quad \forall\, w_S \in V_{Sr},\end{aligned} \tag{3.4.22}$$

and

$$\begin{aligned}&- \langle q_S, \partial_t (p_S(u) - p_S)\rangle_X + \langle a_1(x)\Delta q_S, \Delta (p_S(u) - p_S)\rangle_X \\ &= (y_S(u) - y_S, q_S)_X, \quad \forall\, q_S \in V_{Sl}.\end{aligned} \tag{3.4.23}$$

Then, let

$$w_S = \int_t^1 \omega_1 (y_S(u) - y_S)\, \mathrm{d}s.$$

By employing similar reasoning to that used in Lemma 3.4.4, we can derive an upper bound for the left-hand side of Eq. (3.4.22), as shown below:

$$\begin{aligned}&\langle \partial_t (y_S(u) - y_S), w_S\rangle_X + \langle a_1(x)\Delta (y_S(u) - y_S), \Delta w_S\rangle_X \\ &\geq \|y_S(u) - y_S\|^2_{L^2_{\omega_1}(I_t;L^2(\Omega))} + \frac{1}{2} c_0 \left\| \int_t^1 \omega_1 \Delta (y_S(u) - y_S)\, \mathrm{d}s \right\|_X^2 .\end{aligned} \tag{3.4.24}$$

Regarding the right-hand side of (3.4.22), we utilize the Schwarz inequality and the norm inequality $\|\cdot\|_{L^2(\Omega)} \leq c_\Omega \|\Delta \cdot\|_{L^2(\Omega)}$ to get

$$\begin{aligned}\langle u - u_S, w_S\rangle_X &\leq \|u - u_S\|_X \left\| \int_t^1 \omega_1 (y_S(u) - y_S)\, \mathrm{d}s \right\|_X \\ &\leq c_\Omega \|u - u_S\|_X \left\| \int_t^1 \omega_1 \Delta (y_S(u) - y_S)\, \mathrm{d}s \right\|_X \\ &\leq \frac{1}{2} c_\Omega^2 c_0^{-1} \|u - u_S\|_X^2 + \frac{1}{2} c_0 \left\| \int_t^1 \omega_1 \Delta (y_S(u) - y_S)\, \mathrm{d}s \right\|_X^2 .\end{aligned} \tag{3.4.25}$$

Therefore, combining (3.4.24) and (3.4.25) with the triangle inequality,

$$\|\cdot\|_X \leq \sqrt{2}\, \|\cdot\|_{L^2_{\omega_2}(I_t;L^2(\Omega))},$$

we have

$$\|y_S(u) - y_S\|_{L^2_{\omega_1}(I_t;L^2(\Omega))} \leq \frac{\sqrt{2}}{2} c_\Omega c_0^{-\frac{1}{2}} \|u - u_S\|_X,$$

which implies the error estimate given by (3.4.21).

Similarly, for Eq. (3.4.23), we choose

$$q_S = \int_{-1}^{t} \omega_2(p_S(u) - p_S)\, \mathrm{d}s$$

such that

$$\|p_S(u) - p_S\|_{L^2_{\omega_2}(I_t;L^2(\Omega))} \leq \frac{\sqrt{2}}{2} c_\Omega c_0^{-\frac{1}{2}} \|y_S(u) - y_S\|_X.$$

Then, it has

$$\|p_S(u) - p_S\|_X \leq \sqrt{2}\, \|p_S(u) - p_S\|_{L^2_{\omega_2}(I_t;L^2(\Omega))} \leq c_\Omega c_0^{-\frac{1}{2}} \|y_S(u) - y_S\|_X.$$

This completes the proof of the lemma. □

The subsequent step involves estimating $\|u - u_S\|_X$ as outlined in Lemma 3.4.6 below. To proceed, let $\mathbb{P}^t_M$ and $\mathbb{P}^x_N$ denote the L^2-orthogonal projection operators in time and space, respectively. Furthermore, we introduce a composition operator

$$\Pi_{MN} = \mathbb{P}^t_M \mathbb{P}^x_N,$$

which will be employed subsequently.

Lemma 3.4.6 *There holds*

$$\|u - u_S\|_X \leq \alpha^{-1} \|p - p_S(u)\|_X. \tag{3.4.26}$$

Proof Set $w_S = p_S(u) - p_S$ in (3.4.22) and $q_S = y_S(u) - y_S$ in (3.4.23). Then,

$$\langle u - u_S, p_S(u) - p_S\rangle_X = \langle y_S(u) - y_S, y_S(u) - y_S\rangle_X = \|y_S(u) - y_S\|_X^2.$$

Next, we deduce the following:

$$
\begin{aligned}
&\alpha \|u - u_S\|_X^2 + \|y_S(u) - y_S\|_X^2 \\
&= \langle u - u_S, \alpha u - \alpha u_S \rangle_X + \langle u - u_S, p_S(u) - p_S \rangle_X \\
&= \langle u - u_S, \alpha u + p \rangle_X + \langle u - u_S, p_S(u) - p \rangle_X - \langle u - \Pi_{MN} u, \alpha u_S + p_S \rangle_X \\
&\quad - \langle \Pi_{MN} u - u_S, \alpha u_S + p_S \rangle_X .
\end{aligned}
$$

Since $u_S \in U_{ad}$ satisfies (3.4.10), it follows that $\langle u - u_S, \alpha u + p \rangle_X \leq 0$. Considering the control constraint set U_{ad} for Cases 1–3 (Sect. 3.4.1.1), we can confirm that $\Pi_{MN} u \in U_{ad}^S$ for $u \in U_{ad}$.

Indeed, this assertion holds true for Case 1. Moving to Case 2, if $u \in U_{ad}$, then

$$
\int_{-1}^{1} \int_{\Omega} u \, \mathrm{d}x \, \mathrm{d}t \geq 0.
$$

By Assumption 3.4.2,

$$
\int_{-1}^{1} \int_{\Omega} \Pi_{MN} u \, \mathrm{d}x \, \mathrm{d}t = \langle \Pi_{MN} u, 1 \rangle_X = \langle u, 1 \rangle_X \geq 0.
$$

For Case 3, when $u \in U_{ad}$ we have $\|u\|_X \leq \gamma$ and by Assumption 3.4.2,

$$
\|\Pi_{MN} u\|_X^2 = \langle \Pi_{MN} u, \Pi_{MN} u \rangle_X = \langle u, \Pi_{MN} u \rangle_X \leq \|u\|_X \, \|\Pi_{MN} u\|_X
$$

which implies that $\|\Pi_{MN} u\|_X \leq \|u\|_X \leq \gamma$.

Therefore, according to Eq. (3.4.14),

$$
\langle \Pi_{MN} u - u_S, \alpha u_S + p_S \rangle_X \geq 0.
$$

Furthermore, taking into account that of $\alpha u_S + p_S \in X_S$ and invoking Assumption 3.4.2, we reach the conclusion that

$$
\langle u - \Pi_{MN} u, \alpha u_S + p_S \rangle_X = 0.
$$

To summarize, by applying Hölder's inequality, we deduce from Eq. (3.4.2) that

$$
\alpha \|u - u_S\|_X^2 \leq \langle u - u_S, p_S(u) - p \rangle_X \leq \|u - u_S\|_X \, \|p_S(u) - p\|_X ,
$$

which leads to (3.4.26) immediately. □

Finally, leveraging the auxiliary results established in the preceding lemmas, we proceed to present the comprehensive proof of Theorem 3.4.3 as follows.

***Proof of Theorem* 3.4.3** By employing the triangle inequality and amalgamating the error findings from Lemmas 3.4.4–3.4.6, we obtain

$$
\begin{aligned}
&\|u-u_S\|_X+\|y-y_S\|_X+\|p-p_S\|_X\\
&\le \|u-u_S\|_X+\left(\|y-y_S(u)\|_X+\|y_S(u)-y_S\|_X\right)\\
&\quad+\left(\|p-p_S(u)\|_X+\|p_S(u)-p_S\|_X\right)\\
&\le \left(1+c_0^{-\frac{1}{2}}c_\Omega+c_0^{-1}c_\Omega^2\right)\|u-u_S\|_X+\|y-y_S(u)\|_X+\|p-p_S(u)\|_X\\
&\le \|y-y_S(u)\|_X+\left[1+\alpha^{-1}\left(1+c_0^{-\frac{1}{2}}c_\Omega+c_0^{-1}c_\Omega^2\right)\right]\|p-p_S(u)\|_X\\
&= \|y-y_S(u)\|_X+C_4\,\|p-p_S(u)\|_X\\
&\le \left(1+2c_\Omega c_0^{-\frac{1}{2}}C_4\right)\|y-y_S(u)\|_X\\
&\quad+C_4\left(2\sqrt{2}\,\|\widehat{p}-p\|_{L^2_{\omega_2}(I_t;L^2(\Omega))}+2c_1c_0^{-\frac{1}{2}}\,\|\Delta(\widehat{p}-p)\|_X\right)\\
&\le C_5\,\|y-\widehat{y}\|_{L^2_{\omega_1}(I_t;L^2(\Omega))}+C_6\,\|\Delta(\widehat{y}-y)\|_X+C_7\,\|\widehat{p}-p\|_{L^2_{\omega_2}(I_t;L^2(\Omega))}\\
&\quad+C_8\,\|\Delta(\widehat{p}-p)\|_X,
\end{aligned}
$$

where

$$
C_4=1+\alpha^{-1}\left(1+c_0^{-\frac{1}{2}}c_\Omega+c_0^{-1}c_\Omega^2\right),\quad C_5=2\sqrt{2}\left(1+2c_\Omega c_0^{-\frac{1}{2}}C_4\right),
$$

$$
C_6=\sqrt{2}c_0^{-\frac{1}{2}}c_1\left(1+2c_\Omega c_0^{-\frac{1}{2}}C_4\right),\quad C_7=2\sqrt{2}C_4,\quad C_8=2c_1c_0^{-\frac{1}{2}}C_4.
$$

Thus,

$$
\begin{aligned}
&\|u-u_S\|_X+\|y-y_S\|_X+\|p-p_S\|_X\\
&\le C_5\,\|y-\widehat{y}\|_{L^2_{\omega_1}(I_t;L^2(\Omega))}+C_6\,\|\Delta(\widehat{y}-y)\|_X\\
&\quad+C_7\,\|\widehat{p}-p\|_{L^2_{\omega_2}(I_t;L^2(\Omega))}+C_8\,\|\Delta(\widehat{p}-p)\|_X\,.
\end{aligned}
\tag{3.4.27}
$$

Next, we proceed to estimate these terms $\|y-\widehat{y}\|_{L^2_{\omega_1}(I_t;L^2(\Omega))}$, $\|\Delta(\widehat{y}-y)\|_X$, $\|\widehat{p}-p\|_{L^2_{\omega_2}(I_t;L^2(\Omega))}$ and $\|\Delta(\widehat{p}-p)\|_X$ in the right-hand side of the inequality given by (3.4.27) above.

We begin with the estimation of $\|y-\widehat{y}\|_{L^2_{\omega_1}(I_t;L^2(\Omega))}$. Utilizing Assumptions 3.4.3–3.4.4 alongside the triangle inequality, we derive the following:

$$
\begin{aligned}
&\|y-\widehat{y}\|_{L^2_{\omega_1}(I_t;L^2(\Omega))} = \left\|y-\mathbb{P}^0_{2,N}\mathbb{P}^M_{\omega_1}y\right\|_{L^2_{\omega_1}(I_t;L^2(\Omega))}\\
&\le \left\|y-\mathbb{P}^0_{2,N}y\right\|_{L^2_{\omega_1}(I_t;L^2(\Omega))} + \left\|\mathbb{P}^0_{2,N}y-\mathbb{P}^0_{2,N}\mathbb{P}^M_{\omega_1}y\right\|_{L^2_{\omega_1}(I_t;L^2(\Omega))}\\
&= \left(\int_{-1}^{1}\omega_1\left\|y-\mathbb{P}^0_{2,N}y\right\|^2_{L^2(\Omega)}\,\mathrm{d}t\right)^{\frac{1}{2}} + \left(\int_{-1}^{1}\omega_1\left\|\mathbb{P}^0_{2,N}y-\mathbb{P}^0_{2,N}\mathbb{P}^M_{\omega_1}y\right\|^2_{L^2(\Omega)}\,\mathrm{d}t\right)^{\frac{1}{2}}\\
&\le C_1N^{-\theta}\left(\int_{-1}^{1}\omega_1\|y\|^2_{H^\theta(\Omega)}\,\mathrm{d}t\right)^{\frac{1}{2}} + c_\Omega\left(\int_{-1}^{1}\omega_1\left\|\Delta y-\mathbb{P}^M_{\omega_1}\Delta y\right\|^2_{L^2(\Omega)}\,\mathrm{d}t\right)^{\frac{1}{2}}\\
&\le C_1N^{-\theta}\|y\|_{L^2_{\omega_1}(I_t;H^\theta(\Omega))} + c_\Omega C_2M^{-\sigma}\left\|\partial_t^\sigma y\right\|_{L^2_{\omega^{\sigma,\sigma-1}}(I_t;H^2(\Omega))}\\
&\le C_1N^{-\theta}\|y\|_{L^2_{\omega_1}(I_t;H^\theta(\Omega))} + \sqrt{2}c_\Omega C_2M^{-\sigma}\left\|\partial_t^\sigma y\right\|_{L^2(I_t;H^2(\Omega))}\\
&\le C_1N^{-\theta}\|y\|_{L^2_{\omega_1}(I_t;H^\theta(\Omega))} + \sqrt{2}c_\Omega C_2M^{-\sigma}\|y\|_{H^\sigma(I_t;H^2(\Omega))},
\end{aligned}
$$

In the estimation process, we additionally leveraged the fact that

$$\|\cdot\|_{L^2_{\omega^{\sigma,\sigma-1}}(I_t)} \le \sqrt{2}\,\|\cdot\|_{L^2(I_t)}$$

which holds true because

$$\max_{t\in[-1,1]}\left\{\omega^{\sigma,\sigma-1}\right\} \le 2$$

for the integer $\sigma \ge 1$. Thus

$$\|y-\widehat{y}\|_{L^2_{\omega_1}(I_t;L^2(\Omega))} \le C_1N^{-\theta}\|y\|_{L^2_{\omega_1}(I_t;H^\theta(\Omega))}+\sqrt{2}c_\Omega C_2M^{-\sigma}\|y\|_{H^\sigma(I_t;H^2(\Omega))}.$$

Next, for $\|p-\widehat{p}\|_{L^2_{\omega_2}(I_t;L^2(\Omega))}$, We employ the same technique to derive

$$\|p-\widehat{p}\|_{L^2_{\omega_2}(I_t;L^2(\Omega))} \le C_1N^{-\theta}\|p\|_{L^2_{\omega_2}(I_t;H^\theta(\Omega))}+\sqrt{2}c_\Omega C_3M^{-\sigma}\|p\|_{H^\sigma(I_t;H^2(\Omega))}.$$

Secondly, invoking Assumptions 3.4.3–3.4.4 along with the triangle inequality, we obtain the following derivation:

$$\begin{aligned}
\|\Delta(\widehat{y}-y)\|_X &= \left\|\Delta\left(\mathbb{P}^0_{2,N}\mathbb{P}^M_{\omega_1}y - y\right)\right\|_X \\
&\le \left\|\Delta\left(\mathbb{P}^0_{2,N}y - y\right)\right\|_X + \left\|\Delta\left(\mathbb{P}^0_{2,N}\mathbb{P}^M_{\omega_1}y - \mathbb{P}^0_{2,N}y\right)\right\|_X \\
&\le \left(\int_{-1}^1 \left\|\mathbb{P}^0_{2,N}y - y\right\|^2_{H^2(\Omega)} \mathrm{d}t\right)^{\frac{1}{2}} + \left(\int_{-1}^1 \left\|\mathbb{P}^M_{\omega_1}y - y\right\|^2_{H^2(\Omega)} \mathrm{d}t\right)^{\frac{1}{2}} \\
&\le C_1 N^{2-\theta}\|y\|_{L^2(I_t;H^\theta(\Omega))} + C_2 M^{-\sigma}\left\|\partial_t^\sigma y\right\|_{L^2_{\omega^{\sigma,\sigma-1}}(I_t;H^2(\Omega))} \\
&\le C_1 N^{2-\theta}\|y\|_{L^2(I_t;H^\theta(\Omega))} + \sqrt{2}C_2 M^{-\sigma}\|y\|_{H^\sigma(I_t;H^2(\Omega))},
\end{aligned}$$

and hence

$$\|\Delta(\widehat{y}-y)\|_X \le C_1 N^{2-\theta}\|y\|_{L^2(I_t;H^\theta(\Omega))} + \sqrt{2}C_2 M^{-\sigma}\|y\|_{H^\sigma(I_t;H^2(\Omega))}.$$

Similarly, for the term $\|\Delta(\widehat{p}-p)\|_X$, we have

$$\|\Delta(\widehat{p}-p)\|_X \le C_1 N^{2-\theta}\|p\|_{L^2(I_t;H^\theta(\Omega))} + \sqrt{2}C_3 M^{-\sigma}\|p\|_{H^\sigma(I_t;H^2(\Omega))}.$$

Finally, regarding the results obtained for these terms $\|y-\widehat{y}\|_{L^2_{\omega_1}(I_t;L^2(\Omega))}$, $\|\widehat{p}-p\|_{L^2_{\omega_2}(I_t;L^2(\Omega))}$, $\|\Delta(\widehat{y}-y)\|_X$, and $\|\Delta(\widehat{p}-p)\|_X$, we substitute them into (3.4.27) to obtain

$$\begin{aligned}
&\|u-u_S\|_X + \|y-y_S\|_X + \|p-p_S\|_X \\
&\le C_5\left(C_1 N^{-\theta}\|y\|_{L^2_{\omega_1}(I_t;H^\theta(\Omega))} + \sqrt{2}c_\Omega C_2 M^{-\sigma}\|y\|_{H^\sigma(I_t;H^2(\Omega))}\right) \\
&\quad + C_7\left(C_1 N^{-\theta}\|p\|_{L^2_{\omega_2}(I_t;H^\theta(\Omega))} + \sqrt{2}c_\Omega C_3 M^{-\sigma}\|p\|_{H^\sigma(I_t;H^2(\Omega))}\right) \\
&\quad + C_6\left(C_1 N^{2-\theta}\|y\|_{L^2(I_t;H^\theta(\Omega))} + \sqrt{2}C_2 M^{-\sigma}\|y\|_{H^\sigma(I_t;H^2(\Omega))}\right) \\
&\quad + C_8\left(C_1 N^{2-\theta}\|p\|_{L^2(I_t;H^\theta(\Omega))} + \sqrt{2}C_3 M^{-\sigma}\|p\|_{H^\sigma(I_t;H^2(\Omega))}\right) \\
&\le C_1 \max\{C_5, C_7\} N^{-\theta}\left(\|y\|_{L^2_{\omega_1}(I_t;H^\theta(\Omega))} + \|p\|_{L^2_{\omega_2}(I_t;H^\theta(\Omega))}\right) \\
&\quad + C_1 \max\{C_6, C_8\} N^{2-\theta}\left(\|y\|_{L^2(I_t;H^\theta(\Omega))} + \|p\|_{L^2(I_t;H^\theta(\Omega))}\right) \\
&\quad + \max\left\{\sqrt{2}c_\Omega C_2 C_5 + \sqrt{2}C_2 C_6, \sqrt{2}c_\Omega C_3 C_7 + \sqrt{2}C_3 C_8\right\} \\
&\quad \cdot M^{-\sigma}\left(\|y\|_{H^\sigma(I_t;H^2(\Omega))} + \|p\|_{H^\sigma(I_t;H^2(\Omega))}\right).
\end{aligned}$$

Thanks to

$$\max\{C_5, C_7\} = 2\sqrt{2}\max\left\{1 + 2c_\Omega c_0^{-\frac{1}{2}} C_4, C_4\right\},$$
$$\max\{C_6, C_8\} = \sqrt{2}c_0^{-\frac{1}{2}} c_1 \max\left\{1 + 2c_\Omega c_0^{-\frac{1}{2}} C_4, \sqrt{2}C_4\right\}$$

and

$$\max\left\{\sqrt{2}c_\Omega C_2 C_5 + \sqrt{2}C_2 C_6, \sqrt{2}c_\Omega C_3 C_7 + \sqrt{2}C_3 C_8\right\}$$
$$\leq \sqrt{2}\max\{C_2, C_3\}\max\{C_5 + C_6, C_7 + C_8\},$$

we have

$$\begin{aligned}
&\|u - u_S\|_X + \|y - y_S\|_X + \|p - p_S\|_X \\
&\leq 2\sqrt{2}C_1 \max\left\{1 + 2c_\Omega c_0^{-\frac{1}{2}} C_4, C_4\right\} N^{-\theta}\left(\|y\|_{L^2_{\omega_1}(I_t;H^\theta(\Omega))} + \|p\|_{L^2_{\omega_2}(I_t;H^\theta(\Omega))}\right) \\
&\quad + \sqrt{2}c_0^{-\frac{1}{2}} c_1 C_1 \max\left\{1 + 2c_\Omega c_0^{-\frac{1}{2}} C_4, \sqrt{2}C_4\right\} \\
&\quad\quad \cdot N^{2-\theta}\left(\|y\|_{L^2(I_t;H^\theta(\Omega))} + \|p\|_{L^2(I_t;H^\theta(\Omega))}\right) \\
&\quad + 2\sqrt{2}\max\{C_2, C_3\}\max\left\{2\sqrt{2}\left(1 + 2c_\Omega c_0^{-\frac{1}{2}} C_4\right)\right. \\
&\quad + \sqrt{2}c_0^{-\frac{1}{2}} c_1\left(1 + 2c_\Omega c_0^{-\frac{1}{2}} C_4\right), \\
&\quad\quad \left. 2\sqrt{2}C_4 + 2c_1 c_0^{-\frac{1}{2}} C_4\right\} M^{-\sigma}\left(\|y\|_{H^\sigma(I_t;H^2(\Omega))} + \|p\|_{H^\sigma(I_t;H^2(\Omega))}\right) \\
&= 2\sqrt{2}C_1 \max\left\{1 + 2c_\Omega c_0^{-\frac{1}{2}} C_4, C_4\right\} N^{-\theta}\left(\|y\|_{L^2_{\omega_1}(I_t;H^\theta(\Omega))} + \|p\|_{L^2_{\omega_2}(I_t;H^\theta(\Omega))}\right) \\
&\quad + \sqrt{2}c_0^{-\frac{1}{2}} c_1 C_1 \max\left\{1 + 2c_\Omega c_0^{-\frac{1}{2}} C_4, \sqrt{2}C_4\right\} \\
&\quad\quad \cdot N^{2-\theta}\left(\|y\|_{L^2(I_t;H^\theta(\Omega))} + \|p\|_{L^2(I_t;H^\theta(\Omega))}\right) \\
&\quad + 4\max\{C_2, C_3\}\max\left\{\left(2 + c_0^{-\frac{1}{2}} c_1\right)\left(1 + 2c_\Omega c_0^{-\frac{1}{2}} C_4\right), \left(2 + \sqrt{2}c_1 c_0^{-\frac{1}{2}}\right) C_4\right\} \\
&\quad\quad \cdot M^{-\sigma}\left(\|y\|_{H^\sigma(I_t;H^2(\Omega))} + \|p\|_{H^\sigma(I_t;H^2(\Omega))}\right),
\end{aligned}$$

which completes the proof of the theorem. □

3.4.3 Numerical Examples

In this subsection, we present four numerical examples to demonstrate the precision of the space-time spectral method discussed above. To solve the spectral discrete system given by (3.4.13), we introduce a straightforward and efficient gradient projection algorithm. For a detailed explanation of the algorithm, readers can refer to [158]. The iterative scheme of this projection algorithm can be summarized as follows.

For $n = 0, 1, 2, \cdots$, let

$$\begin{cases} \left\langle u_S^{n+\frac{1}{2}}, v_S \right\rangle_X = \langle u_S^n, v_S\rangle_X - \rho \left\langle J_S'(u_S^n), v_S\right\rangle_X, & \forall\, v_S \in X_S, \\ u_S^{n+1} = \widehat{\Pi}_S \left(u_S^{n+\frac{1}{2}} \right), \end{cases}$$

where $\rho > 0$ is an iteration parameter. Moreover, the projection operator

$$\widehat{\Pi}_S : X_S \to U_{ad}^S$$

satisfies the following property: Given $w_S \in X_S$, the task is to find $\widehat{\Pi}_S\,(w_S) \in U_{ad}^S$ that fulfills

$$\left\langle \widehat{\Pi}_S\,(w_S) - w_S, \widehat{\Pi}_S\,(w_S) - w_S \right\rangle_X = \min_{v_S \in U_{ad}^S} \langle v_S - w_S, v_S - w_S\rangle_X\,,$$

which is equivalent to

$$\left\langle \widehat{\Pi}_S\,(w_S) - w_S, v_S - \widehat{\Pi}_S\,(w_S) \right\rangle_X \geq 0, \quad \forall\, v_S \in U_{ad}^S.$$

Upon comparison with the definition of the discrete projection $\widehat{\Pi}_S\,(w_S) = \Pi_S\,(-\alpha w_S)$, it becomes evident that $\widehat{\Pi}_S\,(w_S) = \Pi_S\,(-\alpha w_S)$ corresponds to Π_S.

Referring to Cases 1–3 presented in Sect. 3.4.1.1, we observe that $\widehat{\Pi}_S\,(w_S) = w_S$ for Case 1. For Case 2,

$$\widehat{\Pi}_S\,(w_S) = -\min\{0, \overline{w_S}\} + w_S,$$

and for Case 3,

$$\widehat{\Pi}_S\,(w_S) = \frac{\gamma w_S}{\max\{\|w_S\|_X\,, \gamma\}}.$$

Below, we outline the primary steps of the gradient projection algorithm.

Algorithm 3 The gradient projection algorithm

1: Set $n = 0$ and choose the initial control $u_S^0 \in U_{ad}^S$.

2: For $n = 0, 1, 2, \cdots$, get $y_S^n \in V_{Sl}$ such that

$$\left(\partial_t y_S^n, w_S\right)_X + \left(a_1(x)\Delta y_S^n, \Delta w_S\right)_X = \left(f + u_S^n, w_S\right)_X, \quad \forall\, w_S \in V_{Sr}.$$

3: Get $p_S^n \in V_{Sr}$ from

$$-\left(q_S, \partial_t p_S^n\right)_X + \left(a_1(x)\Delta q_S, \Delta p_S^n\right)_X = \left(y_S^n - y_d, q_S\right)_X, \quad \forall\, q_S \in V_{Sl}.$$

4: Compute $u_S^{n+\frac{1}{2}} \in X_S$ by

$$u_S^{n+\frac{1}{2}} = u_S^n - \rho\left(\alpha u_S^n + p_S^n\right).$$

5: Get $u_S^{n+1} \in U_{ad}^S$ from

$$u_S^{n+1} = \widehat{\Pi}_S\left(u_S^{n+\frac{1}{2}}\right).$$

6: Stop if the stopping criterion

$$\left\| u_S^{n+1} - u_S^n \right\|_X \leq \text{Tol (the tolerance)}$$

is satisfied. Otherwise, set $n = n + 1$ and go to Step 2.

Next, we apply the aforementioned algorithm to solve four examples. However, given the simplicity of Case 1, we primarily focus on the constrained Cases 2 and 3 in our numerical examples related to the parabolic optimization problem given by (3.4.3)–(3.4.4).

3.4.3.1 One-Dimensional Case in Space

In this subsection, we consider the parabolic optimization problem on the domain $\Omega = (-1, 1)$.

Let $L_n(x)$ be the n-th degree Legendre polynomial (see [204, 253]) and denote

$$c_k = \frac{1}{\sqrt{2(2k+3)^2(2k+5)}}, \quad k = 0, 1, \cdots, N-4,$$
$$\phi_k(x) = c_k\left(L_k(x) - \frac{2(2k+5)}{2k+7}L_{k+2}(x) + \frac{2k+3}{2k+7}L_{k+4}(x)\right),$$
$$k = 0, 1, \cdots, N-4.$$

Upon inspection, we promptly observe that

$$\phi_k(\pm 1) = \frac{\partial \phi_k}{\partial \mathbf{n}}(\pm 1) = 0.$$

Subsequently, to solve the spectral discrete system $\mathcal{Q}_S$ presented in Sect. 3.4.1, we proceed to construct the following function spaces:

$$Q_N = \text{span}\,\{L_i(x)\}_{i=0}^{N}, \quad Q_M = \text{span}\,\{L_j(t)\}_{j=0}^{M}, \quad Y_N = \text{span}\,\{\phi_i(x)\}_{i=0}^{N-4},$$
$$Q_{Ml} = \text{span}\,\{\psi_j(t) := L_j(t) + L_{j+1}(t)\}_{j=0}^{M-1},$$
$$Q_{Mr} = \text{span}\,\{\chi_j(t) := L_j(t) - L_{j+1}(t)\}_{j=0}^{M-1}.$$

Furthermore, we express the spectral approximation solutions y_S, p_S and u_S in the following manners:

$$y_S = \sum_{i=0}^{N-4}\sum_{j=0}^{M-1} y_{ij}\phi_i(x)\psi_j(t), \quad p_S = \sum_{i=0}^{N-4}\sum_{j=0}^{M-1} p_{ij}\phi_i(x)\chi_j(t),$$
$$u_S = \sum_{i=0}^{N}\sum_{j=0}^{M} u_{ij}L_i(x)L_j(t).$$

Based on this, the state equation (3.4.13)(a) can be converted into the following matrix equation:

$$BYG^\top + AYH^\top = F, \tag{3.4.28}$$

where

$$A = (a_{ki})_{(N-3)\times(N-3)}, \quad B = (b_{ki})_{(N-3)\times(N-3)}, \quad G = \left(g_{mj}\right)_{M\times M},$$
$$H = \left(h_{mj}\right)_{M\times M}, \quad Y = \left(y_{ij}\right)_{(N-3)\times M}, \quad F = (f_{km})_{(N-3)\times M},$$
$$a_{ki} = \left(a_1(x)\phi_k'', \phi_i''\right)_\Omega, \quad b_{ki} = (\phi_k, \phi_i)_\Omega, \quad g_{mj} = \left(\chi_m, \psi_j'\right)_{I_t},$$
$$h_{mj} = \left(\chi_m, \psi_j\right)_{I_t}, \quad f_{km} = \langle f + u_S, \phi_k\chi_m\rangle_X,$$

and $(\cdot,\cdot)_\Omega$ and $(\cdot,\cdot)_{I_t}$ denote the inner products in $L^2(\Omega)$ and $L^2(I_t)$, respectively. $(\cdot)^\top$ is the transposed matrix.

During the calculation process, we utilize a highly efficient solution method described in [207] to solve the matrix equation (3.4.28). Similarly, the adjoint state equation (3.4.13)(b) can also be effectively solved.

Example 3.4.7 In the parabolic optimization problem delineated by (3.4.3)–(3.4.4), we designate $\alpha = \frac{3}{4}$. The ensuing control constraint set and the prescribed functions are detailed below:

$$U_{ad} = \{v \in U \mid \|v\|_X \leq \gamma\} \text{ with } \gamma = 1,\ \ a_1(x) = 2 - x^2,$$

$$f = \left(8\pi^2 + 32\pi^4 - 16\pi^4 x^2 - \frac{2\sqrt{3}}{3}\right)\sin^2(\pi x)\sin(\pi t) + \pi\sin^2(\pi x)\cos(\pi t)$$
$$+16\pi^3 x\sin(2\pi x)\sin(\pi t) - \left(4\pi^2 + 16\pi^4 - 8\pi^4 x^2\right)\sin(\pi t),$$

$$y_d = \left(1 + \frac{16\sqrt{3}}{3}\pi^2 + \frac{64\sqrt{3}}{3}\pi^4 - \frac{32\sqrt{3}}{3}\pi^4 x^2\right)\sin^2(\pi x)\sin(\pi t)$$
$$-\frac{2\sqrt{3}}{3}\pi\sin^2(\pi x)\cos(\pi t) + \frac{32\sqrt{3}}{3}\pi^3 x\sin(2\pi x)\sin(\pi t)$$
$$-\frac{8\sqrt{3}}{3}\pi^2\left(1 + 4\pi^2 - 2\pi^2 x^2\right)\sin(\pi t).$$

The exact solutions that satisfy the corresponding optimality conditions are presented in the following:

$$y = \sin^2(\pi x)\sin(\pi t),\ \ u = \frac{2\sqrt{3}}{3}\sin^2(\pi x)\sin(\pi t),\ \ p = -\frac{2\sqrt{3}}{3}\sin^2(\pi x)\sin(\pi t).$$

For the aforementioned example, we initially set $N = 16$ and examine the errors specifically related to the temporal direction. Table 3.5 outlines the numerical outcomes for $\|u - u_S\|_X$, $\|y - y_S\|_X$, and $\|p - p_S\|_X$ with varying values of $M = 4, 6, 8, 10, 12$. Upon reviewing Table 3.5, it becomes evident that as M increases, the errors $\|u - u_S\|_X$, $\|y - y_S\|_X$, and $\|p - p_S\|_X$ decrease. Furthermore, these error results are visually represented in Fig. 3.12a. The error curves depicted in Fig. 3.12a exhibit a linear decrease, indicating an exponential decay in these errors.

Table 3.5 The numerical results of error estimates versus M with $N = 16$ in Example 3.4.7

M	4	6	8	10	12
$\|u - u_S\|_X$	2.6735e-1	2.0432e-2	8.0550e-4	1.9906e-5	4.0128e-7
$\|y - y_S\|_X$	2.5661e-1	1.9019e-2	7.4403e-4	1.8202e-5	3.6027e-7
$\|p - p_S\|_X$	2.7749e-1	2.0434e-2	8.0541e-4	1.9922e-5	4.0364e-7

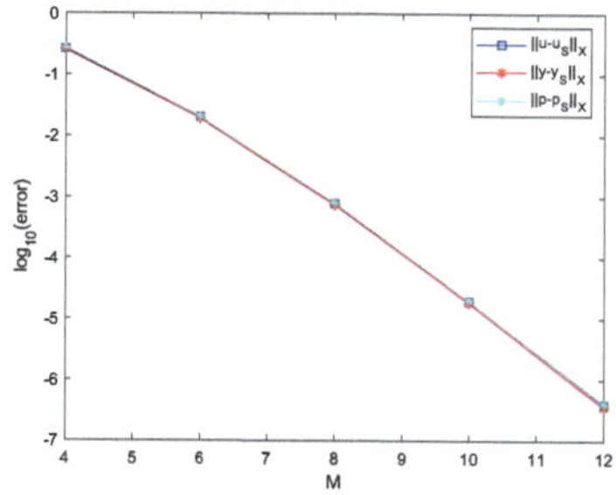

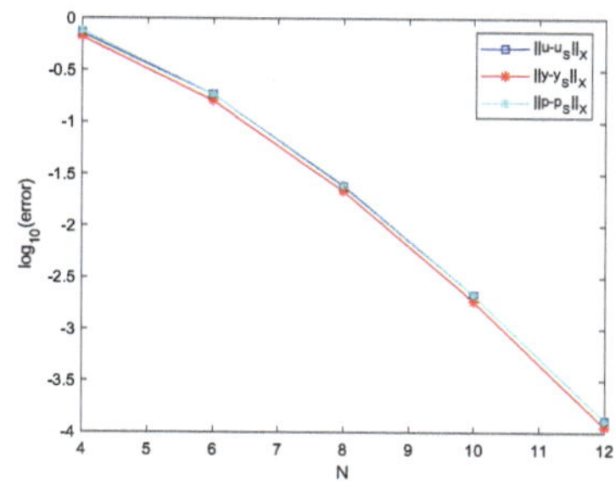

Fig. 3.12 Error estimates in Example 3.4.7. (**a**) $N = 16$. (**b**) $M = 16$

Table 3.6 The numerical results of error estimates versus N with $M = 16$ in Example 3.4.7

N	4	6	8	10	12
$\|u - u_S\|_X$	7.2398e-1	1.8638e-1	2.4561e-2	2.1601e-3	1.2982e-4
$\|y - y_S\|_X$	6.6338e-1	1.6126e-1	2.1630e-2	1.8708e-3	1.1242e-4
$\|p - p_S\|_X$	7.6287e-1	1.8557e-1	2.4975e-2	2.1601e-3	1.2982e-4

Next, we maintain $M = 16$ and explore how the errors $\|u - u_S\|_X$, $\|y - y_S\|_X$, and $\|p - p_S\|_X$ behave in the spatial direction. The corresponding numerical results are presented in Table 3.6 and Fig. 3.12b. These results indicate that the errors in the spatial direction exhibit similar trends to those in the temporal direction. Specifically, as N increases, the errors in the spatial direction not only diminish but also decline exponentially.

These observations suggest that the space-time spectral method introduced in this section offers a precise approximation for the L^2-norm control-constrained optimal control problem governed by the parabolic system.

Clearly, relying solely on this example is insufficient to establish the effectiveness of the space-time spectral method for solving a fourth-order parabolic optimization problem. Moreover, it is widely recognized that spectral accuracy is contingent upon the regularity of continuous solutions. In the case of Example 3.4.7 mentioned earlier, the exact solutions possess ample smoothness. Specifically, their regularity is highly satisfactory. To provide a more comprehensive demonstration of the efficacy of the space-time spectral method, we proceed to examine an instance where the state variable exhibits limited regularity in the temporal direction.

Example 3.4.8 Consider an integral mean control constraint, and enumerate the pertinent data as follows:

$$U_{ad} = \left\{ v \in U \,\middle|\, \int_{-1}^{1} \int_{\Omega} v \,\mathrm{d}x \,\mathrm{d}t \geq 0 \right\}, \quad a_1(x) \equiv 1, \quad \alpha = 1,$$

$$f = \operatorname{sgn}\left(\frac{1}{2} - t\right)(1 + \cos(\pi x))\left(t - \frac{1}{2}\right)^2$$
$$+\frac{1}{3}\pi^4 \cos(\pi x)\left(\frac{27}{8} - \left|t - \frac{1}{2}\right|^3\right) - \sin^2(\pi x)\sin(\pi t),$$

$$y_d = \frac{1}{3}(1 + \cos(\pi x))\left(\frac{27}{8} - \left|t - \frac{1}{2}\right|^3\right) - \pi \sin^2(\pi x)\cos(\pi t)$$
$$-8\pi^4 \cos(2\pi x)\sin(\pi t).$$

By addressing the associated optimality conditions, we derive the precise solutions as outlined below:

$$y = \frac{1}{3}(1 + \cos(\pi x))\left(\frac{27}{8} - \left|t - \frac{1}{2}\right|^3\right), \quad u = \sin^2(\pi x)\sin(\pi t),$$
$$p = -\sin^2(\pi x)\sin(\pi t),$$

where $\operatorname{sgn}(\cdot)$ is the sign function, i.e., $\operatorname{sgn}(t) = 1$ for $t > 0$, $\operatorname{sgn}(t) = -1$ for $t < 0$, and $\operatorname{sgn}(t) = 0$ for $t = 0$.

In Example 3.4.8, the numerical results regarding the errors $\|u - u_S\|_X$, $\|y - y_S\|_X$, and $\|p - p_S\|_X$ are presented in Tables 3.7 and 3.8. These results demonstrate that the errors decrease as we increase the order of spectral approximation, whether we keep N fixed or M. Similarly, an error estimate $\|y - y_S\|_X$ is illustrated in Fig. 3.13, which reveals an algebraic decay of $\|y - y_S\|_X$ under a log-log representation. Consequently, the space-time spectral method proves to be effective for solving the control-constrained parabolic optimization problem given by Eqs. (3.4.1)–(3.4.2), even with limited regularity. Furthermore, it offers a reliable approximation for such control problems.

Table 3.7 The numerical results of error estimates versus M with $N = 16$ in Example 3.4.8

M	2	4	6	8	10	12
$\|u - u_S\|_X$	1.5461e-0	2.4357e-1	1.8796e-2	6.5755e-4	1.5708e-5	9.6694e-5
$\|y - y_S\|_X$	2.3426e-1	8.7262e-3	2.6265e-3	1.0560e-3	2.4378e-4	1.6250e-4
$\|p - p_S\|_X$	1.2673e-0	2.4358e-1	1.7679e-2	6.5756e-4	1.5549e-5	9.6694e-5

Table 3.8 The numerical results of error estimates versus N with $M = 16$ in Example 3.4.8

N	4	6	8	10	12
$\|u - u_S\|_X$	6.5948e-1	1.8336e-1	2.5193e-2	2.1668e-3	1.2954e-4
$\|y - y_S\|_X$	1.2300e-1	5.4459e-3	1.5343e-4	4.5379e-5	4.5297e-5
$\|p - p_S\|_X$	6.5948e-1	1.8336e-1	2.5193e-2	2.1667e-3	1.2953e-4

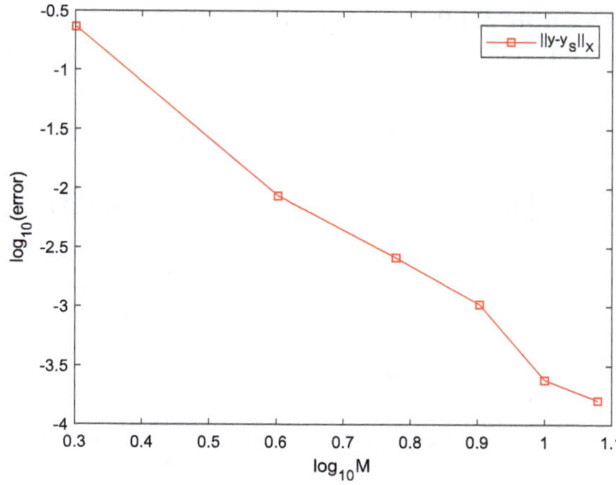

Fig. 3.13 Error estimates versus M with $N = 16$ in Example 3.4.8

Furthermore, as demonstrated thus far, all examples have been one-dimensional in space. Subsequently, we will explore examples that are two-dimensional in space.

3.4.3.2 Two-Dimensional Case in Space

Here, we investigate the control problem on the domain $\Omega = (-1, 1)^2$ involving two spatial variables, x_1 and x_2. Let $L_n(\cdot)$ represent the n-th degree Legendre polynomial. We denote $\phi_k(\cdot)$, Q_{Ml} and Q_{Mr} similarly to Sect. 3.4.3.1. Moreover, we define the following:

$$Q_N = \text{span}\,\{L_i(x_1)L_k(x_2)\}_{i,k=0}^{N}, \quad Y_N = \text{span}\,\{\phi_i(x_1)\phi_k(x_2)\}_{i,k=0}^{N-4}.$$

Next, we express y_S, p_S, and u_S as follows:

$$y_S = \sum_{i=0}^{N-4}\sum_{k=0}^{N-4}\sum_{j=0}^{M-1} y_{ikj}\phi_i(x_1)\phi_k(x_2)\psi_j(t),$$

$$u_S = \sum_{i=0}^{N}\sum_{k=0}^{N}\sum_{j=0}^{M} u_{ikj}L_i(x_1)L_k(x_2)L_j(t),$$

$$p_S = \sum_{i=0}^{N-4}\sum_{k=0}^{N-4}\sum_{j=0}^{M-1} p_{ikj}\phi_i(x_1)\phi_k(x_2)\chi_j(t).$$

Example 3.4.9 In the parabolic optimization problem defined by Eqs. (3.4.3)–(3.4.4), let $\alpha = \frac{3}{4}$. The other known data and functions are provided as follows:

$$U_{ad} = \{v \in U \mid \|v\|_X \le \gamma\} \text{ with } \gamma = 1, \ \ a_1(x) \equiv 1,$$

$$f = \left(64\pi^4 - \frac{4}{3}\right)\sin^2(\pi x_1)\sin^2(\pi x_2)\sin(\pi t) + \pi\sin^2(\pi x_1)\sin^2(\pi x_2)\cos(\pi t)$$
$$-24\pi^4\sin^2(\pi x_1)\sin(\pi t) - 24\pi^4\sin^2(\pi x_2)\sin(\pi t) + 8\pi^4\sin(\pi t),$$

$$y_d = \left(1 + \frac{256}{3}\pi^4\right)\sin^2(\pi x_1)\sin^2(\pi x_2)\sin(\pi t) - \frac{4}{3}\pi\sin^2(\pi x_1)\sin^2(\pi x_2)\cos(\pi t)$$
$$-\frac{96}{3}\pi^4\sin^2(\pi x_1)\sin(\pi t) - \frac{96}{3}\pi^4\sin^2(\pi x_2)\sin(\pi t) + \frac{32}{3}\pi^4\sin(\pi t).$$

The exact solutions that satisfy the corresponding optimality conditions are presented below:

$$y = \sin^2(\pi x_1)\sin^2(\pi x_2)\sin(\pi t), \ \ u = \frac{4}{3}\sin^2(\pi x_1)\sin^2(\pi x_2)\sin(\pi t),$$
$$p = \frac{4}{3}\sin^2(\pi x_1)\sin^2(\pi x_2)\sin(\pi t).$$

Then, Table 3.9 presents the errors $\|u - u_S\|_X$, $\|y - y_S\|_X$, and $\|p - p_S\|_X$ in the temporal direction, with values of $M = 4, 6, 8, 10, 12$, and a fixed $N = 14$. Similarly, Table 3.10 showcases the errors in the spatial direction, with $N = 4, 6, 8, 10, 12$, and a constant $M = 14$. These error results are visually represented in Fig. 3.14. All experimental data in Tables 3.9 and 3.10 demonstrate that these errors converge with high precision as M or N increases. Moreover, Fig. 3.14 illustrates that an exponential convergence rate is achievable in both the temporal and spatial directions. This example confirms that the space-time spectral method provides an accurate approximation for the two-dimensional case of the control problem given by Eqs. (3.4.3)–(3.4.4), subject to L^2-norm control constraints.

Table 3.9 The numerical results of error estimates versus M with $N = 14$ in Example 3.4.9

M	4	6	8	10	12
$\|u - u_S\|_X$	2.8277e-1	2.2323e-2	9.0317e-4	2.4526e-5	9.1509e-6
$\|y - y_S\|_X$	2.2620e-1	1.7289e-2	7.0066e-4	1.8947e-5	6.8776e-6
$\|p - p_S\|_X$	2.9416e-1	2.2327e-2	9.0309e-4	2.4546e-5	9.1127e-6

Table 3.10 The numerical results of error estimates versus N with $M = 14$ in Example 3.4.9

N	4	6	8	10	12
$\|u - u_S\|_X$	9.0104e-1	2.4823e-2	3.7304e-2	3.3514e-3	2.0369e-4
$\|y - y_S\|_X$	6.5356e-1	1.9859e-2	2.8635e-2	2.5156e-3	1.5277e-4
$\|p - p_S\|_X$	8.7113e-1	2.6477e-2	3.8175e-2	3.3540e-3	2.0369e-4

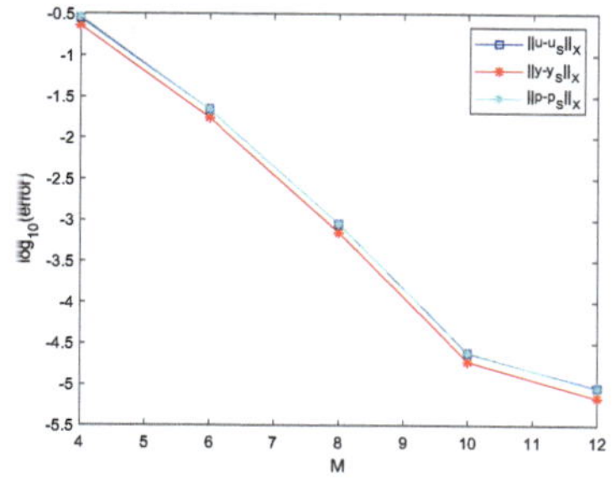

(a)

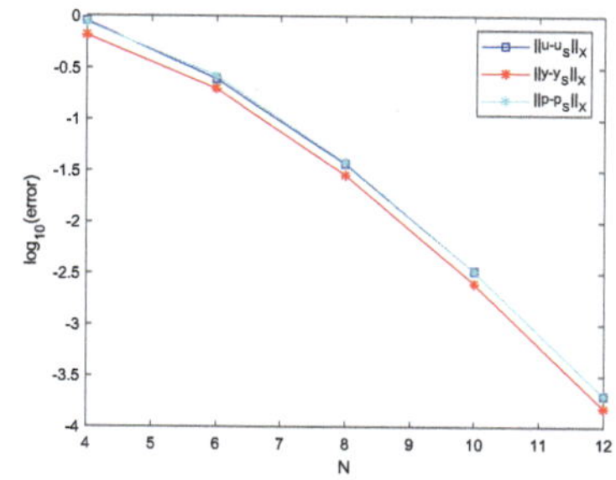

(b)

Fig. 3.14 Error estimates in Example 3.4.9. (**a**) $N = 14$. (**b**) $M = 14$

Lastly, in the subsequent example, we address an integral mean control constraint.

Example 3.4.10 We provide other known data and functions as follows:

$$U_{ad} = \left\{ v \in U \;\middle|\; \int_{-1}^{1} \int_{\Omega} v \,\mathrm{d}x\,\mathrm{d}t \geq 0 \right\}, \quad a_1(x) \equiv 1,$$

$$f = (1+\cos(\pi x_1))(1+\cos(\pi x_2))\left(e^t + \sin(\pi t)\right)$$
$$+\pi^4 (\cos(\pi x_1) + \cos(\pi x_2) + 4\cos(\pi x_1)\cos(\pi x_2))\left(e^t - e^{-1}\right),$$

$$y_d = (1+\cos(\pi x_1))(1+\cos(\pi x_2))\left(e^t - e^{-1} + \pi\cos(\pi t)\right)$$
$$-\pi^4 (\cos(\pi x_1) + \cos(\pi x_2) + 4\cos(\pi x_1)\cos(\pi x_2))\sin(\pi t).$$

The exact solutions satisfying the corresponding optimality conditions are

$$y = (1+\cos(\pi x_1))(1+\cos(\pi x_2))\left(e^t - e^{-1}\right),$$
$$u = -(1+\cos(\pi x_1))(1+\cos(\pi x_2))\sin(\pi t),$$
$$p = (1+\cos(\pi x_1))(1+\cos(\pi x_2))\sin(\pi t).$$

Similarly, the numerical results for errors $\|u - u_S\|_X$, $\|y - y_S\|_X$, and $\|p - p_S\|_X$ in the temporal direction and spatial direction are presented in Tables 3.11 and 3.12, respectively. From these two tables, we observe that the errors decrease rapidly as M or N increases.

In addition, Fig. 3.15 illustrates the error curves for the numerical results of $\|u - u_S\|_X$, $\|y - y_S\|_X$, and $\|p - p_S\|_X$. This figure demonstrates that the errors decay exponentially. Overall, the space-time spectral method proves to be effective for solving the two-dimensional case of the control problem given by Eqs. (3.4.3)–(3.4.4) with the integral mean control constraint.

Table 3.11 The numerical results of error estimates versus M with $N = 14$ in Example 3.4.10

M	4	6	8	10	12
$\|u - u_S\|_X$	9.2474e-1	6.1536e-2	2.4340e-3	5.5967e-5	8.9987e-7
$\|y - y_S\|_X$	4.1858e-2	3.7996e-3	1.6333e-4	3.7134e-6	5.7744e-8
$\|p - p_S\|_X$	8.4912e-1	6.1544e-2	2.3427e-3	5.5097e-5	8.8207e-7

Table 3.12 The numerical results of error estimates versus N with $M = 14$ in Example 3.4.10

N	4	6	8	10	12
$\|u - u_S\|_X$	2.1487e-1	9.4348e-3	2.7567e-4	4.9387e-6	7.1424e-8
$\|y - y_S\|_X$	3.1639e-1	1.3647e-2	3.7899e-4	7.1197e-6	1.0138e-7
$\|p - p_S\|_X$	2.1487e-1	9.2748e-3	2.5742e-4	4.8196e-6	6.9612e-8

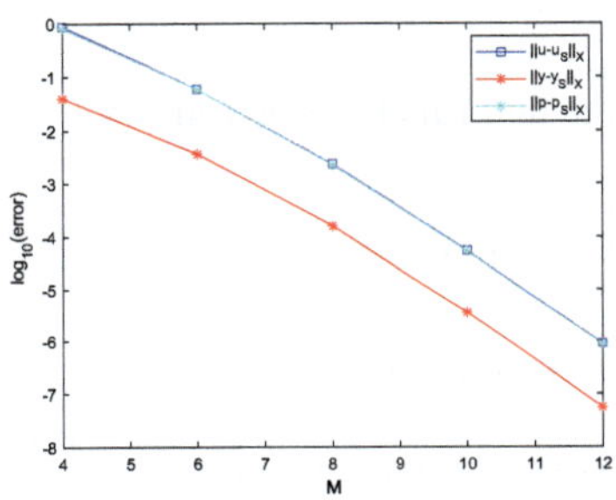

(a)

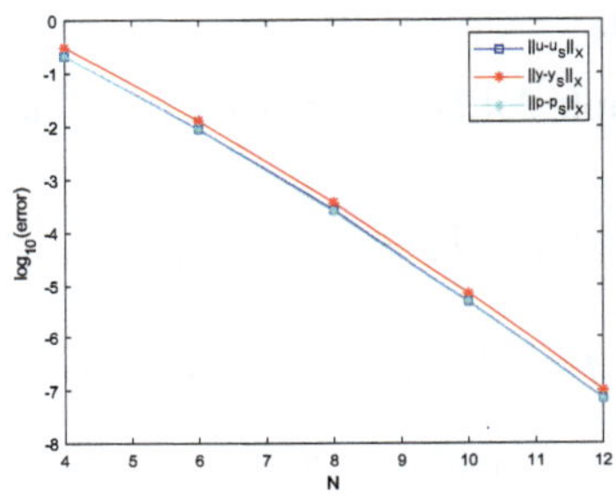

(b)

Fig. 3.15 Error estimates in Example 3.4.10. (**a**) $N = 14$. (**b**) $M = 14$

Part II
Closed-Loop Control

Chapter 4
Dynamic Programming Viscosity Solution Approach

In control theory, three crucial concepts stand out: feedback, optimization, and numerical solution. Feedback and optimization serve as the cornerstone ideas and instruments of control theory, whereas numerical solution occupies a pivotal position in the digital progression of contemporary science. The question of optimization and feedback emerges in virtually every facet of control system design.

However, in the context of continuous control systems, various approaches aimed at finding the numerical solution for optimal control possess their own limitations [202, 232]. The Pontryagin maximum principle offers a necessary condition for optimal control, but its translation into a feedback form remains challenging. Consequently, they inherently operate as open-loop systems.

Often, "those sophisticated necessary conditions rarely give an insight into the structure of the optimal controls" [194]. Furthermore, the open-loop control defined by the Pontryagin maximum principle (when solvable) only addresses the problem for specific initial data, whereas optimal control problems typically require solutions for a wide range of initial conditions. Additionally, examples presented in [146] demonstrate that optimal controls and trajectories corresponding to different initial data may exhibit distinct structures, complicating the search for numerical solutions [173]. Indeed, it is noted in [30] that "the use of the Pontryagin maximum principle to solve concrete problems is very difficult since it contains two different simultaneous mathematical problems: integration of the canonical differential system and, simultaneously, the maximization of the Hamiltonian."

Moreover, in the case of the two-point boundary problem derived from the necessary conditions of optimality, the shooting method encounters difficulties in estimating the initial data to initiate the iterative numerical process [218]. An exception occurs when the control has a valid bound, where the forward-backward sweep iterative method introduced in [112] proves effective for any initial estimate. Nonetheless, given the existence and uniqueness of optimal controls for most practical problems, the Pontryagin maximum principle offers the potential to seek

B. Sun et al., *Maximum Principle and Dynamic Programming Viscosity Solution Approach*, Systems & Control: Foundations & Applications,
https://doi.org/10.1007/978-981-96-5739-1_4

numerical solutions for optimal control, albeit primarily in an open-loop format [80, 81].

Regarding other approaches, the simplification of the original problem often results in a decrease in reliability and accuracy. Additionally, when the degrees of discretization and parameterization are excessively high, the computational workload becomes prominent, and the solving process gives rise to the notorious "curse of dimensionality" [35].

Since Pontryagin's era, it has been established that the value function of an optimal control system, when approached through dynamic programming, satisfies a nonlinear partial differential equation known as the HJB equation. Furthermore, unlike the Pontryagin maximum principle, which focuses on a single extremal at a time, the dynamic programming method simultaneously considers families of extremals. Bellman's dynamic programming offers a pathway to determining the optimal feedback control. Once the HJB equation, which the value function satisfies, is formulated, the feedback control law can be determined by solving this first-order nonlinear partial differential equation. Additionally, according to the viscosity solution theory [16], the value function is typically the unique viscosity solution to the corresponding HJB equation. Evidently, the feedback form of optimal control is more appealing in terms of real-time operation and robustness. Finding the optimal feedback control is widely regarded as the ultimate goal in control theory [116].

However, given the intricate nature of analytically solving dynamic programming equations, it is impractical to obtain closed-form viscosity solutions for the HJB equation in almost all nonlinear optimal control problems, with the exception of linear quadratic problems and a handful of specific examples. This limitation renders numerical solutions as the only viable and practically significant approach to tackling the HJB equation and, consequently, optimal control [76, 128]. In other words, numerical methods become virtually the sole feasible option for solving optimal control problems, barring a few exceptions. Nonetheless, the introduction of viscosity solution concepts has imparted a rigorous mathematical foundation to classical dynamic programming techniques, thereby paving the way for a new avenue of research in optimal feedback control.

Over the past two decades, numerous finite-difference schemes have been developed to address this challenge. These schemes encompass the upwind finite-difference method [233], the method of vanishing viscosity [62], and the parallel algorithm leveraging the domain decomposition technique [78]. However, most of these algorithms are plagued by the infamous "curse of dimensionality" [166], limiting their applicability to only low-dimensional "toy" problems.

One notable exception to the curse of dimensionality constraint is the max-plus method [167] for controlling and estimating nonlinear systems. This approach takes advantage of semiconvex dual operators and max-plus linearity by focusing on the dual-space semigroup linked to the HJB equation. This semigroup proves especially valuable due to its max-plus integral operator form, whose kernel is derived from the originating semigroup. Repeated applications of the dual-space semigroup can yield a numerical solution [166, 167].

Another intriguing study [118] integrates recent findings on the structure of the HJB equation and its simplification to a linear PDE. The curse of dimensionality is bypassed through the use of separated low-rank tensor representations. The central idea involves approximating the solution and its related operators using low-rank tensors. If the problem's components can be accurately modeled within the region, complexity increases based on the approximation's rank rather than the dimensionality.

The curse of dimensionality can also be overcome through effective model reduction, a field where various techniques, such as proper orthogonal decomposition (POD) [74, 91, 142], are commonly employed. By utilizing the POD technique alongside the Galerkin projection approach, the state equation, even if it is a PDE, can be approximated by a low-order ordinary differential equation. Subsequently, the corresponding reduced HJB equation can be numerically solved using a difference scheme.

In this chapter, we deviate from these traditional methods and present an approach rooted in dynamic programming and viscosity solutions. This method is constructed and illustrated through several captivating paradigms stemming from technical or biological origins, aiming to obtain approximate or suboptimal control. Let us delve into an introduction of this methodology.

4.1 DPVS Approach

Consider the following control system:

$$\begin{cases} y'(t) = f(y(t), u(t)), & t \in (0, T],\ T > 0, \\ y(0) = z, \end{cases} \tag{4.1.1}$$

where $y(\cdot) \in \mathbb{R}^n$ is the state, $u(\cdot) \in \mathcal{U}[0, T] = L^\infty([0, T]; U)$ is the admissible control with compact set $U \subset \mathbb{R}^m$, and z is the initial value. Assume that $f : \mathbb{R}^n \times U \to \mathbb{R}^n$ is continuous in its variables.

Given a running cost $L(t, y, u)$ and a terminal cost $\psi(y)$, the optimal control problem for system (4.1.1) is to seek an optimal control $u^*(\cdot) \in \mathcal{U}[0, T]$, such that

$$J(u^*(\cdot)) = \inf_{u(\cdot)\in\mathcal{U}[0,T]} J(u(\cdot)), \tag{4.1.2}$$

where J is the cost functional given by

$$J(u(\cdot)) = \int_0^T L(\tau, y(\tau), u(\tau))\, d\tau + \psi(y(T)).$$

Then, consider the optimal control problem for the following system for any $(t, x) \in [0, T) \times \mathbb{R}^n$:

$$\begin{cases} y'_{t,x}(s) = f(y_{t,x}(s), u(s)), \quad s \in (t, T], \\ y_{t,x}(t) = x, \end{cases}$$

with the cost functional

$$J_{t,x}(u(\cdot)) = \int_t^T L(\tau, y_{t,x}(\tau), u(\tau))\, \mathrm{d}\tau + \psi(y_{t,x}(T)).$$

Define the value function

$$V(t, x) = \inf_{u(\cdot) \in \mathcal{U}[t,T]} J_{t,x}(u(\cdot)), \quad \forall\, (t, x) \in [0, T) \times \mathbb{R}^n$$

with the terminal value

$$V(T, x) = \psi(x)$$

for all $x \in \mathbb{R}^n$.

If V is smooth enough, say $V \in C^1([0, T] \times \mathbb{R}^n)$, then V satisfies the following HJB equation [16]:

$$\begin{cases} V_t(t, x) + \inf_{u \in U} \{f(x, u) \cdot \nabla_x V(t, x) + L(t, x, u)\} = 0, \quad (t, x) \in [0, T) \times \mathbb{R}^n, \\ V(T, x) = \psi(x), \quad x \in \mathbb{R}^n, \end{cases} \tag{4.1.3}$$

where $\nabla_x V(t, x)$ stands for the gradient of V in x.

The next two lemmas highlight the significant role played by the value function in determining the optimal feedback law, as referenced in [16].

Lemma 4.1.1 *Let $V \in C^1([0, T] \times \mathbb{R}^n)$ be the value function. Then if there exists a control $u^*(\cdot) \in \mathcal{U}[0, T]$ such that*

$$\begin{aligned} & f(y^*(t), u^*(t)) \cdot \nabla_x V(t, y^*(t)) + L(t, y^*(t), u^*(t)) \\ = & \inf_{u \in U} \left\{ f(y^*(t), u) \cdot \nabla_x V(t, y^*(t)) + L(t, y^*(t), u) \right\}, \end{aligned}$$

then $u^*(\cdot)$ is an optimal control, where y^* is the state corresponding to u^*.

As usual, we denote the optimal control as

$$u^*(t) \in \arg\inf_{u \in U} \left\{ f(y^*(t), u) \cdot \nabla_x V(t, y^*(t)) + L(t, y^*(t), u) \right\}$$

for almost all $t \in [0, T]$.

Lemma 4.1.2 *Let* $V(t,x) \in C^1([0,T] \times \mathbb{R}^n)$ *be the value function. Then* $(u^*(\cdot), y^*(\cdot))$ *is an optimal control-trajectory pair in feedback form if and only if*

$$V_t(t, y^*(t)) + f(y^*(t), u^*(t)) \cdot \nabla_x V(t, y^*(t)) + L(t, y^*(t), u^*(t)) = 0$$

for almost all $t \in [0, T)$.

By Lemmas 4.1.1 and 4.1.2, we arrive at Theorem 4.1.3, which provides a method for constructing the feedback law through the value function.

Theorem 4.1.3 *Let* $V(t,x) \in C^1([0,T] \times \mathbb{R}^n)$ *be the value function. Suppose* $u(t,x)$ *satisfies*

$$\begin{aligned} &f(x, u(t,x)) \cdot \nabla_x V(t,x) + L(t,x,u(t,x)) \\ &= \inf_{u \in U} \{f(x,u) \cdot \nabla_x V(t,x) + L(t,x,u)\}. \end{aligned}$$

Then

$$u_z^*(t) = u(t, y_z(t))$$

is the feedback law of the optimal control problem (4.1.1)–(4.1.2), *where* $y_z(t)$ *satisfies the following:*

$$y_z'(t) = f(y_z(t), u(t, y_z(t))), \quad y_z(0) = z, \ t \in [0, T].$$

We observe from Theorem 4.1.3 that to determine the feedback control law, both the value function V itself and its gradient $\nabla_x V$ are required.

As we understand it, despite the smoothness of the functions f and L, Eq. (4.1.3) typically lacks a classical solution. Fortunately, under certain fundamental assumptions regarding f and L, the value function V emerges as the sole viscosity solution to (4.1.3). Nevertheless, obtaining an analytical solution for (4.1.3) remains elusive when dealing with general nonlinear functions f and L. Consequently, numerical resolution of (4.1.3) assumes paramount importance, especially in practical applications. Indeed, several difference schemes have been introduced to approximate these viscosity solutions, as documented in [85, 120, 121, 234].

Once a viscosity solution of (4.1.3) has been numerically obtained, we can proceed to construct a numerical solution for the feedback law based on Theorem 4.1.3.

4.2 Algorithm of Finding Optimal Feedback Law

In this subsection, we utilize Theorem 4.1.3 to develop an algorithm aimed at obtaining numerical solutions for the optimal feedback control and the corresponding optimal trajectory pair. The algorithm involves two interconnected discretization steps. Firstly, we discretize the HJB equation (4.1.3) to determine the feedback law. Secondly, we discretize the state equation (4.1.1) to identify the optimal trajectory.

In fact, numerous approximation schemes have emerged for the numerical solution of (4.1.3). Meanwhile, for the numerical solution of the state equation (4.1.1), a lot of classical methods are available, such as the Euler method, the Runge-Kutta method, and the Hamming algorithm [218].

It is worth noting that when employing Theorem 4.1.3 to determine the optimal feedback law, it is the directional derivative $\nabla_x V \cdot f$ rather than the gradient $\nabla_x V$ itself that is required. This aspect significantly eases our pursuit of numerical solutions. The crucial step involves approximating $\nabla_x V \cdot f$ through its natural definition, as outlined below:

$$
\begin{aligned}
&\nabla_x V(t,x)\cdot f(x,u)\\
&=\left[\nabla_x V(t,x)\cdot \eta \frac{f(x,u)}{1+\|f(x,u)\|}\right]\frac{1+\|f(x,u)\|}{\eta}\\
&\approx \frac{V\left(t, x+\eta \dfrac{f(x,u)}{1+\|f(x,u)\|}\right)-V(t,x)}{\eta}(1+\|f(x,u)\|),
\end{aligned}
\tag{4.2.1}
$$

where $\eta > 0$ is a small number and $\|\cdot\|$ denotes the Euclidean norm in $\mathbb{R}^n$. Based on observation (4.2.1), the HJB equation (4.1.3) can be subsequently approximated through a finite-difference scheme, employing both temporal and spatial mesh approximations:

$$
\begin{cases}
\dfrac{V_i^{j+1}-V_i^j}{\Delta t}+\dfrac{V_{i+1}^j-V_i^j}{\eta}\cdot\left(1+\left\|f_i^j\right\|\right)+L_i^j=0,\\
u_i^{j+1}\in \arg\inf\limits_{u\in U}\left\{\dfrac{V_{i+1}^{j+1}-V_i^{j+1}}{\eta}\cdot(1+\|f(x_i,u)\|)+L\left(t_{j+1},x_i,u\right)\right\}
\end{cases}
\tag{4.2.2}
$$

for $i = 0, 1, \cdots, M$ and $j = 0, 1, \cdots, N-1$, where $V_i^j = V(t_j, x_i)$, $f_i^j = f\left(x_i, u_i^j\right)$, $L_i^j = L\left(t_j, x_i, u_i^j\right)$. Meanwhile, it is assumed that

$$
\frac{|\Delta t|}{\eta}\left(1+\max_{i,j}\left\|f_i^j\right\|\right)\le 1,
$$

which is a necessary condition for the stability of the finite-difference scheme [187].

It is worth highlighting that the aforementioned approximation, denoted by (4.2.1), confers significant advantages to the algorithm introduced in this subsection. Attempting to initially derive the viscosity solution of (4.1.3) would likely result in encountering the "curse of dimensionality" for high-dimensional problems, as it necessitates acquiring data for all grids across the entire region. This could explain why numerous studies on the viscosity solution of (4.1.3) tend to be confined to 1-D or 2-D problems, as exemplified in [233]. Conversely, our approach focuses on searching for the optimal control solely along the direction of f, rather than exploring the entire region. Consequently, the novel algorithm requires considerably less data. This concept can also be extended to infinite-dimensional systems, as demonstrated in [102]. This represents the first endeavor to leverage dynamic programming in this manner for finding optimal control.

Based on the preceding discussion, we proceed to devise an algorithm for obtaining numerical solutions of the optimal feedback control-trajectory pairs.

Algorithm 4 Algorithm for obtaining numerical solutions of the optimal feedback control-trajectory pairs

1: Initial partition on time and space. Select two positive integers N and M. Let $t_j = T + j\Delta t,\ j = 0, 1, \cdots, N$ be a backward partition of $[0, T]$, where $\Delta t = -T/N$. For any initial given $\tilde{u} \in U$, let initial state $x_0 = z$ and

$$x_i = x_{i-1} + \eta \frac{f(x_{i-1}, \tilde{u})}{1 + \|f(x_{i-1}, \tilde{u})\|}, \quad i = 1, 2, \cdots, M. \tag{4.2.3}$$

2: Initialization of value function and control. Let

$$\begin{cases} V_i^0 = \psi(x_i), \\ u_i^0 \in \arg\inf\limits_{u \in U} \left\{ \dfrac{V_{i+1}^0 - V_i^0}{\eta} \cdot (1 + \|f(x_i, u)\|) + L(T, x_i, u) \right\}, \\ \qquad i = 0, 1, \cdots, M. \end{cases} \tag{4.2.4}$$

3: Iteration for HJB equation. By (4.2.2) and Step 2, we obtain all

$$\left\{ \left\{ V_i^j \right\}_{i=0}^{M} \right\}_{j=0}^{N} \text{ and } \left\{ \left\{ u_i^j \right\}_{i=0}^{M} \right\}_{j=0}^{N}$$

by

$$\begin{cases} V_i^{j+1} = \left(1 + \dfrac{\Delta t}{\eta}\left(1 + \left\| f_i^j \right\|\right)\right) V_i^j - \dfrac{\Delta t}{\eta}\left(1 + \left\| f_i^j \right\|\right) V_{i+1}^j - \Delta t L_i^j, \\ u_i^{j+1} \in \arg\inf\limits_{u \in U} \left\{ \dfrac{V_{i+1}^{j+1} - V_i^{j+1}}{\eta} \cdot (1 + \|f(x_i, u)\|) + L(t_{j+1}, x_i, u) \right\}. \end{cases}$$

Here,

$$\left(u_0^N, y^0\right) = \left(u_0^N, y(0)\right) = (u(0), z)$$

is the first optimal feedback control-trajectory pair.

4: Iteration for state equation. For $j = 0, 1, 2, \cdots, N-1$, solve the state equation

$$\frac{y^{j+1} - y^j}{-\Delta T} = f\left(y^j, u(t_{N-j})\right),$$

to obtain $y^{j+1} = y(t_{N-j-1})$. Replace $(\tilde{u}, z)$ in Step 1 by $(u(t_{N-j}), y^{j+1})$ and go to Steps 2 and 3 to obtain u_0^{N-j-1}. Then

$$\left(u_0^{N-j-1}, y^{j+1}\right) = \left(u(t_{N-j-1}), y(t_{N-j-1})\right)$$

is the $(j+2)$-th optimal feedback control-trajectory pair. Continue the iteration to obtain all $(u(t_{N-j}), y(t_{N-j}))_{j=0}^{N}$.

After completing Steps 1 through 4, we ultimately obtain all the sought-after optimal feedback control-trajectory pairs.

$$\left(u(t_{N-j}), y(t_{N-j})\right), \quad j = 0, 1, \cdots, N.$$

It is important to emphasize that the primary focus of the aforementioned algorithm is not to solve the HJB equation, nor is it to determine the value function itself. This approach differs significantly from most literature in the field, which tends to concentrate on solving HJB equations (see, for instance, [120, 121, 234]). Instead, our ultimate goal is to identify numerical solutions for both optimal feedback control and the corresponding optimal trajectory. The entire algorithm involves solving the state equation once and the HJB equation N times. Based on our numerical experiments documented in [102, 103] and the example presented in the following section, the mesh point sequence generated in space through the recursive relation (4.2.3) does not introduce oscillations in the space variable within subregions of the given space, even when f changes its sign. This stability is attributed to the more appropriate definition of the directional derivative provided by (4.2.1), which enables us to search for the optimal control along its natural direction, f.

If our sole concern is the solution of the HJB equation (4.1.3), such as computing the values of V within the polyhedral domain $\prod_{i=1}^{n} [a_i, b_i]$ of $\mathbb{R}^n$, it becomes necessary to generate a monotone mesh point sequence in space using the recursive relation (4.2.3). To achieve this, we may need to alter the searching direction intentionally. In such cases, we recommend employing the following approximation for the directional derivative, rather than the approximation given by (4.2.1). Specifically, for every fixed (x, u), $x = \{x^1, x^2, \cdots, x^n\}$, $f = \{f^1, f^2, \cdots, f^n\}$,

$$\begin{aligned}
&\nabla_x V(t,x) \cdot f(x,u) \\
&= \sum_{p=1}^{n} \left[V_{x^p}(t,x) \cdot \eta \frac{f^p(x,u)\,\mathrm{sgn}(f^p(x,u))}{1 + \|f^p(x,u)\|} \right] \frac{1 + \|f^p(x,u)\|}{\eta\,\mathrm{sgn}(f^p(x,u))} \\
&\approx \sum_{p=1}^{n} \frac{V\left(t, x + \eta \dfrac{f^p(x,u)\,\mathrm{sgn}(f^p(x,u))}{1 + \|f^p(x,u)\|} I_p \right) - V(t,x)}{\eta} \frac{1 + \|f^p(x,u)\|}{\mathrm{sgn}(f^p(x,u))},
\end{aligned}$$

where

$$\mathrm{sgn}(\wp) = \begin{cases} 1, & \wp \geq 0, \\ -1, & \wp < 0, \end{cases}$$

I_p is the n-dimension unit vector with the pth component 1. V_{x^p} is the partial derivative of the value function V with respect to x^p.

In this manner, potential oscillations arising from sign changes in the function f can be prevented when utilizing the formula (4.2.3) for spatial mesh partitioning. Consequently, the mesh point sequence in space, generated by the recursive relation (4.2.3), is now adjusted to

$$x_i^p = x_{i-1}^p + \eta \frac{\mathrm{sgn}\left(f^p(x_{i-1}, \tilde{u})\right) f^p\left(x_{i-1}, \tilde{u}\right)}{1 + \|f^p(x_{i-1}, \tilde{u})\|},$$

$$p = 1, 2, \cdots, n, \quad i = 1, 2, \cdots, M,$$

where $x_k = \{x_k^1, x_k^2, \cdots, x_k^n\}$, $k = i-1, i$.

To conclude this section, we emphasize that the solution obtained through the algorithm represents a single closed-loop solution commencing from the initial state, as referenced in [190]. This observation becomes evident during the execution process of the algorithm:

(a) Initially, we evaluate V_i^0, $i = 0, 1, \cdots, M$ using the terminal condition and compute u_i^0 according to Eq. (4.2.4). Given any initial control $\tilde{u}$, we choose an initial state $x^0 = z$ to initiate the computation. By employing the finite-difference scheme (4.2.2) for the HJB equation (4.1.3), we obtain the sets

$$\left\{ \left\{ V_i^j \right\}_{i=0}^{M} \right\}_{j=0}^{N} \text{ and } \left\{ \left\{ u_i^j \right\}_{i=0}^{M} \right\}_{j=0}^{N}$$

(or in the continuous case, $V(t, z)$ and $u(t, z)$). Finally, $\left(u_0^N, x^0\right) = (u(0), z)$ represents the first optimal control-trajectory pair.

(b) Next, we substitute the initial optimal control $u(0)$ and the corresponding state z into the finite-difference scheme for the state equation. This allows us to compute the new state y^1 at time t_{N-1}. Then, we replace the previous control $\tilde{u}$ and state z in Step (a) with the new values $u(0)$ and y^1. Using the difference scheme (4.2.2), we obtain the updated control value u_0^{N-1}. At this point, we have obtained the second optimal control-trajectory pair.

$$\left(u_0^{N-1}, y^1\right) = (u(t_{N-1}), y(t_{N-1})).$$

(c) We continue the computation iteratively, obtaining each subsequent optimal control-trajectory pair $(u(t_{N-j}), y(t_{N-j}))$, $j = 0, 1, \cdots, N$. Once all pairs have been computed, the entire optimal control trajectory has been determined, and the computation is complete.

It is important to note that throughout the execution of the entire algorithm, the finite-difference scheme for the HJB equation is invoked once for every time node update in the difference scheme for the state equation. Upon completion of the algorithm, we have called the state equation (specifically, its corresponding finite-difference scheme) one time and the HJB equation (likewise, its corresponding finite-difference scheme) a total of N times.

This computational process is initiated by the initial state z, and the solution we obtain is specifically tailored to this initial data. To fully construct the feedback map for the optimal control problem, we need to repeat this algorithm and complete all computations for initial states spanning the entire definition domain.

4.3 Optimal Control of HIV/AIDS

In this section, we demonstrate the application of the new algorithm to an optimal control problem involving two types of drug therapies for HIV/AIDS. Through this example, we aim to illustrate the approximate optimal control strategy obtained using the algorithm. By leveraging this paradigm, we endeavor to showcase the validity and effectiveness of the new algorithm, as referenced in [105].

Adopting mathematical methods to understand HIV/AIDS dynamics has been a focal point of numerous studies. For instance, researchers in [178] and [230] have explored how mathematical models can shed light on the complex interactions within HIV infection. Similarly, [130] demonstrates the use of mathematics in comprehending HIV immune dynamics.

To gain insights into the in vivo dynamics of HIV-1 infection, studies like [183] have employed dynamical modeling and parameter estimation techniques. These approaches have revealed crucial aspects of HIV pathogenesis. Furthermore, [178] discusses virus dynamics and investigates the mathematical principles underlying immunology and virology.

The integration of HIV/AIDS education into university curricula has been explored in [59] and [60]. These works analyze HIV/AIDS models from a control engineering perspective, highlighting the importance of interdisciplinary approaches. Using control theory, the question of when to initiate HIV therapy has been examined in [125], concluding that therapy is most effective when the viral load is more manageable.

Optimal chemotherapy strategies for HIV have been derived using the Pontryagin maximum principle in studies such as [37, 83, 132]. These strategies provide valuable insights into effective treatment approaches. Additionally, [31] develops methods for stable control of the HIV-1 population using external feedback control, demonstrating how immune system components can be strengthened against the virus.

An intriguing study on dynamic multi-drug therapies for HIV is presented in [236], proposing a dynamic, albeit not optimal, treatment policy. Employing a model predictive control approach, [254] derives treatment schedules for HIV therapy based on the modified Wodarz-Nowark model, offering a closed-loop control solution.

A neighboring-optimal control policy for HIV is introduced in [217]. This policy builds upon the linearization of the nonlinear model at steady state. Furthermore, [188] proposes a control strategy for HIV dynamics based on linearization and extends it to a linear-quadratic optimal control problem. The resulting controller minimizes the square of the error between actual and desired (equilibrium) values.

For fractional-order HIV-immune systems with memory, [67] discusses the necessary conditions for optimality in a general fractional optimal control problem. The study numerically solves the fractional-order two-point boundary value problem and demonstrates the effects of mathematically optimal therapy.

In [5], researchers fit a nonlinear dynamical mathematical model of HIV infection to longitudinal clinical data for individual patients. They combine a statistically based censored data method with inverse problem techniques to estimate dynamic parameters. The significant works of [4] and [3] investigate optimal control strategies for HIV using the Pontryagin maximum principle. These studies consider optimal structured treatment interruption control problems, employing various approaches to reduce computational burden, such as crude direct search, 5-day segment strategy, and the subperiod method. However, they derive suboptimal structured treatment interruption therapy protocols without explicitly mentioning the HJB equation or a concrete algorithm.

In contrast, our proposed algorithm, as described in the previous sections, offers a systematic and efficient approach to solving optimal control problems in the context of HIV/AIDS treatment. By leveraging the HJB equation and finite-difference schemes, we can obtain approximate optimal control strategies tailored to specific initial conditions and system dynamics. This algorithm holds promise in advancing the field of HIV/AIDS treatment optimization and potentially improving patient outcomes.

Now, let us introduce the HIV/AIDS therapy model under investigation in this section. Let x_1 denote the concentration of uninfected CD4$^+$T cells and x_2 represent the population of free infectious virus particles. For any given $t_f > 0$, the two drug treatment approaches for the HIV/AIDS model can be described by the following system of ordinary differential equations, as referenced in [127, 131]:

$$\begin{cases} x_1'(t) = s_1 - \dfrac{s_2 x_2(t)}{B_1 + x_2(t)} - \mu x_1(t) - k x_2(t) x_1(t) + u_1(t) x_1(t), \quad t \in (0, t_f), \\ x_2'(t) = \dfrac{g(1 - u_2(t)) x_2(t)}{B_2 + x_2(t)} - c x_2(t) x_1(t), \quad t \in (0, t_f), \\ (x_1(0), x_2(0)) = z = (z_1, z_2), \end{cases} \tag{4.3.1}$$

where $u(\cdot) = (u_1(\cdot), u_2(\cdot)) \in \mathcal{U}[0, t_f] = L^\infty((0, t_f); U)$ represents the admissible control set. In this set, u_1 and u_2 correspond to immune boosting and viral suppressing drugs, respectively. Specifically, we can consider IL-2 immunotherapy as the immune-system-enhancing drug, denoted by u_1. The term $u_1(t)x_1(t)$ signifies that the enhancement of the immune system via IL-2 leads to a proportional increase in CD4$^+$T cells at a rate determined by $u_1(t)$. The control constraint set is given by $U = [c_{1\ell}, c_{1r}] \times [c_{2\ell}, c_{2r}]$ where $c_{1\ell}$ and $c_{2\ell}$ are constants for $i = 1, 2$. Within the model equation, the expression

$$s_1 - \frac{s_2 x_2(t)}{B_1 + x_2(t)}$$

accounts for the source and proliferation of uninfected CD4$^+$T cells. This includes contributions from both external sources, such as the thymus and lymph nodes, and internal differentiation of CD4$^+$T cells within the plasma. As the disease progresses, this T-cell source deteriorates, approaching a limiting value of $s_1 - s_2$. Here s_1 and s_2 represent constant sources/production of CD4$^+$T cells, μ denotes the death rate of uninfected CD4$^+$T cells, k signifies the rate at which CD4$^+$T cells become infected by free virus x_2, g represents the input rate of an external viral source, c is the loss rate of the virus, and B_1 and B_2 are half-saturation constants.

This simplified HIV progression model focuses solely on the interaction between uninfected CD4$^+$T cells and free viruses within the plasma. As HIV research evolves, more advanced models have emerged, incorporating additional state variables such as infected CD4$^+$T cell populations. Some of these models, like the one referenced in [230], encompass recent advancements in IL-2 coadministration for HIV therapy. Nonetheless, even in its biological simplicity, our chosen model offers valuable insights into the dynamics of HIV progression within the plasma, based on fundamental assumptions about the interplay between uninfected CD4$^+$T cells and free viruses. Given the abundance of data available for these two populations throughout the disease's progression, model simulations can be readily validated against real-world observations.

The mathematical model (4.3.1) for HIV/AIDS progression, originally presented in [131], delves into the interactions between $CD4^{+}$T cells and HIV particles within the immune system. For a more detailed exploration of its biological underpinnings, readers are encouraged to refer to the original publication. Furthermore, [127] employs the Pontryagin maximum principle to explore optimal control strategies within this model.

In this section, our objective is to introduce a novel algorithm for finding numerical solutions to approximate optimal control problems based on dynamic programming, without relying on exhaustive solutions for the value function. To illustrate our approach, we consider the practical application of drug combinations in virus treatment. The simplicity of the HIV/AIDS model adopted here facilitates a clearer exposition of our methodology. However, it is important to note that our proposed algorithm has the versatility to handle more complex and realistic models, such as those discussed in [192, 193], where drug resistance, immune responses, and HIV latency emerge as critical factors contributing to HIV persistence despite long-term effective therapy.

We now introduce a novel optimal control problem for system (4.3.1). Specifically, our aim is to identify an optimal control $u^*(\cdot) \in \mathcal{U}[0, t_f]$ that minimizes a given objective functional:

$$J(u^*(\cdot)) = \inf_{u(\cdot)\in\mathcal{U}[0,t_f]} J(u(\cdot)), \tag{4.3.2}$$

where J is the cost functional given by

$$J(u(\cdot)) = \int_0^{t_f} \left[A_1 u_1^2(\tau) + A_2 u_2^2(\tau) - x_1(\tau)\right] \mathrm{d}\tau + A_3 x_2^2(t_f), \tag{4.3.3}$$

with positive constants A_1, A_2, A_3 where $A_3 \neq 0$ in (4.3.3), a distinction from [127], our cost functional reflecting the benefits of $CD4^{+}$T cells and the expenses associated with drug therapies. The square of the virus particle count at the end of the treatment period comprises the terminal cost, underscoring the importance of suppressing the final virus count. Quadratic terms are used in the running cost to account for the toxicity of high drug dosages. Our objective is to strike a balance between minimizing drug usage, maximizing the uninfected cell count, and achieving the lowest possible virus particle count at the conclusion of the study period. By incorporating the terminal cost, we effectively prevent a common issue seen in existing literature, where the viral load rebounds simply due to the end of the treatment horizon (as exemplified in Figs. 13 and 15 of [4] and Figs. 3 and 5 of [3]).

Based on (4.1.3), provided that the value function of the optimal control problem (4.3.2) is sufficiently smooth, one can deduce the following HJB equation, which is satisfied by the value function:

$$\begin{cases} V_t(t,x) + \inf_{u\in U} \left\{ V_x(t,x) \cdot F(x,u) + A_1 u_1^2 + A_2 u_2^2 - x_1 \right\} = 0, \\ V(t_f, x) = A_3 x_2^2, \end{cases}$$

where $t \in [0, t_f)$, $x = (x_1, x_2) \in \mathbb{R}^2$. The function $F(x,u) = (F_1(x,u), F_2(x,u))^T$ is defined by

$$F(x,u) = \begin{pmatrix} s_1 - \dfrac{s_2 x_2}{B_1 + x_2} - \mu x_1 - k x_2 x_1 + u_1 x_1 \\ \dfrac{g(1-u_2)x_2}{B_2 + x_2} - c x_2 x_1 \end{pmatrix}.$$

Next, we proceed to generate numerical solutions for the optimal control problem defined by (4.3.1)–(4.3.2). This process strictly adheres to the algorithm outlined in Sect. 4.2. The program is coded in C, while the results are visualized using MATLAB. The parameters employed in our computations are listed below:

$$N = 640,\ M = 5,\ A_1 = 250000,\ A_2 = 75,\ t_f = 40,\ \eta = 0.02,\ z_1 = 400.0,$$
$$z_2 = 3.5,\ c_{1\ell} = 0.0,\ c_{1r} = 0.02,\ c_{2\ell} = 0.0,\ c_{2r} = 0.9,\ s_1 = 2.0,\ s_2 = 1.5,$$
$$\mu = 0.002,\ k = 2.5 \times 10^{-4},\ g = 30.0,\ c = 0.007,\ B_1 = 14.0,\ B_2 = 1.0.$$

The constant A_3 is considered for two numerical values, 100 and 0, solely for the purpose of numerical analysis. Among the parameters and constants enumerated above, several were directly adopted from [131], where model (4.3.1) was originally established.

It is worth noting that the difference in magnitude between A_1 and A_2 serves to balance the terms in the equation. The control bounds for u_1 are set to $[0, 2.0 \times 10^{-2}]$, with its square ranging from $[0, 4.0 \times 10^{-4}]$, while the control bounds for u_2 are $[0, 0.9]$ with its square spanning $[0, 0.81]$. The controls u_1 and u_2 represent drug administration schedules. For instance, a value of $u_1 = 0.02$ indicates that drug u_1 is being administered at its full daily dose. Similarly, when $u_2 = 0.9$, it signifies that drug u_2 is being given at its maximum daily level.

We have computed the approximate optimal control-trajectory pairs for both cases where $A_3 = 100$ and $A_3 = 0$. The results are illustrated in two separate figures. Figure 4.1 displays the numerical solution of the approximate optimal control components u_1^* and u_2^*, along with the corresponding trajectories of CD4$^+$T cells and HIV particles, when $A_3 = 100$. To emphasize the importance of the terminal cost, we have also plotted the case where $A_3 = 0$ in Fig. 4.2, following the same format.

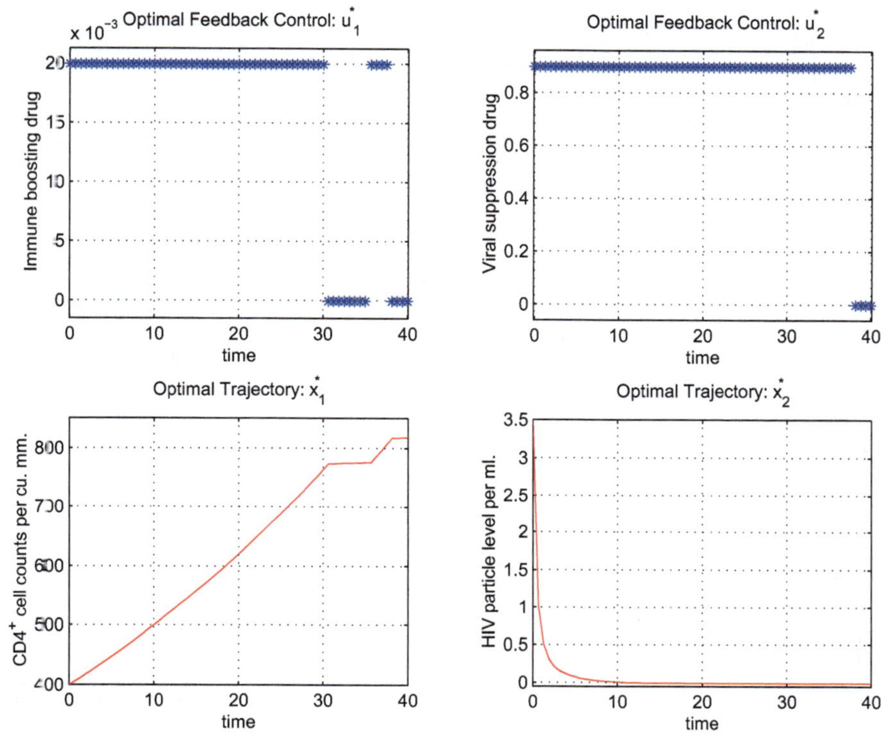

Fig. 4.1 Numerical solution of approximate optimal control-trajectory pair with $A_3 = 100$

It is important to note that while the results we have obtained may not appear to be strictly optimal (as optimal controls should be continuous according to Theorems 6.1 and 6.2 of [84]), they are indeed approximately optimal (or suboptimal) in the sense that they effectively suppress HIV particles and enhance the immune system's function. From a numerical analysis perspective, our results are also quite satisfactory. To verify this, we conducted numerical experiments comparing the cost associated with our approximate optimal control-trajectory pairs to that of arbitrarily chosen admissible controls and their corresponding trajectories.

The arbitrarily selected admissible controls consist of various combinations of two controls. Each control is chosen from one of five different functions:

$$u_1(t) = \begin{cases} 0.02, & \text{Case a}, \\ -0.02t/40 + 0.02, & \text{Case b}, \\ 0.02t/40, & \text{Case c}, \\ 0, & \text{Case d}, \\ 0.02|\sin 30t|, & \text{Case e}, \end{cases} \qquad u_2(t) = \begin{cases} 0.9, & \text{Case A}, \\ -0.9t/40 + 0.9, & \text{Case B}, \\ 0.9t/40, & \text{Case C}, \\ 0, & \text{Case D}, \\ 0.9|\sin 50t|, & \text{Case E}. \end{cases}$$

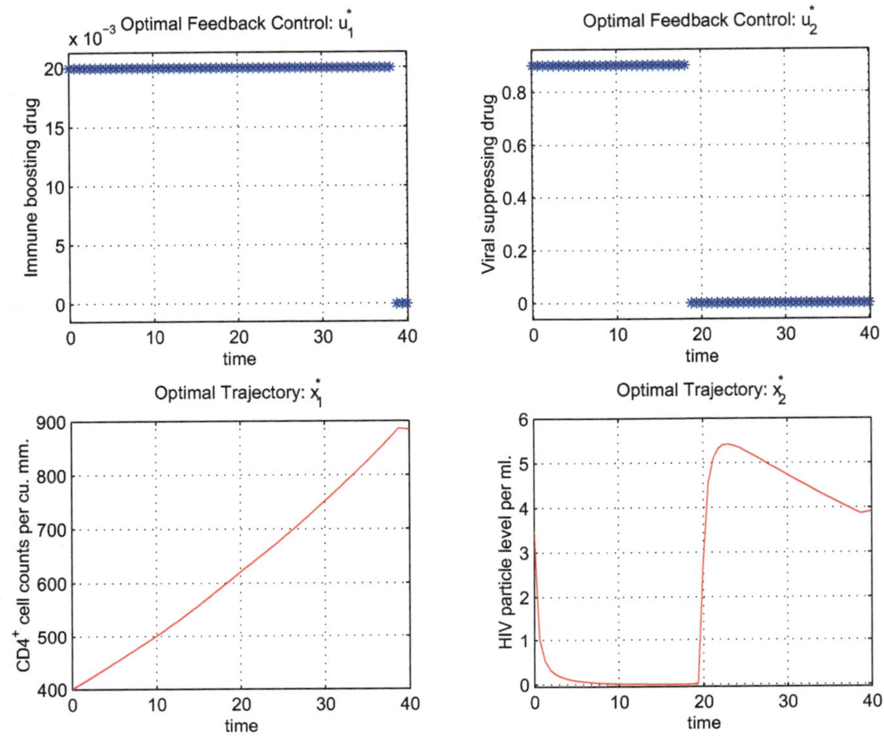

Fig. 4.2 Numerical solution of approximate optimal control-trajectory pair with $A_3 = 0$

Table 4.1 Costs corresponding to different control-trajectory pairs

Control (u_1, u_2)	J values with $A_3 = 100$	J values with $A_3 = 0$
Case (a, A)	– 19037.364346	– 19037.364346
Case (b, B)	– 16091.030761	– 19455.373985
Case (b, C)	– 18927.367136	– 18927.595354
Case (a, D)	– 18365.038145	– 19997.793922
Case (c, B)	– 13315.540793	– 16752.069485
Case (c, C)	– 16250.916681	– 16251.276633
Case (d, A)	– 14479.391625	– 14479.394837
Case (d, D)	– 5464.061856	– 15709.633929
Case (e, E)	– 18057.659758	– 18094.224090
Approximate optimal (u_1^*, u_2^*)	– 19308.169228	– 20177.695872

We subsequently computed the cost functionals (denoted as J values) for these selected controls. The outcomes are presented in Table 4.1.

Remark 4.3.1 It is evident from Table 4.1 that the cost associated with the approximate optimal control-trajectory pair is the lowest compared to all other values obtained under various controls (u_1, u_2) regardless of the value of A_3.

Specifically, when $A_3 = 100$, the cost corresponding to the control (d, D) is the highest. This represents the scenario where no treatment is administered. In the case where $A_3 = 0$, the highest cost occurs with the control (d, A), indicating that the use of the viral suppressing drug u_2 alone, without the immune boosting drug u_1, is not effective. For $A_3 = 100$, the control (a, A) corresponds to the second-lowest cost, representing a full therapy regimen where both drugs are administered at their maximum levels throughout the treatment period. This suggests that full therapy can be a viable treatment option when superior strategies are not available. Notably, in both (d, A) and (d, D) controls, where the immune boosting drug u_1 is absent, the associated costs are significantly high, emphasizing the crucial role of the immune boosting drug in achieving an optimal therapy effect. □

Remark 4.3.2 To demonstrate the effectiveness of our approach, we conducted a comparison with existing references. The results indicate that our findings align with those reported in references [127, 132]. Specifically, the optimal solutions exhibit similar patterns. However, it is important to note that the investigational objects differ among [127, 132] and our work. In reference [132], the model differs from ours, and only one control is applied. Additionally, the cost functionals in [127, 132] do not incorporate a terminal cost. To facilitate a meaningful comparison with these existing results, we calculated the scenario where no terminal cost is considered ($A_3 = 0$) and presented the findings in Fig. 4.3. The patterns observed in our results indeed align with those reported in the references, further validating the effectiveness of our approach.

In [127], it is mentioned that the format of the optimal controls aligns with those reported in papers [31, 35, 60] where only one control is considered. Due to the lack of information regarding the initial values used in [127] (as the optimal control is dependent on these values), we are unable to replicate the exact graph presented in that study. Nonetheless, we conducted a 50-day simulation using the initial value of (400.0, 3.5) as specified in [127]. Our findings reveal that the first control closely resembles the one described in [127], and the pattern of CD4$^+$T cells is consistent with the results reported in that study (refer to Fig. 4.3 for a comparison with existing results). Regarding the second control and HIV particles, our results bear a stronger resemblance to Fig. 4 on page 789 of [132]. □

In conclusion, this section introduces a novel algorithm to derive optimal single closed-loop solutions commencing from the initial state, as outlined in [190], using a dynamic programming approach. The algorithm relies on two key observations: (a) the value function in the considered optimal control problem serves as the viscosity solution for the corresponding HJB equation; and (b) the gradient of this value function appears in the HJB equation in the form of directional derivatives. The algorithm suggests a discretization method for approximating optimal control-trajectory pairs, which is based on a finite-difference scheme in time, by simultaneously solving the HJB and state equations. We have applied this algorithm to an HIV/AIDS optimal control problem, demonstrating its validity and effectiveness through the obtained results. Additionally, this methodology can be extended to other high-dimensional optimal control problems.

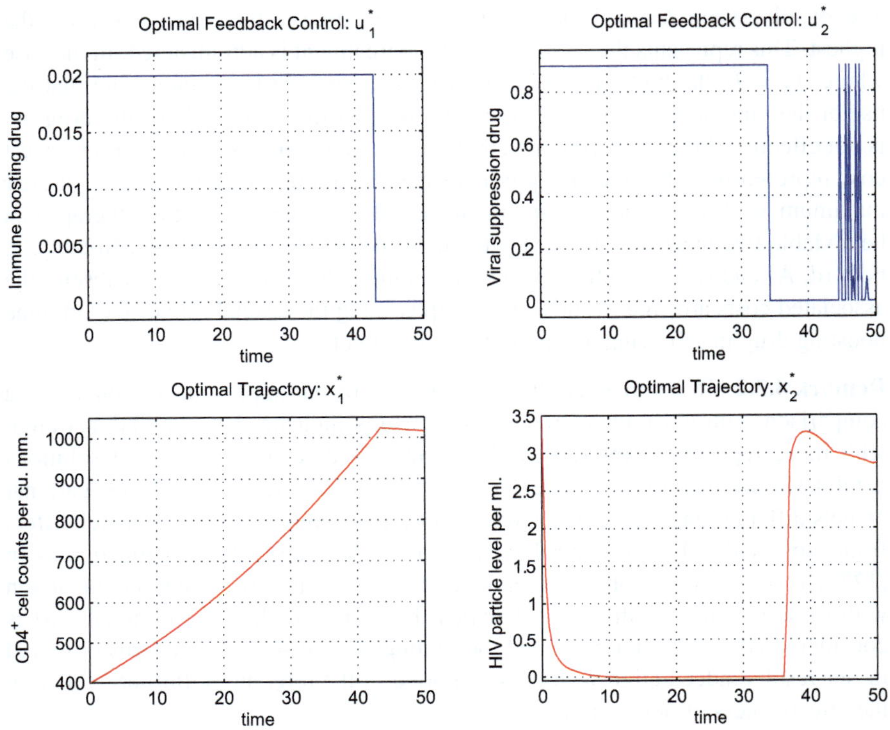

Fig. 4.3 Comparison with existing results

Our numerical analysis of the HIV/AIDS model reveals that variations in initial conditions exert minimal influence on the outcomes. Instead, the incorporation of a terminal cost within the cost functional proves to be a pivotal factor, markedly distinguishing our results from those reported in prior literature.

Lastly, it is important to acknowledge that the obtained solution may represent a local minimum, particularly influenced by the selection of the initial control. To address this concern in practical applications, we recommend comparing the solution against an "arbitrarily chosen" admissible control to ensure its optimality.

4.4 Optimal Feedback Control of Continuous Casting Process

In this section, we examine an optimal control problem related to the continuous casting of steel using the DPVS approach, as described in [103]. This methodology allows us to numerically determine the optimal feedback control, which differs from the approach taken in [171]. In the latter study, the focus was on addressing the optimal control problem of continuous casting within the context of non-differentiable multi-objective optimization, which was solved using the interactive NIMBUS method. Our approach, on the other hand, utilizes the DPVS method to seek a numerical solution for the optimal feedback control, offering a distinct perspective and solution technique for this important industrial process.

4.4.1 Introduction

Continuous casting is a widely utilized technique in the steel industry for the production of various steel grades. This process aims to transform molten steel into a solid structure while minimizing defects. A concise description of the continuous casting process involves the transfer of molten steel into a ladle, which is then poured into a continuous caster (refer to Fig. 4.4). Through this method, the molten steel undergoes a controlled solidification process, resulting in a solid steel structure optimized for further use. The continuous casting process involves several key steps. Initially, molten steel is poured from a ladle into the tundish, which serves as a reservoir. From here, liquid steel flows through a stopper rod and submerged entry nozzle (SEN) system into the mold. This mold, made of water-cooled copper, is where the primary heat extraction occurs.

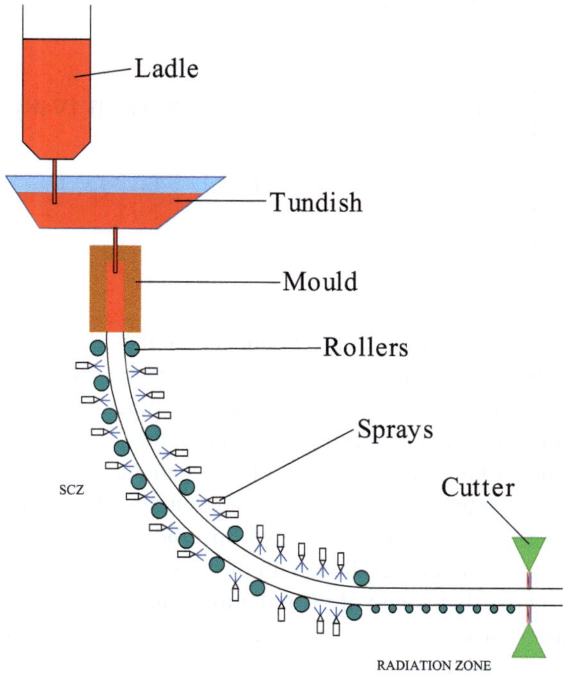

Fig. 4.4 The continuous casting process

Subsequently, the strand enters the secondary cooling zone (SCZ), where heat is further extracted by means of a water spray. In this zone, rollers support the strand as it cools. Following the SCZ, the strand undergoes natural cooling in the radiation zone before being cut and sent for further processing, such as rolling into sheet metal.

Controlling the temperature profile within the SCZ is crucial for enhancing the quality of the cast product. However, improper cooling, such as excessive reheating in this zone, can lead to crack formation on the surface and interior of the strand, as observed in [32]. Therefore, it is essential to adjust the water spray in a way that brings the temperature profile as close as possible to the desired profile at the end of the SCZ, while also ensuring that temperature changes within the SCZ remain within a reasonable range to prevent improper cooling. This consideration gives rise to the optimal control problem in the SCZ that we will address in this section.

4.4.2 *Semi-Discrete Model*

Suppose that the billet has a rectangular cross section, denoted by $\Omega = [0, a] \times [0, b]$, which moves along the z-direction at a constant speed v. Let $P = P(x, y, z)$ represent the temperature at the point (x, y, z). We define a new function $W(x, y, t)$ such that $W(x, y, t) = P(x, y, z)$ where $z = vt$. Then, W satisfies the following nonlinear heat conduction equation with boundary conditions, as referenced in [143]:

$$\begin{cases} \rho(W)[c(W) + Lf(W)]\dfrac{\partial W}{\partial t} = \operatorname{div}\left(K(W)\nabla W\right), \quad (x, y) \in \Omega,\ 0 \le t \le t^*, \\ -K(W)\dfrac{\partial W}{\partial \mathbf{n}} = Q(x, y, t, W), \quad (x, y) \in \Gamma, \\ W(x, y, 0) = W_{\text{mold}}, \end{cases} \tag{4.4.1}$$

where Γ represents the boundary of the rectangular domain Ω. In this context, $c(W)$ denotes the specific heat, $\rho(W)$ represents the density, and $K(W)$ signifies the thermal conductivity, all of which are functions of the temperature W. The pouring temperature at the entrance of the mold is given by W_{mold}. Here, $\mathbf{n}$ represents the outward normal unit vector on the boundary Γ. The time variable $t^* = t_1 + t_2 + t_3$. These intervals correspond to the lengths of different zones traversed by the billet as it moves along the z-axis. Specifically, vt_1 represents the length of the mold, $v(t_2 - t_1)$ denotes the length of the secondary cooling zone (SCZ), and $v(t^* - t_2)$ signifies the length of the radiation zone. The parameter L represents the latent heat associated with phase transitions, while the function $f(W)$ describes the variation of the solid-phase fraction with temperature. It is assumed that all these parameter functions are bounded, positive, and differentiable with respect to W. Q represents the heat flux on the boundary, as described in [32]:

$$Q(x, y, t, W) = \begin{cases} Q_t \left[-\dfrac{3(1-c_w)}{2+c_w} \left(\dfrac{r_d - r}{r_d} \right)^2 + \dfrac{3}{2+c_w} \right], & \text{in the mold;} \\ h\left(W - W_{H_2O}\right) + \sigma\varepsilon\left(W^4 - W_{ext}^4\right), & \text{in the SCZ;} \\ \sigma\varepsilon\left(W^4 - W_{ext}^4\right), & \text{in the radiation zone,} \end{cases} \tag{4.4.2}$$

where W_{H_2O} denotes the temperature of the spray water, W_{ext} represents the ambient temperature in the spray zone, σ is the Stefan-Boltzmann constant, and ε is the emission factor. The heat-transfer coefficient h is a crucial parameter determined by the water spray in the SCZ. As such, it serves as the real control variable in the system. The constant c_w signifies the ratio of the heat flux in the corners of the mold relative to the heat flux at the middle surface. Meanwhile, r_d represents half of the width of the mold. The heat flux Q_t is assumed to follow a specific formulation as described in [32]:

$$Q_t = \frac{6\left(\alpha - \beta\sqrt{t_c}\right)}{1+2c_a}\left(1 + \sqrt{\frac{t}{t_c}}(c_a - 1)\right), \tag{4.4.3}$$

where α and β are constants, and t_c is the dwell time in the mold. c_a is the ratio of the heat flux at the mold exit to the heat flux at the top level of liquid steel (meniscus) in the mold. $r = x$ at $(x, 0, t)$ or (x, b, t) and $r = y$ at $(0, y, t)$ or (a, y, t).

Thanks to the symmetry of the billet's cross section, we can confine our analysis to just one quarter of the area, specifically $\Omega_0 = \left\{(x, y) \mid 0 < x < \frac{a}{2},\ 0 < y < \frac{b}{2}\right\}$. Now, let us briefly explore the semi-discrete modeling approach applied to continuous casting, as detailed in [110]. In this framework, we discretize the spatial dimensions while preserving continuity in the time domain. To this end, we introduce a grid with fixed spacings Δx and Δy and represent the temperature at each grid point as $T_{ij}(t) = T(i\Delta x, j\Delta y, t),\ 0 \le i \le m,\ 0 \le j \le n,\ m \ge n$. Based on this discretization, [110] derived a semi-discrete approximation equation for (4.4.1).

$$\begin{cases} \dfrac{dT(t)}{dt} = \mathcal{F}(T(t))(\mathcal{A}T(t) + \mathcal{B}U(t)), \\ T(0) = g(W_{mold}), \end{cases} \tag{4.4.4}$$

where

$$T = (T_0, T_1, \dots, T_n)^\top,$$

$$T_j = \left(T_{mj}, T_{(m-1)j}, \dots, T_{jj}, \dots, T_{jn}\right)^\top, \quad j = 0, 1, 2, \dots, n,$$

$$U = (u_{1m}, \dots, u_{10}, u_{20}, \dots, u_{2n})^\top,$$

$$\begin{cases} u_{1i}(t) = Q\left(i\Delta x, 0, t, g^-(T_{i0}(t))\right), \\ u_{2j}(t) = Q\left(0, j\Delta y, t, g^-(T_{0j}(t))\right), \\ \quad i = 0, 1, \dots, m, \;\; j = 0, 1, \dots, n. \end{cases} \tag{4.4.5}$$

$$\mathcal{A} = \begin{pmatrix} A_0 & C_0 & 0 & \cdots & 0 & 0 & 0 \\ B_1 & A_1 & C_1 & \cdots & 0 & 0 & 0 \\ & & & \cdots & & & \\ 0 & 0 & 0 & \cdots & B_{n-1} & A_{n-1} & C_{n-1} \\ 0 & 0 & 0 & \cdots & 0 & B_n & A_n \end{pmatrix}, \quad \mathcal{B} = \begin{pmatrix} B \\ 0 \\ \cdots \\ 0 \end{pmatrix}.$$

And g^- is the inverse function of the Kirchhoff transform [143]:

$$T = g(W) = \int_{W^0}^{W} K(\rho)\,\mathrm{d}\rho \tag{4.4.6}$$

for a given W^0. $F^*(T_j) = \operatorname{diag}\left\{F(T_{ij})\right\}$, $\mathcal{F}(T) = \operatorname{diag}\left\{F^*(T_j)\right\}$ are diagonal matrices and

$$F(T) = \frac{K(g^-(T))}{\rho(g^-(T))[c(g^-(T)) + Lf(g^-(T))]}. \tag{4.4.7}$$

For $1 \le j \le n-1$,

$$A_j = \left(-\frac{2}{\Delta x^2} - \frac{2}{\Delta y^2}\right) I_{m+n-2j+1} + \left[A_{j1}, A_{j2}\right],$$

$$A_{j1} = \begin{pmatrix} 0 & \dfrac{2}{\Delta x^2} & 0 & \cdots & 0 & \mathbf{0_{1,m-j+1}} \\ \dfrac{1}{\Delta x^2} & 0 & \dfrac{1}{\Delta x^2} & \cdots & 0 & 0 \\ \cdots & \cdots & \cdots & \cdots & \cdots & \cdots \\ \mathbf{0}_{m-j+1,1} & 0 & 0 & \cdots & \dfrac{1}{\Delta x^2} & 0 \\ 0 & 0 & 0 & \cdots & 0 & \dfrac{1}{\Delta y^2} \\ \cdots & \cdots & \cdots & \cdots & \cdots & \cdots \\ 0 & 0 & 0 & \cdots & 0 & 0 \end{pmatrix},$$

$$A_{j2} = \begin{pmatrix} \mathbf{0_{1,m-j+2}} & 0 & 0 & \cdots & 0 & 0 \\ 0 & 0 & 0 & \cdots & 0 & 0 \\ \cdots & \cdots & \cdots & \cdots & \cdots & \cdots \\ \dfrac{1}{\Delta y^2} & 0 & 0 & \cdots & 0 & 0 \\ 0 & \dfrac{1}{\Delta y^2} & 0 & \cdots & 0 & 0 \\ \dfrac{1}{\Delta y^2} & 0 & \dfrac{1}{\Delta y^2} & \cdots & 0 & 0 \\ \cdots & \cdots & \cdots & \cdots & \cdots & \cdots \\ 0 & 0 & 0 & \cdots & \dfrac{2}{\Delta y^2} & 0 \end{pmatrix},$$

$$B = \begin{pmatrix} -\dfrac{2}{\Delta y} I_m & 0 & 0 & 0 \\ 0 & -\dfrac{2}{\Delta y} & -\dfrac{2}{\Delta x} & 0 \\ 0 & 0 & 0 & -\dfrac{2}{\Delta x} I_n \end{pmatrix},$$

$$A_n = -\left(\frac{2}{\Delta x^2} + \frac{2}{\Delta y^2}\right) I_{m-n+1} + \begin{pmatrix} 0 & \frac{2}{\Delta x^2} & 0 & \cdots & 0 & 0 & 0 \\ \frac{2}{\Delta x^2} & 0 & \frac{2}{\Delta x^2} & \cdots & 0 & 0 & 0 \\ \cdots & \cdots & \cdots & \cdots\cdots & \cdots & \cdots \\ 0 & 0 & 0 & \cdots & 0 & \frac{2}{\Delta x^2} & 0 \end{pmatrix},$$

$$C_0 = \begin{pmatrix} \frac{2}{\Delta y^2} I_{m-1} & 0 & 0 \\ 0 & \frac{2}{\Delta y^2} & 0 \\ 0 & 0 & 0 \\ 0 & \frac{2}{\Delta x^2} & 0 \\ 0 & 0 & \frac{2}{\Delta x^2} I_{n-1} \end{pmatrix},$$

$$C_j = \begin{pmatrix} \frac{1}{\Delta y^2} I_{m-j-1} & 0 & 0 \\ 0 & \frac{1}{\Delta y^2} & 0 \\ 0 & 0 & 0 \\ 0 & \frac{1}{\Delta x^2} & 0 \\ 0 & 0 & \frac{1}{\Delta x^2} I_{n-j-1} \end{pmatrix},$$

$$B_j = \begin{pmatrix} \frac{1}{\Delta y^2} I_{m-j} & 0 & 0 & 0 & 0 \\ 0 & \frac{1}{\Delta y^2} & 0 & \frac{1}{\Delta x^2} & 0 \\ 0 & 0 & 0 & 0 & \frac{1}{\Delta x^2} I_{n-j} \end{pmatrix},$$

$$B_n = \begin{pmatrix} \frac{2}{\Delta y^2} I_{m-n} & \mathbf{0} \\ \mathbf{0} & \frac{2}{\Delta y^2} \; 0 \; \frac{2}{\Delta x^2} \end{pmatrix},$$

where $\mathbf{0}_{i,j}$ represents an entry of 0 at the position (i, j) within matrices, while I_k denotes the $k \times k$ identity matrix.

It becomes apparent that Eq. (4.4.4) constitutes a standard lumped control system, wherein T serves as the state variable and U as the input. This aspect emerging from the semi-discrete model of the continuous casting process constitutes one of its advantages when compared to the infinite-dimensional formulation represented by (4.4.1). A notable observation is that the control variable U exclusively appears within the SCZ. More precisely, its influence is confined to the mold region:

$$\begin{cases} u_{1i}(t) = Q_t\left[-\dfrac{3(1-c_w)}{2+c_w}\left(\dfrac{a/2 - i\Delta x}{a/2}\right)^2 + \dfrac{3}{2+c_w}\right], \\ u_{2j}(t) = Q_t\left[-\dfrac{3(1-c_w)}{2+c_w}\left(\dfrac{b/2 - j\Delta y}{b/2}\right)^2 + \dfrac{3}{2+c_w}\right], \\ \qquad i = 0, 1, \ldots, m, \quad j = 0, 1, \ldots, n. \end{cases} \tag{4.4.8}$$

In the SCZ,

$$\begin{aligned} U &= \sigma\varepsilon\left(\left(g^-(T_0)\right)^4 - W_{\text{ext}}^4\right) + \left(g^-(T_0) - W_{\text{H}_2\text{O}}\right) h, \\ h &= (h_{1m}, \ldots, h_{10}, h_{20}, \ldots, h_{2n})^T, \\ &\begin{cases} h_{1i}(t) = h(i\Delta x, 0, t), & i = 0, 1, \ldots, m, \\ h_{2j}(t) = h(0, j\Delta y, t), & j = 0, 1, \ldots, n. \end{cases} \end{aligned} \tag{4.4.9}$$

Note that $\left(g^-(T_0)\right)^4 - W_{\text{ext}}^4$ denotes the column vector

$$\left(\left(g^-(T_{i0})\right)^4 - W_{\text{ext}}^4, \left(g^-(T_{0j})\right)^4 - W_{\text{ext}}^4\right)^\top.$$

Therefore, in the SCZ, Eq. (4.4.4) becomes

$$\begin{cases} \dfrac{dT(t)}{dt} = \mathcal{F}(T(t))\Big[\mathcal{A}T(t) + \mathcal{B}\sigma\varepsilon\left(\left(g^-(T_0(t))\right)^4 - W_{\text{ext}}^4\right) \\ \qquad + \mathcal{B}\left(g^-(T_0(t)) - W_{\text{H}_2\text{O}}\right) h(t)\Big] \quad \text{for almost all } t \in (t_1, t_2), \\ T(t_1) = S^0, \end{cases} \tag{4.4.10}$$

where h is the control variable determined by the water spray in the SCZ. Furthermore, $g^-\left(S^0\right)$ represents the temperature profile of the end section of the mold. The value of S^0 can be determined by solving Eqs. (4.4.4) and (4.4.8) because the pouring temperature W_{mold} is usually known. In the radiation zone,

$$U = \sigma\varepsilon\left(\left(g^-(T_0)\right)^4 - W_{\text{ext}}^4\right).$$

For any bounded measurable function h, the solution to (4.4.10) is understood to be the solution of the following integral equation [214, p.345]:

$$\begin{aligned} T(t) = S^0 + \int_{t_1}^{t} \mathcal{F}(T(\rho))\Big[\mathcal{A}T(\rho) + \mathcal{B}\sigma\varepsilon\left(\left(g^-(T_0(\rho))\right)^4 - W_{\text{ext}}^4\right) \\ + \mathcal{B}\left(g^-(T_0(\rho)) - W_{\text{H}_2\text{O}}\right)h(\rho)\Big]\,\mathrm{d}\rho \quad \text{for all } t \in [t_1, t_2]. \end{aligned} \tag{4.4.11}$$

Theorem 4.4.1 *For any bounded measurable $h(\cdot)$, the following two assertions hold:*

(i) *There exists a unique continuous solution T to Eq.* (4.4.11) *with respect to (t_1, S^0) in $[t_1, t_2]$.*

(ii) $\left\|T^1(t) - T^2(t)\right\| \le C_2\left[\left\|S^1 - S^2\right\| + \int_{t_1}^{t_2}\left\|h^1(\rho) - h^2(\rho)\right\|\,\mathrm{d}\rho\right]$ *for all $t \in [t_1, t_2]$, where C_2 is a constant and T^i is the solution to* (4.4.11) *corresponding to h^i and initial condition $T^i(t_1) = S^i$, $i = 1, 2$.*

Proof Based on certain assumptions regarding the parameter functions in (4.4.1), it can be established that, for any given t, the function appearing on the right-hand side of (4.4.10) is differentiable with respect to T and remains bounded within any compact set of T and $t \in [t_1, t_2]$. By invoking Theorem 36 on Page 347 and Proposition C.3.4 on Page 351 of [214], it follows that (4.4.11) admits a unique local continuous solution. A straightforward argument demonstrates that this local solution can be extended to cover the entire interval $[t_1, t_2]$. This concludes the proof of (i).

Moving on to (ii), by leveraging the assumptions made about the parameter functions in (4.4.1) along with the results obtained in (i), it becomes straightforward to deduce that

$$\begin{aligned} &\left\|T^1(t) - T^2(t)\right\| \\ &\le M_1\left(\left\|S^1 - S^2\right\| + \int_{t_1}^{t_2}\left\|h^1(\rho) - h^2(\rho)\right\|\,\mathrm{d}\rho\right) + M_2\int_{t_1}^{t}\left\|T^1(\rho) - T^2(\rho)\right\|\,\mathrm{d}\rho \end{aligned}$$

for some constants M_i, $i = 1, 2$. The assertion (ii) then follows from the Gronwall inequality. □

Now, let us illustrate the numerical solutions of Eq. (4.4.4) through an example involving a low-carbon billet caster. This example will be crucial for the optimal control computations discussed in the subsequent Sect. 4.4.5. Consider a caster with a section measuring 10 cm × 10 cm, where the mold and the SCZ have lengths of 0.7 m and 5 m, respectively.

Let W represent the temperature profile, as described in Eq. (4.4.1). Furthermore, we discretize this profile by defining $W_{ij}(t) = W(i\Delta x, j\Delta y, t),\ i = 0, 1, \cdots, m,\ j = 0, 1, \cdots, n$. Given that we are only considering one quarter of the section, we can impose the following boundary conditions:

$$m\Delta x = \frac{a}{2}, \quad n\Delta y = \frac{b}{2}.$$

Let

$$\begin{cases} W(t) = (W_0(t), W_1(t), \ldots, W_n(t))^\top, \\ W_j(t) = \left(W_{mj}(t), \ldots, W_{jj}(t), \ldots, W_{jn}(t)\right)^\top, \quad j = 0, 1, 2, \ldots, n. \end{cases} \tag{4.4.12}$$

In this context, $W_j(t)$ represents the temperature profile at the grid points of the jth layer of the section at a given time t. For the purpose of our analysis, we consider one quarter of the section, specifically the region [0, 5 cm] × [0, 5 cm]. We discretize this region using a grid with spacing $\Delta x = \Delta y = 1$ cm, which implies that $m = n = 5$ in the discretized equation (4.4.12). Within this discretized region, there are a total of 36 grid points. The corner point W_{00} is expected to have the lowest temperature due to the water spraying from two sides onto this point. Conversely, the center point W_{55} is anticipated to have the highest temperature as it is located at the center of the section. The remaining parameters relevant to this analysis are listed in Table 4.2. These parameters include various thermal properties, dimensions, and boundary conditions that are essential for accurately modeling the temperature distribution within the casting section.

The numerical solutions to Eq. (4.4.4) are obtained using the classical Runge-Kutta method. This method is applied to all 36 grid points within the discretized

Table 4.2 Parameters (=P-) used for the numerical simulation

P-	Value	P-	Value
a	0.1 m	b	0.1 m
Δx	0.01 m	W_{H_2O}, W_{ext}	25°C
Δy	0.01 m	σ	$\frac{5.67\times10^{-8}W}{m^2 K^4}$
t_0	21 s	ε	0.8
t_f	171 s	c_a, c_w	0.5
v	$\frac{1}{30}$ m/s	α	968.1 $KW \cdot m^{-2}$
β	$\frac{80.1\,KW}{m^2 \cdot s^{1/2}}$	t_c	21 s
W^0	34.286°C	W_m	1580°C

domain. After obtaining the solutions in terms of T, we transform these values back to W using the relationship $W = g^{-}(T)$.

To numerically solve Eq. (4.4.4), we require the nonlinear functions $F(T)$ from (4.4.7) and $g(W)$ from (4.4.6). The function $g(W)$ can be determined by solving the following ordinary differential equation:

$$\frac{dT}{dW} = K(W), \quad T\left(W^0\right) = 0 \tag{4.4.13}$$

And $F(T)$ can be constructed by utilizing the relationship between enthalpy derivative and temperature, as well as the relationship between thermal conductivity and temperature

$$F(T) = \frac{K(g^{-}(T))}{H'(g^{-}(T))}, \tag{4.4.14}$$

where

$$H = H(W) = \int_{W^0}^{W} \rho(s)[c(s) + Lf(s)] \, ds$$

is the enthalpy-temperature relation pertaining to the specific steel being considered. Evidently, to determine both $g(W)$ and $F(W)$, we require $K(W)$ and $H(W)$, which can be acquired from the look-up table (refer to Table 4.3) and subsequent interpolation. The outcomes obtained for these functions are illustrated in Fig. 4.5.

Figure 4.6 illustrates the numerical solutions for the temperature profile within the mold. It comprises a total of 36 curves, each representing the temperature profile at a specific grid point. The lowest curve depicts the temperature variation at the corner point of the mold, while the topmost curve reflects the temperature changes at the center point of the mold.

Table 4.3 Look-up table for K and H

W[°C]	$K[W/\text{m}°\text{C}]$	$H[\text{J/m}^3]$
34.286	60.0	1.0
502.9	–	2.0e9
708.5	31.85	3.14e9
800.0	25.38	4.257e9
900.0	26.488	–
1000.0	27.596	5.0e9
1200.0	–	6.0e9
1300.0	30.92	–
1400.0	–	6.857e9
1508.6	33.23	7.286e9
155	202.38	9.571e9
1600.0	205.44	9.857e9

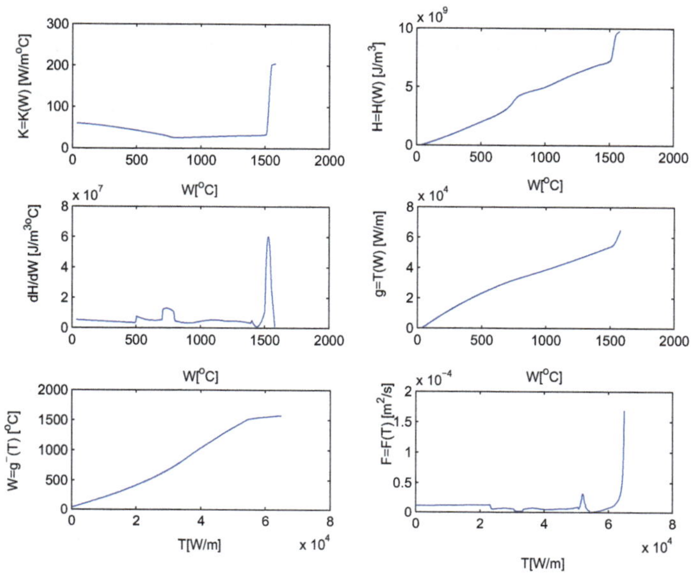

Fig. 4.5 Six functions required in the simulation

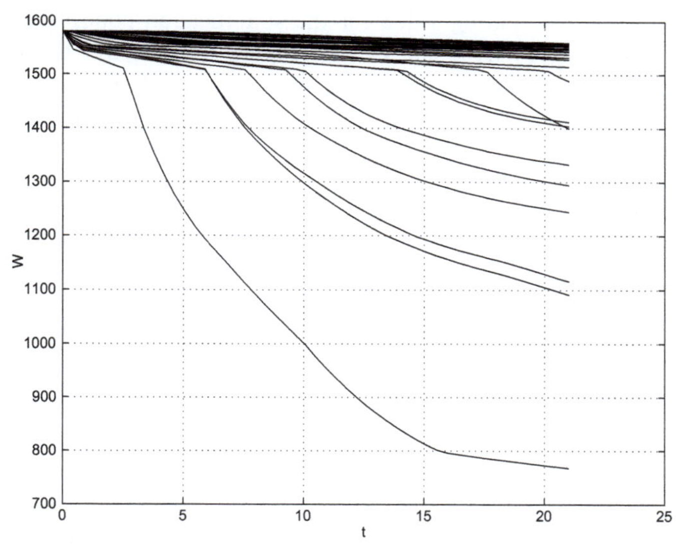

Fig. 4.6 Temperature profile in the mold

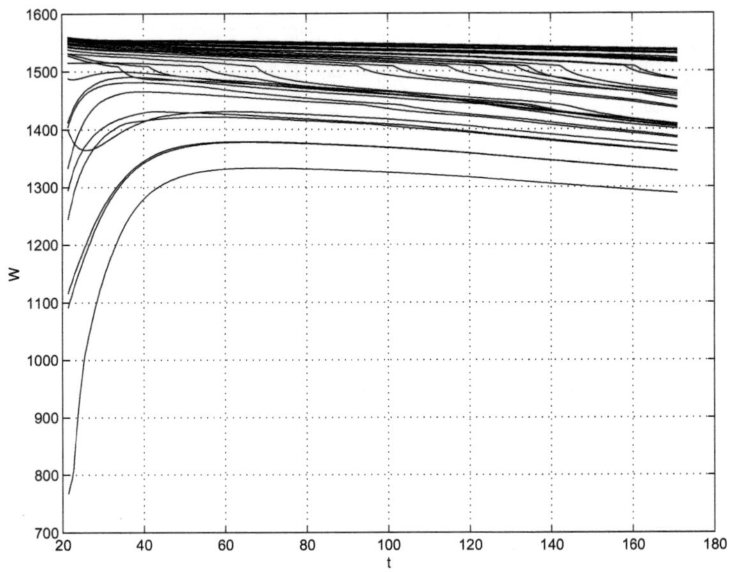

Fig. 4.7 Temperature profile in the SCZ for $h = 0$

Using $T(t_1)$ as the starting value for the solidification control zone, we can proceed to solve Eqs. (4.4.10) and (4.4.9). Figure 4.7 displays the temperature profile in the SCZ without any control measures applied (i.e., when $h = 0$). Meanwhile, Fig. 4.8 demonstrates the outcome when the control components $h(t)$ are set to be

$$h_{1i}(t),\ h_{2j}(t) = \begin{cases} 700, & 0.7 \leq vt \leq 1.7, \\ 500, & 1.7 < vt \leq 2.7, \\ 300, & 2.7 < vt \leq 3.7, \\ 200, & 3.7 < vt \leq 4.7, \\ 700, & 4.7 < vt \leq 5.7, \end{cases} \quad i = 0, 1, \ldots, m,\ j = 0, 1, \ldots, n. \tag{4.4.15}$$

It is observed that in the absence of control (i.e., when $h = 0$), the outer layer experiences reheating at the onset of the SCZ. However, when h is applied as defined in (4.4.15), despite an initial temperature decrease in the outer layer at the start of the SCZ, it undergoes reheating later within the same zone. The temperature differences between the connected layers for both scenarios are presented in Figs. 4.9 and 4.10, respectively.

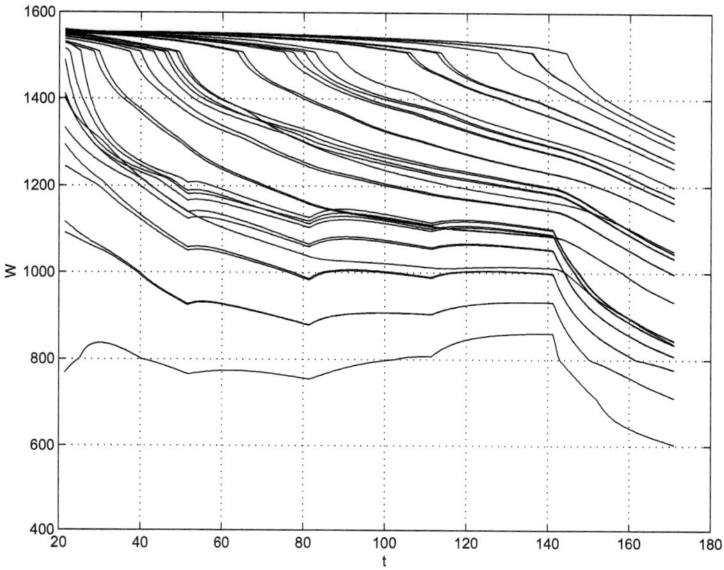

Fig. 4.8 Temperature profile in the SCZ for h given in (4.4.15)

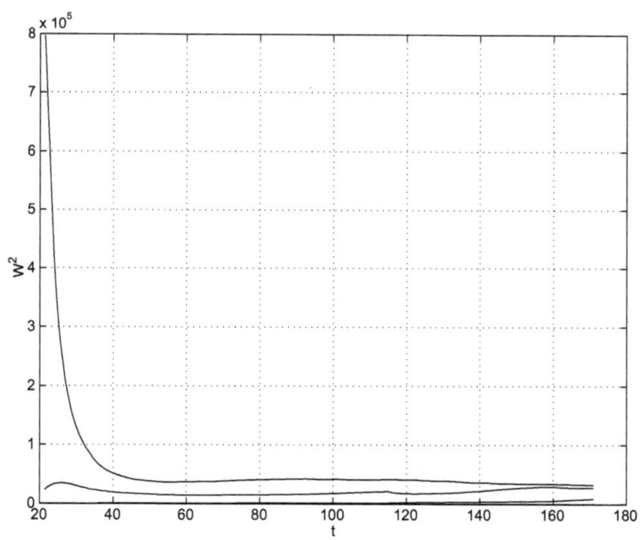

Fig. 4.9 $\|W_j - W_{j-1}\|^2$, $j = 1, ..., 5$ for $h = 0$

4.4.3 Optimal Control Formulation and DPP

In this subsection, we formulate an optimal control problem specific to continuous casting, aimed at preventing crack formation in the solidification control zone.

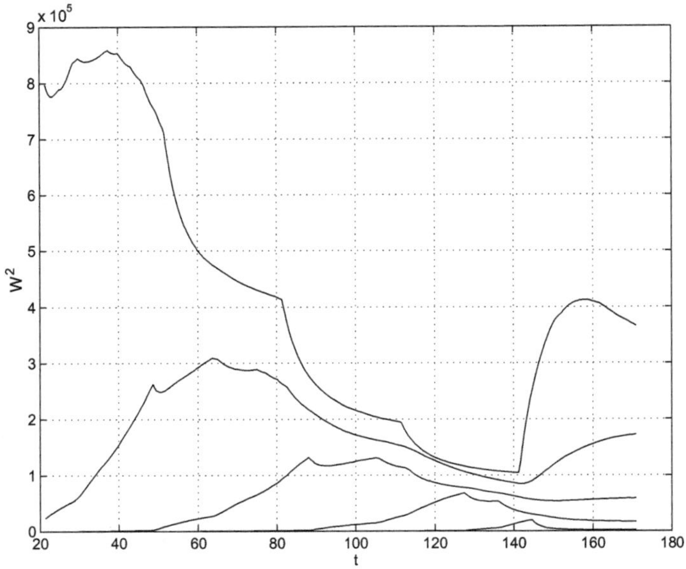

Fig. 4.10 $\|W_j - W_{j-1}\|^2$, $j = 1, ..., 5$ for h given in (4.4.15)

Through the transformation $W = g^{-}(T)$ (as seen in (4.4.6)), we frame this optimal control problem in terms of T. Given an ideal temperature profile $W^* = \left(W_0^*, W_1^*, \ldots, W_n^*\right)^T$ for one quarter-end section of the SCZ, the challenge is to determine an optimal control vector $h^* = \left(h_{1m}^*, \ldots, h_{10}^*, h_{20}^*, \ldots, h_{2n}^*\right)^T$ such that

$$J\left(t_1, S^0, h^*\right) = \inf_{h(\cdot)\in\mathcal{U}(t_1,t_2)} J\left(t_1, S^0, h\right) \tag{4.4.16}$$

subject to (4.4.10), where

$$\begin{aligned} &J\left(t_1, S^0, h\right) \\ &= \int_{t_1}^{t_2}\left[\|h(t)\|^2 + \sum_{j=1}^{n}\max\left(G_j\left(g^{-}(T(t))\right), 0\right)\right]\mathrm{d}t + \left\|g^{-}(T(t_2)) - W^*\right\|^2 \\ &= \int_{t_1}^{t_2}\left[\sum_{i=0}^{m}h_{1i}^2(t) + \sum_{j=0}^{n}h_{2j}^2(t) + \sum_{j=1}^{n}\max\left(G_j\left(g^{-}(T(t))\right), 0\right)\right]\mathrm{d}t \\ &\quad + \sum_{j=1}^{n}\left[\sum_{i=j+1}^{m}\left|g^{-}(T_{ij}(t_2)) - W_{ij}^*\right|^2 + \sum_{k=j}^{n}\left|g^{-}(T_{jk}(t_2)) - W_{jk}^*\right|^2\right] \end{aligned} \tag{4.4.17}$$

with being $c_j > 0$ are given constants for

$$
\begin{aligned}
&G_j\left(g^-(T(t))\right)\\
&=\left\|g^-(T_j(t))-g^-(T_{j-1}(t))\right\|^2-c_j\\
&=\sum_{i=j}^{m}\left|g^-(T_{ij}(t))-g^-(T_{i(j-1)}(t))\right|^2+\left|g^-\left(T_{(j-1)(j-1)}(t)\right)-g^-\left(T_{jj}(t)\right)\right|^2\\
&\quad+\sum_{k=j}^{n}\left|g^-(T_{jk}(t))-g^-(T_{(j-1)k}(t))\right|^2-c_j,\quad j=1,2,\cdots,n.
\end{aligned}
\tag{4.4.18}
$$

Additionally,

$$
\begin{aligned}
&\mathcal{U}(t_1,t_2)=L^\infty(t_1,t_2;\mathbf{E}),\\
&\mathbf{E}=\big\{h=(h_{1i},h_{2j})\mid 0\le h_{1i}\le d_{1i},\ 0\le h_{2j}\le d_{2j},\\
&\quad i=0,1,2,\cdots,m,\ j=0,1,2,\cdots,n\big\}\subset\mathbb{R}^{m+n+2}
\end{aligned}
\tag{4.4.19}
$$

where $d_{1i},\ d_{2j}$ are given constants.

Noteworthily,

$$
\sum_{j=1}^{n}\max\left(G_j\left(g^-\left(T(t)\right)\right),0\right)
$$

the term in the cost functional (4.4.17) can be interpreted as a relaxed state constraint [172]. Its purpose is to ensure that the temperature profile differences between two connected layers remain within a specified range as much as possible.

Let $\langle\cdot,\cdot\rangle$ represent the Euclidean inner product and $\|\cdot\|$ represent the corresponding induced Euclidean norm in the Euclidean space of the appropriate dimension. We define the state space as $\mathbb{R}^{(m+1)(n+1)}$. An admissible control h is one that satisfies the condition $h(\cdot)\in\mathcal{U}(t_1,t_2)$.

The optimization problem formulated above aims to achieve four objectives: (a) optimally control the rapid reheating in the initial part of the SCZ, (b) align the temperature profile $g^-(T(t_2))$ of the SCZ's end section as closely as possible to a predefined desired temperature profile W^*, (c) maintain the temperature differences between adjacent layers within a specified range, and (d) ensure that the amount of water spray remains within a reasonable limit. These objectives, to some extent, provide assurances that quality issues such as crack formation on both the surface and interior of the strand are prevented. Now define the value function as follows:

$$
V(\tau,S)=\inf_{h(\cdot)\in\mathcal{U}(\tau,t_2)}J(\tau,S,h)
\tag{4.4.20}
$$

for all $(\tau, S) \in \Theta = [t_1, t_2] \times \mathbb{R}^{(m+1)(n+1)}$. Here by $J(\tau, S, h)$, we understand to be the cost functional (4.4.17) and (4.4.18) with (4.4.10) where the condition $T(t_1) = S^0$ is replaced by $T(\tau) = S$ and t_1 is replaced by τ. Moreover, $\mathcal{U}(\tau, t_2)$ is similarly defined as $\mathcal{U}(t_1, t_2)$.

Theorem 4.4.2 *The value function V is continuous with respect to $(\tau, S) \in \Theta$.*

Proof For any $S^1,\ S^2 \in \mathbb{R}^{(m+1)(n+1)}$, $\tau^1,\ \tau^2 \in [t_1, t_2]$ and any given $\delta > 0$, suppose $\tau^2,\ S^2$ are fixed. Let $\tau^m = \max(\tau^1, \tau^2)$. By Theorem 4.4.1, one can find a smooth $h \in \mathcal{U}(t_1, t_2)$ such that

$$
\begin{aligned}
& V(\tau^2, S^2) \\
& \geq \int_{\tau^2}^{t_2} \left[\|h(\rho)\|^2 + \sum_{j=1}^{n} \max\left(G_j \left(g^-(T^2(\rho)) \right), 0 \right) \right] \mathrm{d}\rho \\
& \quad + \left\| g^- \left(T^2(t_2) \right) - W^* \right\|^2 - \delta,
\end{aligned}
$$

where T^2 is the solution to (4.4.11) produced by h and the initial condition $T^2(\tau^2) = S^2$. Suppose that T^1 is the solution to (4.4.11) produced by the same h and the initial condition $T^1(\tau^1) = S^1$. Then,

$$
\begin{aligned}
& V\left(\tau^1, S^1\right) - V\left(\tau^2, S^2\right) \\
\leq & \int_{\tau^1}^{t_2} \left[\|h(\rho)\|^2 + \sum_{j=1}^{n} \max\left(G_j\left(g^-\left(T^1(\rho)\right)\right), 0\right) \right] \mathrm{d}\rho + \left\| g^-\left(T^1(t_2)\right) - W^* \right\|^2 \\
& - \int_{\tau^2}^{t_2} \left[\|h(\rho)\|^2 + \sum_{j=1}^{n} \max\left(G_j\left(g^-\left(T^2(\rho)\right)\right), 0\right) \right] \mathrm{d}\rho \\
& - \left\| g^-\left(T^2(t_2)\right) - W^* \right\|^2 + \delta \\
\leq & \int_{\tau^1}^{\tau^2} \|h(\rho)\|^2 \, \mathrm{d}\rho + \int_{\tau^1}^{\tau^m} \sum_{j=1}^{n} \max\left(G_j\left(g^-\left(T^1(\rho)\right)\right), 0\right) \mathrm{d}\rho \\
& + \int_{\tau^m}^{t_2} \sum_{j=1}^{n} \max\left(G_j\left(g^-\left(T^1(\rho)\right)\right), 0\right) \mathrm{d}\rho + \left\| g^-\left(T^1(t_2)\right) - W^* \right\|^2 \\
& - \int_{\tau^m}^{t_2} \sum_{j=1}^{n} \max\left(G_j\left(g^-\left(T^2(\rho)\right)\right), 0\right) \mathrm{d}\rho - \left\| g^-\left(T^2(t_2)\right) - W^* \right\|^2 + \delta
\end{aligned}
$$

$$\leq \int_{\tau^m}^{t_2} \sum_{j=1}^{n} \left[\frac{G_j\left(g^-\left(T^1(\rho)\right)\right) + \left|G_j\left(g^-\left(T^1(\rho)\right)\right)\right|}{2} \right.$$
$$\left. - \frac{G_j\left(g^-\left(T^2(\rho)\right)\right) + \left|G_j\left(g^-\left(T^2(\rho)\right)\right)\right|}{2} \right] \mathrm{d}\rho$$
$$+\mathcal{M}\left|\tau^1 - \tau^2\right| + \left\|g^-\left(T^1(t_2)\right) - W^*\right\|^2 - \left\|g^-\left(T^2(t_2)\right) - W^*\right\|^2 + \delta$$

for some constant $\mathcal{M}$. Since T^1 is continuous, it follows that

$$T^1(\tau^2) - S^2 = T^1(\tau^2) - T^1(\tau^1) + S^1 - S^2 \to 0 \quad \text{as } \tau^1 \to \tau^2, S^1 \to S^2.$$

By Theorem 4.4.1,

$$T^1(t) - T^2(t) \to 0 \quad \text{for all } t \in [\tau^m, t_2] \text{ as } \tau^1 \to \tau^2, S^1 \to S^2.$$

This, combined with the dominant convergence theorem, leads to

$$\overline{\lim_{\tau^1 \to \tau^2, S^1 \to S^2}} \left[V\left(\tau^1, S^1\right) - V\left(\tau^2, S^2\right)\right] \leq 0.$$

The inverse inequality

$$\underline{\lim_{\tau^1 \to \tau^2, S^1 \to S^2}} \left[V\left(\tau^1, S^1\right) - V\left(\tau^2, S^2\right)\right] \geq 0$$

can be proved similarly. Therefore, $V(\tau, S)$ is continuous in (τ, S). This completes the proof of the theorem. □

Theorem 4.4.3 (DPP) *For any initial condition* $(\tau, S) \in \Theta$ *and* $r \in [\tau, t_2]$,

$$V(\tau, S) = \inf_{h(\cdot)\in\mathcal{U}(\tau,r)} \left\{ \int_{\tau}^{r} \left[\|h(t)\|^2 + \sum_{j=1}^{n} \max\left(G_j\left(g^-(T(t))\right), 0\right) \right] \mathrm{d}t + V(r, T(r)) \right\}. \tag{4.4.21}$$

Proof Firstly, based on the definition of the value function in (4.4.20), for any given $\delta > 0$, one can select an admissible control $h^1(\cdot) \in \mathcal{U}(r, t_2)$ that satisfies

$$\int_{r}^{t_2} \left[\left\|h^1(t)\right\|^2 + \sum_{j=1}^{n} \max\left(G_j\left(g^-\left(T^1(t)\right)\right), 0\right) \right] \mathrm{d}t + \left\|g^-\left(T^1(t_2)\right) - W^*\right\|^2$$
$$\leq V(r, T(r)) + \delta \tag{4.4.22}$$

where the state T is produced by the admissible control $h \in \mathcal{U}(\tau, r)$ with the initial condition $T(\tau) = S$, and T^1 is the state corresponding to $h^1(\cdot)$ such that $T^1(r) = T(r)$. Such a control $h^1(\cdot)$ is referred to as δ-optimal.

Define an admissible control $\tilde{h}(\cdot) \in \mathcal{U}(\tau, t_2)$ by

$$\tilde{h}(t) = \begin{cases} h(t), & \tau \le t \le r, \\ h^1(t), & r < t \le t_2, \end{cases}$$

and let $\tilde{T}(\cdot)$ be the state corresponding to $\tilde{h}(\cdot)$. In view of (4.4.22), it has

$$\begin{aligned} & V(\tau, S) \le J(\tau, S, \tilde{h}) \\ &= \int_\tau^{t_2} \left[\left\|\tilde{h}(t)\right\|^2 + \sum_{j=1}^n \max\left(G_j\left(g^-\left(\tilde{T}(t)\right)\right), 0\right) \right] dt + \left\|g^-\left(\tilde{T}(t_2)\right) - W^*\right\|^2 \\ &\le \int_\tau^r \left[\|h(t)\|^2 + \sum_{j=1}^n \max\left(G_j\left(g^-(T(t))\right), 0\right) \right] dt + V(r, T(r)) + \delta. \end{aligned}$$

Secondly, given the same $\delta > 0$, select an admissible control $h(\cdot) \in \mathcal{U}(\tau, t_2)$ which satisfies the condition for any $r \in [\tau, t_2]$,

$$\begin{aligned} \delta + V(\tau, S) &\ge J(\tau, S, h) \\ &= \int_\tau^r \left[\|h(t)\|^2 + \sum_{j=1}^n \max\left(G_j\left(g^-(T(t))\right), 0\right) \right] dt + J(r, T(r), h) \\ &\ge \int_\tau^r \left[\|h(t)\|^2 + \sum_{j=1}^n \max\left(G_j\left(g^-(T(t))\right), 0\right) \right] dt + V(r, T(r)). \end{aligned}$$

Thus, (4.4.21) follows from the arbitrariness of δ. □

4.4.4 HJB Equation and Viscosity Solution

Beginning with the DPP, we can deduce the corresponding HJB equation, which is presented in Theorem 4.4.4 below.

Theorem 4.4.4 *If $V(\tau, S) \in C^1(\Theta)$, then the value function V satisfies the HJB equation of the following:*

$$\begin{cases} -V_\tau(\tau, S) + \mathcal{H}(\tau, S, D_S V(\tau, S)) = 0, & \forall\, (\tau, S) \in [t_1, t_2) \times \mathbb{R}^{(m+1)(n+1)}, \\ V(t_2, S) = \left\| g^-(S) - W^* \right\|^2, & \forall\, S \in \mathbb{R}^{(m+1)(n+1)}, \end{cases} \tag{4.4.23}$$

where the Hamiltonian $\mathcal{H}$ is given by

$$\begin{aligned} &\mathcal{H}(\tau, S, D_S V(\tau, S)) \\ &= \sup_{h \in \mathbf{E}} \Bigg\{ \Big\langle -\mathcal{F}(S) \Big[\mathcal{A}S + \mathcal{B}\sigma\varepsilon \left((g^-(S_0))^4 - W_{\text{ext}}^4 \right) + \mathcal{B}\left(g^-(S_0) - W_{\text{H}_2\text{O}} \right) h \Big], \\ &\qquad D_S V(\tau, S) \Big\rangle - \|h\|^2 - \sum_{j=1}^{n} \max\left(G_j(g^-(S)), 0 \right) \Bigg\}, \\ &\qquad\qquad \forall\, (\tau, S, D_S V(\tau, S)) \in \Theta \times \mathbb{R}^{(m+1)(n+1)} \end{aligned} \tag{4.4.24}$$

with $S = (S_0, S_1, \cdots, S_n)^\top$ and $D_S V$ being the partial Fréchet gradient of $V(\tau, \cdot)$.

Proof For any $0 < \delta < t_2 - \tau$ and for any specified constant control $h \in \mathbf{E}$, according to the DPP given in (4.4.21),

$$V(\tau, S) \le \int_\tau^{\tau+\delta} \left[\|h\|^2 + \sum_{j=1}^{n} \max\left(G_j(g^-(T(t))), 0 \right) \right] \mathrm{d}t + V(\tau + \delta, T(\tau + \delta))$$

where T is the state corresponding to h and initial condition $T(\tau) = S$. Therefore,

$$\begin{aligned} 0 \le \frac{1}{\delta} \int_\tau^{\tau+\delta} &\left[\|h\|^2 + \sum_{j=1}^{n} \max\left(G_j\left(g^-(T(t)) \right), 0 \right) \right] \mathrm{d}t \\ &+ \frac{V(\tau + \delta, T(\tau + \delta)) - V(\tau, T(\tau))}{\delta}. \end{aligned}$$

Passing to the limit as $\delta \to 0+$ gives

$$-V_\tau(\tau, S) + \mathcal{H}(\tau, S, D_S V(\tau, S)) \le 0. \tag{4.4.25}$$

On the other hand, for any given $\eta > 0$, $0 < \delta < t_2 - \tau$, and $h \in \mathbf{E}$, according to Theorem 4.4.1, one can construct a smooth control function $\hat{h}(\cdot) \in \mathcal{U}(\tau, t_2)$ satisfying $\hat{h}(\tau) = h$ and such that

$$V(\tau, S) + \eta\delta \geq \int_{\tau}^{\tau+\delta} \left[\left\| \hat{h}(t) \right\|^2 + \sum_{j=1}^{n} \max\left(G_j \left(g^- \left(\hat{T}(t) \right) \right), 0 \right) \right] \mathrm{d}t$$
$$+ V\left(\tau + \delta, \hat{T}(\tau + \delta) \right),$$

where $\hat{T}(\cdot)$ is the state corresponding to $\hat{h}(\cdot)$ and the initial condition $\hat{T}(\tau) = S$. Therefore,

$$\begin{aligned}
\eta \geq & \frac{1}{\delta} \int_{\tau}^{\tau+\delta} \left[\left\| \hat{h}(t) \right\|^2 + \sum_{j=1}^{n} \max\left(G_j \left(g^- \left(\hat{T}(t) \right) \right), 0 \right) \right] \mathrm{d}t \\
& + \frac{V\left(\tau + \delta, \hat{T}(\tau + \delta) \right) - V(\tau, S)}{\delta} \\
= & \frac{1}{\delta} \int_{\tau}^{\tau+\delta} \left[\left\| \hat{h}(t) \right\|^2 + \sum_{j=1}^{n} \max\left(G_j \left(g^- \left(\hat{T}(t) \right) \right), 0 \right) \right] \mathrm{d}t + V_\tau(\tau, S) \\
& + \Big\langle D_S V(\tau, S), \frac{1}{\delta} \int_{\tau}^{\tau+\delta} \mathcal{F}(\hat{T}(t)) \Big[\mathcal{A}\hat{T}(t) + \mathcal{B}\sigma\varepsilon \left(\left(g^- \left(\hat{T}_0(t) \right) \right)^4 - W_{\mathrm{ext}}^4 \right) \\
& + \mathcal{B} \left(g^- \left(\hat{T}_0(t) \right) - W_{\mathrm{H_2O}} \right) \hat{h}(t) \Big] \mathrm{d}t \Big\rangle + \frac{1}{\delta} \circ \left(|\delta| + \left\| \hat{T}(\tau + \delta) - S \right\| \right) \\
\geq & \frac{1}{\delta} \int_{\tau}^{\tau+\delta} -\mathcal{H}(t, T(t), D_S V(t, T(t))) \, \mathrm{d}t + V_\tau(\tau, S) \\
& + \frac{1}{\delta} \circ \left(|\delta| + \left\| \hat{T}(\tau + \delta) - S \right\| \right).
\end{aligned}$$

Passing to the limit as $\delta \to 0+$ once again leads to the conclusion that $-V(\tau, S) + \mathcal{H}(\tau, S, D_S V(\tau, S)) \geq -\eta$ and hence

$$-V_\tau(\tau, S) + \mathcal{H}(\tau, S, D_S V(\tau, S)) \geq 0. \tag{4.4.26}$$

The proof is finalized by amalgamating the results obtained in (4.4.25) and (4.4.26). □

Next, we present a sufficient condition for optimality.

Theorem 4.4.5 *Let $Y \in C^1(\Theta)$ satisfy (4.4.23) and let V be the value function. Then*

(i) $Y(\tau, S) \leq V(\tau, S), \ \forall(\tau, S) \in \Theta$.
(ii) *If there exists a* $h^*(\cdot) \in \mathcal{U}(t_1, t_2)$ *such that*

$$\begin{aligned}
&\Big\langle \mathcal{F}(T^*(t)) \Big[\mathcal{A}T^*(t) + \mathcal{B}\sigma\varepsilon \left(\left(g^-\left(T_0^*(t)\right)\right)^4 - W_{\text{ext}}^4 \right) \\
&\quad + \mathcal{B}\left(g^-\left(T_0^*(t)\right) - W_{\text{H}_2\text{O}}\right) h^*(t) \Big], D_S Y(t, T^*(t)) \Big\rangle \\
&\quad + \|h^*(t)\|^2 + \sum_{j=1}^{n} \max\left(G_j\left(g^-(T^*(t))\right), 0\right) \\
&= -\mathcal{H}\left(t, T^*(t), D_S Y(t, T^*(t))\right),
\end{aligned} \tag{4.4.27}$$

which, in usual way, one writes

$$\begin{aligned}
h^*(t) \in \arg\inf_{h \in \mathbf{E}} &\Big\{ \Big\langle \mathcal{F}\left(T^*(t)\right) \Big[\mathcal{A}T^*(t) + \mathcal{B}\sigma\varepsilon \left(\left(g^-\left(T_0^*(t)\right)\right)^4 - W_{\text{ext}}^4 \right) \\
&+ \mathcal{B}\left(g^-\left(T_0^*(t)\right) - W_{\text{H}_2\text{O}}\right) h \Big], D_S Y\left(t, T^*(t)\right) \Big\rangle \\
&+ \|h\|^2 + \sum_{j=1}^{n} \max\left(G_j\left(g^-\left(T^*(t)\right)\right), 0\right) \Big\},
\end{aligned} \tag{4.4.28}$$

for almost all $t \in [t_1, t_2]$*, where* T^* *is the state corresponding to* h^* *and initial condition* $T^*(t_1) = S^0$*. As a result,* $h^*(\cdot)$ *is an optimal control.*

Proof For any control function $h(\cdot) \in \mathcal{U}(\tau, t_2)$, by employing the dynamic programming equation stated in (4.4.23), we obtain

$$\begin{aligned}
&Y(t_2, T(t_2)) \\
&= Y(\tau, S) + \int_{\tau}^{t_2} \left[\frac{\partial}{\partial \tau} Y(t, T(t)) + \left\langle \frac{dT(t)}{dt}, D_S Y(t, T(t)) \right\rangle \right] \mathrm{d}t \\
&= Y(\tau, S) + \int_{\tau}^{t_2} \Big[\frac{\partial}{\partial \tau} Y(t, T(t)) + \Big\langle \mathcal{F}(S)\Big[\mathcal{A}S + \mathcal{B}\sigma\varepsilon \left(\left(g^-(S_0)\right)^4 - W_{\text{ext}}^4 \right) \\
&\quad + \mathcal{B}\left(g^-(S_0) - W_{\text{H}_2\text{O}}\right) h\Big], D_S Y(t, T(t)) \Big\rangle \Big] \mathrm{d}t \\
&\geq Y(\tau, S) - \int_{\tau}^{t_2} \left[\|h(t)\|^2 + \sum_{j=1}^{n} \max\left(G_j\left(g^-(T(t))\right), 0\right) \right] \mathrm{d}t.
\end{aligned} \tag{4.4.29}$$

$Y(\tau, S) \le J(\tau, S, h)$ and (i) then follows.

For the second assertion, suppose $h^*(\cdot) \in \mathcal{U}(t_1, t_2)$ satisfies (4.4.27). Substituting $(h^*(\cdot), T^*(\cdot))$ into (4.4.29) yields an equality. Specifically, $Y\left(t_1, S^0\right) = J\left(t_1, S^0, h^*\right)$. By (i), we have $V\left(t_1, S^0\right) = J\left(t_1, S^0, h^*\right)$, indicating that $h^*(\cdot)$ is an optimal control. This completes the proof. □

Theorem 4.4.6

(i) *The sufficient condition for optimality stated in (4.4.27) becomes necessary if the value function V is smooth. Hence,* $(T^*(\cdot), h^*(\cdot))$ *is an optimal control-trajectory pair if and only if*

$$V_t(t, T^*(t)) + \Big\langle \mathcal{F}(T^*(t))\Big[\mathcal{A}T^*(t) + \mathcal{B}\sigma\varepsilon\left(\left(g^-\left(T_0^*(t)\right)\right)^4 - W_{\text{ext}}^4\right) + \mathcal{B}\left(g^-\left(T_0^*(t)\right) - W_{\text{H}_2\text{O}}\right)h^*(t)\Big], D_S V(t, T^*(t))\Big\rangle + \|h^*(t)\|^2 + \sum_{j=1}^{n} \max\left(G_j\left(g^-(T^*(t))\right), 0\right) = 0 \quad \textit{for almost all } t \in [t_1, t_2]. \tag{4.4.30}$$

(ii) *Let V be the value function. Then for any control-trajectory pair* $(h^*(\cdot), T^*(\cdot))$, $T^*(t_1) = S^0$, *the function*

$$t \to V(t, T^*(t)) - \int_t^{t_2} \left[\left\|h^*(\rho)\right\|^2 + \sum_{j=1}^{n} \max\left(G_j\left(g^-\left(T^*(\rho)\right)\right), 0\right)\right] \mathrm{d}\rho \tag{4.4.31}$$

is nondecreasing over $[t_1, t_2]$.

Moreover, $(h^*(\cdot), T^*(\cdot))$ *is an optimal control-trajectory pair if and only if the above function is constant on* $[t_1, t_2]$.

Proof This proof follows a standard approach, and thus, we omit the detailed steps here. □

Now we introduce the definition of a viscosity solution to the HJB equation given in (4.4.23).

Definition 4.4.7 Let $u(\tau, S) \in C(\Theta)$. We say that

(i) u is a viscosity supersolution to the first equation of (4.4.23) in Θ if

$$-\varphi_\tau(\tau, S) + \mathcal{H}\left(\tau, S, D_S\varphi(\tau, S)\right) \ge 0$$

for any $\varphi \in C^1(\Theta)$ such that $u - \varphi$ has a local minimum, relative to Θ at $S \in \mathbb{R}^{(m+1)(n+1)}$.

(ii) u is a viscosity subsolution to the first equation of (4.4.23) in Θ if

$$-\varphi_\tau(\tau, S) + \mathcal{H}(\tau, S, D_S\varphi(\tau, S)) \le 0$$

for any $\varphi \in C^1(\Theta)$ such that $u - \varphi$ has a local maximum, relative to Θ at $S \in \mathbb{R}^{(m+1)(n+1)}$.

(iii) u is a viscosity solution to the first equation of (4.4.23) in Θ if it is simultaneously a viscosity subsolution and supersolution in Θ.

Theorem 4.4.8 *The value function V is a viscosity solution of the HJB equation (4.4.23) in Θ.*

Proof Let $\varphi \in C^1(\Theta)$ and $(\tau, S) \in \Theta$ be a local maximum point of $V - \varphi$. Let $h \in \mathbf{E}$ be an arbitrary constant control and T be the state corresponding to h and initial condition $T(\tau) = S$. Then for any given $\delta > 0$, $(r, T(r)) \in \mathbb{B}((\tau, S), \delta)$, the ball centered at (τ, S) with radius δ in Θ, for all sufficiently small r. Hence,

$$\varphi(\tau, S) - \varphi(r, T(r)) \le V(\tau, S) - V(r, T(r))$$

for all r small enough. By the DPP (4.4.21), we get

$$\varphi(\tau, S) - \varphi(r, T(r)) \le \int_\tau^r \left[\|h\|^2 + \sum_{j=1}^n \max\left(G_j\left(g^-(T(t))\right), 0\right) \right] dt.$$

Divide by $r - \tau > 0$ on both sides above and let $r \to \tau$, to obtain

$$-\varphi_\tau(\tau, S) - \left\langle \mathcal{F}(S)\left[\mathcal{A}S + \mathcal{B}\sigma\varepsilon\left(\left(g^-(S_0)\right)^4 - W_{\text{ext}}^4\right) + \mathcal{B}\left(g^-(S_0) - W_{\text{H}_2\text{O}}\right)h\right],\right.$$
$$\left. D_S\varphi(\tau, S)\right\rangle \le \|h\|^2 + \sum_{j=1}^n \max\left(G_j\left(g^-(S)\right), 0\right).$$

This shows that

$$-\varphi_\tau(\tau, S) + \mathcal{H}(\tau, S, D_S V(\tau, S)) \le 0$$

for any $h \in \mathbf{E}$. Therefore, V is a viscosity subsolution to (4.4.23) in $(\tau, S) \in \Theta$.

Next assume that $(\tau, S) \in \Theta$ is a local minimum point of $V - \varphi$, that is, for a given $\delta > 0$, $V(\tau, S) - V(r, T(r)) \le \varphi(\tau, S) - \varphi(r, T(r))$ for all $(r, T(r)) \in \mathbb{B}((\tau, S), \delta)$. For any given $\rho > 0$ and $h \in \mathbf{E}$, by the DPP (4.4.21) and Theorem 4.4.1, there exists a smooth $\overline{h} \in \mathcal{U}(\tau, t_2)$ depending on ρ and r such that $\overline{h}(\tau) = h$ and

$$V(\tau, S) \ge \int_\tau^r \left[\left\|\overline{h}(t)\right\|^2 + \sum_{j=1}^n \max\left(G_j\left(g^-\left(\overline{T}(t)\right)\right), 0\right) \right] dt + V\left(r, \overline{T}(r)\right) - r\rho,$$

where $\overline{T}$ is the state corresponding to $\overline{h}$ and initial condition $\overline{T}(\tau) = S$. Therefore,

$$\varphi(\tau, S) - \varphi(r, \overline{T}(r)) \geq \int_{\tau}^{r} \left[\|\overline{h}\|^2 + \sum_{j=1}^{n} \max \left(G_j \left(g^- \left(\overline{T}(t) \right) \right), 0 \right) \right] \mathrm{d}t - r\rho.$$

Divide by $r - \tau > 0$ and let $r \to \tau$ to get

$$-\varphi_\tau(\tau, S) - \Big\langle \mathcal{F}(S) \Big[\mathcal{A}S + \mathcal{B}\sigma\varepsilon \left(\left(g^- (S_0) \right)^4 - W_{\text{ext}}^4 \right) + \mathcal{B} \left(g^- (S_0) - W_{\text{H}_2\text{O}} \right) \overline{h} \Big],$$

$$D_S\varphi(\tau, S) \Big\rangle \geq \|h\|^2 + \sum_{j=1}^{n} \max \left(G_j \left(g^- \left(\overline{T}(\tau) \right) \right), 0 \right) - \rho.$$

Since ρ is arbitrary, the last inequality reads

$$-\varphi_\tau(\tau, S) + \mathcal{H}(\tau, S, D_S V(\tau, S)) \geq 0,$$

that is, V is a viscosity supersolution to (4.4.23). Therefore, V is a viscosity solution to (4.4.23) in Θ. The proof is complete. □

To conclude this subsection, let us revisit the discussion on optimality conditions. It is widely recognized that assertion (i) of Theorem 4.4.6 leads to the classical verification theorem. This theorem plays a crucial role in assessing the optimality of a given control-trajectory pair and, more significantly, in constructing the optimal feedback control, as described in [249]. Assuming that the value function $V(\tau, S)$ is smooth, it follows that $V(\tau, S)$ is a classical solution to the HJB equation stated in Theorem 4.4.4:

$$-V_\tau(\tau, S) + \mathcal{H}(\tau, S, D_S V(\tau, S)) = 0.$$

Therefore, the optimality condition given in (4.4.30) is equivalent to (4.4.27). It is important to note that (4.4.27) states that $h^*(\cdot) \in \mathcal{U}(t_1, t_2)$ is an optimal control with the initial state S^0 if and only if (4.4.28) holds. Furthermore, (4.4.28) provides a formula for finding the optimal feedback control.

However, it is crucial to acknowledge that (4.4.28) holds only under the assumption that the value function is differentiable. This assumption may not always be valid. Nonetheless, it serves as the foundation for our numerical approach to solving for the optimal feedback control, which will be discussed in the next subsection.

4.4.5 Finite Difference Scheme for the Numerical Solution of Optimal Feedback Control

The initial step involves discretizing the HJB equation given in (4.4.23). To accomplish this, we define a sequence of time points $\tau_j = t_2 + j\Delta\tau,\ j = 0, 1, \cdots, N$, where $\Delta\tau = \frac{t_1 - t_2}{N}$ and N is an integer. For a given positive value of η, we approximate the Fréchet partial derivative using the following method:

$$\begin{aligned}\langle D_S V(\tau, S), f\rangle &= \left\langle D_S V(\tau, S), \eta \frac{f}{\|f\|}\right\rangle \frac{\|f\|}{\eta} \\ &\approx \left[V\left(\tau, S + \eta\frac{f}{\|f\|}\right) - V(\tau, S)\right]\frac{\|f\|}{\eta}\end{aligned} \tag{4.4.32}$$

where

$$\begin{aligned}f &= f(h, S) \\ &= -\mathcal{F}(S)\left[\mathcal{A}S + \mathcal{B}\sigma\varepsilon\left(\left(g^-(S_0)\right)^4 - W_{\text{ext}}^4\right) + \mathcal{B}\left(g^-(S_0) - W_{\text{H}_2\text{O}}\right)h\right].\end{aligned}$$

Let $f_i^j = f\left(h_i^j, S^{(i)}\right)$ and $f_i = f\left(h, S^{(i)}\right)$. For the initial state $S^{(0)} = S^0$ set

$$S^{(i)} = S^{(i-1)} + \frac{f_{i-1}}{\|f_{i-1}\|}\eta, \quad i = 1, 2, \cdots, M. \tag{4.4.33}$$

We approximate the HJB equation (4.4.23) using a difference scheme, as described in [85], to obtain

$$\begin{cases}\dfrac{V_i^{j+1} - V_i^j}{\Delta\tau} + \dfrac{V_i^j - V_{i-1}^j}{\eta}\left\|f_i^j\right\| + \left\|h_i^j\right\|^2 + \displaystyle\sum_{p=1}^{n}\max\left(G_p\left(g^-\left(S^{(i)}\right)\right), 0\right) = 0, \\ h_i^{j+1} = \displaystyle\arg\inf_{h\in\mathbf{E}}\left\{\dfrac{V_i^{j+1} - V_{i-1}^{j+1}}{\eta}\|f_i\| + \|h\|^2 + \sum_{p=1}^{n}\max\left(G_p\left(g^-\left(S^{(i)}\right)\right), 0\right)\right\}\end{cases} \tag{4.4.34}$$

for all $i = 1, 2, \cdots, M$ and $j = 0, 1, \cdots, N$, where $V_i^j \approx V\left(\tau_j, S^{(i)}\right)$. Furthermore, we once again impose the following necessary condition for the stability of the aforementioned difference scheme given in (4.4.34), as stated in [149]:

$$\frac{|\Delta\tau|}{\eta}\max_{1\le i\le M}\|f_i\| \le 1. \tag{4.4.35}$$

In summary, we have the following difference algorithm for solving the HJB equation.

Algorithm 5 The algorithm for solving the HJB equation

1: Initialization. Set

$$\begin{cases} V_i^0 = V\left(t_2, S^{(i)}\right) = \left\|g^-\left(S^{(i)}\right) - W^*\right\|^2, \quad S^{(i)} = S^{(i-1)} + \dfrac{f_{i-1}}{\|f_{i-1}\|}\eta, \\ h_i^0 \in \arg\inf\limits_{h\in\mathbf{E}} \left\{ \dfrac{\left\|g^-\left(S^{(i)}\right) - W^*\right\|^2 - \left\|g^-\left(S^{(i-1)}\right) - W^*\right\|^2}{\eta}\|f_i\| \right. \\ \qquad \left. + \|h\|^2 + \displaystyle\sum_{p=1}^{n} \max\left(G_p\left(g^-\left(S^{(i)}\right)\right), 0\right) \right\}, \quad i = 1, 2, \cdots, M. \end{cases} \tag{4.4.36}$$

Here the formula for h_i^0 comes from

$$\begin{aligned} h(t) \in \arg\inf_{h\in\mathbf{E}} \Big\{ &\langle f(h, T(t)), D_S V(t, T(t))\rangle + \|h\|^2 \\ &+ \sum_{p=1}^{n} \max\left(G_p\left(g^-(T(t))\right), 0\right) \Big\}. \end{aligned} \tag{4.4.37}$$

2: Iteration. By (4.4.34),

$$\begin{cases} V_i^{j+1} = \left(1 - \dfrac{\Delta\tau}{\eta}\left\|f_i^j\right\|\right) V_i^j + \dfrac{\Delta\tau}{\eta}\left\|f_i^j\right\| V_{i-1}^j - \Delta\tau \left\|h_i^j\right\|^2 \\ \qquad -\Delta\tau \displaystyle\sum_{p=1}^{n} \max\left(G_p\left(g^-\left(S^{(i)}\right)\right), 0\right), \\ h_i^{j+1} = \arg\inf\limits_{h\in\mathbf{E}} \left\{ \dfrac{V_i^{j+1} - V_{i-1}^{j+1}}{\eta}\|f_i\| + \|h\|^2 + \displaystyle\sum_{p=1}^{n} \max\left(G_p\left(g^-\left(S^{(i)}\right)\right), 0\right) \right\} \end{cases} \tag{4.4.38}$$

for all $i = 1, 2, \cdots, M$ and $j = 0, 1, \cdots, N-1$.

From (4.4.37), the optimal feedback control is

$$h_{S^0}^*(t) \in \arg\inf_{h\in\mathbf{E}} \Big\{ \big\langle f(h, T^*(t)), D_S V(t, T^*(t)) \big\rangle + \|h\|^2 + \sum_{p=1}^{n} \max\left(G_p\left(g^-(T^*(t))\right), 0\right) \Big\} \tag{4.4.39}$$

where $T^*(\cdot)$ denotes the optimal trajectory of the system, subject to the initial condition $T^*(t_1) = S^0$. Because (4.4.39) relies on T^*, it becomes imperative to solve (4.4.10). Here, we numerically determine the solution of (4.4.10) using the classical Runge-Kutta method.

The following steps outline the procedure for obtaining the optimal feedback control.

Steps for Finding the Optimal Feedback Control

Step 1. Invoke the algorithm for solving the HJB equation to obtain the feedback control function $h(t_1)$. Next, substitute $\left(h(t_1), S^{(0)}\right)$ into (4.4.10) to derive the optimal trajectory $T^*(\ell_1)$, where $\ell_1 = t_1 + \Delta\ell$, Δ and $\ell = \frac{t_2-t_1}{J}$ for a given integer J.

Step 2. Replace the optimal trajectory $T^*(\ell_1)$ obtained in the previous step as the new initial data $S^{(0)}$ for the next iteration. Then, re-invoke the algorithm for solving the HJB equation to determine the updated feedback control function $(h(\ell_1), T^*(\ell_1))$. Afterward, substitute the pair $(h(\ell_1), T^*(\ell_1))$ into (4.4.10) to compute the optimal trajectory $T^*(\ell_2)$, at the next time point, where $\ell_2 = t_1 + 2\Delta\ell$.

Step 3. Repeat the aforementioned process until all feedback control functions $h(\ell_k)$ and their corresponding optimal trajectories $T^*(\ell_k)$ have been obtained, where $\ell_k = t_1 + k\Delta\ell$ and $k = 0, 1, \cdots, J$. In other words,

$$h_{S^0}^*(t) = \{h(t_1), h(\ell_1), h(\ell_2), \cdots, h(t_2)\} \tag{4.4.40}$$

which is the optimal feedback control.

Now that we have obtained all the necessary feedback control functions and optimal trajectories, we are ready to proceed with finding numerical solutions for the optimal control problem given by (4.4.16), (4.4.10), and (4.4.17). We will employ the schemes (4.4.36) and (4.4.38), along with the classical Runge-Kutta method, for this purpose. All the parameters required for the computation are listed in Table 4.2. Furthermore, let us proceed with

$$\begin{cases} c_1 = 4.427 \times 10^5, & c_2 = 3.0829 \times 10^5, \\ c_3 = 1.2743 \times 10^5, & c_4 = 6.6941 \times 10^4, \\ c_5 = 1.4287 \times 10^4 \end{cases} \tag{4.4.41}$$

and d_{1i}, $d_{2j} = 700$, representing the maximum value of control as defined by (4.4.15) and (4.4.19). The temperature profile observed at $t_2 = 171s$ when applying

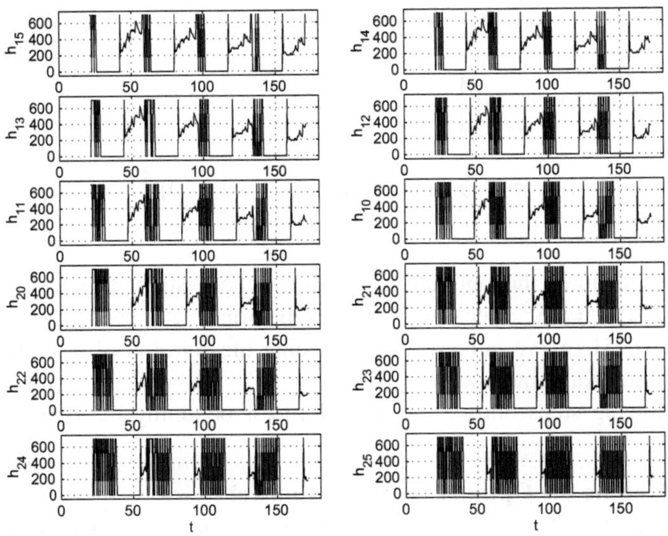

Fig. 4.11 Optimal feedback control h^*

the control (4.4.15) is regarded as the ideal temperature profile, denoted as W^*. Consistent with Sect. 4.4.2, we set $m = n = 5$.

The computation was performed using Visual C++ 6.0, and the numerical results were plotted using MATLAB 6.1. Figure 4.11 illustrates the obtained optimal feedback control at a total of 11 grid points: $\left(h_{1i}^*(t), h_{2j}^*(t)\right)$, $i, j = 0, 1, 2, 3, 4, 5$, $21 \le t \le 171$ (notice that h_{10} and h_{20}, for the time range $21 \le t \le 171$ (note that h_{10} and h_{20} correspond to the same grid point). With this optimal control in place, we obtained the corresponding optimal temperature profile $W^{**} = g^-(T^*)$ for the SCZ, which is shown in Fig. 4.12. Figure 4.13 displays the temperature differences between adjacent layers. From these two figures, it is evident that the reheating on the front section of the SCZ is effectively controlled, and the temperature differences between different layers remain mostly within the range observed in Fig. 4.9, where no control was applied.

The remaining task involves conducting a numerical experiment to verify the optimality of the obtained numerical solutions for the optimal feedback control. This is achieved by comparing the cost functional $J\left(t_1, S^0, h^*\right)$ corresponding to the obtained optimal control-trajectory pair $J\left(t_1, S^0, h\right)$ associated with an arbitrarily chosen admissible control h and its corresponding trajectory. Specifically, we aim to confirm whether the following inequality holds true:

$$J\left(t_1, S^0, h^*\right) \le J\left(t_1, S^0, h\right). \tag{4.4.42}$$

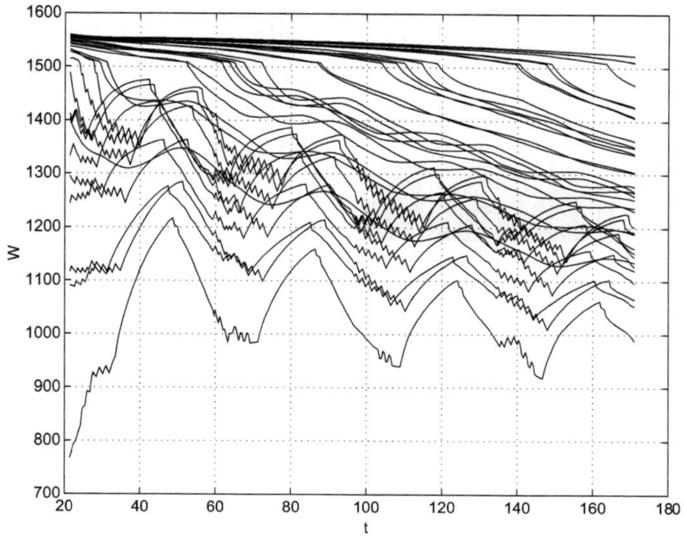

Fig. 4.12 Optimal temperature profile $W^{**} = g^{-}(T^{*})$

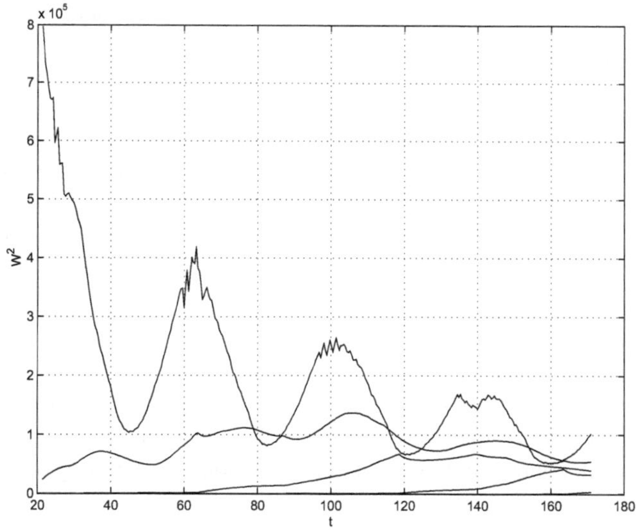

Fig. 4.13 $\left\| W_j^{**} - W_{j-1}^{**} \right\|^2, \; j = 1, ..., 5$

Figure 4.14 presents various admissible controls, including the one given by (4.4.15), denoted as case 8. We have computed the corresponding cost functionals $J\left(t_1, S^0, h\right)$ for all these controls. The results are summarized in Table 4.4. As can be seen from Table 4.4, the cost functional associated with the optimal feedback control h^*, is $J\left(t_1, S^0, h^*\right) = 23830013.958020$ (labeled as case 9), which

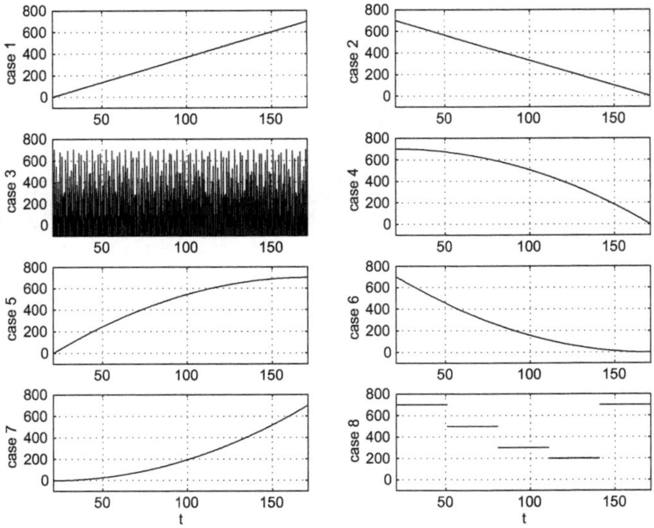

Fig. 4.14 Several different arbitrarily chosen admissible controls

Table 4.4 Different admissible controls h and their corresponding costs J

Case of h	Corresponding J
case 1	25032343.051020
case 2	34980529.161010
case 3	25179601.149153
case 4	56672703.455952
case 5	40723454.557675
case 6	24165907.148881
case 7	25318719.513016
case 8	53515593.114259
case 9	23830013.958020

is clearly lower than all other cost functionals $J\left(t_1, S^0, h\right)$. These comparisons provide strong evidence that we have indeed obtained numerical solutions for the optimal feedback control in the context of continuous steel casting.

4.5 Optimal Birth Feedback Control of Population Dynamics

In this section, we explore the optimal control problem for an age-structured population dynamics of the McKendrick type using the DPVS approach, as outlined in [102]. This investigation involves an infinite-dimensional bilinear control problem. Significant research has been conducted on partial differential equation models that describe population dynamics.

Studies such as [50] and [51] have investigated optimal birth control problems for age-structured population systems of the McKendrick type and established maximum principles for these problems. The optimal control problem for a Gurtin-MacCamy-type system, which describes the evolution of an age-structured population, is examined in [13], where necessary optimality conditions are provided.

Additionally, [9] looks into an optimal harvesting problem for a nonlinear age-dependent population system. For certain approximating problems, optimal controllers in feedback form are given using the dynamic programming method. Meanwhile, [10] considers the optimal harvesting problem for the linear Lokta-McKendrick model with periodic vital rates and a periodic forcing term that sustains oscillations. The study presents necessary optimality conditions and develops a numerical algorithm to approximate the optimal control and optimal harvest.

The model discussed in this section focuses on the population evolution of human beings with birth control, as developed in [213]. We derive the HJB equation satisfied by the value function and show that the value function is the viscosity solution of this HJB equation. Through the value function and its gradient, we determine the optimal birth feedback control.

Furthermore, we design a finite-difference scheme to obtain the numerical solution for the optimal birth feedback control. The optimality of the obtained control is verified numerically by comparing it with different controls under the same constraints. All the data used in our computations are sourced from the 1989 census of China's population, as documented in [66].

4.5.1 Problem Formulation and DPP

The McKendrick-type model for age-structured population dynamics, as established in [213], constitutes a first-order partial differential equation featuring a nonlocal boundary condition, which is characterized by

$$\begin{cases} \dfrac{\partial p(r,t)}{\partial t} + \dfrac{\partial p(r,t)}{\partial r} = -\mu(r)p(r,t), \quad 0 < r < a_m,\ t > 0, \\ p(r,0) = p_0(r), \quad 0 \le r \le a_m, \\ p(0,t) = \beta(t)\displaystyle\int_{a_1}^{a_2} b(r)p(r,t)\,\mathrm{d}r, \quad t \ge 0, \end{cases} \tag{4.5.1}$$

where $p(r,t)$ represents the age density distribution of a closed population at a specific time t and age r. $\mu(r)$ denotes the relative mortality of the population, expressed as a nonnegative measurable function satisfying

$$\int_0^r \mu(\rho)\,\mathrm{d}\rho < \infty \quad \text{for } r < a_m \text{ and } \quad \int_0^{a_m} \mu(\rho)\,\mathrm{d}\rho = \infty. \tag{4.5.2}$$

The highest age ever attained by individuals within the population is designated as a_m. The function $b(r) = k(r)h(r)$ plays a crucial role, where $k(r)$ signifies the female-to-population ratio and $h(r)$ characterizes the fertility pattern of females satisfying

$$\int_{a_1}^{a_2} h(r)\,\mathrm{d}r = 1.$$

It is a reasonable assumption that the function $b(r)$ is bounded and measurable within the fecundity period $[a_1, a_2]$ where $0 < a_1 < a_2 < a_m$. Outside this fecundity period, $b(r)$ is zero. The initial age distribution of the population is denoted by $p_0(r)$. Additionally, $\beta(t)$ represents the specific fertility rate of females at time t, which serves as a macro-level control for population birth rates.

Let $\mathbf{H} = L^2(0, a_m)$ be the state space with the usual inner product $\langle\cdot,\cdot\rangle$ and the inner product induced norm $\|\cdot\|$. For any $t > s \geq 0$, let

$$\mathcal{U}[s,t] = \left\{\beta(\tau) \in [\beta_0, \beta_1] \subset \mathbb{R}^+ \mid \tau \in [s,t],\ \beta(\tau) \text{ is measurable on } [s,t]\right\}.$$

Given $T > 0$ and $p_0 \in \mathbf{H}$, the optimal control problem is to find an optimal control $\beta^*(\cdot) \in \mathcal{U}[0,T]$ such that

$$\begin{cases} J(\beta^*) = \inf\limits_{\beta(\cdot)\in\mathcal{U}[0,T]} J(\beta), \\ J(\beta) = \displaystyle\int_0^T \int_0^{a_m} L(p(r,t), \beta(t), r, t)\,\mathrm{d}r\,\mathrm{d}t + \int_0^{a_m} f_0(r, p(r,T))\,\mathrm{d}r, \end{cases} \tag{4.5.3}$$

where $p(r,t)$ is the solution to Eq. (4.5.1) corresponding to $\beta(\cdot)$, and L, f_0 satisfy

$$\begin{cases} \left|\displaystyle\int_0^{a_m} f_0(r, p(r))\,\mathrm{d}r\right|, \quad \left|\displaystyle\int_0^{a_m} L(p(r), \beta, r, t)\,\mathrm{d}r\right| \\ \quad \leq C_1 + C_2\|p\|, \quad \forall\,(t,\beta) \in [0,T]\times[\beta_0,\beta_1],\ p \in \mathbf{H}, \\ \left|\displaystyle\int_0^{a_m} [f_0(r, p_1(r)) - f_0(r, p_2(r))]\,\mathrm{d}r\right|, \\ \left|\displaystyle\int_0^{a_m} [L(p_1(r), \beta, r, t) - L(p_2(r), \beta, r, t)]\,dr\right| \leq C_3\|p_1 - p_2\|, \\ \qquad \forall\,(t,\beta) \in [0,T]\times[\beta_0,\beta_1],\ p_1,\ p_2 \in \mathbf{H}, \\ \displaystyle\int_0^{a_m} L(p(r), \beta, r, t)\,\mathrm{d}r \text{ is continuous in } (t, p, \beta) \in \mathbb{R}^+ \times \mathbf{H} \times [\beta_0, \beta_1] \end{cases} \tag{4.5.4}$$

for some constants C_i, $i = 1, 2, 3$.

Define the time-dependent operators $\mathbb{A}_\beta(t)$ as follows:

$$\begin{cases} \mathbb{A}_\beta(t)\phi(r) = -\phi'(r) - \mu(r)\phi(r), \quad \forall\, \phi \in D(\mathbb{A}_\beta(t)), \\ D(\mathbb{A}_\beta(t)) = \left\{\phi(r) \;\middle|\; \phi,\ \mathbb{A}_\beta(t)\phi \in \mathbf{H},\ \phi(0) = \beta(t)\displaystyle\int_{a_1}^{a_2} b(r)\phi(r)\,\mathrm{d}r\right\}. \end{cases} \tag{4.5.5}$$

Then, Eq. (4.5.1) can be reformulated as a first-order evolution equation in $\mathbf{H}$ as

$$\begin{cases} \dfrac{\partial p(r,t)}{\partial t} = \mathbb{A}_\beta(t)p(r,t), \\ p(r,0) = p_0(r). \end{cases} \tag{4.5.6}$$

From this point onward, we will also employ $\mathbb{A}_\beta$ to denote the operator $\mathbb{A}_\beta(t)$ when $\beta(t) \equiv \beta$ independent of time t, and the semigroup generated by $\mathbb{A}_\beta$ will be denoted as $\mathbb{T}_\beta(t)$. It is seen that $\mathbb{A}_\beta(t) = \mathbb{A}_{\beta(t)}$.

Define a family of evolution operators $\mathbb{T}(t,s;\beta),\ 0 \le s \le t < \infty$ by

$$\mathbb{T}(t,s;\beta)\phi(r) = \begin{cases} \phi(r-t+s)\, e^{-\int_{r-t+s}^{r}\mu(\rho)\,\mathrm{d}\rho}, \quad r \ge t-s, \\ \beta(t-r)\displaystyle\int_{a_1}^{a_2} b(\tau)\phi(\tau-t+s+r)\, e^{-\int_{\tau-t+s+r}^{\tau}\mu(\rho)\,\mathrm{d}\rho}\,\mathrm{d}\tau \\ e^{-\int_0^r \mu(\rho)\,\mathrm{d}\rho}, \quad r < t-s, \end{cases}$$

$\forall \phi \in \mathbf{H},\ 0 \le t-s \le a_1$ and

$$\mathbb{T}(t,s;\beta) = \mathbb{T}\left(t, s + \left[\frac{t-s}{a_1}\right]a_1; \beta\right) \prod_{n=1}^{\left[\frac{t-s}{a_1}\right]} \mathbb{T}(s+na_1, s+(n-1)a_1; \beta) \tag{4.5.7}$$

as $t-s > a_1$, where $[x]$ denotes the maximal integer not exceeding x and $\mathbb{T}(t,s;\beta)$ is uniquely determined by $\{\beta(\tau) \mid \tau \in [s,t]\}$.

The following Theorem 4.5.1 is taken from [109] and its validity can be easily ascertained through a simple and direct computation.

Theorem 4.5.1

(i) *For each* $\beta(\cdot) \in \mathcal{U}[s,t]$, $\mathbb{T}(t,s;\beta)$ *is a bounded linear operator on* $\mathbf{H}$ *with*

$$\mathbb{T}(s,s;\beta) = \mathbb{I},\ \mathbb{T}(t,s;\beta) = \mathbb{T}(t,t_0;\beta)\mathbb{T}(t_0,s;\beta), \quad \forall\, 0 \le s \le t_0 \le t < \infty,$$

where $\mathbb{I}$ *stands for the identity operator on* $\mathbf{H}$.

(ii) $\mathbb{T}(\cdot,\cdot;\beta)\phi$ *is (strongly) continuous for each* $\phi \in \mathbf{H}$ *uniformly in* $\beta(\cdot) \in \mathcal{U}[s,t]$.

(iii) *Let* $\mathbb{T}_{\beta_1}(t)$ *be the* C_0*-semigroup generated by* $\mathbb{A}_{\beta_1}$ *on* **H**. *Then* [213]

$$\|\mathbb{T}(t,s;\beta)\| \le \|\mathbb{T}_{\beta_1}(t)\| \le Me^{\omega(t-s)} \quad \text{for some } M > 1 \text{ and } \omega \in \mathbb{R}.$$

(iv) *The family of evolution operators* $\{\mathbb{T}(t,s;\beta)\}$ *defined by* (4.5.7) *has the iterative relation*

$$\mathbb{T}(t,s;\beta)\phi(r) = \begin{cases} \phi(r-t+s)\, e^{-\int_{r-t+s}^{r} \mu(\rho)\,\mathrm{d}\rho}, & r \ge t-s, \\ \beta(t-r)\displaystyle\int_{a_1}^{a_2} b(\tau)\mathbb{T}(t-r,s;\beta)\phi(\tau)\,\mathrm{d}\tau\; e^{-\int_0^r \mu(\rho)\,\mathrm{d}\rho}, & r < t-s. \end{cases} \tag{4.5.8}$$

(v) *If* $\beta(t) \in C^1(s,\infty)$, *then* $\mathbb{T}(t,s;\beta)$ *is an evolution system, i.e.,*

$$\frac{d}{dt}\mathbb{T}(t,s;\beta)p_0 = \mathbb{A}_\beta(t)\mathbb{T}(t,s;\beta)p_0, \quad \forall\, p_0 \in D(\mathbb{A}_\beta(0)).$$

(vi) *For any* $t > s \ge 0$ *and any* $\beta_i(\cdot) \in L^2(s,t)$ $(i = 1,2)$,

$$\|\mathbb{T}(t,s;\beta_1)p_0 - \mathbb{T}(t,s;\beta_2)p_0\| \le \|b\|\|p_0\|\|\beta_1(\cdot) - \beta_2(\cdot)\|_{L^2(s,t)}. \tag{4.5.9}$$

Theorem 4.5.2 *Let* $\mathbb{A}_\beta^*$ *be the adjoint operator of* $\mathbb{A}_\beta$ *in* **H**. *Then there exists an operator* $\mathbb{C}$ *on* **H** *that is bounded, linear, self-adjoint, and positive definite such that for each* $\beta \in [\beta_0, \beta_1]$, $\mathbb{A}_\beta^*\mathbb{C}$ *is a bounded linear operator on* **H** *and*

$$\sup_{\beta \in [\beta_0,\beta_1]} \left\|\mathbb{A}_\beta^*\mathbb{C}\right\| \le 1 + \beta_1\|b\|\sqrt{a_m}. \tag{4.5.10}$$

Furthermore, the set $\mathcal{D}^*$ *defined by*

$$\mathcal{D}^* = D(\mathbb{A}_\beta^*) = \left\{\phi(r) \;\middle|\; \phi(r),\ \phi'(r) - \mu(r)\phi(r) \in \mathbf{H},\ \phi(a_m) = 0\right\} \tag{4.5.11}$$

is independent of $\beta \in [\beta_0, \beta_1]$ *and dense in* **H**.

Proof Let the operator $\mathbb{D}$ be defined as

$$\begin{cases} \mathbb{D}\phi(r) = \phi'(r) - \mu(r)\phi(r), \\ D(\mathbb{D}) = \{\phi(r) \mid \phi,\ \mathbb{D}\phi \in \mathbf{H},\ \phi(a_m) = 0\}. \end{cases} \tag{4.5.12}$$

Then

$$-\phi(0) = \int_0^{a_m} \frac{d}{dr}\left[\phi(r)e^{-\int_0^r \mu(\rho)\,\mathrm{d}\rho}\right] \mathrm{d}r$$
$$= \int_0^{a_m} \left[\phi'(r) - \mu(r)\phi(r)\right] e^{-\int_0^r \mu(\rho)\,\mathrm{d}\rho}\,\mathrm{d}r = \left\langle \mathbb{D}\phi, e^{-\int_0^r \mu(\rho)\,\mathrm{d}\rho}\right\rangle.$$

It is straightforward to verify that

$$\mathbb{A}_\beta^*\phi(r) = \phi'(r) - \mu(r)\phi(r) + \beta b(r)\phi(0) = \mathbb{D}\phi(r) - \beta b(r)\left\langle \mathbb{D}\phi, e^{-\int_0^r \mu(\rho)\,\mathrm{d}\rho}\right\rangle. \tag{4.5.13}$$

So $D(\mathbb{A}_\beta^*) = D(\mathbb{D})$. Furthermore, let $\mathbb{C} = (\mathbb{I} + \mathbb{D}^*\mathbb{D})^{-\frac{1}{2}}$. Then it follows from Theorem 13.13 of [195] that $\mathbb{C}$ is a linear bounded self-adjoint positive operator and $\|\mathbb{D}\mathbb{C}\| \le 1$. By (4.5.13),

$$\left\|\mathbb{A}_\beta^*\mathbb{C}\right\| \le 1 + \beta_1 \|b\| \sqrt{a_m}.$$

This is the desired result. □

Theorem 4.5.3 *Let $\mathcal{D}^*$ be defined as* (4.5.11) *and let $q(\cdot) \in \mathcal{D}^*$. Then $\langle q(\cdot), p(\cdot, t)\rangle$ is differentiable almost everywhere on* $[0, T]$ *and*

$$\frac{d}{dt}\langle q(\cdot), p(\cdot, t)\rangle = \left\langle \mathbb{A}_\beta^*(t) q(\cdot), p(\cdot, t)\right\rangle, \quad t \in [0, T] \text{ a.e.}, \tag{4.5.14}$$

where $p(r, t) = \mathbb{T}(t, 0; \beta)p_0(r)$ for any $p_0 \in \mathbf{H}$ and $\beta(\cdot) \in \mathcal{U}[0, T]$. $\mathbb{A}_\beta^*(t) = \mathbb{A}_{\beta(t)}^*$ is the adjoint operator of $\mathbb{A}_\beta(t)$.

Proof Let $t \in (0, T)$. For simplicity, we only consider the right derivative. Let $\delta > 0,\ t + \delta < a_1$. Then from (4.5.8), it has

$$p(r, t+\delta)$$
$$= \begin{cases} p(r-\delta, t)\, e^{-\int_{r-\delta}^r \mu(\rho)\,\mathrm{d}\rho}, & r \ge \delta, \\ \beta(\delta + t - r) \displaystyle\int_{a_1}^{a_2} b(\tau) p(\tau - r + \delta, t)\, e^{-\int_{\tau-r+\delta}^{\tau} \mu(\rho)\,\mathrm{d}\rho}\,\mathrm{d}\tau\, e^{-\int_0^r \mu(\rho)\,\mathrm{d}\rho}, & \\ & r < \delta. \end{cases}$$

So

$$
\begin{aligned}
&\langle q(\cdot), p(\cdot, t+\delta) - p(\cdot, t)\rangle \\
&= \int_{\delta}^{a_m} q(r)\left[p(r-\delta, t)\, e^{-\int_{r-\delta}^{r} \mu(\rho)\,\mathrm{d}\rho} - p(r,t)\right] \mathrm{d}r \\
&\quad + \int_0^{\delta} q(r)\Big[\beta(\delta + t - r) \int_{a_1}^{a_2} b(\tau) p(\tau - r + \delta, t) \\
&\quad \cdot e^{-\int_{\tau-r+\delta}^{\tau} \mu(\rho)\,\mathrm{d}\rho}\, \mathrm{d}\tau\, e^{-\int_0^{r} \mu(\rho)\,\mathrm{d}\rho} - p(r,t)\Big] \mathrm{d}r \\
&= \int_0^{a_m-\delta} p(r,t)\left[q(r+\delta)\, e^{-\int_r^{r+\delta} \mu(\rho)\,\mathrm{d}\rho} - q(r)\right] \mathrm{d}r - \int_0^{\delta} p(r,t) q(r)\, \mathrm{d}r \\
&\quad - \int_{a_m-\delta}^{a_m} q(r) p(r,t)\, \mathrm{d}r + \int_0^{\delta} q(r) p(r,t)\, \mathrm{d}r \\
&\quad - \int_{\delta+t}^{t} \Big[q(\delta + t - r)\, e^{-\int_0^{\delta+t-r} \mu(\rho)\,\mathrm{d}\rho} \beta(r) \\
&\quad \cdot \int_{a_1}^{a_2} b(\tau) p(\tau + r - t, t)\, e^{-\int_{\tau+r-t}^{\tau} \mu(\rho)\,\mathrm{d}\rho}\, \mathrm{d}\tau\Big] \mathrm{d}r \\
&= \int_0^{a_m-\delta} p(r,t)\left[q(r+\delta)\, e^{-\int_r^{r+\delta} \mu(\rho)\,\mathrm{d}\rho} - q(r)\right] \mathrm{d}r - \int_{a_m-\delta}^{a_m} q(r) p(r,t)\, \mathrm{d}r \\
&\quad + \int_t^{t+\delta} \Big[q(\delta + t - r)\, e^{-\int_0^{\delta+t-r} \mu(\rho)\,\mathrm{d}\rho}\, \beta(r) \\
&\quad \cdot \int_{a_1}^{a_2} b(\tau) p(\tau + r - t, t)\, e^{-\int_{\tau+r-t}^{\tau} \mu(\rho)\,\mathrm{d}\rho}\, \mathrm{d}\tau\Big] \mathrm{d}r
\end{aligned}
$$

and hence

$$
\begin{aligned}
&\lim_{\delta \downarrow 0} \left\langle q(\cdot), \frac{p(\cdot, t+\delta) - p(\cdot, t)}{\delta} \right\rangle \\
&= \int_0^{a_m} p(r,t)\left[q'(r) - \mu(r) q(r)\right] \mathrm{d}r + q(0)\beta(t) \int_{a_1}^{a_2} b(\tau) p(\tau, t)\, \mathrm{d}\tau \\
&= \left\langle \mathbb{A}_{\beta}^{*}(t) q(\cdot), p(\cdot, t) \right\rangle,
\end{aligned}
$$

for all such t that the above limit exists, which are dense in $[0, T]$. □

By invoking Theorem 4.5.3, we consider $p(r,t) = \mathbb{T}(t, 0; \beta) p_0(r)$ as the (weak) solution of Eq. (4.5.1) in the sense specified by (4.5.14). This solution is obtained by integrating (4.5.1) along the characteristics.

Furthermore, the value function $V(t, p_0)$ for the optimal control problem is defined as follows:

$$V(t, p_0) = \inf_{\beta(\cdot)\in\mathcal{U}[t,T]} \left\{ \int_t^T \int_0^{a_m} L(p(r,\tau), \beta(\tau), r, \tau)\, dr\, d\tau + \int_0^{a_m} f_0(r, p(r,T))\, dr \right\} \tag{4.5.15}$$

where $p(r,\tau) = \mathbb{T}(\tau, t; \beta)p_0(r)$ is the solution of (4.5.1) corresponding to $\beta(\cdot) \in \mathcal{U}[t, T]$.

Theorem 4.5.4 (DPP) *For any* $0 < \delta < T - t$,

$$\begin{cases} V(t, p_0) \\ = \inf_{\beta(\cdot)\in\mathcal{U}[t,t+\delta]} \left\{ \int_t^{t+\delta} \int_0^{a_m} L(p(r,\tau), \beta(\tau), r, \tau)\, dr\, d\tau + V(t+\delta, p(\cdot, t+\delta)) \right\}, \\ V(T, p_0) = \int_0^{a_m} f_0(r, p_0(r))\, dr \end{cases} \tag{4.5.16}$$

where $p(r,\tau) = \mathbb{T}(\tau, t; \beta)p_0(r)$.

Proof Let

$$W(t) = \inf_{\beta(\cdot)\in\mathcal{U}[t,t+\delta]} \left\{ \int_t^{t+\delta} \int_0^{a_m} L(p(r,\tau), \beta(\tau), r, \tau)\, dr\, d\tau + V(t+\delta, p(\cdot, t+\delta)) \right\}.$$

It is obvious that $V(t, p_0) \le W(t)$. For any given $\varepsilon > 0$, there exists a $\hat{\beta}(\cdot) \in \mathcal{U}[t, T]$ such that

$$\int_t^T \int_0^{a_m} L\left(\hat{p}(r,\tau), \hat{\beta}(\tau), r, \tau\right) dr\, d\tau + \int_0^{a_m} f_0\left(r, \hat{p}(r,T)\right) dr \le V(t, p_0) + \varepsilon$$

where $\hat{p}(r,\tau) = \mathbb{T}\left(\tau, t; \hat{\beta}\right) p_0(r)$. Since

$$V\left(t+\delta, \hat{p}(\cdot, t+\delta)\right) \le \int_{t+\delta}^T \int_0^{a_m} L\left(\hat{p}(r,\tau), \hat{\beta}(\tau), r, \tau\right) dr\, d\tau + \int_0^{a_m} f_0\left(r, \hat{p}(r,T)\right) dr,$$

it follows that

$$
\begin{aligned}
W(t) &\leq \int_t^{t+\delta} \int_0^{a_m} L\left(\hat{p}(r,\tau), \hat{\beta}(\tau), r, \tau\right) \,\mathrm{d}r \,\mathrm{d}\tau + V\left(t+\delta, \hat{p}(\cdot, t+\delta)\right) \\
&\leq \int_t^{T} \int_0^{a_m} L\left(\hat{p}(r,\tau), \hat{\beta}(\tau), r, \tau\right) \,\mathrm{d}r \,\mathrm{d}\tau + \int_0^{a_m} f_0\left(r, \hat{p}(r,T)\right) \mathrm{d}r \\
&\leq V(t, p_0) + \varepsilon.
\end{aligned}
$$

By the arbitrariness of ε, we obtain that $W(t) \leq V(t, p_0)$. Therefore, $W(t) = V(t, p_0)$. □

Theorem 4.5.5 *Given M and ω as defined in Theorem 4.5.1, along with the constants C_i, $i = 1, 2, 3$ specified in (4.5.4), the following statements hold true:*

(i) $V(t, p_0) \leq (1+T)C_1 + MC_2\left(1+\omega^{-1}\right) e^{\omega T} \|p_0\|$, *for all* $t \in [0, T]$, $p_0 \in \mathbf{H}$.
(ii) $|V(t, p_1) - V(t, p_2)| \leq \left(1+\omega^{-1}\right) MC_3\, e^{\omega T} \|p_1 - p_2\|$, *for all* $t \in [0, T]$, p_1, $p_2 \in \mathbf{H}$.
(iii) *For any fixed p_0, $V(t, p_0)$ is continuous in t.*

Proof For any $\beta(\cdot) \in \mathcal{U}[t, T]$,

$$
\begin{aligned}
|V(t, p_0)| &\leq \int_t^T \left(C_1 + C_2\|\mathbb{T}(\tau, t; \beta)p_0\|\right) \mathrm{d}\tau + C_1 + C_2\|\mathbb{T}(T, t; \beta)p_0\| \\
&\leq C_1 T + \frac{M}{\omega} C_2\, e^{\omega T} \|p_0\| + C_1 + MC_2\, e^{\omega T} \|p_0\| \\
&= (1+T)C_1 + \left(1 + \frac{1}{\omega}\right) MC_2\, e^{\omega T} \|p_0\|.
\end{aligned}
$$

This is (i). For any $\varepsilon > 0$, take $\hat{\beta}(\cdot) \in \mathcal{U}[t, T]$ such that

$$
\begin{aligned}
V(t, p_2) \geq & \int_t^T \int_0^{a_m} L\left(\hat{p}_2(r,\tau), \hat{\beta}(\tau), r, \tau\right) \mathrm{d}r \,\mathrm{d}\tau \\
&+ \int_0^{a_m} f_0\left(r, \hat{p}_2(r, T)\right) \mathrm{d}r - \varepsilon,
\end{aligned}
$$

where $\hat{p}_2(r,\tau) = \mathbb{T}\left(\tau, t; \hat{\beta}\right) p_2(r)$. Let $\hat{p}_1(r,\tau) = \mathbb{T}\left(\tau, t; \hat{\beta}\right) p_1(r)$. Then

$$
\begin{aligned}
& V(t, p_1) - V(t, p_2) \\
\leq & \int_t^T \int_0^{a_m} L\left(\hat{p}_1(r,\tau), \hat{\beta}(\tau), r, \tau\right) \mathrm{d}r \,\mathrm{d}\tau + \int_0^{a_m} f_0\left(r, \hat{p}_1(r, T)\right) \mathrm{d}r
\end{aligned}
$$

$$
\begin{aligned}
&- \int_t^T \int_0^{a_m} L\left(\hat{p}_2(r,\tau), \hat{\beta}(\tau), r, \tau\right) \,\mathrm{d}r\, \mathrm{d}\tau - \int_0^{a_m} f_0\left(r, \hat{p}_2(r,T)\right) \mathrm{d}r + \varepsilon \\
&\le \int_t^T \left| \int_0^{a_m} \left[L\left(\hat{p}_1(r,\tau), \hat{\beta}(\tau), r, \tau\right) - L\left(\hat{p}_2(r,\tau), \hat{\beta}(\tau), r, \tau\right)\right] \mathrm{d}r \right| d\tau \\
&\quad + \left| \int_0^{a_m} \left[f_0\left(r, \hat{p}_1(r,T)\right) - f_0\left(r, \hat{p}_2(r,T)\right)\right] \mathrm{d}r \right| + \varepsilon \\
&\le C_3 \int_t^T \left\| \hat{p}_1(\cdot,\tau) - \hat{p}_2(\cdot,\tau)\right\| \,\mathrm{d}\tau + C_3 \left\| \hat{p}_1(\cdot,T) - \hat{p}_2(\cdot,T)\right\| + \varepsilon \\
&\le \frac{M}{\omega} C_3\, e^{\omega T} \|p_1 - p_2\| + C_3 M e^{\omega T} \|p_1 - p_2\| + \varepsilon \\
&= \left(1 + \omega^{-1}\right) M C_3\, e^{\omega T} \|p_1 - p_2\| + \varepsilon.
\end{aligned}
$$

Therefore, (ii) has been proven. It is worth noting that in the final line above, we utilized the fact from (iii) of Theorem 4.5.1 which states that

$$
\begin{aligned}
&\left\| \hat{p}_2(\cdot,\tau) - \hat{p}_2(\cdot,\tau) \right\| \\
&\le \left\| \mathbb{T}\left(\tau, t; \hat{\beta}\right) \right\| \|p_1 - p_2\| \le M e^{\omega(\tau - t)} \|p_1 - p_2\|, \quad \forall\, \tau \in [t, T].
\end{aligned}
$$

Finally, for given $p_0 \in \mathbf{H}$ and any $t_1 > t_2 \in [0, T]$, there exists a $\overline{\beta}(\cdot) \in \mathcal{U}[t_2, T]$ such that

$$
V(t_2, p_0) \ge \int_{t_2}^T \int_0^{a_m} L\left(\overline{p}_2(r,t), \overline{\beta}(t), r, t\right) \,\mathrm{d}r\, \mathrm{d}t + \int_0^{a_m} f_0\left(r, \overline{p}_2(r,T)\right) \mathrm{d}r - \varepsilon,
$$

where $\overline{p}_2(r,t) = \mathbb{T}\left(t, t_2; \overline{\beta}\right) p_0(r)$. Let $\overline{p}_1(r,t) = \mathbb{T}\left(t, t_1; \overline{\beta}\right) p_0(r)$. Then

$$
\begin{aligned}
&V(t_1, p_0) - V(t_2, p_0) \\
&\le \int_{t_1}^T \int_0^{a_m} L\left(\overline{p}_1(r,t), \overline{\beta}(t), r, t\right) \,\mathrm{d}r\, \mathrm{d}t + \int_0^{a_m} f_0\left(r, \overline{p}_1(r,T)\right) \mathrm{d}r \\
&\quad - \int_{t_2}^T \int_0^{a_m} L\left(\overline{p}_2(r,t), \overline{\beta}(t), r, t\right) \,\mathrm{d}r\, \mathrm{d}t - \int_0^{a_m} f_0\left(r, \overline{p}_2(r,T)\right) \mathrm{d}r + \varepsilon \\
&= - \int_{t_2}^{t_1} \int_0^{a_m} L\left(\overline{p}_2(r,t), \overline{\beta}(t), r, t\right) \,\mathrm{d}r\, \mathrm{d}t
\end{aligned}
$$

$$+\int_{t_1}^{T}\int_0^{a_m}\left[L\left(\overline{p}_1(r,t),\overline{\beta}(t),r,t\right)-L\left(\overline{p}_2(r,t),\overline{\beta}(t),r,t\right)\right]\,dr\,\mathrm{d}t$$

$$+\int_0^{a_m}\left[f_0\left(r,\overline{p}_1(r,T)\right)-f_0\left(r,\overline{p}_2(r,T)\right)\right]\,\mathrm{d}r+\varepsilon$$

$$\le\left(C_1+C_2M\,e^{\omega T}\|p_0\|\right)(t_1-t_2)+C_3\int_{t_1}^{T}\left\|\mathbb{T}\left(t,t_1;\overline{\beta}\right)p_0-\mathbb{T}\left(t,t_2;\overline{\beta}\right)p_0\right\|\,\mathrm{d}t$$

$$+C_3\left\|\mathbb{T}\left(T,t_1;\overline{\beta}\right)p_0-\mathbb{T}\left(T,t_2;\overline{\beta}\right)p_0\right\|+\varepsilon$$

$$\le\left(C_1+C_2M\,e^{\omega T}\|p_0\|\right)(t_1-t_2)+C_3(1+T)M\,e^{\omega T}\left\|\mathbb{T}\left(t_1,t_2;\overline{\beta}\right)p_0-p_0\right\|+\varepsilon.$$

Hence, as $t_1-t_2\to 0$, $V(t_1,p_0)-V(t_2,p_0)\to 0$, provided that

$$\left\|\mathbb{T}\left(t_1,t_2;\overline{\beta}\right)p_0-p_0\right\|\to 0.$$

Now suppose $t_1-t_2<a_1$. We then have

$$\left\|\mathbb{T}\left(t_1,t_2;\overline{\beta}\right)p_0-p_0\right\|^2=\int_{t_1-t_2}^{a_m}\left[p_0(r-t_1+t_2)\,e^{-\int_{r-t_1+t_2}^{r}\mu(\rho)\,\mathrm{d}\rho}-p_0(r)\right]^2\,\mathrm{d}r$$

$$+\int_0^{t_1-t_2}\left[\overline{\beta}(t_1-r)\int_{a_1}^{a_2}b(\tau)p_0(\tau-t_1+t_2+r)\,e^{-\int_{\tau-t_1+t_2+r}^{\tau}\mu(\rho)\,\mathrm{d}\rho}\,\mathrm{d}\tau\right.$$

$$\left.\cdot e^{-\int_0^r\mu(\rho)\,\mathrm{d}\rho}-p_0(r)\right]^2\,\mathrm{d}r$$

$$\le 2\int_{t_1-t_2}^{a_m}\left[p_0(r-t_1+t_2)-p_0(r)\right]^2\,\mathrm{d}r$$

$$+2\int_{t_1-t_2}^{a_m}\left[\left(e^{-\int_{r-t_1+t_2}^{r}\mu(\rho)\,\mathrm{d}\rho}-1\right)p_0(r-t_1+t_2)\right]^2\,\mathrm{d}r$$

$$+2\beta_1^2\|b\|(a_2-a_1)^{\frac{1}{2}}\|p_0\||t_1-t_2|+2\int_0^{t_1-t_2}|p_0(r)|^2\,\mathrm{d}r\to 0$$

uniformly for $\overline{\beta}$ as $t_1-t_2\to 0$. So (iii) is valid. The proof is complete. □

Remark 4.5.6 The proof of Theorem 4.5.5 (iii) demonstrates that for any given $\|\mathbb{T}(t_1,t_2;\beta)p_0-p_0\|$ converges uniformly to zero for all $\beta(\cdot)\in\mathcal{U}[t_2,t_1]$ as $t_1-t_2\to 0$. We will frequently rely on this fact in subsequent discussions.

4.5.2 HJB Equation and Viscosity Solution

For brevity in notation, we rewrite (4.5.6) as follows:

$$\begin{cases} \dfrac{d\mathbf{P}(t)}{dt} = \mathbb{A}_\beta(t)\mathbf{P}(t), \\ \mathbf{P}(0) = P, \end{cases} \tag{4.5.17}$$

where $\mathbf{P}(t) = p(\cdot, t)$, $P = p_0(\cdot)$. The value function is rewritten as

$$V(t, P) = \inf_{\beta(\cdot)\in\mathcal{U}[t,T]} \int_t^T f(\tau, \mathbf{P}(\tau), \beta(\tau))\ d\tau + \psi(\mathbf{P}(T))$$

where

$$\begin{cases} \mathbf{P}(\tau) = \mathbb{T}(\tau, t; \beta)P, \\ f(\tau, \mathbf{P}(\tau), \beta(\tau)) = \displaystyle\int_0^{a_m} L(p(r, \tau), \beta(\tau), r, \tau)\ dr, \\ \psi(\mathbf{P}(T)) = \displaystyle\int_0^{a_m} f_0(r, p(r, T))\ dr. \end{cases}$$

Theorem 4.5.7 *If $V(t, P) \in C^1([0, T] \times \mathbf{H})$, then V satisfies the following HJB equation:*

$$\begin{cases} V_t(t, P) + \inf\limits_{\beta\in[\beta_0,\beta_1]} \left\{\langle V_p(t, P), \mathbb{A}_\beta P\rangle + f(t, P, \beta)\right\} = 0, \\ V(T, P) = \psi(P), \quad \forall\, t \in [0, T], \quad P \in \bigcap\limits_{\beta\in[\beta_0,\beta_1]} D(\mathbb{A}_\beta). \end{cases} \tag{4.5.18}$$

Proof For any $0 < \delta < T - t$ and $\beta \in [\beta_0, \beta_1]$, the DPP implies that

$$V(t, P) \le \int_t^{t+\delta} f\left(\tau, \mathbb{T}_\beta(\tau - t)P, \beta\right)\ d\tau + V\left(t + \delta, \mathbb{T}_\beta(t + \delta)P\right). \tag{4.5.19}$$

If $P \in D(\mathbb{A}_\beta)$, since $\mathbb{T}_\beta(t)P = P$, it follows that

$$P - \mathbb{T}_\beta(t + \delta)P = P - \left[P + \delta\mathbb{A}_\beta P + o(\delta)\right] = -\delta\mathbb{A}_\beta P + o(\delta).$$

Hence,

$$V(t, P) - V\left(t + \delta, \mathbb{T}_\beta(t + \delta)P\right) = -V_t(t, P)\delta - \langle V_p(t, P), \mathbb{A}_\beta P\rangle\,\delta + o(\delta),$$

from which (4.5.19) can be rewritten as

$$- V_t(t, P) - \langle V_p(t, P), \mathbb{A}_\beta P\rangle \le \frac{1}{\delta}\int_t^{t+\delta} f\left(\tau, \mathbb{T}_\beta(\tau - t)P, \beta\right)\, d\tau + \frac{o(\delta)}{\delta}.$$

Passing to the limit as $\delta \to 0$ gives

$$- V_t(t, P) - \langle V_p(t, P), \mathbb{A}_\beta P\rangle \le f(t, P, \beta).$$

So

$$V_t(t, P) + \inf_{\beta\in[\beta_0,\beta_1]} \left\{\langle V_p(t, P), \mathbb{A}_\beta P\rangle + f(t, P, \beta)\right\} \ge 0.$$

On the other hand, for any given $\varepsilon > 0$, there exists a $\hat{\beta}(\cdot) \in \mathcal{U}[t, t+\delta]$ such that

$$V(t, P) \ge \int_t^{t+\delta} f\left(\tau, \hat{\mathbf{P}}(\tau), \hat{\beta}(\tau)\right)\, d\tau + V\left(t+\delta, \hat{\mathbf{P}}(t+\delta)\right) - \varepsilon \tag{4.5.20}$$

where $\hat{\mathbf{P}}(\tau) = \mathbb{T}\left(\tau, t; \hat{\beta}\right) P$. By (vi) of Theorem 4.5.1, we may assume without loss of generality that $\hat{\beta}(\cdot) \in C^1[t, t+\delta] \cap \mathcal{U}[t, t+\delta]$ and $\hat{\beta}(t) = \beta$ for any given $\beta \in [\beta_1, \beta_2]$. If $P \in \bigcap_{\beta\in[\beta_0,\beta_1]} D(\mathbb{A}_\beta)$, then $\hat{\mathbf{P}}(t+\delta)$ is differentiable with respect to δ and

$$\frac{d}{d\delta}\hat{\mathbf{P}}(t+\delta)\Big|_{\delta=0} = \mathbb{A}_{\hat{\beta}}(t)P.$$

This together with (4.5.20) gives

$$- V_t(t, P) - \left\langle V_p(t, P), \mathbb{A}_{\hat{\beta}}(t)P\right\rangle \ge f\left(t, P, \hat{\beta}(t)\right) - \varepsilon,$$

i.e.,

$$V_t(t, P) + \left\langle V_p(t, P), \mathbb{A}_{\hat{\beta}(t)}P\right\rangle + f\left(t, P, \hat{\beta}(t)\right) \le \varepsilon.$$

So

$$V_t(t, P) + \inf_{\beta\in[\beta_0,\beta_1]} \left\{\left\langle V_p(t, P), \mathbb{A}_\beta P\right\rangle + f(t, P, \beta)\right\} \le \varepsilon$$

and hence

$$V_t(t, P) + \inf_{\beta\in[\beta_0,\beta_1]} \left\{\left\langle V_p(t, P), \mathbb{A}_\beta P\right\rangle + f(t, P, \beta)\right\} \le 0.$$

This completes the proof of the theorem. □

Remark 4.5.8 We shall denote

$$\mathcal{D} = \bigcap_{\beta \in [\beta_0, \beta_1]} D(\mathbb{A}_\beta)$$
$$\supset \left\{ \phi(r), \phi'(r) + \mu(r)\phi(r) \in \mathbf{H} \;\middle|\; \phi(0) = \int_{a_1}^{a_2} b(r)\phi(r)\, \mathrm{d}r = 0 \right\}. \tag{4.5.21}$$

From now on, we consistently assume that $\mathbb{A}_\beta(t)$ is dissipative for all β belonging to the interval $[\beta_0, \beta_1]$. This assumption is justifiable because, instead of (4.5.17), we can alternatively consider

$$\begin{cases} \dfrac{d\mathbf{Q}(t)}{dt} = [\mathbb{A}_\beta(t) - C]\mathbf{Q}(t), \\ \mathbf{Q}(0) = P \end{cases} \tag{4.5.22}$$

and then $\mathbb{A}_\beta(t) - C$ with $C = \frac{1}{2}\|b\|^2$ is dissipative for each $\beta \in [\beta_0, \beta_1]$ because

$$\mathrm{Re}\langle \mathbb{A}_\beta(t)P, P\rangle \leq \frac{1}{2}\|b\|^2\langle P, P\rangle.$$

Since $\mathbf{Q}(t) = \mathbf{P}(t)\, e^{-Ct}$, the value function becomes

$$V(t, P) = \inf_{\beta(\cdot)\in\mathcal{U}[t,T]} \left\{ \int_t^T f\left(\tau, \mathbf{Q}(\tau)\, e^{C\tau}, \beta(\tau)\right) \mathrm{d}\tau + \psi\left(e^{CT}\mathbf{Q}(T)\right) \right\}$$

where $\mathbf{Q}(\tau) = e^{-Ct}\mathbb{T}(\tau, t; \beta)P$.

Furthermore, we define the solution of the HJB equation (4.5.18) in the "viscosity sense" [247]. Let Ω be an open set of $\mathbf{H}$ and set

$$\mathrm{USC}([0, T] \times \Omega) = \{\text{upper-semicontinuous mappings } u : [0, T] \times \Omega \to \mathbb{R}\},$$

$$\mathrm{LSC}([0, T] \times \Omega) = \{\text{lower-semicontinuous mappings } u : [0, T] \times \Omega \to \mathbb{R}\}.$$

Definition 4.5.9 The function $U(t, P) \in \mathrm{USC}([0, T] \times \Omega)$ (respectively the function $U(t, P) \in \mathrm{LSC}([0, T] \times \Omega)$) is a subsolution (respectively, supersolution) of (4.5.18) on $[0, T] \times \Omega$ if for every test function $\Phi = \varphi + g,\ \varphi,\ \varphi_p \in C([0, T] \times \Omega; \mathbb{R}),\ g \in C^1(\Omega; \mathbb{R})$ satisfying condition (4.5.23) and condition (4.5.24) below:

$$\begin{cases} \mathrm{Range}(\varphi_p) \subset \mathcal{D}^*, \\ \text{the mapping } (t, P) \to \left\langle \mathbb{A}_\beta^*\varphi_p(t, P), P \right\rangle \\ \text{from } [0, T] \times \Omega \text{ to } \mathbb{R} \text{ is equicontinuous in } \beta \in [\beta_0, \beta_1]; \end{cases} \tag{4.5.23}$$

$$\begin{cases} \text{there exists a } \tilde{h} : [0, \infty) \to \mathbb{R} \text{ such that} \\ \tilde{h} \text{ is nondecreasing, } \tilde{h}'(0) = 0\text{and} \\ g(P) = \tilde{h}(\|P\|), \ \forall \ P \in \mathbf{H} \end{cases} \tag{4.5.24}$$

and for the local maximum point (respectively, the minimum point) (t, P) of $U - \Phi$ (respectively $U + \Phi$), we have

$$\varphi_t(t, P) + \inf_{\beta \in [\beta_0, \beta_1]} \left\{ \left\langle \mathbb{A}_\beta^* \varphi_P(t, P), P \right\rangle + f(t, P, \beta) \right\} \geq 0. \tag{4.5.25}$$

(respectively,

$$-\varphi_t(t, P) + \inf_{\beta \in [\beta_0, \beta_1]} \left\{ -\left\langle \mathbb{A}_\beta^* \varphi_P(t, P), P \right\rangle + f(t, P, \beta) \right\} \leq 0.) \tag{4.5.26}$$

$U(t, P) \in C([0, T] \times \Omega)$ is a viscosity solution of (4.5.18) if it is both a subsolution and a supersolution on $[0, T] \times \Omega$.

Theorem 4.5.10 *The value function $V(t, P)$ is a viscosity solution of the HJB equation* (4.5.18).

Proof Assume that (t, P) is a local maximum point of $V - \Phi$ for the test function Φ. Then there exists a $\delta > 0$ such that

$$V(\tau, Q) - \Phi(\tau, Q) \leq V(t, P) - \Phi(t, P), \text{ for all } |\tau - t|, \ \|Q - P\| < \delta,$$

i.e.,

$$\Phi(\tau, Q) - \Phi(t, P) \geq V(\tau, Q) - V(t, P), \text{ for all } |\tau - t|, \ \|Q - P\| < \delta.$$

By Remark 4.5.6, there exists an $\varepsilon > 0$ such that for all $\beta \in [\beta_0, \beta_1]$,

$$\begin{aligned} \Phi\left(t + \delta_0, \mathbb{T}_\beta(\delta_0) P\right) - \Phi(t, P) &\geq V(t + \delta_0, \mathbb{T}_\beta(\delta_0) P) - V(t, P) \\ &\geq -\int_t^{t+\delta_0} f(\tau, \mathbb{T}_\beta(\tau - t) P, \beta) \, \mathrm{d}\tau \quad \text{as } 0 < \delta_0 < \varepsilon. \end{aligned}$$

The inequality mentioned previously is derived from Theorem 4.5.4. It is worth noting that, under the conditions specified in (4.5.23) and (4.5.24), we have

$$\begin{aligned} \varphi(\tau + \delta_0, \mathbb{T}_\beta(\delta_0) Q) - \varphi(\tau, Q) &= \int_0^{\delta_0} \frac{d}{d\rho} \varphi(\tau + \rho, \mathbb{T}_\beta(\rho) Q) \, \mathrm{d}\rho \\ = \int_0^{\delta_0} \varphi_t(\tau + \rho, \mathbb{T}_\beta(\rho) Q) \, \mathrm{d}\rho &+ \int_0^{\delta_0} \left\langle \mathbb{A}_\beta^* \varphi_P(\tau + \rho, \mathbb{T}_\beta(\rho) Q), \mathbb{T}_\beta(\rho) Q \right\rangle \mathrm{d}\rho, \end{aligned} \tag{4.5.27}$$

which holds for all τ, δ_0 and all Q by first considering those $Q \in D(\mathbb{A}_\beta)$ and then for $Q \in \mathbf{H}$ via the density argument. Also (4.5.24) implies that

$$g(\mathbb{T}_\beta(\delta_0)Q) - g(Q) \leq 0 \tag{4.5.28}$$

for all $Q \in \mathbf{H}$ and all $\beta \in [\beta_0, \beta_1]$ by first for $Q \in D(\mathbb{A}_\beta)$ and then via the density argument. Putting things all together, we have

$$\begin{aligned} -\int_t^{t+\delta_0} f(\tau, \mathbb{T}_\beta(\tau - t)P, \beta)\, \mathrm{d}\tau &\leq \Phi(t+\delta_0, \mathbb{T}_\beta(\delta_0)P) - \Phi(t, P) \\ &\leq \varphi(t+\delta_0, \mathbb{T}_\beta(\delta_0)P) - \varphi(t, P) \\ &= \int_0^{\delta_0} \varphi_t(t+\tau, \mathbb{T}_\beta(\tau)P)\, \mathrm{d}\tau + \int_0^{\delta_0} \left\langle \mathbb{A}_\beta^* \varphi_P\left(t+\tau, \mathbb{T}_\beta(\tau)P\right), \mathbb{T}_\beta(\tau)P \right\rangle \mathrm{d}\tau \end{aligned}$$

for all $0 < \delta_0 < \varepsilon$. Dividing δ_0 on both sides above and letting δ_0 go to zero, we obtain

$$\varphi_t(t, P) + \left\langle \mathbb{A}_\beta^* \varphi_P(t, P), P \right\rangle + f(t, P, \beta) \geq 0.$$

And hence, (4.5.25) holds by the arbitrariness of β.

Next, assume that (t, P) is a local minimum point of $V + \Phi$ for the test function Φ. Then there exists an $\varepsilon > 0$ such that

$$-\Phi(\tau, \mathbf{P}(\tau)) + \Phi(t, P) \leq V(\tau, \mathbf{P}(\tau)) - V(t, P) \tag{4.5.29}$$

for all $|\tau - t| < \varepsilon$ and all admissible control $\beta(\cdot) \in \mathcal{U}[t, t+\varepsilon]$, in which $\mathbf{P}(\tau) = \mathbb{T}(\tau, t; \beta)P$. Let $0 < \delta_0 < \varepsilon$. By Theorem 4.5.4,

$$\begin{aligned} V(t, P) &= \inf_{\beta(\cdot)\in\mathcal{U}[t,t+\delta_0]} \left\{ \int_t^{t+\delta_0} f\left(\tau, \mathbf{P}(\tau), \beta(\tau)\right) \mathrm{d}\tau + V(t+\delta_0, \mathbf{P}(t+\delta_0)) \right\} \\ &\geq \inf_{\beta(\cdot)\in\mathcal{U}[t,t+\delta_0]} \left\{ \int_t^{t+\delta_0} f(\tau, \mathbf{P}(\tau), \beta(\tau))\, \mathrm{d}\tau + V(t, P) - \Phi(t+\delta_0, \mathbf{P}(t+\delta_0)) \right. \\ &\qquad \left. + \Phi(t, P) \right\}. \end{aligned}$$

So

$$\inf_{\beta(\cdot)\in\mathcal{U}[t,t+\delta_0]} \left\{ \int_t^{t+\delta_0} f(\tau, \mathbf{P}(\tau), \beta(\tau))\, \mathrm{d}\tau - \Phi(t+\delta_0, \mathbf{P}(t+\delta_0)) + \Phi(t, P) \right\} \leq 0 \tag{4.5.30}$$

holds for all $0 < \delta_0 < \varepsilon$. As stated in (4.5.27) and (4.5.28), we have the following

$$\varphi(t+\delta_0, \mathbf{P}(t+\delta_0)) - \varphi(t, P) = \int_0^{\delta_0} \varphi_t(t+\tau, \mathbf{P}(t+\tau))\, \mathrm{d}\tau + \int_0^{\delta_0} \Big\langle \mathbb{A}_\beta^*(t)\varphi_p(t+\tau, \mathbf{P}(t+\tau)), \mathbf{P}(t+\tau)\Big\rangle\, \mathrm{d}\tau$$

and

$$g(\mathbf{P}(t+\delta_0)) - g(P) \le 0$$

for all $\beta(\cdot) \in \mathcal{U}[t, t+h]$. (4.5.30) then implies that

$$\inf_{\beta(\cdot)\in\mathcal{U}[t,t+\delta_0]} \left\{ \int_t^{t+\delta_0} f(\tau, \mathbf{P}(\tau), \beta(\tau))\, \mathrm{d}\tau - \varphi(t+\delta_0, \mathbf{P}(t+\delta_0)) + \varphi(t, P) \right\} \le 0$$

or

$$\inf_{\beta(\cdot)\in\mathcal{U}[t,t+\delta_0]} \left\{ \int_t^{t+\delta_0} f(\tau, \mathbf{P}(\tau), \beta(\tau))\, \mathrm{d}\tau - \int_0^{\delta_0} \varphi_t(t+\tau, \mathbf{P}(t+\tau))\, d\tau - \int_0^{\delta_0} \Big\langle \mathbb{A}_\beta^*(t)\varphi_p(t+\tau, \mathbf{P}(t+\tau)), \mathbf{P}(t+\tau)\Big\rangle\, \mathrm{d}\tau \right\} \le 0. \tag{4.5.31}$$

Supposing the contrary,

$$-\varphi_t(t, P) + \inf_{\beta\in[\beta_0,\beta_1]} \left\{ -\Big\langle \mathbb{A}_\beta^*\varphi_p(t, P), P\Big\rangle + f(t, P, \beta) \right\} > 2\vartheta$$

for some $\vartheta > 0$. Then for all $\beta \in [\beta_0, \beta_1]$,

$$-\varphi_t(t, P) - \Big\langle \mathbb{A}_\beta^*\varphi_p(t, P), P\Big\rangle + f(t, P, \beta) > 2\vartheta.$$

By Remark 4.5.6, there exists a $\delta > 0$, $\delta < \varepsilon$, such that

$$-\varphi_t(t, \mathbf{P}(t+\tau)) - \Big\langle \mathbb{A}_\beta^*(t)\varphi_p(t, \mathbf{P}(t+\tau)), \mathbf{P}(t+\tau)\Big\rangle + f(t+\tau, \mathbf{P}(t+\tau), \beta(\tau)) > \vartheta$$

for all $\tau < \delta$ and all $\beta(\cdot) \in \mathcal{U}[t, t+\delta]$. It contradicts (4.5.31). So V is a supersolution. The proof is complete. □

4.5.3 Optimal Feedback Control

A necessary condition, which also characterizes optimal trajectory-control pairs, can be formulated through the utilization of the value function $V(t, P)$.

Theorem 4.5.11 *Let $V(t, P)$ be the value function. Then for any trajectory-control pair $(\mathbf{P}^*(t), \beta^*(t))$, $\mathbf{P}^*(t) = \mathbb{T}(t, 0; \beta^*) P$, the function*

$$t \to V(t, \mathbf{P}^*(t)) - \int_t^T f\left(\tau, \mathbf{P}^*(\tau), \beta^*(\tau)\right) d\tau \tag{4.5.32}$$

is nondecreasing in $[0, T]$. Moreover, $(\mathbf{P}^*(\cdot), \beta^*(\cdot))$ is an optimal pair if and only if the above function is constant on $[0, T]$. Consequently, if $V(t, P) \in C^1([0, T] \times \mathbf{H})$, $\beta^*(\cdot) \in C^1[0, T]$, $\mathbf{P}^*(0) \in D\left(\mathbb{A}_{\beta^*(0)}\right)$, then $(\mathbf{P}^*(\cdot), \beta^*(\cdot))$ is an optimal pair if and only if

$$\begin{cases} V_t(t, \mathbf{P}^*(t)) + \left\langle V_p(t, \mathbf{P}^*(t)), \mathbb{A}_{\beta^*(t)} \mathbf{P}^*(t)\right\rangle + f(t, \mathbf{P}^*(t), \beta^*(t)) = 0, \\ V(T, \mathbf{P}^*(T)) = \psi(\mathbf{P}^*(T)), \quad \forall\, t \in [0, T]. \end{cases} \tag{4.5.33}$$

Proof The first assertion is a direct outcome of the DPP. The final assertion arises from differentiating (4.5.32) with respect to time t and applying (v) of Theorem 4.5.1.

Regarding the second assertion, if $(\mathbf{P}^*(t), \beta^*(t))$ is not optimal, then there necessarily exists another trajectory-control pair such that

$$\int_0^T f\left(\tau, \mathbf{P}(\tau), \beta(\tau)\right) d\tau + \psi(\mathbf{P}(T)) < \int_0^T f\left(\tau, \mathbf{P}^*(\tau), \beta^*(\tau)\right) d\tau + \psi(\mathbf{P}^*(T))$$

and hence

$$\begin{aligned} V(0, P) &= V(0, \mathbf{P}^*(0)) = V(0, \mathbf{P}(0)) \\ &\le \int_0^T f(\tau, \mathbf{P}(\tau), \beta(\tau))\, d\tau + \psi(\mathbf{P}(T)) \\ &< \int_0^T f(\tau, \mathbf{P}^*(\tau), \beta^*(\tau))\, d\tau + \psi(\mathbf{P}^*(T)) \\ &= \int_0^T f(\tau, \mathbf{P}^*(\tau), \beta^*(\tau))\, d\tau + V(T, \mathbf{P}^*(T)), \end{aligned}$$

i.e.,

$$V(0, \mathbf{P}^*(0)) - \int_0^T f(\tau, \mathbf{P}^*(\tau), \beta^*(\tau))\, d\tau < V(T, \mathbf{P}^*(T)).$$

Thus,

$$V(t, \mathbf{P}^*(t)) - \int_t^T f(\tau, \mathbf{P}^*(\tau), \beta^*(\tau))\, \mathrm{d}\tau$$

is not constant on $[0, T]$. On the other hand, if it is not constant on $[0, T]$, the first assertion implies that

$$V(0, P) < \int_0^T f(\tau, \mathbf{P}^*(\tau), \beta^*(\tau))\, \mathrm{d}\tau + \psi(\mathbf{P}^*(T)).$$

Let

$$\varepsilon = \int_0^T f(\tau, \mathbf{P}^*(\tau), \beta^*(\tau))\, \mathrm{d}\tau + \psi(\mathbf{P}^*(T)) - V(0, P) > 0.$$

By the definition of the value function, there exists a $(\mathbf{P}(t), \beta(t))$ such that

$$V(0, P) > \int_0^T f(\tau, \mathbf{P}(\tau), \beta(\tau))\, \mathrm{d}\tau + \psi(\mathbf{P}(T)) - \varepsilon.$$

Hence,

$$\int_0^T f(\tau, \mathbf{P}(\tau), \beta(\tau))\, \mathrm{d}\tau + \psi(\mathbf{P}(T)) < \int_0^T f(\tau, \mathbf{P}^*(\tau), \beta^*(\tau))\, \mathrm{d}\tau + \psi(\mathbf{P}^*(T)),$$

which implies that $(\mathbf{P}^*(t), \beta^*(t))$ is not optimal, proving the required result. □

The final assertion of Theorem 4.5.11 will lead to the classical verification theorem, which holds a pivotal role in validating the optimality of a given admissible pair. Furthermore, it serves as a crucial tool in the construction of optimal feedback controls, as highlighted in [97, 249]. Assuming the smoothness of the value function $V(t, P)$, since $V(t, P)$ is a viscosity solution of the HJB equation, we have

$$V_t(t, P) + \inf_{\beta \in [\beta_0, \beta_1]} \left\{ \langle V_P(t, P), \mathbb{A}_\beta P \rangle + f(t, P, \beta) \right\} = 0,$$

then the condition of optimality (4.5.33) is equivalent to

$$\begin{aligned} &\left\langle V_P(t, \mathbf{P}^*(t)), \mathbb{A}_{\beta^*(t)} \mathbf{P}^*(t) \right\rangle + f(t, \mathbf{P}^*(t), \beta^*(t)) \\ &= \inf_{\beta \in [\beta_0, \beta_1]} \left\{ \langle V_P(t, \mathbf{P}^*(t)), \mathbb{A}_\beta \mathbf{P}^*(t) \rangle + f(t, \mathbf{P}^*(t), \beta) \right\} \\ &:= \mathcal{H}\left(\mathbf{P}^*(t), V_P(t, \mathbf{P}^*(t))\right) \quad \text{for } t \in [0, T] \text{ a.e.} \end{aligned} \tag{4.5.34}$$

Let

$$\begin{aligned}\mathcal{S}(Z) &:= \underset{\beta\in[\beta_0,\beta_1]}{\arg\min}\left\{\langle V_z(t,Z),\mathbb{A}_\beta Z\rangle + f(t,Z,\beta)\right\}\\ &= \{\beta\in[\beta_0,\beta_1] \mid \mathcal{H}(Z,V_z(t,Z)) = \langle V_z(t,Z),\mathbb{A}_\beta Z\rangle + f(t,Z,\beta)\}.\end{aligned} \tag{4.5.35}$$

According to (4.5.34), $\beta^*(\cdot)\in\mathcal{U}[0,T]$ is an optimal control for the initial state p_0 if and only if

$$\beta^*(t)\in S(\mathbf{P}^*(t)) \quad \text{for } t\in[0,T] \text{ a.e.}, \tag{4.5.36}$$

where $\mathbf{P}^*(t) = \mathbb{T}(t,0;\beta^*)p_0$. Indeed, (4.5.36) serves as the formula for determining the optimal feedback control.

It is worth noting that (4.5.36) holds true only when the value function is differentiable. However, this is not always the case in terms of analyticity. Nonetheless, it remains a crucial formula for numerically constructing the optimal feedback control, which will be put to test in the subsequent subsection.

4.5.4 *Finite Difference Scheme for the Numerical Solution of Optimal Feedback Control*

In this subsection, we will confine our discussion to the following linear quadratic optimal control problem:

$$\begin{aligned}J(\beta^*) &= \inf_{\beta(\cdot)\in\mathcal{U}[0,T]} J(\beta)\\ &= \inf_{\beta(\cdot)\in\mathcal{U}[0,T]} \frac{1}{2}\left\{\int_0^T \left[\beta(t)-\bar{\beta}(t)\right]^2 dt + \int_0^{a_m} [p(r,T)-\bar{p}(r)]^2\, dr\right\},\end{aligned} \tag{4.5.37}$$

where $\bar{\beta}(t)$ represents the mean critical fertility rate of females in an ideal society, and $\bar{p}(r)$ signifies the stable age density distribution of the population with a zero growth rate. In alignment with (4.5.3) and (4.5.18), we have

$$L(p(r,t),\beta(t),r,t) = \frac{1}{2}\left[\beta(t)-\bar{\beta}(t)\right]^2, \quad f_0(r,p(r,T)) = \frac{1}{2}[p(r,T)-\bar{p}(r)]^2,$$

$$f(t,p(\cdot,t),\beta(t)) = \frac{a_m}{2}\left[\beta(t)-\bar{\beta}(t)\right]^2,$$

$$\psi(p(\cdot,T)) = \int_0^{a_m}\frac{1}{2}[p(r,T)-\bar{p}(r)]^2\, dr,$$

and the HJB equation can be rewritten as

$$\begin{cases} V_t(t,p) + \inf\limits_{\beta\in[\beta_0,\beta_1]} \left\{ \langle V_p(t,p), \mathbb{A}_\beta p\rangle + \frac{a_m}{2}\left[\beta - \bar{\beta}(t)\right]^2 \right\} = 0, \\ V(T,p) = \int_0^{a_m} \frac{1}{2}[p(r) - \bar{p}(r)]^2\,dr, \quad \forall\, t\in[0,T],\ p \in \bigcap\limits_{\beta\in[\beta_0,\beta_1]} D(\mathbb{A}_\beta). \end{cases} \tag{4.5.38}$$

Now, we proceed to numerically solve the equation mentioned above. To accomplish this, we set $t_j = T + j\Delta t,\ \ j = 0, 1, 2, \cdots, N$, where $\Delta t = -\frac{T}{N}$ and N is an integer. Given a small $\varepsilon > 0$, we employ the following approximation:

$$\begin{aligned} \langle V_p(t,p), \mathbb{A}_\beta p\rangle &= \left\langle V_p(t,p), \varepsilon\frac{\mathbb{A}_\beta p}{\|\mathbb{A}_\beta p\|}\right\rangle \frac{\|\mathbb{A}_\beta p\|}{\varepsilon} \\ &\approx \left[V\left(t, p + \varepsilon\frac{\mathbb{A}_\beta p}{\|\mathbb{A}_\beta p\|}\right) - V(t,p)\right]\frac{\|\mathbb{A}_\beta p\|}{\varepsilon}. \end{aligned}$$

For the initial state p_0 let $p_i = p_{i-1} + \frac{\mathbb{A}_\beta p_{i-1}}{\|\mathbb{A}_\beta p_{i-1}\|}\varepsilon,\ i = 1, 2, \cdots, M$. Approximate (4.5.38) by the difference equation:

$$\begin{cases} \dfrac{V_i^{j+1} - V_i^j}{\Delta t} + \dfrac{V_i^j - V_{i-1}^j}{\varepsilon}\left\|\mathbb{A}_{\beta_i^j} p_i\right\| + \dfrac{a_m}{2}\left[\beta_i^j - \bar{\beta}^j\right]^2 = 0, \\ \beta_i^{j+1} \in \arg\inf\limits_{\beta\in[\beta_0,\beta_1]} \left\{ \dfrac{V_i^{j+1} - V_{i-1}^{j+1}}{\varepsilon}\|\mathbb{A}_\beta p_i\| + \dfrac{a_m}{2}\left[\beta - \bar{\beta}^{j+1}\right]^2 \right\} \end{cases} \tag{4.5.39}$$

for $i = 1, 2, \cdots, M$ and $j = 0, 1, 2, \cdots, N$, where $V_i^j \approx V(t_j, p_i)$. Once again, referring to [149], the following necessary condition for the stability of the difference scheme will be assumed:

$$\frac{|\Delta t|}{\varepsilon} \max_{1\le i\le M} \|\mathbb{A}_\beta p_i\| \le 1. \tag{4.5.40}$$

Set $\alpha_i^j = \frac{\Delta t}{\varepsilon}\left\|\mathbb{A}_{\beta_i^j} p_i\right\|$. We present the algorithm for numerically solving (4.5.38) as follows:

Algorithm 6 Algorithm for solving the HJB equation

1: Initialization. Set

$$\begin{cases} V_i^0 = V(T, p_i) = \psi(p_i), \ p_i = p_{i-1} + \dfrac{\mathbb{A}_\beta p_{i-1}}{\|\mathbb{A}_\beta p_{i-1}\|}\varepsilon, \\ \beta_i^0 \in \underset{\beta\in[\beta_0,\beta_1]}{\arg\inf} \left\{\dfrac{\psi(p_i) - \psi(p_{i-1})}{\varepsilon}\|\mathbb{A}_\beta p_i\| + \dfrac{a_m}{2}\left[\beta - \bar{\beta}^0\right]^2\right\} \\ \quad := \underset{\beta\in[\beta_0,\beta_1]}{\arg\inf} \left\{H(\beta, p)\right\}, \quad i = 1, 2, \cdots, M. \end{cases} \tag{4.5.41}$$

Here the formula for β_i^0 comes from

$$\beta(t, p(\cdot)) \in \underset{\beta\in[\beta_0,\beta_1]}{\arg\inf} \left\{\langle V_p(t, p), \mathbb{A}_\beta p\rangle + \frac{a_m}{2}\left[\beta - \bar{\beta}(t)\right]^2\right\}. \tag{4.5.42}$$

2: Iteration. By (4.5.39),

$$\begin{cases} V_i^{j+1} = \left(1 - \alpha_i^j\right) V_i^j + \alpha_i^j V_{i-1}^j - \dfrac{a_m}{2}\left(\beta_i^j - \bar{\beta}^j\right)^2 \Delta t, \\ \beta_i^{j+1} \in \underset{\beta\in[\beta_0,\beta_1]}{\arg\inf} \left\{\dfrac{V_i^{j+1} - V_{i-1}^{j+1}}{\varepsilon}\|\mathbb{A}_\beta p_i\| + \dfrac{a_m}{2}\left[\beta - \bar{\beta}^{j+1}\right]^2\right\} \end{cases} \tag{4.5.43}$$

for $i = 1, 2, \cdots, M$ and $j = 0, 1, 2, \cdots, N-1$.

It is seen that

$$\beta_i^0 = \begin{cases} \beta_1, \ H(\beta_1, p) = \min\left\{H(\beta_1, p), \ H(\beta_0, p), \ H\left(\bar{\beta}^0, p\right)\right\}, \\ \beta_0, \ H(\beta_0, p) = \min\left\{H(\beta_1, p), \ H(\beta_0, p), \ H\left(\bar{\beta}^0, p\right)\right\}, \\ \bar{\beta}^0, \ H(\bar{\beta}^0, p) = \min\left\{H(\beta_1, p), \ H(\beta_0, p), \ H\left(\bar{\beta}^0, p\right)\right\} \end{cases} \tag{4.5.44}$$

for $i = 1, 2, \cdots, M$. The similar formula can be also obtained for β_i^j, $i = 1, 2, \cdots, M$, $j = 1, 2, \cdots, N-1$.

From (4.5.42), the optimal feedback control is

$$\beta_{p_0}^*(t, p^*(\cdot, t)) \in \underset{\beta\in[\beta_0,\beta_1]}{\arg\inf} \left\{\langle V_p(t, p^*(\cdot, t)), \mathbb{A}_\beta p^*(\cdot, t)\rangle + \frac{a_m}{2}\left[\beta - \bar{\beta}(t)\right]^2\right\} \tag{4.5.45}$$

where $p^*(\cdot, t)$ is the optimal trajectory of the system. Since the optimal trajectory referenced in (4.5.45) is crucial to our analysis, it becomes imperative to solve

(4.5.1) to determine this trajectory. To this end, we employ the difference scheme to approximate the solution of the initial value problem stated in (4.5.1) (see [213] for further details). Adopting the equally spaced grid method, we have

$$\begin{cases} \dfrac{p_{i,j}-p_{i-1,j}}{\Delta s}+\dfrac{p_{i,j}-p_{i,j-1}}{\Delta s}=-\mu_i p_{i,j}, & 1\le j\le J,\ 1\le i\le 2K, \\ p_{i,0}=p_0(i\Delta s), \quad 0\le i\le 2K, & \\ p_{0,j}=\beta_j\Delta s\displaystyle\sum_{i=a_1}^{a_2} b_i p_{i,j}, \quad 1\le j\le J & \end{cases} \tag{4.5.46}$$

in which $b_i = b(i\Delta s),\ \beta_j = \beta(j\Delta s),\ i = a_1, \cdots, a_2,\ 2K\Delta s = a_m,\ J\Delta s = T$. It follows that

$$\begin{cases} p_{i,j}=\dfrac{p_{i-1,j}+p_{i,j-1}}{2+\mu_i\Delta s}, \quad 1\le j\le J,\ 1\le i\le 2K, \\ p_{i,0}=p_0(i\Delta s), \quad 0\le i\le 2K, \\ p_{0,j}=\beta_j\displaystyle\sum_{i=a_1}^{a_2} b_i p_{i,j}\Delta s, \quad 1\le j\le J. \end{cases} \tag{4.5.47}$$

Next, we will provide detailed steps for finding the optimal feedback control. The focus is on solving the difference scheme (4.5.47) to obtain the optimal trajectory. Additionally, throughout this process, the algorithm for solving the HJB equation will be frequently invoked to derive the corresponding feedback control function.

Steps for Finding the Optimal Feedback Control

Step 1: Use the algorithm for solving the HJB equation to obtain the feedback control function $\beta(t, p_0)$. Then, substitute $\beta(t, p_0)$ into Eq. (4.5.47) to determine the optimal trajectory $p^*(\cdot, s_1)$, $s_1 = \Delta s$.

Step 2: Using $p^*(\cdot, s_1)$ as the initial data p_0 from the previous step, reapply the algorithm for solving the HJB equation to derive the feedback control $\beta(s_1, p^*(\cdot, s_1))$. Next, substitute $\beta(s_1, p^*(\cdot, s_1))$ into Eq. (4.5.47) to calculate the optimal trajectory $p^*(\cdot, s_2)$ where $s_2 = 2\Delta s$.

Step 3: By repeatedly following the aforementioned processes, we obtain the complete set of feedback control functions $\beta\left(s_j, p^*(\cdot, s_j)\right)$ for $s_j = j\Delta s$ and $j = 0, 1, \cdots, J$. In other words,

$$\begin{aligned} \beta^*_{p_0}(t, p^*(\cdot, t)) = \{&\beta(0, p_0(\cdot)),\ \beta\left(s_1, p^*(\cdot, s_1)\right),\ \beta\left(s_2, p^*(\cdot, s_2)\right), \cdots, \\ &\beta\left(T, p^*(\cdot, T)\right)\}, \end{aligned} \tag{4.5.48}$$

which is the optimal feedback control.

Table 4.5 The age structure of an ideal society

Age r	Proportion	Age r	Proportion
0	0.013	46–50	0.063
1–5	0.065	51–55	0.062
6–10	0.065	56–60	0.061
11–15	0.065	61–65	0.057
16–20	0.065	66–70	0.053
21–25	0.065	71–75	0.047
26–30	0.065	76–80	0.037
31–35	0.065	81–85	0.024
36–40	0.064	>85	0.003
41–45	0.064		

We are now ready to compute the numerical solution for the linear quadratic optimal control problem concerning the Chinese population, utilizing the formulas from (4.5.41), (4.5.43), and (4.5.47). The starting point of our analysis is the data obtained from the Chinese population sampling census conducted in 1989, as referenced in [66]. These data have been visualized using MATLAB 6.1 and are presented in Fig. 4.18 (depicting the age structure of females), Fig. 4.19 (showing relative mortality rates), and Fig. 4.20 (illustrating the age structure of the total population).

To establish an ideal population structure, we refer to the proportions outlined in Table 4.5, sourced from [213]. We multiply these proportions by the total ideal population size, which is set at $N_{\text{sum}} = 1{,}400{,}000$. Regarding the fertility pattern $h(r)$, we approximate it using the Gamma density distribution curve from statistics, as described in [213]. For this approximation, we assume that the peak fertility age is 24. The remaining parameters for our analysis are as follows:

$$\begin{cases} T = 25, \ \ \beta_0 = 1, \ \ \beta_1 = 3, \ \ \bar{\beta}(t) = 2, \\ a_1 = 15, \ a_2 = 49, \ a_m = 99, \ \varepsilon = 0.01. \end{cases}$$

In addition, due to the limitation of computer memory, the data in Figs. 4.18 and 4.20 and N_{sum} are divided by 10,000.

All numerical computations have been executed using Visual C++ 6.0, and the resulting data has been visualized with MATLAB 6.1. Figure 4.15 depicts the value function $V(t, p^*(\cdot, t))$, while Fig. 4.16 illustrates the optimal feedback control $\beta^*_{p_0}(t, p^*(\cdot, t))$.

To conclude this section, we must verify the optimality of our numerical solution for the optimal feedback control. This verification is accomplished by comparing the cost functional $J\left(\beta^*_{p_0}(t, p^*(\cdot, t))\right)$ of the obtained optimal control-trajectory pair

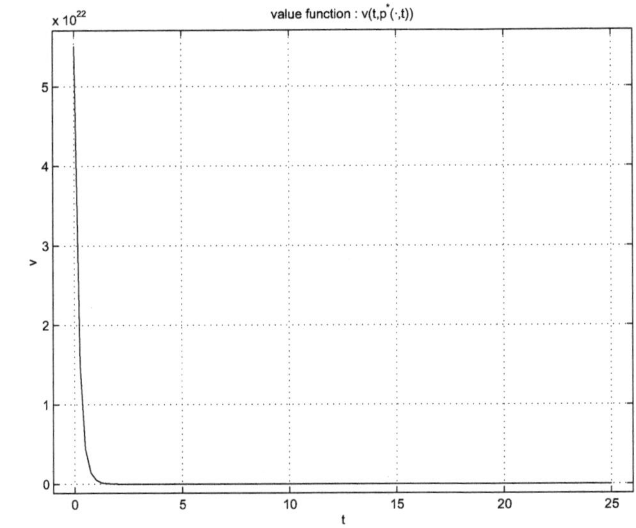

Fig. 4.15 The value function $V(t, p^*(\cdot, t))$

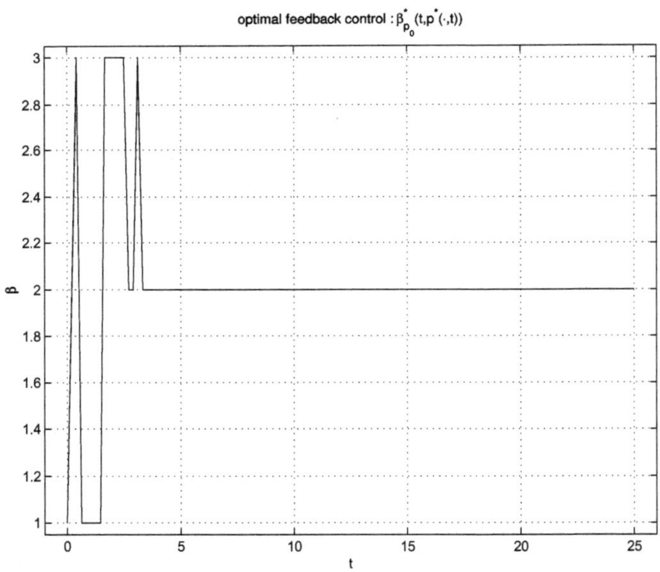

Fig. 4.16 The optimal feedback control $\beta^*_{p0}(t, p^*(\cdot, t))$

with that of an arbitrarily chosen control $\beta_{p0}(t)$, $J(\beta_{p0}(t))$ under the same initial conditions. Specifically, we evaluate:

$$J\left(\beta^*_{p0}(t, p^*(\cdot, t))\right) \leq J(\beta_{p0}(t)). \tag{4.5.49}$$

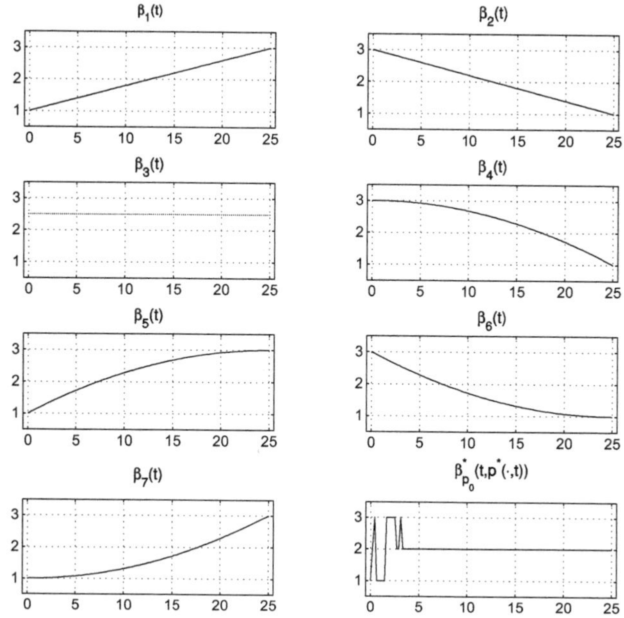

Fig. 4.17 Seven different arbitrarily chosen control β_i and the optimal feedback control $\beta^*_{p0}(t, p^*(\cdot, t))$

Table 4.6 Different β and its corresponding cost $J(\beta)$

β	$J(\beta)$	β	$J(\beta)$
$\beta_1(t)$	11641608005.975365	$\beta_5(t)$	11641608172.639233
$\beta_2(t)$	11641608005.975365	$\beta_6(t)$	11641608172.639233
$\beta_3(t)$	11641607901.862444	$\beta_7(t)$	11641608172.639233
$\beta_4(t)$	11641608172.639233	$\beta^*_{p0}(t, p^*(\cdot, t))$	11641607717.814667

Referring to Fig. 4.17, we compute these costs $J(\beta_i)$ $(i = 1, 2, \cdots, 7)$ for seven different controls $\beta_i \in \mathcal{U}[0, T], i = 1, 2, \cdots, 7$.

Additionally, we have calculated the cost value $J\left(\beta^*_{p0}(t, p^*(\cdot, t))\right)$ corresponding to the optimal feedback control. These results are summarized in Table 4.6.

Examining Table 4.6, it is evident that the cost associated with the optimal feedback control $\beta^*_{p0}(t, p^*(\cdot, t))$, is $J\left(\beta^*_{p0}(t, p^*(\cdot, t))\right) = 11641607717.814667$, which is significantly lower than the costs $J(\beta_i)$ for all other controls $i = 1, 2, \cdots, 7$. This confirms that we have indeed obtained the numerical solution for the optimal birth feedback control for the Chinese population from 1989 to 2014 (Figs. 4.18, 4.19, and 4.20).

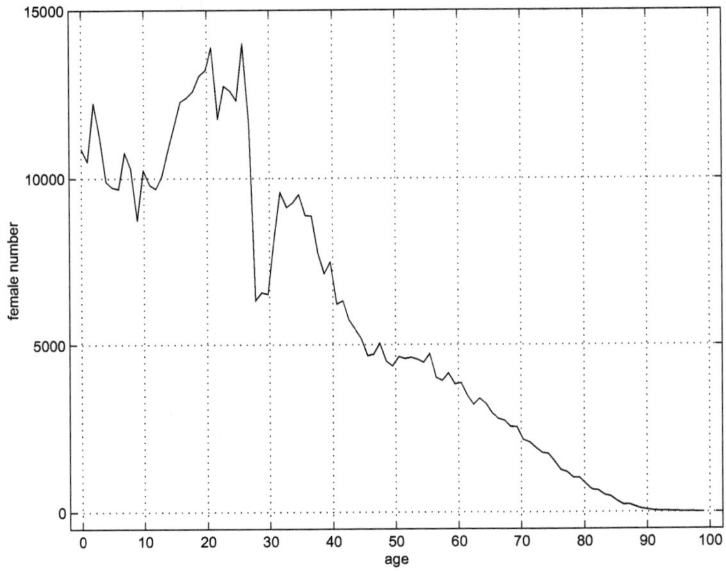

Fig. 4.18 The age structure of females of the Chinese population in 1989

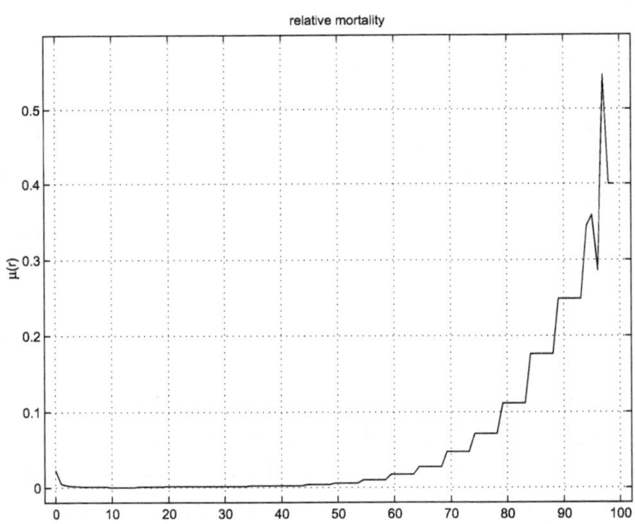

Fig. 4.19 The relative mortality of the Chinese population in 1989

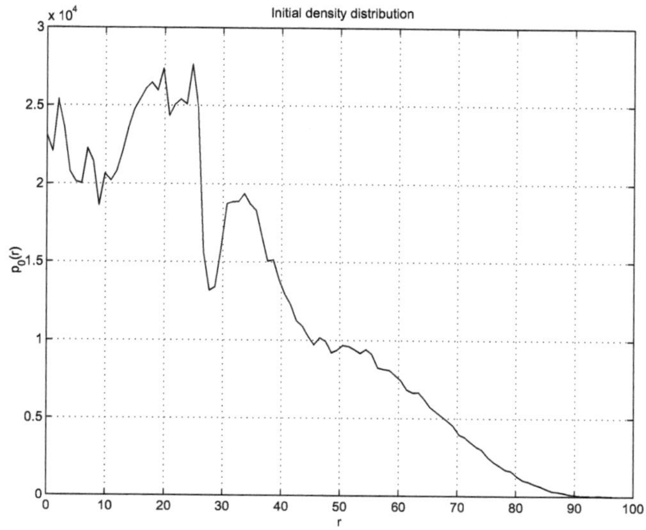

Fig. 4.20 The age-structure of the Chinese population in 1989

Chapter 5
Multiple Algorithms for DPVS Approach

As previously emphasized, determining the optimal feedback control law holds fundamental significance in optimal control theory, often described as the holy grail due to its pivotal role. The Bellman dynamic programming approach stands out for its ability to generate optimal controls in feedback form, greatly enhancing their appeal in terms of online implementation and robustness.

The introduction of viscosity solutions provides a rigorous mathematical framework for the classical dynamic programming approach. However, the curse of dimensionality poses significant computational challenges for practical implementations of numerical algorithms in optimal control design, especially when the DPVS approach tackles high-dimensional problems. It is evident that the dynamic programming approach requires substantial effort to develop effective numerical approximation schemes. One reason, applicable even in low dimensions, is the low regularity of viscosity solutions, which are often only Lipschitz continuous or even discontinuous. Apart from a few exceptional cases, obtaining analytical solutions for the HJB equation is not feasible due to its analytical complexity. Additionally, to overcome the curse of dimensionality, we must devise efficient algorithms that prioritize low memory allocation and, potentially, incorporate rules to minimize CPU time.

In this chapter, we systematically explore the numerical solutions for optimal feedback control, focusing primarily on the DPVS approach. Our objective is to devise a genuinely effective algorithm for numerically solving optimal control problems and demonstrate its convergence. We devote particular focus to distributed parameter systems, given their extensive practical applications.

B. Sun et al., *Maximum Principle and Dynamic Programming Viscosity Solution Approach*, Systems & Control: Foundations & Applications,
https://doi.org/10.1007/978-981-96-5739-1_5

5.1 A Direction Derivative Algorithm

In this section, we revisit the algorithm presented in Sect. 4.2 for determining numerical solutions for the optimal feedback control-trajectory, referring to it as the direction derivative algorithm [104]. This algorithm has been distilled from our numerical investigations into practical optimal control problems conducted in [102, 103]. Its foundation lies in the HJB equation and the theory of viscosity solutions, making it both constructive and applicable to both finite-dimensional [103] and infinite-dimensional systems [102]. To illustrate its effectiveness, we present an example where both the value function and the feedback law admit closed-form solutions. This enables us to assess the accuracy of the numerical solutions generated by our algorithm. It is worth noting that in this particular example, the value function constitutes a viscosity solution, rather than a classical solution, to its corresponding HJB equation [242].

Now, let us consider the state equation that governs the system dynamics.

$$\begin{cases} y'(t) = u(t), & t \in (0, 1), \\ y(0) = z, \end{cases} \tag{5.1.1}$$

for which the state space is $\mathbb{R}$. The control constraint $U = [-1, 1]$ and hence $\mathcal{U}[0, 1] = L^\infty([0, 1]; U)$. The cost functional is

$$J(u(\cdot)) = -y^2(1). \tag{5.1.2}$$

Now, let us introduce the value function associated with this problem. Given any point $(t, x) \in [0, 1) \times \mathbb{R}$, we consider the following system:

$$\begin{cases} y'_{t,x}(s) = u(s), & s \in [t, 1], \\ y_{t,x}(t) = x. \end{cases} \tag{5.1.3}$$

Accordingly, the cost functional is defined as follows:

$$J_{t,x}(u(\cdot)) = -y_{t,x}^2(1).$$

Define the value function

$$V(t, x) = \inf_{u(\cdot)\in\mathcal{U}[t,1]} J_{t,x}(u(\cdot)). \tag{5.1.4}$$

A simple calculation gives

$$
\begin{aligned}
V(t,x) &= \begin{cases} -[x+(1-t)]^2, & x \geq 0, \\ -[x-(1-t)]^2, & x \leq 0, \end{cases} \quad t \in [0,1], \\
&= -[|x| + (1-t)]^2, \quad x \in \mathbb{R},\ t \in [0,1].
\end{aligned}
\tag{5.1.5}
$$

It is observed that V is continuous in its variables but does not belong to $C^1([0,1] \times \mathbb{R})$. The corresponding HJB equation is

$$
\begin{cases} V_t(t,x) + \inf\limits_{|u|\leq 1} \{V_x(t,x) \cdot u\} = 0, & (t,x) \in [0,1) \times \mathbb{R}, \\ V(1,x) = -x^2, \quad x \in \mathbb{R}. \end{cases}
\tag{5.1.6}
$$

From [242], we know that the value function V is the unique viscosity solution to the HJB equation (5.1.6). Furthermore, the optimal feedback control and the corresponding trajectory pair are analytically given as

$$
u^*_{t,x}(s) = \operatorname{sgn}\left(y^*_{t,x}(s)\right) \tag{5.1.7}
$$

where

$$
\operatorname{sgn}(y) = \begin{cases} 1, & y > 0, \\ \{\pm 1\}, & y = 0, \\ -1, & y < 0, \end{cases}
$$

and

$$
y^*_{t,x}(s) = \begin{cases} x + (s-t), & x \geq 0, \\ x - (s-t), & x \leq 0. \end{cases} \tag{5.1.8}
$$

We apply the algorithm described in Sect. 4.2 systematically, obtaining numerical solutions for the optimal feedback control-trajectory pairs [104]. We initialize the values from − 1 to 1 with a step size of 0.2. The parameters used are $N = M = 100$ and $\eta = 0.02$, whose definitions are detailed in Sect. 4.2. All computations are executed using Visual C++ 6.0, and the numerical results are visualized using MATLAB 6.0.

In Fig. 5.1, the numerical solutions for the optimal trajectories are presented. The solid lines represent the analytical solutions given by (5.1.8). Notably, the analytical and numerical solutions align perfectly. Similarly, Fig. 5.2 illustrates the computed numerical solutions for the optimal feedback controls, where solid lines depict the analytical solutions specified by (5.1.7). Once again, there is a seamless agreement between the analytical and numerical solutions.

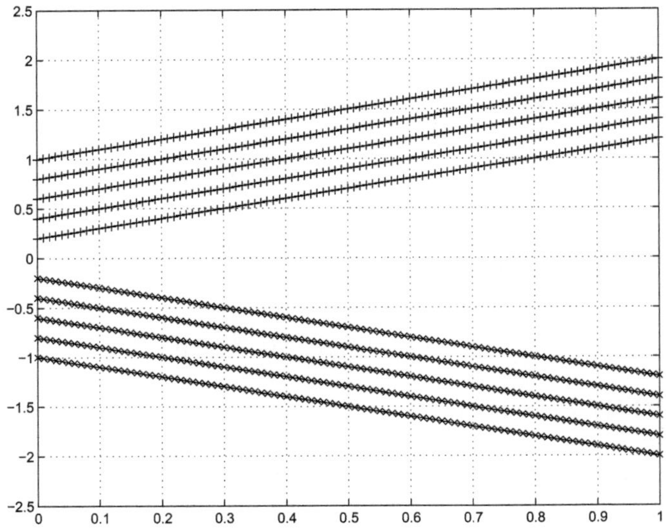

Fig. 5.1 Numerical solutions of the optimal trajectories

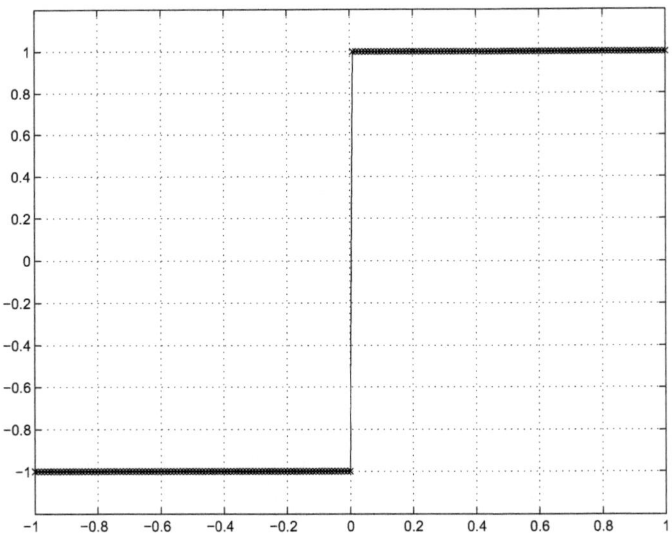

Fig. 5.2 Numerical solutions of the optimal feedback laws

5.2 An Interpolation Algorithm

In this section, we focus on the feedback design for the numerical solution of the optimal control problem within a finite time horizon, leveraging the DPVS approach. To formulate the optimal feedback control, we employ the upwind finite-difference scheme to solve the HJB equation.

Unlike traditional algorithms, our approach involves interpolating the numerical control function sequentially. This interpolation allows us to approximate the optimal feedback control-trajectory pair without the need to repeatedly solve the HJB equation, significantly enhancing computational efficiency, as described in [227].

Furthermore, this design is capable of handling optimal control problems involving multiple controls. Notably, the interpolation methods used here are not confined to one-dimensional linear interpolation; various other techniques, such as multidimensional linear interpolation, cubic spline interpolation, nearest-neighbor interpolation, and cubic-Hermite interpolation, can also be effectively incorporated into this novel design.

5.2.1 *DPVS Approach and Interpolation Algorithm*

Let $T > 0$ and let $U \subset \mathbb{R}^m$ be a closed, bounded, and convex set. We denote by $\mathcal{U}[0, T] = L^\infty([0, T]; U)$ the set of admissible controls. Now, we consider the following control system:

$$\begin{cases} \dfrac{d}{dt}y(t) = f(t, y(t), u(t)), & t \in (0, T], \\ y(0) = y_0, \end{cases} \tag{5.2.1}$$

where $y : [0, T] \to \mathbb{R}^n$ represents the state, the control $u(\cdot) \in \mathcal{U}[0, T]$, and $y_0 \in \mathbb{R}^n$ is a given initial value. Assume that the mapping $f : [0, T] \times \mathbb{R}^n \times U \to \mathbb{R}^n$ is continuous.

Now, consider the cost functional defined as follows:

$$J(u(\cdot)) = \int_0^T L(s, y(s), u(s))\, ds + h(y(T)), \tag{5.2.2}$$

where both the running cost $L : [0, T] \times \mathbb{R}^n \times \mathbb{R}^m \to \mathbb{R}$ and the terminal cost $h : \mathbb{R}^n \to \mathbb{R}$ are continuous. The optimal control problem aims to minimize the cost functional (5.2.2) over all admissible controls $\mathcal{U}[0, T]$, i.e.,

$$\inf_{u \in \mathcal{U}[0,T]} J(u(\cdot)) \tag{5.2.3}$$

subject to the control system (5.2.1).

In order to introduce the DPP, we consider a family of dynamic optimal control problems. Specifically, for each $t \in [0, T]$, we minimize the cost functional over the set of admissible controls $\mathcal{U}[t, T] = L^\infty([t, T]; U)$. The cost functional is defined as

$$J(u(\cdot); t, x) = \int_t^T L(s, y(s), u(s))\, \mathrm{d}s + h(y(T)),$$

subject to

$$\begin{cases} \dfrac{d}{ds} y(s) = f(s, y(s), u(s)), & s \in (t, T], \\ y(t) = x, \end{cases}$$

where $(t, x) \in [0, T) \times \mathbb{R}^n$ is the initial time and state. Define the value function by

$$\begin{aligned} & v(t, x) = \inf_{u(\cdot)\in\mathcal{U}[t,T]} J(u(\cdot); t, x), \quad (t, x) \in [0, T) \times \mathbb{R}^n, \\ & v(T, x) = h(x), \quad x \in \mathbb{R}^n, \end{aligned} \tag{5.2.4}$$

which represents the associated minimal cost functional. This functional satisfies the DPP:

$$v(t, x) = \inf_{u(\cdot)\in\mathcal{U}[t,\tau]} \left\{ \int_t^\tau L(s, y(s; t, x), u(s))\, \mathrm{d}s + v(\tau, y(\tau; t, x)) \right\},$$

for any $\tau \in [t, T)$ [16].

Then, if the value function v possesses sufficient smoothness, specifically $v \in C^1([0, T] \times \mathbb{R}^n)$, the HJB equation can be derived from the DPP as follows:

$$\begin{aligned} & -\frac{\partial}{\partial t} v(t, x) - \inf_{u\in U} \{ f(t, x, u) \cdot \nabla_x v(t, x) + L(t, x, u) \} = 0, \\ & \qquad\qquad (t, x) \in [0, T) \times \mathbb{R}^n, \\ & v(T, x) = h(x), \quad x \in \mathbb{R}^n, \end{aligned} \tag{5.2.5}$$

where ∇_x denotes the gradient with respect to x. By solving the HJB equation (5.2.5), we proceed to the feedback synthesis, which involves constructing the optimal control $u^*(\cdot)$ corresponding to the optimal state $y^*(\cdot)$. For any mapping $\bar{u}(t, x)$ satisfying

$$\bar{u}(t, x) \in \arg\inf_{u\in U} \{ f(t, x, u) \cdot \nabla_x v(t, x) + L(t, x, u) \},$$

we set

$$u^*(t) = \bar{u}\left(t, y^*(t)\right), \tag{5.2.6}$$

for almost all $t \in [0, T]$, where $y^*(t)$ satisfies

$$\frac{d}{dt} y^*(t) = f\left(t, y^*(t), \bar{u}(t, y^*(t))\right), \quad t \in (0, T], \tag{5.2.7}$$

with $y^*(0) = y_0$. Then we have

$$\begin{aligned} & f(t, y^*(t), u^*(t)) \cdot \nabla_x v(t, y^*(t)) + L(t, y^*(t), u^*(t)) \\ &= \inf_{u \in U} \left\{ f\left(t, y^*(t), u\right) \cdot \nabla_x v(t, y^*(t)) + L(t, y^*(t), u) \right\} \end{aligned}$$

for almost all $t \in [0, T]$. Thus, $u^*(\cdot)$ constitutes an optimal control for the problem stated in (5.2.3), as established in [16]. In this context, $u^*(\cdot)$ appearing in (5.2.6) is referred to as the control function or the optimal feedback mapping. When the mapping $\bar{u}$ is known and incorporated into (5.2.7), the resulting system (5.2.7) operates as a closed-loop system.

The DPVS approach indicates that once a viscosity solution of the HJB equation (5.2.5) is obtained numerically, we can determine the numerical solution for the feedback law through the aforementioned feedback synthesis.

Next, we revisit the upwind finite-difference scheme for solving the HJB equation (5.2.5). This scheme is highly suitable and has been effectively employed in numerous examples, as documented in [91, 103, 105]. Most importantly, its convergence has been rigorously established in the subsequent Sect. 5.5.

Take a polygonal region $[0, T] \times \prod_{k=1}^{n} [a_k, b_k]$ as the computational domain. Set the time step and denote $t_j = T + j\Delta t, j = 0, 1, 2, \ldots, N$, which divide the time domain $[0, T]$ into N equal parts. Similarly, for each $k = 1, 2, \cdots, n$, divide $[a_k, b_k]$ $[a_k, b_k]$ into M_k equal parts. Denote

$$\Delta x_k = \frac{b_k - a_k}{M_k}, \quad k = 1, 2, \cdots, n,$$

the space steps. And let $i = (i_1, i_2, \cdots, i_n)$, $ik^+ = (i_1, \ldots, i_{k-1}, i_k + 1, i_{k+1}, \ldots, i_n)$, $ik^- = (i_1, \cdots, i_{k-1}, i_k - 1, i_{k+1}, \cdots, i_n)$, where $i_k = 0, 1, \cdots, M_k$, $k = 1, 2, \ldots, n$. Denote the point $x_i = (a_1 + i_1\Delta x_1, a_2 + i_2\Delta x_2, \ldots, a_n + i_n\Delta x_n)$. Then the upwind finite-difference scheme of the HJB equation (5.2.5) can be written as

$$\begin{aligned} & \frac{v_i^{j+1} - v_i^j}{\Delta t} + \sum_{k=1}^{n} \left(\frac{1+\operatorname{sign} f_{k,i}^j}{2} \frac{v_{ik^+}^j - v_i^j}{\Delta x_k} + \frac{1-\operatorname{sign} f_{k,i}^j}{2} \frac{v_i^j - v_{ik^-}^j}{\Delta x_k} \right) f_{k,i}^j + L_i^j = 0, \\ & u_i^{j+1} = \arg\inf_{u \in U} \left(\sum_{k=1}^{n} f_k(t_{j+1}, x_i, u) \frac{v_i^{j+1} - v_{ik^-}^{j+1}}{\Delta x_k} + L(t_{j+1}, x_i, u) \right), \end{aligned} \tag{5.2.8}$$

for $i_k = j+1, j+2, \ldots, M_k - j - 1, k = 1, 2, \ldots, n$ and $j = 0, 1, \ldots, N-1$. Here, the sign z denotes the sign of z. And u_i^j, v_i^j are the approximations of $\bar{u}(t_j, x_i)$ and $v(t_j, x_i)$, respectively. Then, denote

$$f_{k,i}^j = f_k\left(t_j, x_i, u_i^j\right), \quad L_i^j = L\left(t_j, x_i, u_i^j\right).$$

In what follows, we formulate a feedback algorithm rooted in the HJB equation for (5.2.3). The specific steps are outlined below.

Algorithm 7 A feedback algorithm rooted in the HJB equation

1: Initial partition on time and space. Divide the computational domain into a mesh as described in Sect. 5.2.1 above. Denote $\left(u^j, y^j\right)$ as the optimal control-trajectory pair at grid points $t^j = -j\Delta t$ for $j = 0, 1, \ldots, N$.

2: Compute the approximations of value function and control function. Based on the numerical scheme (5.2.8), set $\gamma_k = \frac{\Delta t}{\Delta x_k}$ for $k = 1, 2, \ldots, n$.

1). Using the same spatial step sizes Δx_k for $k = 1, 2, \ldots, n$, expand the initial solution domain $\prod_{k=1}^{n} [a_k, b_k]$ to a new larger domain:

$$\prod_{k=1}^{n} [a_k - (N+1)\Delta x_k, b_k + (N+1)\Delta x_k]$$

by adding $2(N+1)$ nodes. (This is a coarse expansion, and more precise extension steps can be found in [233]). After adding these nodes, renumber the points on the new expanded region:

$$\prod_{k=1}^{n} [a_k - (N+1)\Delta x_k, b_k + (N+1)\Delta x_k].$$

Denote $ii = (ii_1, ii_2, \ldots, ii_n)$, $iik^+ = (ii_1, \ldots, ii_{k-1}, ii_k + 1, ii_{k+1}, \ldots, ii_n)$, $iik^- = (i_1, \ldots, ii_{k-1}, ii_k - 1, ii_{k+1}, \ldots, ii_n)$, where $ii_k = 0, 1, \ldots, M_k + 2(N+1)$ and $k = 1, 2, \ldots, n$. Define $x_{ii} = (a_1 + (ii_1 - N - 1)\Delta x_1,\ a_2 + (ii_2 - N - 1)\Delta x_2, \ldots,\ a_n + (ii_n - N - 1)\Delta x_n)$.

2). Initialize $\left\{v_{ii}^0\right\}_{ii_k=0}^{M_k+2(N+1)}$ and $\left\{u_{ii}^0\right\}_{ii_k=0}^{M_k+2(N+1)}$ as follows:

$$v_{ii}^0 = h(x_{ii}),$$
$$u_{ii}^0 = \arg\inf_{u \in U}\left(\sum_{k=1}^{n} f_k(t_0, x_{ii}, u)\frac{v_{ii}^0 - v_{iik^-}^0}{\Delta x_k} + L(t_0, x_{ii}, u)\right).$$

3). For $j = 0$ to $N-1$, compute v_{ii}^{j+1} and u_{ii}^{j+1} as follows. For $ii_k = j+1$ to $M_k + 2N - j + 1$, $k = 1, 2, \ldots, n$, set

$$v_{ii}^{j+1} = \left(1 + \sum_{k=1}^{n} \gamma_k \left|f_{k,ii}^{j}\right|\right) v_{ii}^{j} - \sum_{k=1}^{n} \frac{1 + \text{sign} f_{k,ii}^{j}}{2} \gamma_k f_{k,ii}^{j} v_{iik^+}^{j}$$

$$+ \sum_{k=1}^{n} \frac{1 - \text{sign} f_{k,ii}^{j}}{2} \gamma_k f_{k,ii}^{j} v_{iik^-}^{j} - \Delta t L_{ii}^{j},$$

$$u_{ii}^{j+1} = \arg\inf_{u \in U} \left(\sum_{k=1}^{n} f_k(t_{j+1}, x_{ii}, u) \frac{v_{ii}^{j+1} - v_{iik^-}^{j+1}}{\Delta x_k} + L(t_{j+1}, x_{ii}, u) \right).$$

4). Select the numerical solutions for the value function and the control function over the domain $[0, T] \times \prod_{k=1}^{n} [a_k, b_k]$. Denote these solutions by

$$\left\{ \left\{ v_i^j \right\}_{i_k=0}^{M_k} \right\}_{j=0}^{N} \quad \text{and} \quad \left\{ \left\{ u_i^j \right\}_{i_k=0}^{M_k} \right\}_{j=0}^{N},$$

respectively, for $k = 1, 2, \ldots, n$. In this notation, v_i^j represents the approximation of the value function $v(t_j, x_i)$ and u_i^j represents the approximation of the control function $\bar{u}(t_j, x_i)$.

3: Compute the optimal pairs $\left\{(u^j, y^j)\right\}_{j=0}^{N}$. For initial setting, let $y^0 = y_0$. Set $j = 0$.

1). Compute the jth optimal feedback control u^j by utilizing the numerical solution $\left\{u_i^{N-j}\right\}_{i_k=0}^{M_k}$ obtained in Step 2, for $k = 1, 2, \ldots, n$. To determine the control value at point y^j, select an interpolation method such as linear interpolation (or alternatively, cubic spline interpolation, nearest-neighbor interpolation, or cubic-Hermite interpolation). Employ the known control values y^j to perform the interpolation. The interpolated value, denoted as $\tilde{u}(t^j, y^j)$, serves as an approximation of the true optimal feedback control $\left\{u_i^{N-j}\right\}_{i_k=0}^{M_k}$. Subsequently, set

$$u^j = \tilde{u}(t^j, y^j).$$

2). Compute the $(j+1)$-th optimal trajectory y^{j+1}. Discretize the state equation using the forward difference method:

$$y^{j+1} = y^j - f(t^j, y^j, u^j)\Delta t.$$

Using the known values of u^j and y^j, we solve the aforementioned finite-difference equation to derive the $(j+1)$-th optimal trajectory point y^{j+1}.

3). Iterate to the next time instant by incrementing j as $j = j + 1$. If j exceeds N, terminate the procedure. Otherwise, proceed to the first step in Step 3. Continue iterating until all optimal feedback control-trajectory pairs $\{(u^j, y^j)\}_{j=0}^N$ have been obtained.

Remark 5.2.1 In the first step of Step 3, various interpolation methods can be employed, including polynomial interpolation, average interpolation, Newton interpolation, Lagrange interpolation, and others as referenced in [203]. Likewise, in the second step of Step 3, alternative difference methods can be utilized to discretize the state equation.

5.2.2 *Simulations*

Five numerical examples are presented to demonstrate the efficacy of the proposed algorithm in this subsection. The simulations are executed using MATLAB on a desktop computer equipped with a Core i5-6600M processor and 8GB of RAM.

Example 5.2.2 Consider the following control problem [249]:

$$\begin{cases} \dot{y}(t) = y(t)u(t), \quad t \in (0, 1], \\ y(0) = y_0, \\ J(u(\cdot)) = -y(1), \\ \min\limits_{u(\cdot)\in\mathcal{U}} J(u(\cdot)), \end{cases}$$

where $\mathcal{U} = L^\infty([0, 1]; U)$, $U = [0, 1]$, $y_0 \in \mathbb{R}$ is a given initial value. Its corresponding HJB equation is

$$-\frac{\partial}{\partial t}v(t, x) - \inf_{u\in[0,1]} \{xu \cdot \nabla_x v(t, x)\} = 0, \ (t, x) \in [0, 1) \times \mathbb{R},$$
$$v(1, x) = -x, \ x \in \mathbb{R}.$$

Its exact value function v is

$$v(t, x) = \begin{cases} -xe^{1-t}, \ x > 0, \\ -x, \quad\quad x \le 0, \end{cases}$$

and the control function is

$$\bar{u}(t, x) = \begin{cases} 1, \ x > 0, \\ 0, \ x \le 0. \end{cases} \tag{5.2.9}$$

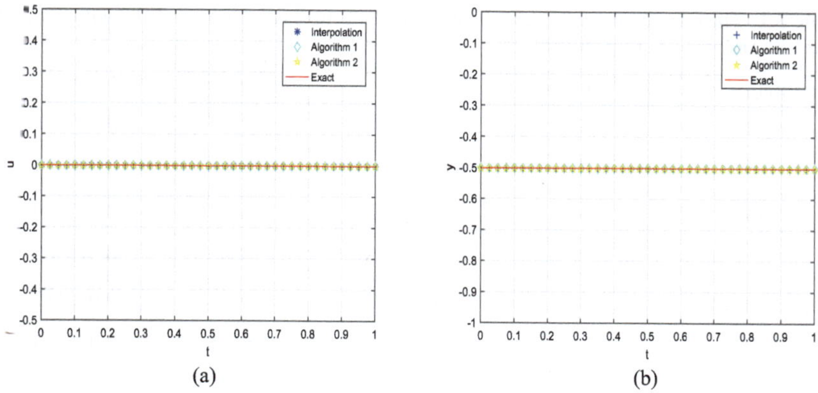

Fig. 5.3 Numerical and exact solutions in Example 5.2.2 with initial state $y_0 = -\frac{1}{2}$. (**a**) Optimal feedback control. (**b**) Optimal trajectory

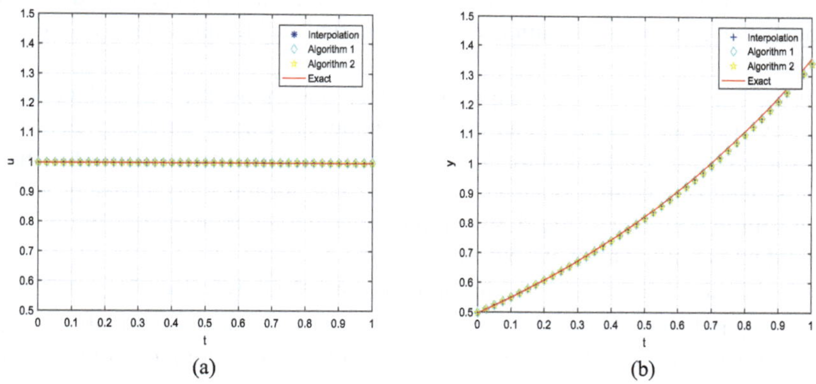

Fig. 5.4 Numerical and exact solutions in Example 5.2.2 with initial state $y_0 = \frac{1}{2}$. (**a**) Optimal feedback control. (**b**) Optimal trajectory

Here, we consider the region $D = [0, 1]\times\left[-\frac{3}{2}, \frac{3}{2}\right]$, specifically, where $0 \leq t \leq 1$ and $-\frac{3}{2} \leq x \leq \frac{3}{2}$. The t-axis is discretized with $N = 41$ nodes, while the x-axis is discretized with $M = 31$ nodes. The time step is set as $\Delta t = -\frac{1}{40}$ and the space step is $\Delta x = \frac{1}{10}$. Utilizing linear interpolation, we numerically solve this example via the interpolation algorithm outlined in Sect. 5.2.1. The results are subsequently visualized in Figs. 5.3 and 5.4.

Furthermore, to demonstrate the efficacy of the presented interpolation algorithm, two existing algorithms are also employed to simultaneously solve Example 5.2.2. The respective results are showcased in Figs. 5.3 and 5.4. Additionally, a comparative analysis of the numerical simulations is provided in Table 5.1.

Specifically, one of the selected algorithms involves the minimization of the Hamiltonian, as detailed in Algorithm 3.2 of [91]. For simplicity, we refer to this

Table 5.1 The maximum norm error estimates and CPU times for Example 5.2.2 with initial state $y_0 = -\frac{1}{2}$ and $y_0 = \frac{1}{2}$, respectively

Algorithm	Initial state	Control	Trajectory	CPU(s)
Interpolation	$y_0 = -\frac{1}{2}$	6.6000e-05	3.3001e-05	0.005615
	$y_0 = \frac{1}{2}$	6.6000e-05	0.016695	0.005484
Algorithm A1	$y_0 = -\frac{1}{2}$	6.6000e-05	3.3001e-05	0.015370
	$y_0 = \frac{1}{2}$	6.6000e-05	0.016695	0.016507
Algorithm A2	$y_0 = -\frac{1}{2}$	6.6000e-05	3.3001e-05	0.025182
	$y_0 = \frac{1}{2}$	6.6000e-05	0.016695	0.025733

as Algorithm A1. The other chosen algorithm, designated as Algorithm A2, also obtains the optimal feedback control by minimizing the Hamiltonian function. However, it differs from Algorithm A1 in that the value function is determined through the convex combination of the numerical value function, as outlined in [140].

When $y_0 = -\frac{1}{2}$, it is straightforward to determine the optimal control as $u^*(t) \equiv 0$ for all $t \in [0, 1]$ and the corresponding optimal trajectory as $y^*(t) \equiv -\frac{1}{2}$ for all $t \in [0, 1]$. Figure 5.3a illustrates that the numerical solutions for optimal feedback control obtained via the three algorithms align closely with the exact solution. Similarly, Fig. 5.3b demonstrates that the numerical solution for the optimal trajectory matches its exact counterpart.

For the case where $y_0 = \frac{1}{2}$, the optimal control is given by $u^*(t) \equiv 1$ for all $t \in [0, 1]$, and the optimal trajectory is described by $y^*(t) = \frac{1}{2}e^t$ for all $t \in [0, 1]$. Figure 5.4 verifies that both the numerical solutions for optimal feedback control and trajectory are in good agreement with their respective exact solutions when the initial state is set to $y_0 = \frac{1}{2}$.

For these three algorithms, Table 5.1 provides a comparison of the computed errors in the maximum norm between the numerical solutions and the exact solutions for both the optimal feedback control u^* and the optimal trajectory y^* with initial state $y_0 = -\frac{1}{2}$ and $\frac{1}{2}$. The small error estimates indicate that the numerical solutions obtained via the interpolation algorithm, as well as Algorithms A1 and A2, closely approximate the exact solutions.

Furthermore, Table 5.1 also includes a comparison of the CPU times required by each algorithm. These times represent the computational duration after obtaining the numerical control function and numerical value function using the upwind finite-difference method. Upon comparing the CPU times, it becomes evident that the interpolation algorithm requires significantly less time than Algorithms A1 and A2, despite achieving similar error estimations. This finding underscores the effectiveness of linear interpolation in approximating the optimal feedback control, even when dealing with a discontinuous control function such as (5.2.9). Overall, the numerical results demonstrate the feasibility and efficiency of the interpolation algorithm in determining the optimal feedback control.

Example 5.2.3 Consider the following control problem [243]:

$$\begin{cases} \dot{y}(t) = y(t) + u(t), \quad t \in (0, 1], \\ y(0) = 1, \\ J(u(\cdot)) = \frac{1}{2}\int_0^1 u^2(t)\,\mathrm{d}t + \frac{1}{2}y^2(1), \\ \min\limits_{u(\cdot)\in\mathcal{U}} J(u(\cdot)), \end{cases}$$

where $\mathcal{U} = L^\infty([0, 1]; U)$, $U = \mathbb{R}$. Its corresponding HJB equation can be read as

$$-\frac{\partial}{\partial t}v(t, x) - \inf_{u\in\mathbb{R}}\left\{(x + u)\cdot\nabla_x v(t, x) + \frac{1}{2}u^2\right\} = 0,\ (t, x) \in [0, 1) \times \mathbb{R},$$
$$v(1, x) = \frac{1}{2}x^2,\ x \in \mathbb{R}. \tag{5.2.10}$$

The exact solution to the HJB equation (5.2.10) is

$$v(t, x) = \frac{x^2}{1 + e^{2(t-1)}},$$

and the control function is

$$\bar{u}(t, x) = \frac{-2x}{1 + e^{2(t-1)}}. \tag{5.2.11}$$

With (5.2.6), (5.2.7), and (5.2.11), we respectively obtain the optimal feedback control and the optimal trajectory as

$$u^*(t) = \frac{-2e^{-t}}{1 + e^{-2}}, \quad y^*(t) = \frac{e^{-t} + e^{t-2}}{1 + e^{-2}}.$$

Now we proceed with the numerical solution of Example 5.2.3. We select the computational domain as $[0, 1] \times [0, 1]$, where $0 \le t \le 1$ represents the time axis and $0 \le x \le 1$ represents the space axis. For the discretization, we choose a uniform mesh along the t-axis with $N = 41$ nodes, giving a time step of $\Delta t = -\frac{1}{40}$, and a uniform mesh along the x-axis with $M = 21$ nodes, resulting in a space step of $\Delta x = \frac{1}{20}$. In this example, we employ linear interpolation.

The obtained numerical solutions are presented in Fig. 5.5. This figure demonstrates that both the numerical solutions for the optimal feedback control and the trajectory align closely with their respective exact solutions.

Furthermore, in Table 5.2, we provide the maximum errors between the numerical solutions and the exact solutions for both the optimal feedback control and the

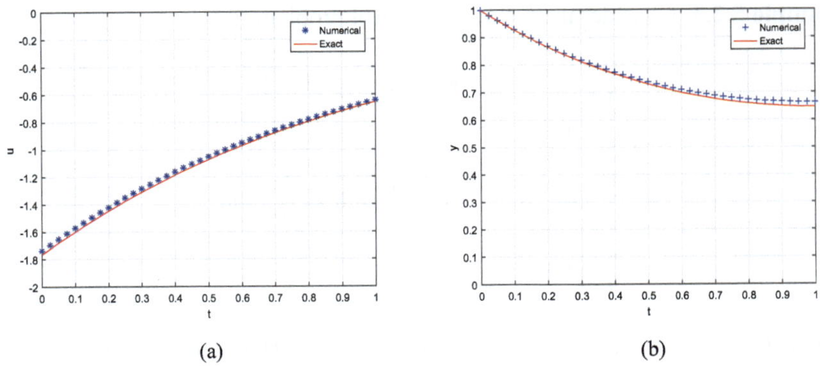

Fig. 5.5 Numerical and exact solutions in Example 5.2.3 with initial state $y_0 = 1$. (**a**) Optimal feedback control. (**b**) Optimal trajectory

Table 5.2 The maximum norm error estimates for Example 5.2.3 with initial state $y_0 = 1$

Error	Control	Trajectory
$y_0 = 1$	0.025896	0.017297

trajectory. These errors are 0.025896 and 0.017297, respectively, indicating a high degree of accuracy. This further confirms the effectiveness of the algorithm.

In addition, in Examples 5.2.2 and 5.2.3, the control problems are all one-dimensional problems. Next, we present other two optimal control examples with two space dimensions.

Example 5.2.4 Consider the following two-dimensional control problem [233] with two state variables and one control.

$$\begin{cases} \begin{pmatrix} \dot{x}(t) \\ \dot{y}(t) \end{pmatrix} = \begin{pmatrix} x(t) \\ y(t) \end{pmatrix} u(t), \quad t \in (0, 1], \\ x(0) = x_0, \quad y(0) = y_0, \\ J(u(\cdot)) = -x(1) - y(1), \\ \min\limits_{u(\cdot)\in\mathcal{U}} J(u(\cdot)), \end{cases}$$

where $\mathcal{U} = L^\infty([0, 1]; U)$, $U = [0, 1]$, $(x_0, y_0) \in \mathbb{R}^2$. Its corresponding HJB equation is

$$\begin{aligned} -\frac{\partial}{\partial t} v(t, x, y) - \inf_{u\in[0,1]} \left\{ xu \cdot v_x(t, x, y) + yu \cdot v_y(t, x, y) \right\} = 0, \\ (t, x, y) \in [0, 1) \times \mathbb{R}^2, \\ v(1, x, y) = -(x + y), \quad (x, y) \in \mathbb{R}^2. \end{aligned} \tag{5.2.12}$$

In this example, the exact solution of the value function is

$$v(t,x,y)=\begin{cases}-(x+y)e^{1-t}, & x+y>0,\\ -(x+y), & x+y\le 0,\end{cases}$$

and the control function is

$$\bar{u}(t,x,y)=\begin{cases}1, & x+y>0,\\ 0, & x+y\le 0.\end{cases} \tag{5.2.13}$$

From (5.2.6), (5.2.7), and (5.2.13), when initial states $(x_0, y_0)=\left(-\frac{1}{2},-\frac{1}{2}\right)$, the optimal feedback control is $u^*(t)\equiv 0,\ t\in[0,1]$ and the optimal trajectory is $(x^*(t), y^*(t))\equiv\left(-\frac{1}{2},-\frac{1}{2}\right),\ t\in[0,1]$. When the initial states $(x_0,y_0)=\left(\frac{1}{2},\frac{1}{2}\right)$, the optimal feedback control is given by $u^*(t)\equiv 1$ and the optimal trajectory is given by

$$(x^*(t), y^*(t))=\left(\frac{1}{2}e^t,\frac{1}{2}e^t\right),\quad t\in[0,1].$$

Next, we use the interpolation algorithm to solve the HJB equation (5.2.12) in the range $[0,1]\times\left[-\frac{3}{2},\frac{3}{2}\right]^2$, i.e., $0\le t\le 1,\ -\frac{3}{2}\le x\le\frac{3}{2},\ -\frac{3}{2}\le y\le\frac{3}{2}$. Here we still choose a uniform mesh on the t-axis, x-axis, and y-axis. The nodes N on the t-axis is set as 41, i.e., $\Delta t=-\frac{1}{40}$. The nodes on x- and y-axis are set as $M_1=M_2=31$, i.e., $\Delta x=\Delta y=\frac{1}{10}$. Again, the linear interpolation is chosen to be applied in the algorithm.

From Figs. 5.6 and 5.7, it is evident that, despite varying initial values, the numerical solutions exhibit strong agreement with the exact solutions for both the optimal feedback control and the trajectory.

Example 5.2.4 demonstrates that linear interpolation remains a valuable tool for approximating the optimal feedback control, even in the context of two-dimensional control problems. Furthermore, the results indicate that the discontinuity present in the control function (5.2.13) does not pose a significant obstacle in determining the optimal feedback control. This finding underscores the versatility and robustness of the interpolation-based approach in handling complex control scenarios (Table 5.3).

Furthermore, all three examples discussed above focus on systems with a single control variable and solely utilize linear interpolation within the algorithm. To exhibit the efficacy and versatility of the interpolation-based algorithm even more convincingly, we present a comprehensive analysis in Example 5.2.5.

In this example, we consider an optimal control problem that involves two state variables and two control variables. This added complexity provides a more rigorous testbed for assessing the performance of the interpolation techniques. To ensure a comprehensive evaluation, we employ four different interpolation methods: linear

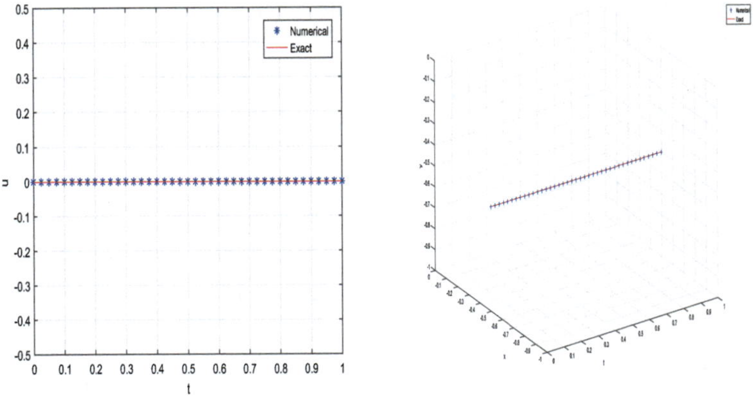

(a) (b)

Fig. 5.6 Numerical and exact solutions in Example 5.2.4 with initial states $(x_0, y_0) = \left(-\frac{1}{2}, -\frac{1}{2}\right)$. (**a**) Optimal feedback control. (**b**) Optimal trajectory

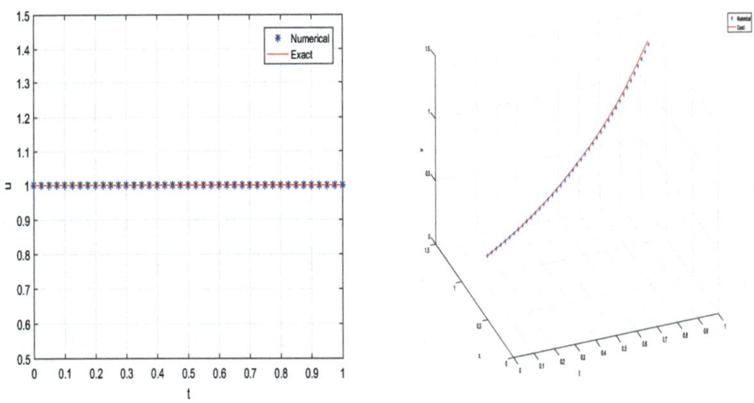

(a) (b)

Fig. 5.7 Numerical and exact solutions in Example 5.2.4 with initial states $(x_0, y_0) = \left(\frac{1}{2}, \frac{1}{2}\right)$. (**a**) Optimal feedback control. (**b**) Optimal trajectory

Table 5.3 The maximum norm error estimates for Example 5.2.4 with initial states $(x_0, y_0) = \left(-\frac{1}{2}, -\frac{1}{2}\right)$ and $(x_0, y_0) = \left(\frac{1}{2}, \frac{1}{2}\right)$, respectively

Errors	Control	Trajectory
$(x_0, y_0) = \left(-\frac{1}{2}, -\frac{1}{2}\right)$	6.6000e-05	4.6671e-05
$(x_0, y_0) = \left(\frac{1}{2}, \frac{1}{2}\right)$	6.6000e-05	0.023611

interpolation, cubic spline interpolation, nearest-neighbor interpolation, and cubic-Hermite interpolation.

By incorporating a broader range of interpolation techniques, we aim to showcase the flexibility and adaptability of our approach in handling complex optimal control problems. The results obtained through this investigation not only reinforce the effectiveness of the algorithm but also highlight its potential for use in a wide array of practical applications where multiple states and controls are involved.

Example 5.2.5 Consider the following optimal control problem with two state variables and two controls.

$$\begin{cases} \begin{pmatrix} \dot{x}_1(t) \\ \dot{x}_2(t) \end{pmatrix} = \begin{pmatrix} x_1(t) + u_1(t) \\ u_2(t) \end{pmatrix}, \quad t \in (0, 1], \\ x_1(0) = x_{10}, \quad x_2(0) = x_{20}, \\ J(u_1(\cdot), u_2(\cdot)) = \frac{1}{2} \int_0^1 \left[u_1^2(t) + u_2^2(t) + x_2^2(t) \right] \mathrm{d}t + \frac{1}{2} x_1^2(1), \\ \min\limits_{(u_1(\cdot), u_2(\cdot)) \in \mathcal{U}} J(u_1(\cdot), u_2(\cdot)), \end{cases} \tag{5.2.14}$$

where $(x_{10}, x_{20}) \in \mathbb{R}^2, \mathcal{U} = L^\infty([0, 1]; U), U = \{(p, q) \in \mathbb{R}^2 \mid |p| \leq 1, |q| \leq 1, |p + q^2| \leq 1\}$. Denote by $|\cdot|$ the absolute value. The HJB equation of (5.2.14) can be written as

$$\begin{aligned} -\frac{\partial}{\partial t} v(t, x_1, x_2) - \inf_{|u_1|, |u_2|, |u_1 + u_2^2| \leq 1} & \Big\{ (x_1 + u_1) \cdot v_{x_1}(t, x_1, x_2) \\ + u_2 \cdot v_{x_2}(t, x_1, x_2) + \frac{1}{2} u_1^2 + \frac{1}{2} u_2^2 + \frac{1}{2} x_2^2 \Big\} &= 0, \quad (t, x_1, x_2) \in [0, 1) \times \mathbb{R}^2, \\ v(1, x_1, x_2) = \frac{1}{2} x_1^2, & \quad (x_1, x_2) \in \mathbb{R}^2. \end{aligned} \tag{5.2.15}$$

Here we choose the computational region $[0, 1] \times [-1, 1]^2$, i.e., $0 \leq t \leq 1$, $-1 \leq x_1 \leq 1$, and $-1 \leq x_2 \leq 1$. The number of nodes N on the t-axis is chosen as 41. And the number of nodes on the x_1- and x_2-axis are set as $M_1 = M_2 = 41$. Then the time step $\Delta t = -\frac{1}{40}$ and the space steps $\Delta x_1 = \Delta x_2 = \frac{1}{20}$. To visualize the numerical results, we have plotted the value function v, two control functions $\bar{u}_1$ and $\bar{u}_2$, and the combined function $\bar{u}_1 + \bar{u}_2^2$ at various cross sections in Figs. 5.8, 5.9, and 5.10.

Examining Figs. 5.8b–d, 5.9b–d, and 5.10b–d, it becomes evident that both the numerical controls $\bar{u}_1$ and $\bar{u}_2$ and $\bar{u}_1 + \bar{u}_2^2$ satisfy the given constraints.

Moreover, in this example, the exact value function v and control functions $(\bar{u}_1, \bar{u}_2)$ for the HJB equation (5.2.15) are not known in the literature. Despite the presence of numerous constraints on the control variables, including nonlinear ones,

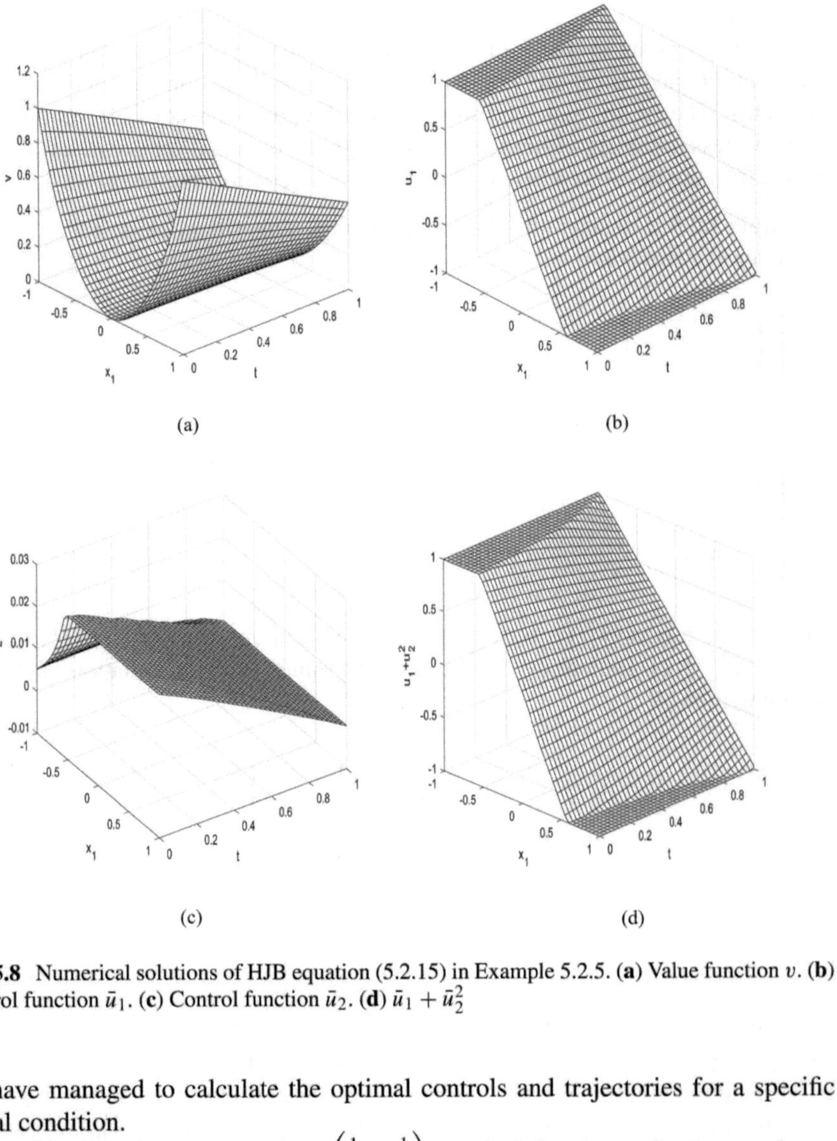

Fig. 5.8 Numerical solutions of HJB equation (5.2.15) in Example 5.2.5. (**a**) Value function v. (**b**) Control function $\bar{u}_1$. (**c**) Control function $\bar{u}_2$. (**d**) $\bar{u}_1 + \bar{u}_2^2$

we have managed to calculate the optimal controls and trajectories for a specific initial condition.

Specifically, when $(x_{10}, x_{20}) = \left(\frac{1}{2}, -\frac{1}{2}\right)$, we find that the optimal controls are given by

$$u_1^*(t) = \frac{-e^{-t}}{1+e^{-2}}, \quad u_2^*(t) = \frac{e^{-t} - e^{t-2}}{2(1+e^{-2})}$$

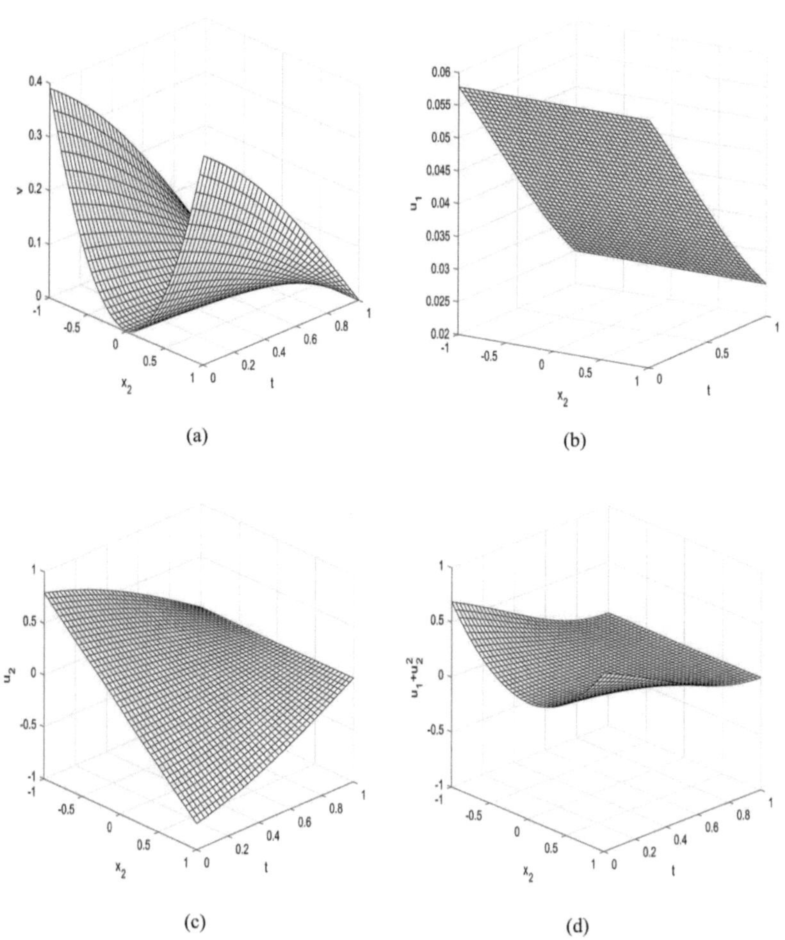

Fig. 5.9 Numerical solutions of HJB equation (5.2.15) in Example 5.2.5 at $x_1 = 0$. (**a**) Value function v. (**b**) Control function $\bar{u}_1$. (**c**) Control function $\bar{u}_2$. (**d**) $\bar{u}_1 + \bar{u}_2^2$

for all $t \in [0, 1]$. Similarly, the optimal trajectories are

$$x_1^*(t) = \frac{e^{-t} + e^{t-2}}{2(1+e^{-2})}, \quad x_2^*(t) = -\frac{e^{-t} + e^{t-2}}{2(1+e^{-2})}$$

for all $t \in [0, 1]$.

To demonstrate the versatility of our approach, we have employed both linear interpolation and cubic spline interpolation to approximate the optimal feedback control. The respective numerical results are presented in Figs. 5.11 and 5.12.

Examining these figures, it is evident that both interpolation methods yield numerical solutions that are in good agreement with the exact solutions.

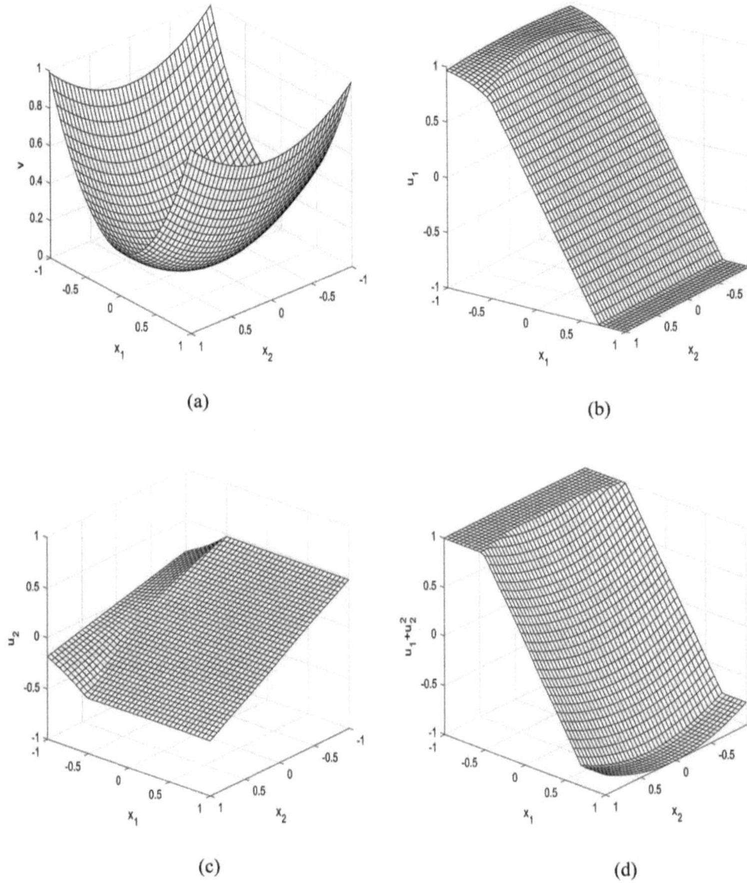

Fig. 5.10 Numerical solutions of the HJB equation (5.2.15) in Example 5.2.5 at $t = \frac{1}{2}$. (**a**) Value function v. (**b**) Control function $\bar{u}_1$. (**c**) Control function $\bar{u}_2$. (**d**) $\bar{u}_1 + \bar{u}_2^2$

To investigate the impact of various interpolation methods on the interpolation algorithm, we provide error results for four interpolation cases in this example. Table 5.4 outlines the computational errors, measured in maximum norms, between the numerical solutions, and the exact solutions. These errors encompass both the optimal control (u_1^*, u_2^*) and the optimal trajectory (x_1^*, x_2^*) with initial state $x_{10} = \frac{1}{2}$ and $x_{20} = -\frac{1}{2}$.

In the case of linear interpolation, the computational errors are 0.038706 and 0.02681 for the optimal control and optimal trajectory, respectively. As for cubic spline interpolation, the errors are 0.038706 and 0.026801. In both nearest-neighbor and cubic-Hermite interpolation cases, we also observe very small errors. Although the errors in the nearest-neighbor case are slightly larger than the others, all methods exhibit a good approximation to the exact solutions.

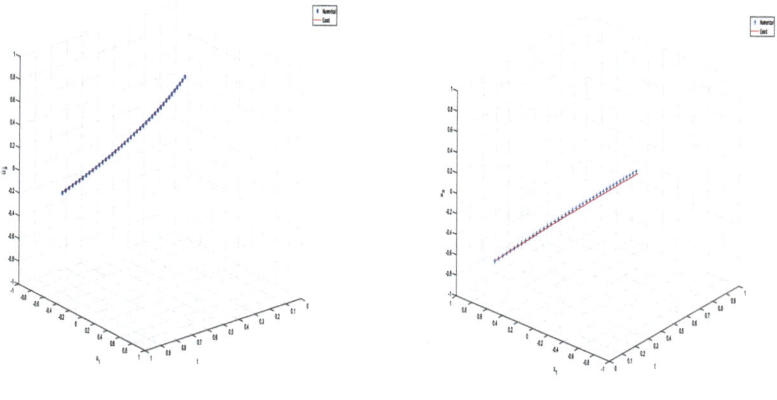

(a) (b)

Fig. 5.11 Numerical and exact solutions in Example 5.2.5 with initial states $(x_{10}, x_{20}) = \left(\frac{1}{2}, -\frac{1}{2}\right)$ (linear interpolation case). (**a**) Optimal feedback control. (**b**) Optimal trajectory

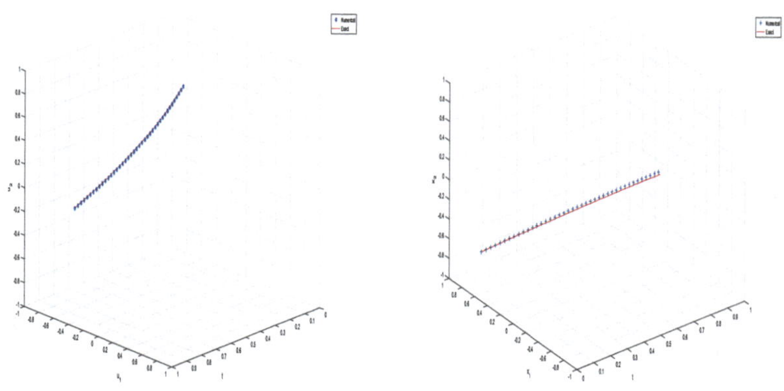

(a) (b)

Fig. 5.12 Numerical and exact solutions in Example 5.2.5 with initial states $(x_{10}, x_{20}) = \left(\frac{1}{2}, -\frac{1}{2}\right)$ (cubic spline interpolation case). (**a**) Optimal feedback control. (**b**) Optimal trajectory

Table 5.4 The maximum norm error estimates for Example 5.2.5 with initial states $(x_{10}, x_{20}) = \left(\frac{1}{2}, -\frac{1}{2}\right)$ in four interpolation cases

Error	Control	Trajectory
Linear	0.038706	0.02681
Cubic spline	0.038706	0.026801
Nearest-neighbor	0.069388	0.030588
Cubic-Hermite	0.038706	0.026804

Comparatively, referring to Table 5.4, we find that nearest-neighbor interpolation performs the least well among four interpolation methods. The other three interpolations show almost equivalent performance. These small error estimates confirm that the numerical solutions obtained via the interpolation algorithm are very close to the exact solutions.

The experimental results of Example 5.2.5 once again demonstrate the exceptional effectiveness of the interpolation algorithm.

Let us emphasize that in the four examples discussed above, all optimal controls are continuous, and we do not consider cases where optimal control exhibits jumps. Moving forward, we will employ the interpolation algorithm to tackle a control problem where the optimal control has jumps. Our aim is to demonstrate that the interpolation algorithm is equally viable for addressing such problems.

Example 5.2.6 Consider the following optimal control problem with two state variables and one control.

$$\begin{cases} \begin{pmatrix} \dot{x}_1(t) \\ \dot{x}_2(t) \end{pmatrix} = \begin{pmatrix} x_2(t) \\ -x_1(t) + \dfrac{1}{4}u(t) \end{pmatrix}, \quad t \in (0, 2\pi], \\ x_1(0) = 0, \quad x_2(0) = 0, \\ J(u(\cdot)) = x_2(2\pi), \\ \min\limits_{(u(\cdot))\in\mathcal{U}} J(u(\cdot)), \end{cases} \tag{5.2.16}$$

where $\mathcal{U} = L^\infty([0, 2\pi]; U)$, $U = [-1, 1]$. The HJB equation of (5.2.16) can be written as

$$-\frac{\partial}{\partial t}v(t, x_1, x_2) - \inf_{|u|\leq 1}\left\{x_2 \cdot v_{x_1}(t, x_1, x_2) + \left(-x_1 + \frac{1}{4}u\right) \cdot v_{x_2}(t, x_1, x_2)\right\} = 0,$$
$$(t, x_1, x_2) \in [0, 2\pi) \times \mathbb{R}^2,$$
$$v(2\pi, x_1, x_2) = x_2, \quad x_2 \in \mathbb{R}.$$

Here we choose the computational region $(t, x_1, x_2) \in [0, 2\pi] \times \left[-\frac{3}{2}, \frac{3}{2}\right]^2$. The number of nodes N on t-axis is chosen as 41. And the number of nodes on thex_1- and x_2-axis are set as $M_1 = M_2 = 11$. Then the time step $\Delta t = -\frac{\pi}{20}$ and the space steps $\Delta x_1 = \Delta x_2 = \frac{3}{10}$. In this case, for the given initial state $(x_{10}, x_{20}) = (0, 0)$, the unique optimal control $u^*(\cdot)$ and the corresponding optimal trajectory $(x_1^*(\cdot), x_2^*(\cdot))$

of system (5.2.16) can be obtained analytically by

$$u^*(t) = \begin{cases} -1, & 0 \le t < \dfrac{\pi}{2}, \\ 1, & \dfrac{\pi}{2} \le t < \dfrac{3\pi}{2}, \\ -1, & \dfrac{3\pi}{2} \le t \le 2\pi, \end{cases} \tag{5.2.17}$$

and

$$\begin{pmatrix} x_1^*(t) \\ x_2^*(t) \end{pmatrix} = \begin{cases} \dfrac{1}{4}\begin{pmatrix} \cos t - 1 \\ -\sin t \end{pmatrix}, & 0 \le t < \dfrac{\pi}{2}, \\ \dfrac{1}{4}\begin{pmatrix} \cos t - 2\sin t + 1 \\ -2\cos t - \sin t \end{pmatrix}, & \dfrac{\pi}{2} \le t < \dfrac{3\pi}{2}, \\ \dfrac{1}{4}\begin{pmatrix} \cos t - 4\sin t - 1 \\ -4\cos t - \sin t \end{pmatrix}, & \dfrac{3\pi}{2} \le t \le 2\pi. \end{cases}$$

From (5.2.17), it is evident that the optimal control is discontinuous, featuring two jumps within the interval $[0, 2\pi]$. For this particular example, we implement the computation using the aforementioned interpolation algorithm. Specifically, we opt for linear interpolation coupled with the Heun method in the second step in Step 3.

The simulation results, depicted in Fig. 5.13, reveal a strong correlation between the numerical solutions and the exact solutions for both optimal control and

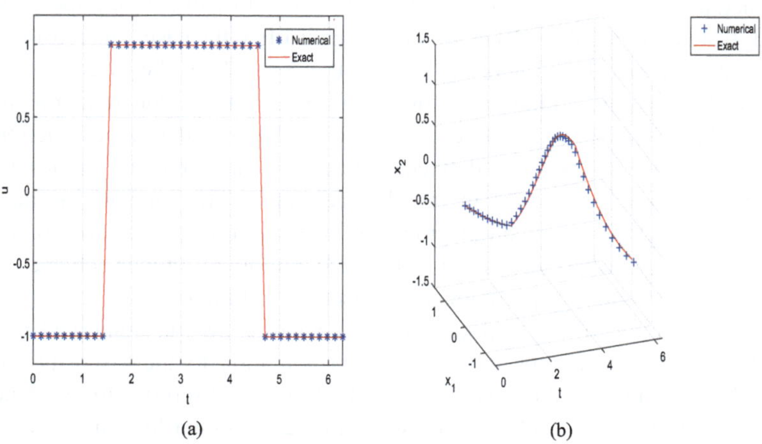

Fig. 5.13 Numerical and exact solutions in Example 5.2.6 (linear interpolation case). (**a**) Optimal feedback control. (**b**) Optimal trajectory

trajectory. The computational errors, measured in maximum norms, between the numerical solutions and the exact solutions for optimal control u^* and optimal trajectory (x_1^*, x_2^*) are 4.8000e5 and 0.0911, respectively.

These small error estimates provide compelling evidence that the interpolation algorithm is equally effective in finding numerical optimal controls for solving control problems where the optimal control exhibits jumps.

5.3 Time Difference and Time-Space Approximation Schemes

This section focuses on the general continuous-time, finite-dimensional deterministic system under a cost functional with a finite horizon. Our objective is to compute approximate solutions for the optimal feedback control. Initially, we employ the DPP to derive the evolutive HJB equation, which characterizes the value function of the optimal control problem.

Subsequently, we introduce two numerical schemes to solve the HJB equation. The first scheme involves a time difference approximation, while the second incorporates both time and space approximations. For each scheme, we establish convergence proofs, demonstrating that the solution obtained from the discrete scheme converges to the viscosity solution of the HJB equation.

Furthermore, we show that the optimal control determined by the discrete system's dynamic programming constitutes a minimizing sequence for the optimal feedback control of its continuous counterpart.

To illustrate the effectiveness of our approach, we present an example utilizing the time-space algorithm. The results obtained confirm the scheme's efficacy in approximating the optimal feedback control [107].

In fact, as we have already observed, when facing difficulties (or impossibilities) in analytically solving the HJB equation, which is often the case, one must resort to numerical solutions. It is important to note that in seeking the viscosity solution of the HJB equation, additional information beyond the value function itself is required. Specifically, we need access to the gradient of the value function. For numerical solutions, specially defined differences can serve as proxies for the gradient, which may not exist in the classical sense. This substitution has been validated through our simulation practices in Sects. 4.4 and 4.5 for some large inertial systems. Furthermore, as shown in [85], simple difference schemes can be employed to numerically approximate the viscosity solution.

The construction of the optimal feedback control law via numerical methods typically involves two steps. Firstly, an approximation to the viscosity solution of the HJB equation is obtained, either through a time-discretization scheme or a time-space discretization scheme. Subsequently, the optimal feedback control can be constructed based on this approximation, provided that the convergence of the approximation is established.

For finite-dimensional optimal control problems with infinite horizon costs, there are several results available in the literature regarding the approximation of the viscosity solution and the optimal feedback control. We refer the reader to [16, 42, 43, 242] for further details. However, the finite horizon case poses additional challenges due to the involvement of time in the HJB equation. Despite some progress made in approximating the viscosity solution (see, for instance, [18] and [77, 85]), the convergence of the approximate solution often relies on the introduction of a discount factor in the cost functional [77]. Currently, there are no convergence results available specifically for the optimal controls in this context.

In this section, we focus on a general deterministic system with a finite horizon cost functional, without the inclusion of a discount factor. We directly discretize the HJB equation using a time-discretization scheme. For this scheme, we prove certain properties introduced in [18], which are crucial for the convergence of the approximate solution of the HJB equation, as demonstrated in [85]. Leveraging piecewise linear interpolation, we devise an algorithm to approximate the viscosity solution. Based on this approximation, we then construct an "almost optimal" feedback control law associated with the nonautonomous finite horizon deterministic problem.

For a given $T > 0$, consider the following finite-dimensional control system by

$$\begin{cases} y'(t) = f(y(t), t, u(t)), & t \in (0, T], \\ y(0) = z, \end{cases} \tag{5.3.1}$$

where $u(\cdot) \in \Delta \triangleq L^\infty([0, T]; \mathbb{U})$ is the control, $\mathbb{U} \subset \mathbb{R}^m$ is a compact subset, $z \in \mathbb{R}^n$ is the initial value, and $y(t) \in \mathbb{R}^n$ is the state at time t. The cost functional J is given by

$$J(u(\cdot)) = \int_0^T L(y(t), t, u(t))\, dt + \psi(y(T)), \tag{5.3.2}$$

where L defines the running cost and ψ represents the final cost. The optimal control problem aims to find $u^*(\cdot) \in \Delta$ such that

$$J\left(u^*(\cdot)\right) = \inf_{u(\cdot)\in\Delta} J(u(\cdot)). \tag{5.3.3}$$

As previously stated, the dynamic programming method involves considering the following family of optimal control problems: each system starts from $(x, s) \in \mathbb{R}^n \times [0, T)$ such that

$$\begin{cases} y'(t) = f(y(t), t, u(t)), & t \in (s, T], \\ y(s) = x, \end{cases} \tag{5.3.4}$$

where $u(\cdot) \in \Delta_s \triangleq L^\infty([s, T]; \mathbb{U})$. The objective is to find $u_*(\cdot) \in \Delta_s$ that minimizes the cost functional of the following:

$$J_{x,s}(u(\cdot)) = \int_s^T L(y(t), t, u(t))\,\mathrm{d}t + \psi(y(T)). \tag{5.3.5}$$

The value function w for the optimal control problem of (5.3.4)–(5.3.6) is given by

$$w(x, s) = J_{x,s}(u_*(\cdot)) = \inf_{u(\cdot)\in\Delta_s} J_{x,s}(u(\cdot)). \tag{5.3.6}$$

Therefore, if $w \in C^1(\mathbb{R}^n \times [0, T])$, then it satisfies the HJB equation

$$\begin{cases} -\dfrac{\partial w}{\partial s}(x, s) - \inf\limits_{u\in\mathbb{U}}\{\nabla_x w(x, s) \cdot f(x, s, u) + L(x, s, u)\} = 0, \\ \qquad\qquad (x, s) \in \mathbb{R}^n \times [0, T), \\ w(x, T) = \psi(x), \quad x \in \mathbb{R}^n. \end{cases} \tag{5.3.7}$$

Below, we define the concept of a viscosity solution and enumerate several properties of the value function.

Definition 5.3.1 A function $w \in C(\mathbb{R}^n \times [0, T])$ is a viscosity subsolution of (5.3.7) if for any $x \in \mathbb{R}^n$, $w(x, T) \le \psi(x)$, and for any test function $v \in C^1(\mathbb{R}^n \times [0, T))$, if $w - v$ attains strictly a local maximum at $(x^*, s^*) \in \mathbb{R}^n \times [0, T)$, then

$$-v_s(x^*, s^*) - \inf_{u\in\mathbb{U}}\left\{v_x(x^*, s^*) \cdot f(x^*, s^*, u) + L(x^*, s^*, u)\right\} \le 0.$$

Similarly, a function $w \in C(\mathbb{R}^n \times [0, T])$ is a viscosity supersolution of (5.3.7) if for any $x \in \mathbb{R}^n$, $w(x, T) \ge \psi(x)$, and for any test function $v \in C^1(\mathbb{R}^n \times [0, T))$, if $w - v$ attains strictly a local minimum at $(x^*, s^*) \in \mathbb{R}^n \times [0, T)$, then

$$-v_s(x^*, s^*) - \inf_{u\in\mathbb{U}}\left\{v_x(x^*, s^*) \cdot f(x^*, s^*, u) + L(x^*, s^*, u)\right\} \ge 0.$$

w is called a viscosity solution of (5.3.7), if w is both viscosity subsolution and viscosity supersolution of (5.3.7).

The following Theorem 5.3.2 is a direct consequence of [63, 242].

Theorem 5.3.2 *For system* (5.3.1)–(5.3.3), *assume that*

$$
\begin{cases}
\|f(x,t,u)-f(y,s,u)\| \le L_f(\|x-y\|+|t-s|), \\
\qquad \forall\, x,y\in\mathbb{R}^n,\ t,s\in[0,T],\ u\in\mathbb{U}, \\
\|f(x,t,u)\|\le M_f, \quad f(x,t,\cdot)\in C(\mathbb{U}), \\
\qquad \forall\, x\in\mathbb{R}^n,\ t\in[0,T],\ u\in\mathbb{U}, \\
|L(x,t,u)-L(y,s,u)|\le L_L(\|x-y\|+|t-s|), \\
\qquad \forall\, x,y\in\mathbb{R}^n,\ t,s\in[0,T],\ u\in\mathbb{U}, \\
|L(x,t,u)|\le M_L,\ L(x,t,\cdot)\in C(\mathbb{U}), \\
\qquad \forall\, x\in\mathbb{R}^n,\ t\in[0,T],\ u\in\mathbb{U}, \\
|\psi(x)-\psi(y)|\le L_\psi\|x-y\|, \quad \forall\, x,y\in\mathbb{R}^n, \\
|\psi(x)|\le M_\psi, \quad \forall\, x\in\mathbb{R}^n,
\end{cases}
\tag{5.3.8}
$$

where $L_f, M_f, L_L, M_L, L_\psi, M_\psi$ *are positive constants. Then, the function* w *defined by* (5.3.6) *belongs to* BUC($\mathbb{R}^n\times[0,T]$), *the space of bounded uniformly continuous functions on* $\mathbb{R}^n\times[0,T]$. *Furthermore, it is the unique viscosity solution of the HJB equation* (5.3.7). *Furthermore, there exists a constant* $L_w>0$ *such that for any* $(x,t),(y,s)\in\mathbb{R}^n\times[0,T]$, *the following holds:*

$$|w(x,t)-w(y,s)|\le L_w(\|x-y\|+|t-s|).$$

Our approach to constructing approximation schemes for the viscosity solution and optimal feedback control stems from the time-discretization dynamic programming principle. Initially, we devise a time finite-difference scheme for the HJB equation, demonstrating that it indeed aligns with the time-discretization DPP inherent to the corresponding discrete control system. Subsequently, a comprehensive time-space discretization scheme is formulated through systematic space triangulation for a bounded polyhedral domain.

The primary advancements of this section are twofold: (a) To the best of our knowledge, the devised time and time-space approximation schemes are the inaugural ones that adhere to the time-discretization DPP in the context of finite horizon problems lacking a discount factor. Moreover, the approximate viscosity solution is proven to converge to its continuous counterpart from an operational standpoint. (b) A precise estimation is established concerning the disparity between the time-approximate viscosity solution and the time-space-approximate viscosity solution derived via an apt space discretization. This estimation is pivotal in establishing the optimality of the time-space discretization control system and ensuring the convergence of the "almost optimal" feedback control derived from the time-space-approximate viscosity solution.

Our methodology unfolds as follows. In the subsequent subsection, we formulate a time-discretization scheme for the HJB equation (5.3.7) and substantiate the convergence of the approximate viscosity solution to that of the corresponding HJB equation. The time-space discretization along with the convergence proof is delineated in Sect. 5.3.2. Section 5.3.3 is dedicated to synthesizing "almost optimal" feedback control laws for both time-discretization and time-space discretization systems. It is shown that each of these feedback control laws constitutes a minimizing sequence for an optimal feedback control law tailored to the original continuous control system. Additionally, relevant convergence rates are provided. Finally, in Sect. 5.3.4, an algorithm is streamlined, and numerical results are presented to corroborate the efficacy of the algorithm.

5.3.1 Time-Discretization Scheme

For each given large positive integer N, subdivide $[0, T]$ into N equal subintervals. Let

$$s_j = jh, \quad j = 0, 1, \cdots, N, \quad h = \frac{T}{N} < 1.$$

When $s = s_{j+1}$, $j = 0, 1, ..., N-1$, we discretize the time derivative term in HJB equation (5.3.7) by the time backward finite difference:

$$w_s(x, s_{j+1}) \cong \frac{w(x, s_{j+1}) - w(x, s_j)}{h},$$

and discretize the gradient term by the finite difference:

$$\nabla_x w(x, s_{j+1}) \cdot f(x, s_{j+1}, u) \cong \frac{w(x + hf(x, s_{j+1}, u), s_{j+1}) - w(x, s_{j+1})}{h}.$$

We then obtain the following time-discretization HJB equation:

$$\begin{cases} w_h(x, s_j) = \min\limits_{u \in \mathbb{U}} \left\{ hL(x, s_{j+1}, u) + w_h\left(x + hf(x, s_{j+1}, u), s_{j+1}\right)\right\}, \\ \qquad\qquad\qquad x \in \mathbb{R}^n, \ j = 0, 1, ..., N-1, \\ w_h(x, T) = \psi(x), \quad x \in \mathbb{R}^n. \end{cases}$$

Modify the above scheme for all $s \in [0, T)$, to obtain

$$\begin{cases} w_h(x,s) = \min_{u \in \mathbb{U}} \left\{ (s_{j+1}-s)L(x,s,u) + w_h\left(x + (s_{j+1}-s)f(x,s,u), s_{j+1}\right) \right\}, \\ \qquad\qquad x \in \mathbb{R}^n,\ s \in [s_j, s_{j+1}),\ j = 0,1,\cdots,N-1, \\ w_h(x,T) = \psi(x), \quad x \in \mathbb{R}^n. \end{cases} \tag{5.3.9}$$

Before proceeding with the estimations for scheme (5.3.9), we need to introduce some notations. We define a set I and a family of functions $\{g_h(\cdot) \mid h \in I\}$, as follows:

$$I \triangleq \left\{ h \mid h = T/N,\ N \in \mathbb{N},\ N \geq [T]+1 \right\},$$

and

$$g_h(s) = \begin{cases} \left[\dfrac{s}{h}\right], & s \in [0,T), \\ N-1, & s = T, \end{cases}$$

where $[a] \in \mathbb{N}$ denotes the largest integer that is not greater than $a \in \mathbb{R}$.

To simplify notation, we also utilize $j_{h,s}$ as a shorthand for $g_h(s)$ for all $s \in [0, T]$. By the definition of g_h, $j_{h,s}$ is the unique integer such that $s \in [s_{j_{h,s}}, s_{j_{h,s}+1})$. It is evident that $I \subset (0, 1)$ and $j_{h,s} \in \{0, 1, \cdots, N-1\}$ for all $s \in [0, T]$.

It is worth noting that N, given by $N = \frac{T}{h}$, increases as h decreases within the set I, indicating a finer discretization.

Proposition 5.3.1 *Assuming that all the hypotheses stated in Theorem* 5.3.2 *are valid, we can assert the existence of a unique solution denoted as* w_h *to* (5.3.9)*. Furthermore, for any arbitrary* $x, y \in \mathbb{R}^n$, $j = 0, 1, \cdots, N-1$, *the subsequent estimates are indeed accurate:*

(i) $|w_h(x,s_j)| \leq (N-j)hM_L + M_\psi$.
(ii) $|w_h(x,s_j) - w_h(y,s_j)| \leq C_j\|x-y\|$, $C_j \triangleq \left(TL_L + L_\psi\right)(1+hL_f)^{N-j} \leq C_{max} \triangleq \left(TL_L + L_\psi\right)e^{L_f}$.
(iii) $|w_h(x,s_{j+1}) - w_h(x,t)| \leq \left(M_L + C_{j+1}M_f\right)|s_{j+1}-t|$ *for any* $t \in [s_j, s_{j+1})$.
(iv) $|w_h(x,s) - w_h(x,t)| \leq \left[M_L + L_L + C_{j+1}(M_f + L_f)\right]|s-t|$ *for any* $s, t \in [s_j, s_{j+1})$.
(v) $\{w_h \mid h \in I\}$ *is uniformly bounded and equi-continuous in* $\mathbb{R}^n \times [0, T]$. *Precisely,*
$|w_h(x,s)| \leq TM_L + M_\psi$ *for any* $s \in [0,T]$,
$|w_h(x,s) - w_h(y,s)| \leq \left[L_L + (1+L_f)C_{max}\right]\|x-y\|$ *for any* $s \in [0,T]$,
$|w_h(x,s) - w_h(x,t)| \leq \left[M_L + L_L + C_{max}(M_f + L_f)\right]|s-t|$ *for any* $s, t \in [0, T]$.

Proof The proof of the existence and uniqueness of w_h is considered straightforward and thus omitted. We proceed now to establish the properties (i)–(v).

(i) Indeed, we can approach this systematically. In fact, by directly referring to (5.3.9), it becomes evident that for every $x \in \mathbb{R}^n$,

$$\begin{aligned}
&|w_h(x, s_N)| = |\psi(x)| \le M_\psi,\\
&|w_h(x, s_{N-1})| \le hM_L + |w_h(x + hf(x, s_{N-1}, u), s_N)| \le hM_L + M_\psi,\\
&|w_h(x, s_{N-2})| \le hM_L + |w_h(x + hf(x, s_{N-2}, u), s_{N-1})| \le 2hM_L + M_\psi,\\
&\quad \cdots\\
&|w_h(x, s_j)| \le hM_L + |w_h(x + hf(x, s_j, u), s_{j+1})| \le (N-j)hM_L + M_\psi,\\
&\qquad\qquad j = N-3, N-4, \cdots, 1, 0,
\end{aligned}$$

proving (i).

(ii) This can be proved in a manner analogous to (i). Specifically, for any $x, y \in \mathbb{R}^n$,

$$|w_h(x, s_N) - w_h(y, s_N)| = |\psi(x) - \psi(y)| \le L_\psi \|x - y\| \triangleq \overline{C}_N \|x - y\|,$$

$$\begin{aligned}
&|w_h(x, s_{N-1}) - w_h(y, s_{N-1})|\\
&\le \max_{u\in\mathbb{U}} \Big| [w_h(x + hf(x, s_{N-1}, u), s_N) - w_h(y + hf(y, s_{N-1}, u), s_N)]\\
&\qquad + h[L(x, s_{N-1}, u) - L(y, s_{N-1}, u)]\Big|\\
&\le \max_{u\in\mathbb{U}} \left\{\overline{C}_N \|x - y + h[f(x, s_{N-1}, u) - f(y, s_{N-1}, u)]\| + hL_L\|x - y\|\right\}\\
&\le \left[\overline{C}_N(1 + hL_f) + hL_L\right] \|x - y\| \triangleq \overline{C}_{N-1}\|x - y\|,
\end{aligned}$$

and

$$\begin{aligned}
&\left|w_h(x, s_j) - w_h(y, s_j)\right|\\
&\le \max_{u\in\mathbb{U}} \Big| \left[w_h\left(x + hf(x, s_j, u), s_{j+1}\right) - w_h\left(y + hf(y, s_j, u), s_{j+1}\right)\right]\\
&\qquad + h\left[L(x, s_j, u) - L(y, s_j, u)\right]\Big|\\
&\le \max_{u\in\mathbb{U}} \left\{\overline{C}_{j+1} \left\|x - y + h\left[f(x, s_j, u) - f(y, s_j, u)\right]\right\| + hL_L\|x - y\|\right\}\\
&\le \left[\overline{C}_{j+1}(1 + hL_f) + hL_L\right] \|x - y\| \triangleq \overline{C}_j\|x - y\|,\\
&\qquad j = N-2, N-3, \cdots, 1, 0,
\end{aligned}$$

that is,

$$\overline{C}_N = L_\psi, \quad \overline{C}_j = hL_L + (1 + hL_f)\overline{C}_{j+1}, \quad j = N-1, N-2, \cdots, 1, 0.$$

Solving the above backward difference equation gives (ii):

$$\begin{aligned}
\overline{C}_j &= \sum_{l=0}^{N-j-1} (1 + hL_f)^l hL_L + (1 + hL_f)^{N-j} L_\psi \\
&\le (N-j)hL_L(1 + hL_f)^{N-j} + (1 + hL_f)^{N-j} L_\psi \\
&\le \left[TL_L + L_\psi\right](1 + hL_f)^{N-j} \triangleq C_j \le C_{max} \triangleq (TL_L + L_\psi)e^{L_f}.
\end{aligned}$$

(iii) For any $j \in \{0, 1, \cdots, N-1\}$, if $t \in [s_j, s_{j+1})$, it follows from (ii) that for any $x \in \mathbb{R}^n$,

$$\begin{aligned}
\left|w_h(x, s_{j+1}) - w_h(x, t)\right| &\le \max_{u \in \mathbb{U}} \Big|w_h(x, s_{j+1}) - (s_{j+1} - t)L(x, t, u) \\
&\qquad - w_h\left(x + (s_{j+1} - t)f(x, t, u), s_{j+1}\right)\Big| \\
&\le [M_L + C_{j+1}M_f]|s_{j+1} - t|.
\end{aligned}$$

This is (iii).
Now we prove (iv). Similar to (iii), for any $j \in \{0, 1, \cdots, N-1\}$, if $s, t \in [s_j, s_{j+1})$, we have, for any $x \in \mathbb{R}^n$, that

$$\begin{aligned}
&|w_h(x, s) - w_h(x, t)| \\
&\le \max_{u \in \mathbb{U}} \Big|(t - s)L(x, s, u) + (s_{j+1} - t)\left[L(x, s, u) - L(x, t, u)\right] \\
&\quad + w_h\left(x + (s_{j+1} - s)f(x, s, u), s_{j+1}\right) \\
&\quad - w_h\left(x + (s_{j+1} - t)f(x, t, u), s_{j+1}\right)\Big| \\
&\le (M_L + hL_L)|s - t| + C_{j+1}(M_f + hL_f)|s - t|, \\
&\le \left[M_L + L_L + C_{j+1}(M_f + L_f)\right]|s - t|.
\end{aligned}$$

(iv) is proved.

(v) Firstly, we claim that $\{w_h \mid h \in I\}$ is uniformly bounded in $\mathbb{R}^n \times [0, T]$. Actually, when $s = T$, $|w_h(x, T)| = |\psi(x)| \le M_\psi, \ \forall\, x \in \mathbb{R}^n$, and for any $(x, s) \in \mathbb{R}^n \times [0, T)$,

$$\begin{aligned}
|w_h(x, s)| &\le (s_{j_{h,s}+1} - s)M_L + \left|w_h\left(x + (s_{j_{h,s}+1} - s)f(x, s, u), s_{j_{h,s}+1}\right)\right| \\
&\le \left[1 + (N - j_{h,s} - 1)\right]hM_L + M_\psi \le TM_L + M_\psi.
\end{aligned}$$

Next, when $s = T$, it has

$$|w_h(x, T) - w_h(y, T)| = |\psi(x) - \psi(y)| \le L_\psi \|x - y\|, \quad \forall\, x, y \in \mathbb{R}^n,$$

and for any $s \in [0, T)$,

$$\begin{aligned}
&|w_h(x, s) - w_h(y, s)| \\
&\le \max_{u \in \mathbb{U}} \Big\{ \big| w_h\left(x + (s_{j_{h,s}+1} - s) f(x, s, u), s_{j_{h,s}+1}\right) \\
&\quad - w_h\left(y + (s_{j_{h,s}+1} - s) f(y, s, u), s_{j_{h,s}+1}\right) \big| \\
&\quad + \left(s_{j_{h,s}+1} - s\right) |L(x, s, u) - L(y, s, u)| \Big\} \\
&\le h L_L \|x - y\| + C_{j_{h,s}+1} \left(\|x - y\| + h L_f \|x - y\|\right) \\
&\le \left[L_L + (1 + L_f) C_{max}\right] \|x - y\|, \quad \forall\, x, y \in \mathbb{R}^n.
\end{aligned}$$

Therefore, for every $s \in [0, T]$, the family of functions $\{w_h(\cdot, s) \mid h \in I\}$ is equi-continuous in $\mathbb{R}^n$. The remaining task is to demonstrate that for any fixed $x \in \mathbb{R}^n$, the family $\{w_h(x, \cdot) \mid h \in I\}$ is also equi-continuous in $[0, T]$. We shall approach this as follows.

By virtue of (iii)–(iv), it suffices to examine the scenario where s and t belong to distinct intervals within $[0, T]$. Specifically, by (iii)–(iv), we only need to consider the case that $s, t \in [0, T]$ are in different intervals, that is, $s \in [s_{j_{h,s}}, s_{j_{h,s}+1})$ and $t \in [s_{j_{h,t}}, s_{j_{h,t}+1})$ with $0 \le j_{h,s} < j_{h,t} \le N - 1$. For any $x \in \mathbb{R}^n$, from (iii) to (iv), we have

$$\begin{aligned}
&|w_h(x, s) - w_h(x, t)| \\
&= \Bigg| w_h(x, s) - w_h(x, s_{j_{h,s}+1}) \\
&\quad + \sum_{l=j_{h,s}+1}^{j_{h,t}-1} [w_h(x, s_l) - w_h(x, s_{l+1})] + w_h(x, s_{j_{h,t}}) - w_h(x, t) \Bigg| \\
&\le \left| w_h(x, s) - w_h(x, s_{j_{h,s}+1}) \right| + \sum_{l=j_{h,s}+1}^{j_{h,t}-1} |w_h(x, s_l) - w_h(x, s_{l+1})| \\
&\quad + \left| w_h(x, s_{j_{h,t}}) - w_h(x, t) \right| \\
&\le C^* \left(|s - s_{j_{h,s}+1}| + \sum_{l=j_{h,s}+1}^{j_{h,t}-1} |s_l - s_{l+1}| + |s_{j_{h,t}} - t| \right) \le C^* |s - t|,
\end{aligned}$$

where $C^* = \max_{l=0,1,\cdots,N-1} \{M_L + C_l M_f,\ M_L + L_L + C_{l+1}(M_f + L_f)\} = M_L + L_L + C_{max}(M_f + L_f)$.

Therefore, $\{w_h \mid h \in I\}$ is uniformly bounded and equi-continuous in $\mathbb{R}^n \times [0, T]$. This completes the proof of the proposition. □

Prior to establishing the convergence of w_h, it is necessary to introduce some additional concepts and present a preliminary result.

For any $s \in [0, T)$, let $p_h(s) = s_{j_{h,s}+1} - s$. Define operator $F_h(s, p_h(s)) : C(\mathbb{R}^n) \to C(\mathbb{R}^n)$ by

$$F_h(s, p_h(s))[\phi(\cdot)](x) = \min_{u\in\mathbb{U}} \{p_h(s)L(x, s, u) + \phi(x + p_h(s)f(x, s, u))\}, \quad \forall\, \phi \in C(\mathbb{R}^n). \tag{5.3.10}$$

For simplicity in notation, we shall also employ F_h to represent $F_h(s, p_h(s))$. With this notation, the time-discretization scheme (5.3.9) can be reformulated as

$$\begin{cases} w_h(x, s) = F_h\left[w_h(\cdot, s_{j+1})\right](x), \\ \qquad x \in \mathbb{R}^n,\ s \in [s_j, s_{j+1}),\ j = 0, 1, \cdots, N-1, \\ w_h(x, T) = \psi(x), \quad x \in \mathbb{R}^n. \end{cases} \tag{5.3.11}$$

Lemma 5.3.3 below outlines the characteristics of F_h that will be pertinent in subsequent discussions.

Lemma 5.3.3 *Let F_h be defined by* (5.3.10). *Then*

(i) *Monotonicity: For any $(x, s) \in \mathbb{R}^n \times [0, T)$, $\phi_1, \phi_2 \in C(\mathbb{R}^n)$, if for any $u \in \mathbb{U}$,*

$$\phi_1(x + p_h(s)f(x, s, u)) \le \phi_2(x + p_h(s)f(x, s, u)),$$

then

$$F_h[\phi_1](x) \le F_h[\phi_2](x).$$

(ii) *Constant translation property:*

$$F_h[\phi + c] = F_h[\phi] + c, \quad \forall\, \phi \in C(\mathbb{R}^n),\ c \in \mathbb{R}.$$

(iii) *Consistency: For any $(x, t) \in \mathbb{R}^n \times [0, T)$ and any test function $v \in C^1(\mathbb{R}^n \times [0, T))$,*

$$\lim_{\substack{(y,s)\to(x,t)\\ h\to 0+}} p_h(s)^{-1}\left\{F_h\left[v(\cdot, s_{j_{h,s}+1})\right](y) - v(y, s)\right\}$$
$$= v_t(x, t) + \inf_{u\in\mathbb{U}} \{v_x(x, t) \cdot f(x, t, u) + L(x, t, u)\},$$

where $p_h(s) = s_{j_{h,s}+1} - s$.

Proof Given that (i) and (ii) follow directly from the definition, we shall focus on proving (iii). Consider the following:

$$\begin{aligned}
&F_h(s, p_h(s))\left[v(\cdot, s_{j_{h,s}+1})\right](y) - v(y, s)\\
&= \min_{u\in\mathbb{U}} \left\{p_h(s)L(y, s, u) + v\left(y + p_h(s)f(y, s, u), s_{j_{h,s}+1}\right) - v(y, s_{j_{h,s}+1})\right.\\
&\quad \left. + v(y, s_{j_{h,s}+1}) - v(y, s)\right\}\\
&= p_h(s)\cdot\left[v_s(y, \bar{s}) + \min_{u\in\mathbb{U}}\left\{v_y(\bar{y}, s_{j_{h,s}+1})\cdot f(y, s, u) + L(y, s, u)\right\}\right],
\end{aligned}$$

where $\bar{s} = s + \theta_1 \cdot p_h(s)$, $\bar{y} = y + \theta_2 \cdot p_h(s) f(y, s, u)$ for some $\theta_1, \theta_2 \in (0, 1)$. As $h \to 0+$, there is $(y, s) \to (x, t)$, and therefore, $p_h(s) \to 0+$, $(\bar{y}, \bar{s}) \to (x, t)$, $s_{j_{h,s}+1} \to t$. The conclusion follows from the fact that $v \in C^1(\mathbb{R}^n \times [0, T))$. □

Theorem 5.3.4 *Suppose that all hypotheses of Theorem* 5.3.2 *hold. Let w_h be the solution of* (5.3.9) *(or* (5.3.11)*) and w be the viscosity solution of* (5.3.7)*. Then for any compact subset $Q \subset \mathbb{R}^n$,*

$$\lim_{h\to 0+} w_h = w \quad \textit{uniformly on } Q \times [0, T].$$

Proof Since, according to (v) of Proposition 5.3.1, the family $\{w_h \mid h \in I\}$ consists of uniformly bounded and equi-continuous functions on $\mathbb{R}^n \times [0, T]$, we can invoke the Arzelà-Ascoli theorem. This theorem guarantees that for any compact subset $Q \subset \mathbb{R}^n$, there exists a subsequence of $\{w_h \mid h \in I\}$ (which we continue to denote by $\{w_h \mid h \in I\}$ for simplicity) such that $\{w_h \mid h \in I\}$ converges uniformly to a function w^* on $Q \times [0, T]$ as $h \to 0+$. Our aim is to demonstrate that this limit function w^* is indeed a viscosity solution of (5.3.7).

For any test function $v \in C^1(\mathbb{R}^n \times [0, T))$, suppose that $(x^*, s^*) \in \mathbb{R}^n \times [0, T)$ is a strict local maximum point of $w^* - v$. Then there exists a closed ball $B = B(x^*, s^*)$ of $\mathbb{R}^n \times [0, T)$ centered at (x^*, s^*) such that

$$w^*\left(x^*, s^*\right) - v\left(x^*, s^*\right) > w^*(x, s) - v(x, s), \quad \forall\, (x, s) \in B\backslash\left(x^*, s^*\right).$$

As a result, there exists a sequence $\{(x_h, s_h)\}$ within B such that each point $(x_h, s_h) \in B$ is a maximum point for the difference $w_h - v$ on B, and

$$x_h \to x^* \text{ and } s_h \to s^* \text{ as } h \to 0+.$$

Note that $s_h \in \left[s_{j_{h,s_h}}, s_{j_{h,s_h}+1}\right)$. Define $p_h = s_{j_{h,s_h}+1} - s_h$. Obviously, $p_h > 0$, and as long as $h > 0$ small enough, $j_{h,s_h} < N - 1$ and $(x_h + p_h f(x_h, s_h, u), s_h + p_h) \in B$ for all $u \in \mathbb{U}$.

With the aforementioned preparations in place, for any given $u \in \mathbb{U}$, we have

$$\begin{aligned} & w_h(x_h, s_h) - v(x_h, s_h) \\ & \geq w_h(x_h + p_h f(x_h, s_h, u), s_h + p_h) - v(x_h + p_h f(x_h, s_h, u), s_h + p_h), \end{aligned}$$

that is,

$$\begin{aligned} & v(x_h + p_h f(x_h, s_h, u), s_h + p_h) - v(x_h, s_h) \\ & \geq w_h(x_h + p_h f(x_h, s_h, u), s_h + p_h) - w_h(x_h, s_h) \end{aligned}$$

or

$$\phi_1(x_h + p_h f(x_h, s_h, u)) \geq \phi_2(x_h + p_h f(x_h, s_h, u)), \quad \forall\, u \in \mathbb{U},$$

where $\phi_1, \phi_2 \in C^1(\mathbb{R}^n)$ are defined as

$$\begin{aligned} \phi_1(x) &\triangleq v(x, s_h + p_h) - v(x_h, s_h), \\ \phi_2(x) &\triangleq w_h(x, s_h + p_h) - w_h(x_h, s_h), \quad x \in \mathbb{R}^n. \end{aligned}$$

Using the operator $F_h(s_h, p_h)$ on ϕ_1 and ϕ_2 and considering the monotonicity and constant translation property of F_h as stated in Lemma 5.3.3, we obtain the following:

$$F_h\left[v(\cdot, s_h + p_h)\right](x_h) - v(x_h, s_h) \geq F_h\left[w_h(\cdot, s_h + p_h)\right](x_h) - w_h(x_h, s_h) = 0.$$

Dividing both sides of the previous equation by p_h, we then take the limit as $h \to 0+$ and apply (iii) from Lemma 5.3.3 to derive the following:

$$v_s\left(x^*, s^*\right) + \inf_{u \in \mathbb{U}} \left\{ v_x\left(x^*, s^*\right) \cdot f\left(x^*, s^*, u\right) + L\left(x^*, s^*, u\right) \right\} \geq 0,$$

which means that w^* is a viscosity subsolution of (5.3.7) by the fact that $w_h(\cdot, T) = \psi(\cdot)$ on $\mathbb{R}^n$.

Following the same reasoning, we can demonstrate that w^* is also a viscosity supersolution of (5.3.7). Consequently, w^* is indeed a viscosity solution of (5.3.7). The desired result then follows from the uniqueness of the viscosity solution of (5.3.7). □

5.3.2 Space-Discretization Scheme

In this subsection, we turn our attention to the spatial approximation of w_h. For simplicity, we adopt a standard assumption from [16] that $\Omega \subset \mathbb{R}^n$ is a closed and bounded polyhedron satisfying certain conditions for all sufficiently small values of h:

$$x + hf(x, s, u) \in \Omega, \quad \forall\, (x, s, u) \in \Omega \times [0, T] \times \mathbb{U}. \tag{5.3.12}$$

We construct a regular triangulation of Ω using a finite number of simplices $\{T_i\}$ such that their union covers the entire domain, i.e., $\Omega = \cup_i T_i$. Here, x_j, $j = 1, 2, \cdots, M$ represent the nodes or vertices of this triangulation. Define the following:

$$k = \max_i\{\text{diam}(T_i)\}, \;\; S = \{s_j \mid j = 0, 1, \cdots, N\}, \;\; G = \{x_j \mid j = 1, 2, \cdots, M\},$$

where $\text{diam}(T_i)$ denotes the diameter of the set T_i. It is clear that

$$\lim_{k \to 0+} \text{dist}(x, G) = 0, \quad \forall\, x \in \Omega.$$

For $(x_i, s_j) \in G \times S$, define

$$\begin{cases} v_h^k(x_i, s_j) = \min\limits_{u \in \mathbb{U}} \left\{ hL(x_i, s_j, u) + \sum\limits_{l=1}^{M} \lambda_i^l(u, s_j) v_h^k(x_l, s_{j+1}) \right\}, \\ \qquad\qquad\qquad\qquad i = 1, 2, \cdots, M, \;\; j = 0, 1, \cdots, N-1, \\ v_h^k(x_i, T) = \psi(x_i), \quad i = 1, 2, \cdots, M, \end{cases} \tag{5.3.13}$$

where $\left\{\lambda_i^l(u, s_j) \in [0, 1] \mid l = 1, 2, \cdots, M\right\}$ is the set of coefficients in the unique convex linear combination

$$x_i + hf(x_i, s_j, u) = \sum_{l=1}^{M} \lambda_i^l(u, s_j) x_l \in \Omega$$

(by assumption (5.3.12)) with $\sum\limits_{l=1}^{M} \lambda_i^l(u, s_j) = 1$. This convex linear combination is constructed using the vertices of the simplex where $x_i + hf(x_i, s_j, u)$ is located.

In this manner, v_h^k is unambiguously defined on the domain $G \times S$. Furthermore, analogous to the results obtained in Proposition 5.3.1, there exists a unique solution $v_h^k \in B(G \times S)$, which satisfies the equation given in (5.3.13). Here, $B(S_0)$ represents the set of all bounded functions defined on the region S_0.

Based on the scheme presented in (5.3.13), we introduce a piecewise linear function as follows: $w_h^k \in B(\Omega \times [0, T])$ by

$$\begin{cases} w_h^k(x,s) = \sum_{i=1}^{M} \lambda_i(x) \left[\mu(s) v_h^k(x_i, s_j) + (1-\mu(s)) v_h^k(x_i, s_{j+1}) \right], \\ \qquad x \in \Omega, \ s \in [s_j, s_{j+1}), \ j = 0, \cdots, N-1, \\ w_h^k(x,T) = \sum_{i=1}^{M} \lambda_i(x) \psi(x_i), \quad x \in \Omega, \end{cases} \tag{5.3.14}$$

where $\mu(s) \triangleq \frac{s_{j+1}-s}{h} \in (0, 1]$ when $s \in [s_j, s_{j+1})$, and

$$\{\lambda_i(x) \in [0, 1] \mid i = 1, 2, \cdots, M\}$$

is the set of coefficients in the unique convex linear combination $x = \sum_{i=1}^{M} \lambda_i(x) x_i$ with $\sum_{i=1}^{M} \lambda_i(x) = 1$. This convex linear combination is determined based on the vertices of the simplex within which the point x is situated. It is evident that $w_h^k = v_h^k$ on the domain $G \times S$.

Theorem 5.3.5 *Suppose that all hypotheses of Theorem 5.3.2 hold. Let w_h, w_h^k be the solutions to (5.3.9) and (5.3.14) on $\Omega \times S$, respectively. Then for any $(x, s_j) \in \Omega \times S$, there holds*

$$\left| w_h(x, s_j) - w_h^k(x, s_j) \right| \le C \left(\frac{k}{h} + k \right), \tag{5.3.15}$$

where $C \triangleq \max\{T C_{max}, L_\psi\}$ is independent of h, k.

Proof For any $(x, s_j) \in \Omega \times S$, suppose that x lies in some simplex T_{i_x}. Let $x = \sum_{i \in I_x} \lambda_i(x) x_i$ be the unique convex linear combination which is produced in terms of the vertices of T_{i_x}, where

$$I_x \triangleq \{i \mid x_i \in T_{i_x}, \ i = 1, 2, \cdots, M\}.$$

Then,

$$\|x - x_i\| \le k \quad \text{for any } i \in I_x.$$

Next, by definition (5.3.14), for any $u \in \mathbb{U}$ and $j \in \{0, 1, \cdots, N-1\}$, it has

$$w_h^k\left(\sum_{l=1}^{M} \lambda_i^l(u, s_j)x_l, s_{j+1}\right) = \sum_{l=1}^{M} \lambda_i^l(u, s_j) w_h^k(x_l, s_{j+1}),$$

where $\left\{\lambda_i^l(u, s_j) \in [0, 1] \mid 1 \le l \le M\right\}$ is the set of coefficients in the unique convex linear combination $x_i + hf(x_i, s_j, u) = \sum_{l=1}^{M} \lambda_i^l(u, s_j)x_l$ with $\sum_{l=1}^{M} \lambda_i^l(u, s_j) = 1$. Once again, this convex linear combination is formulated in terms of the vertices of the simplex where $x_i + hf(x_i, s_j, u)$ lies.

It follows from (ii) of Proposition 5.3.1 that

$$\left|w_h(x, s_j) - w_h(x_i, s_j)\right| \le C_{max}\|x - x_i\|, \quad \forall\, i \in I_x,\ j = 0, 1, \cdots, N-1.$$

When $j = N$, it has

$$\begin{aligned} \left|w_h(x, s_N) - w_h^k(x, s_N)\right| &= \left|\psi(x) - \sum_{i \in I_x} \lambda_i(u, s_N)\psi(x_i)\right| \\ &\le \sum_{i \in I_x} \lambda_i(u, s_N)|\psi(x) - \psi(x_i)| \le L_\psi k. \end{aligned} \tag{5.3.16}$$

For any $j \in \{0, 1, \cdots, N-1\}$, based on the previously justified facts and the definitions provided in (5.3.9) and (5.3.13)–(5.3.14), we can iteratively deduce that

$$\begin{aligned} &\left|w_h(x, s_j) - w_h^k(x, s_j)\right| \\ &\le \sum_{i \in I_x} \lambda_i(x)\left\{\left|w_h(x, s_j) - w_h(x_i, s_j)\right| + \left|w_h(x_i, s_j) - w_h^k(x_i, s_j)\right|\right\} \\ &\le C_{max}k + \sum_{i \in I_x} \lambda_i(x)\left|w_h(x_i, s_j) - w_h^k(x_i, s_j)\right| \\ &= C_{max}k + \sum_{i \in I_x} \lambda_i(x)\Bigg|\min_{u \in \mathbb{U}}\left\{hL(x_i, s_j, u) + w_h(y_u, s_{j+1})\right\} \\ &\qquad - \min_{u \in \mathbb{U}}\left\{hL(x_i, s_j, u) + \sum_{l=1}^{M} \lambda_i^l(u, s_j) w_h^k(x_l, s_{j+1})\right\}\Bigg| \end{aligned}$$

$$
\begin{aligned}
&\le C_{max}k + \sum_{i\in I_x} \lambda_i(x) \max_{u\in\mathbb{U}} \left| w_h(y_u, s_{j+1}) - \sum_{l=1}^{M} \lambda_i^l(u, s_j) w_h^k(x_l, s_{j+1}) \right| \\
&= C_{max}k + \sum_{i\in I_x} \lambda_i(x) \max_{u\in\mathbb{U}} \left| w_h(y_u, s_{j+1}) - w_h^k\left(\sum_{l=1}^{M} \lambda_i^l(u, s_j) x_l, s_{j+1} \right) \right| \\
&= C_{max}k + \sum_{i\in I_x} \lambda_i(x) \max_{u\in\mathbb{U}} \left| w_h(y_u, s_{j+1}) - w_h^k(y_u, s_{j+1}) \right| \\
&\le C_{max}k + \sup_{z_1\in\Omega} \left| w_h(z_1, s_{j+1}) - w_h^k(z_1, s_{j+1}) \right| \\
&\le 2C_{max}k + \sup_{z_1\in\Omega} \sup_{z_2\in\Omega} \left| w_h(z_2, s_{j+2}) - w_h^k(z_2, s_{j+2}) \right| \\
&= 2C_{max}k + \sup_{z_2\in\Omega} \left| w_h(z_2, s_{j+2}) - w_h^k(z_2, s_{j+2}) \right| \\
&\le \cdots \\
&\le (N-j)C_{max}k + \sup_{z_{N-j}\in\Omega} \left| w_h(z_{N-j}, s_N) - w_h^k(z_{N-j}, s_N) \right| \\
&\le (N-j)C_{max}k + L_\psi k \le TC_{max}\frac{k}{h} + L_\psi k \\
&\le C\left(\frac{k}{h} + k\right), \quad \text{where } C \triangleq \max\{TC_{max}, L_\psi\}.
\end{aligned}
\tag{5.3.17}
$$

Notice that in the preceding discussion, the convex linear combination

$$
y_u \triangleq x_i + hf(x_i, s_j, u) = \sum_{l=1}^{M} \lambda_i^l(u, s_j) x_l
$$

is generated in the same manner as outlined in (5.3.13). The proof is subsequently completed by referencing (5.3.16) and (5.3.17). □

Theorem 5.3.6 *Suppose that all hypotheses of Theorem* 5.3.2 *hold. Let* $k = \mathcal{O}(h^{1+\gamma})$ *for some constant* $\gamma > 0$, *and let* w *be the viscosity solution of* (5.3.7), w_h^k *the solution of* (5.3.14). *Then,*

$$
\lim_{h\to 0+} w_h^k = w \quad \textit{uniformly on } \Omega \times [0, T].
$$

Proof Firstly, when $s = T$, utilizing (5.3.15), we can deduce that for any $x \in \Omega$, there holds

$$\begin{aligned}\left|w_h^k(x,T) - w(x,T)\right| &\leq \left|w_h^k(x,T) - w_h(x,T)\right| + |w_h(x,T) - w(x,T)| \\ &= \left|w_h^k(x,T) - w_h(x,T)\right| \leq C\left(\frac{k}{h} + k\right) \\ &= \mathcal{O}\left(h^\gamma + h^{1+\gamma}\right) \to 0 \quad \text{as } h \to 0+.\end{aligned} \tag{5.3.18}$$

Next, for any $s \in [0, T)$, define $\mu(s) = \frac{s_{j_{h,s}+1} - s}{h}$. Then,

$$\mu(s) \in (0, 1] \text{ and } s = \mu(s)s_{j_{h,s}} + (1 - \mu(s))s_{j_{h,s}+1}.$$

By definition (5.3.14), for any $x \in \Omega$,

$$\begin{aligned}&\left|w_h^k(x,s) - w(x,s)\right| \\ &\leq \mu(s)\left|w_h^k(x, s_{j_{h,s}}) - w(x,s)\right| + (1 - \mu(s))\left|w_h^k(x, s_{j_{h,s}+1}) - w(x,s)\right|,\end{aligned} \tag{5.3.19}$$

and (we treat only $s_{j_{h,s}}$ term)

$$\begin{aligned}&\left|w_h^k(x, s_{j_{h,s}}) - w(x,s)\right| \\ &\leq \left|w_h^k(x, s_{j_{h,s}}) - w_h(x, s_{j_{h,s}})\right| + \left|w_h(x, s_{j_{h,s}}) - w_h(x,s)\right| \\ &\quad + |w_h(x,s) - w(x,s)| \\ &\triangleq I_1 + I_2 + I_3.\end{aligned} \tag{5.3.20}$$

By (5.3.15), the term

$$I_1 \triangleq \left|w_h^k(x, s_{j_{h,s}}) - w_h(x, s_{j_{h,s}})\right| \leq C\left(\frac{k}{h} + k\right) = \mathcal{O}\left(h^\gamma + h^{1+\gamma}\right) \to 0$$

as $h \to 0+$.

Utilizing (iv) of Proposition 5.3.1 and Theorem 5.3.4, we can deduce that the terms I_2 and I_3 converge uniformly to 0 on $\Omega \times [0, T]$ as $h \to 0+$. The proof is subsequently finalized by referencing (5.3.18) through (5.3.20). □

5.3.3 Approximation of Optimal Feedback Control

In this subsection, we initially establish the corresponding DPP for the time-discretization scheme. Leveraging the approximations of the viscosity solution discussed in Sects. 5.3.1 and 5.3.2, we are able to design the optimal feedback controls for the associated discrete control systems. Subsequently, the convergence of the discrete optimal feedback control to its continuous counterpart is addressed by referring to the infinite horizon case examined in [16].

5.3.3.1 Time-Discretization Case

We firstly consider the time-discretization case. Recall that for a given large positive integer N, the interval $[0, T]$ is partitioned into N equal subintervals, with $s_j = jh,\ j = 0, 1, \cdots, N$, where $h = \frac{T}{N} < 1$. Additionally, recall that for any $s \in [0, T)$, $j_{h,s}$ is the unique integer such that $s \in [s_{j_{h,s}}, s_{j_{h,s}+1})$. Let

$$
\begin{aligned}
t_0 = s, \quad & t_l = s_{j_{h,s}+l}, \quad l = 1, 2, \cdots, N - j_{h,s},\\
h_0 = s_{j_{h,s}+1} - s, \quad & h_l = h, \quad l = 1, 2, \cdots, N - j_{h,s} - 1.
\end{aligned}
$$

For any $x \in \mathbb{R}^n$, consider the time-discretization counterpart for control system (5.3.4) which starts from (x, s):

$$
\begin{cases}
y_{l+1} = y_l + h_l f(y_l, t_l, u_l), \quad l = 0, 1, \cdots, N - j_{h,s} - 1,\\
y_0 = x,
\end{cases} \tag{5.3.21}
$$

where $u_l = u_h(t_l),\ y_l = y_{t_l}, u_h(\cdot) \in \Delta_s$. The cost functional $J^h_{x,s}$ and value function v_h to the discrete system (5.3.21) are defined, respectively, by

$$
J^h_{x,s}(u_h(\cdot)) = \sum_{l=0}^{N-j_{h,s}-1} h_l L(y_l, t_l, u_l) + \psi(y_{N-j_{h,s}}), \tag{5.3.22}
$$

and

$$
v_h(x, s) = \inf_{u_h(\cdot)\in\Delta_s} J^h_{x,s}(u_h(\cdot)). \tag{5.3.23}
$$

Lemma 5.3.7 (Time-Discretization DPP) *Suppose that all hypotheses of Theorem 5.3.2 hold. Let v_h be defined as* (5.3.23). *Then for $i = 1, 2, \cdots, N - j_{h,s} - 1$,*

$$
v_h(x, s) = \min_{u_0,\cdots,u_{i-1}\in\mathbb{U}} \left\{ \sum_{l=0}^{i-1} h_k L(y_l, t_l, u_l) + v_h(y_i, t_i) \right\}, \tag{5.3.24}
$$
$$
\forall\ (x, s) \in \mathbb{R}^n \times [0, T).
$$

Proof The proof follows a similar approach to that of Lemma 2.1 in [77] and is therefore omitted here for brevity. □

Corollary 5.3.1 *Let* $i = 1$ *in Lemma* 5.3.7. *Then* (5.3.24) *reduces to the time-discretization scheme* (5.3.9) *in Sect.* 5.3.1*:*

$$\begin{cases} v_h(x,s) = \min_{u\in\mathbb{U}}\{(s_{j+1}-s)L(x,s,u) + v_h(x+(s_{j+1}-s)f(x,s,u), s_{j+1})\}, \\ \qquad\qquad x\in\mathbb{R}^n,\ s\in[s_j,s_{j+1}),\ j=0,1,\cdots,N-1, \\ v_h(x,T) = \psi(x), \quad x\in\mathbb{R}^n. \end{cases}$$

By Corollary 5.3.1, $v_h \equiv w_h$ on $\mathbb{R}^n \times [0,T]$. Therefore, in the subsequent analysis, we will employ w_h in place of v_h. Let us now consider the following subset of $\mathbb{U}$:

$$\begin{aligned} A_1(x,s) = \min_{\|\cdot\|}\Big\{u\in\mathbb{U} \mid w_h(x,s) &= (s_{j+1}-s)L(x,s,u) \\ &+ w_h\left(x+(s_{j+1}-s)f(x,s,u), s_{j+1}\right)\Big\}, \\ x\in\mathbb{R}^n,\ s&\in[s_j,s_{j+1}),\ j=0,1,\cdots,N-1. \end{aligned} \tag{5.3.25}$$

We observe that $A_1(x,s)$ represents a subset of controls possessing minimal energy, which adhere to the time-discretization scheme given by (5.3.9). Based on the assumption stated in (5.3.8), it follows that for any $(x,s) \in \mathbb{R}^n \times [0,T)$, the set $A_1(x,s)$ is nonempty. It is worth noting that $A_1(x,s)$ may contain multiple elements, but we have the liberty to select any element $a_1(x,s)$ of $A_1(x,s) \subset \mathbb{U} \subset \mathbb{R}^m$ for the purpose of determining the minimal energy control function denoted by m_h^*:

$$m_h^*(x,s) = a_1(x,s), \quad x\in\mathbb{R}^n,\ s\in[s_j,s_{j+1}),\ j=0,1,\cdots,N-1. \tag{5.3.26}$$

In this context, the control function m_h^* is well-defined on the specified domain $\mathbb{R}^n \times [0,T)$.

The time-discretization analogue for the control system given by (5.3.1) constitutes a particular instance of the control system described by (5.3.21), wherein $x = z$ along with $s = j_{h,s} = 0$,

$$\begin{cases} y^{j+1} = y^j + hf\left(y^j, s_j, u_h^*(s_j)\right), \quad j=0,1,\cdots,N-1, \\ y^0 = z, \end{cases} \tag{5.3.27}$$

in which $z\in\mathbb{R}^n$ is given and the optimal feedback control law u_h^* is taken as

$$u_h^*(t) = u_j^* \triangleq m_h^*\left(y^j, s_j\right), \quad t\in[s_j,s_{j+1}),\ j=0,1,\cdots,N-1. \tag{5.3.28}$$

Here, to differentiate the solution of (5.3.21) originating from the initial point (x, s), we employ the notation $y^j \approx y(s_j)$ to represent the solution of the system given by (5.3.27).

Now, we proceed to present two convergence results.

Theorem 5.3.8 (Time-Discretization Optimal Feedback Control) *Suppose that all the hypotheses stated in Theorem* 5.3.2 *are valid. Let* u_h^* *be defined as per* (5.3.28) *and let* w_h *represent the solution of* (5.3.9). *Then, for any* $z \in \mathbb{R}^n$,

$$w_h(z,0) \le J_{z,0}^h\left(u_h(\cdot)\right), \quad \forall\, u_h(\cdot) \in \Delta,$$

$$w_h(z,0) = J_{z,0}^h\left(u_h^*(\cdot)\right),$$

where $J_{z,0}^h$ *is defined by* (5.3.22).

Proof The first assertion is supported by the following argument:

$$\begin{aligned} w_h(z,0) &= \min_{u\in\mathbb{U}} \left\{hL(z,0,u) + w_h\left(z + hf(z,0,u), s_1\right)\right\} \\ &\le hL(z,0,u_0) + w_h(y_1, s_1) \\ &= hL(y_0, s_0, u_0) + \min_{u\in\mathbb{U}} \left\{hL(y_1, s_1, u) + w_h\left(y_1 + hf(y_1, s_1, u), s_2\right)\right\} \\ &\le hL(y_0, s_0, u_0) + hL(y_1, s_1, u_1) + w_h(y_2, s_2) \\ &\le \cdots \\ &\le \sum_{l=0}^{N-1} hL(y_l, s_l, u_l) + w_h(y_N, s_N) = J_{z,0}^h\left(u_h(\cdot)\right), \end{aligned} \tag{5.3.29}$$

where $u_l = u_h(s_l)$, $l = 0, 1, \cdots, N-1$, and y_l is the solution of (5.3.21) with $x = z$ and $s = j_{h,s} = 0$.

For the second assertion, we choose $u_h(\cdot) = u_h^*(\cdot)$. Then $u_l = u_l^* = m_h^*\left(y^l, s_l\right)$ and $y_l = y^l$, $l = 0, 1, \cdots, N-1$, where y^l is the solution of (5.3.27). By the definition of $m_h^*\left(y^l, s_l\right)$, all inequalities in (5.3.29) become equalities. The result then follows. □

Theorem 5.3.9 (Minimizing Sequence of Continuous Optimal Feedback Control) *Suppose that all hypotheses of Theorem* 5.3.2 *hold. Let* u_h^* *be defined as* (5.3.28). *Then for any given* $z \in \mathbb{R}^n$,

$$\lim_{h\to 0+} J_{z,0}\left(u_h^*(\cdot)\right) = \inf_{u(\cdot)\in\Delta} J_{z,0}\left(u(\cdot)\right) \equiv w(z,0), \tag{5.3.30}$$

where $J_{z,0}$ *is defined by* (5.3.5).

Proof We assert that the statement given by (5.3.30) is tantamount to saying that

$$\lim_{h\to 0+} J_{z,0}\left(u_h^*(\cdot)\right) = \lim_{h\to 0+} J_{z,0}^h\left(u_h^*(\cdot)\right) \equiv \lim_{h\to 0+} w_h(z,0) = w(z,0). \quad (5.3.31)$$

Actually,

$$\begin{aligned} &\left|J_{z,0}\left(u_h^*(\cdot)\right) - J_{z,0}^h\left(u_h^*(\cdot)\right)\right| \\ \leq &\left|\int_0^T L\left(y^*(t), t, u_h^*(t)\right) \mathrm{d}t + \psi\left(y^*(T)\right) - \sum_{j=0}^{N-1} hL\left(y^j, s_j, u_j^*\right) - \psi\left(y^N\right)\right| \\ \leq &\left|\int_0^T L\left(y^*(t), t, u_h^*(t)\right) \mathrm{d}t - \sum_{j=0}^{N-1} hL\left(y^j, s_j, u_j^*\right)\right| + \left|\psi(y^*(T)) - \psi\left(y^N\right)\right| \\ \triangleq\ & I_4 + I_5, \end{aligned} \quad (5.3.32)$$

where $y^*(\cdot)$ is the solution of system (5.3.1) with control $u_h^*(\cdot)$ defined by (5.3.28), $u_j^* = m_h^*\left(y^j, s_j\right)$ is determined by (5.3.25)–(5.3.26) and y^j is the solution of system (5.3.27). The proof will be accomplished if we can show that

$$\left\|y^*(t) - y^j\right\| \leq C_T h, \quad t \in [s_j, s_{j+1}],\ \forall\, j = 0, 1, \cdots, N-1. \quad (5.3.33)$$

In fact,

$$\begin{aligned} I_4 &\leq \sum_{j=0}^{N-1} \int_{s_j}^{s_{j+1}} \left|L\left(y^*(t), t, u_j^*\right) - L\left(y^j, s_j, u_j^*\right)\right| \mathrm{d}t \\ &\leq \sum_{j=0}^{N-1} \int_{s_j}^{s_{j+1}} L_L\left(\left\|y^*(t) - y^j\right\| + |t - s_j|\right) \mathrm{d}t \leq T L_L (C_T + 1) h, \end{aligned} \quad (5.3.34)$$

and

$$I_5 = \left|\psi\left(y^*(T)\right) - \psi\left(y^N\right)\right| \leq L_\psi \left\|y^*(T) - y^N\right\| \leq L_\psi C_T h, \quad (5.3.35)$$

where C_T is a constant independent of h. (5.3.31) then follows from (5.3.32) and (5.3.34)–(5.3.35).

Now, we verify the validity of (5.3.33). To this end, we define

$$z^*(t) = y^j, \quad t \in [s_j, s_{j+1}),\ j = 0, 1, \cdots, N-1.$$

Since

$$\begin{cases} y^*(t) = y^*(s_j) + \int_{s_j}^{t} f\left(y^*(\tau), \tau, u_j^*\right) \mathrm{d}\tau, & t \in [s_j, s_{j+1}], \\ y^{j+1} = y^j + hf\left(y^j, s_j, u_j^*\right), & j = 0, 1, \cdots, N-1, \end{cases}$$

and

$$\begin{aligned} &\|y^*(s_j) - y^j\| \\ &\le \left\|y^*(s_{j-1}) - y^{j-1}\right\| + \int_{s_{j-1}}^{s_j} \left|f\left(y^*(\tau), \tau, u_{j-1}^*\right) - f\left(y^{j-1}, s_{j-1}, u_{j-1}^*\right)\right| \mathrm{d}\tau \\ &\le \left\|y^*(s_{j-1}) - y^{j-1}\right\| + L_f \int_{s_{j-1}}^{s_j} \left\|y^*(\tau) - y^{j-1}\right\| \mathrm{d}\tau + L_f h^2, \quad 1 \le j \le N, \end{aligned}$$

we have, for any $t \in [s_j, s_{j+1}),\ j \in \{0, 1, \cdots, N-1\}$, that

$$\begin{aligned} &\|y^*(t) - z^*(t)\| \\ &= \left\|y^*(t) - y^j\right\| \le \left\|y^*(t) - y^{j+1}\right\| + M_f h \\ &\le \left\|y^*(s_j) - y^j\right\| + \int_{s_j}^{t} \left|f\left(y^*(\tau), \tau, u_j^*\right) - f\left(y^j, s_j, u_j^*\right)\right| \mathrm{d}\tau + 2M_f h \\ &\le \left\|y^*(s_j) - y^j\right\| + L_f \int_{s_j}^{t} \left(\|y^*(\tau) - y_j\| + |\tau - s_j|\right) \mathrm{d}\tau + 2M_f h \\ &\le \left\|y^*(s_j) - y^j\right\| + L_f \int_{s_j}^{t} \|y^*(\tau) - z^*(\tau)\| \mathrm{d}\tau + L_f h^2 + 2M_f h \\ &\le \left\|y^*(s_{j-1}) - y^{j-1}\right\| + L_f \int_{s_{j-1}}^{t} \|y^*(\tau) - z^*(\tau)\| \mathrm{d}\tau + 2L_f h^2 + 2M_f h \\ &\le \cdots \\ &\le L_f \int_{s_0}^{t} \|y^*(\tau) - z^*(\tau)\| \mathrm{d}\tau + (j+1)L_f h^2 + 2M_f h \\ &\le L_f \int_{0}^{t} \|y^*(\tau) - z^*(\tau)\| \mathrm{d}\tau + \overline{C}_T h, \end{aligned} \tag{5.3.36}$$

where $\overline{C}_T = TL_f + 2M_f$. Apply the Gronwall inequality to (5.3.36), to obtain (5.3.33):

$$\|y^*(t) - z^*(t)\| \le C_T h, \quad \forall\, t \in [s_j, s_{j+1}],\ j = 0, 1, \cdots, N-1$$

where

$$C_T = \overline{C}_T e^{TL_f} = (TL_f + 2M_f)e^{TL_f}.$$

This completes the proof of the theorem. □

5.3.3.2 Time-Space Discretization Case

Now, let us turn our attention to the time-space discretization scenario. Recall from Sect. 5.3.2 that $\Omega \subset \mathbb{R}^n$ is assumed to be a closed and bounded polyhedron, satisfying the condition that for all sufficiently small values of h,

$$x + hf(x, s, u) \in \Omega, \quad \forall\, (x, s, u) \in \Omega \times [0, T] \times \mathbb{U}.$$

For $(x_i, s_j) \in G \times S$, $v_h^k(x_i, s_j)$ is defined by (5.3.13), while for $(x, s) \in \Omega \times [0, T]$, $w_h^k(x, s)$ is defined by (5.3.14), and $w_h^k = v_h^k$ on $G \times S$.

Using the relation given by (5.3.14), we can reformulate (5.3.13) for each point $(x_i, s_j) \in G \times S$ as follows:

$$\begin{cases} w_h^k(x_i, s_j) = \min\limits_{u \in \mathbb{U}} \left\{ hL(x_i, s_j, u) + w_h^k\left(x_i + hf(x_i, s_j, u), s_{j+1}\right) \right\}, \\ \qquad\qquad i = 1, 2 \cdots, M, \ j = 0, 1, \cdots, N-1, \\ w_h^k(x_i, T) = \psi(x_i), \quad i = 1, 2 \cdots, M. \end{cases} \tag{5.3.37}$$

Based on (5.3.9) and (5.3.37), we are inspired to define P and P^k as follows:

$$\begin{aligned} P(x, s_j, u) &\triangleq hL(x, s_j, u) + w_h\left(x + hf(x, s_j, u), s_{j+1}\right), \\ &\qquad (x, s_j, u) \in \Omega \times S\backslash\{T\} \times \mathbb{U}, \end{aligned} \tag{5.3.38}$$

and

$$\begin{aligned} P^k(x, s_j, u) &\triangleq hL(x, s_j, u) + w_h^k\left(x + hf(x, s_j, u), s_{j+1}\right), \\ &\qquad (x, s_j, u) \in \Omega \times S\backslash\{T\} \times \mathbb{U}. \end{aligned} \tag{5.3.39}$$

Notice that the value of w_h^k appearing in the definition of P^k can be unambiguously determined by referring to the relations provided by (5.3.14) and (5.3.13).

The subsequent Corollary 5.3.2 arises as a direct result of Theorem 5.3.5.

Corollary 5.3.2 *Suppose that all the hypotheses stated in Theorem* 5.3.2 *are valid. Let* C *denote the constant referenced in Theorem* 5.3.5. *Then,*

$$\left|P^k - P\right| \le C\left(\frac{k}{h} + k\right) \to 0 \quad \textit{uniformly on } \Omega \times S\backslash\{T\} \times \mathbb{U} \textit{ as } k \to 0+.$$

Next, based on the assumptions given by (5.3.8), for any point $x \in \Omega$ and $j \in \{0, 1, \cdots, N-1\}$, we can deduce that there exist at least one control $m_h^*(x, s_j) \in \mathbb{U}$ and one control $m_k^*(x, s_j) \in \mathbb{U}$ such that the following conditions are satisfied:

$$P\left(x, s_j, m_h^*(x, s_j)\right) = \min_{u \in \mathbb{U}} P(x, s_j, u) \equiv w_h(x, s_j) \tag{5.3.40}$$

by (5.3.9) and

$$P^k\left(x, s_j, m_k^*(x, s_j)\right) = \min_{u \in \mathbb{U}} P^k(x, s_j, u). \tag{5.3.41}$$

Consider the following subset of $\mathbb{U}$:

$$A_2(x, s_j) = \min_{\|\cdot\|}\left\{v \in \mathbb{U} \,\middle|\, P^k\left(x, s_j, v\right) = \min_{u \in \mathbb{U}} P^k(x, s_j, u)\right\}. \tag{5.3.42}$$

This is a collection of controls that possess minimal energy while adhering to the constraint given by (5.3.41). The fact that $A_2(x, s_j)$ is non-empty is ensured by the assumptions stated in (5.3.8). It is worth noting that $A_2(x, s_j)$ might contain more than one element, but for our purposes, we can arbitrarily select a single element $a_2(x, s_j) \in A_2(x, s_j)$ to specify the minimal energy control function m_k^* by

$$m_k^*(x, s_j) = a_2(x, s_j), \quad x \in \Omega, \ j = 0, 1, \cdots, N-1. \tag{5.3.43}$$

In this manner, the control m_k^* is unambiguously defined on the domain $\Omega \times S\backslash\{T\}$.

Having established this, we are now in a position to introduce a feedback control law denoted by u_k^* for time-space discretization system:

$$u_k^*(s) = u_{k,j}^* \triangleq m_k^*\left(y^j, s_j\right), \quad s \in [s_j, s_{j+1}), \ j = 0, 1, \cdots, N-1, \tag{5.3.44}$$

where, similarly, y^j represents an approximation of $y(s_j)$ which is the solution of the subsequent system controlled by u_k^* as determined by (5.3.44) in the absence of diffusion:

$$\begin{cases} y^{j+1} = y^j + hf\left(y^j, s_j, u_k^*(s_j)\right), & j = 0, 1, \cdots, N-1, \\ y_0 = z, \end{cases} \tag{5.3.45}$$

where $z \in \Omega$ is a given point. Equation (5.3.45) serves as the discrete analog for the control system described by (5.3.1).

Theorem 5.3.10 (Minimizing Sequence of Time-Discretization Optimal Feedback Control) *Assuming that all the hypotheses stated in Theorem* 5.3.2 *are valid, and let u_k^* be defined as in* (5.3.44). *Then, for any $z \in \Omega$, we have the following conclusion:*

$$\left| J_{z,0}^h \left(u_k^*(\cdot)\right) - w_h(z,0) \right| \leq 4TC\frac{k}{h^2} \to 0 \quad \text{as } k \to 0+, \tag{5.3.46}$$

where $J_{z,0}^h$ is defined by (5.3.22) *and C is the constant in Theorem* 5.3.5.

Proof From (5.3.39), we have, for all $j = 0, 1, \cdots, N-1$ and $x \in \Omega$, that

$$\begin{aligned} &hL\left(x, s_j, m_k^*(x, s_j)\right) \\ &= -w_h^k\left(x + hf(x, s_j, m_k^*(x, s_j)), s_{j+1}\right) + P^k\left(x, s_j, m_k^*(x, s_j)\right), \end{aligned} \tag{5.3.47}$$

where $m_k^*(x, s_j)$ is determined according to the relations given by (5.3.42)–(5.3.43). Now, let us set $x = y$ in (5.3.47), where y^j represents the solution of the system described by (5.3.45). By summing up the expression over j from 0 to $N-1$, we obtain the following result:

$$\begin{aligned} &\sum_{j=0}^{N-1} hL\left(y^j, s_j, u_{k,j}^*\right) \\ &= \sum_{j=0}^{N-1} \left[-w_h^k\left(y^j + hf\left(y^j, s_j, u_{k,j}^*\right), s_{j+1}\right) + P^k\left(y^j, s_j, u_{k,j}^*\right)\right], \end{aligned} \tag{5.3.48}$$

where $u_{k,j}^* = m_k^*\left(y^j, s_j\right)$.

Based on the definitions of $u_j^* = m_h^*\left(y^j, s_j\right)$ given in (5.3.28), which is characterized by (5.3.25)–(5.3.26), and $u_{k,j}^* = m_k^*\left(y^j, s_j\right)$ provided in (5.3.28), which is described by (5.3.42)–(5.3.43), we can deduce the following:

$$P\left(y^j, s_j, u_{k,j}^*\right) - P\left(y^j, s_j, u_j^*\right) \geq 0,$$

$$P^k\left(y^j, s_j, u_{k,j}^*\right) - P^k\left(y^j, s_j, u_j^*\right) \leq 0.$$

Taking into account the aforementioned facts, on the one hand, by combining (5.3.40), (5.3.48), Theorem 5.3.5, and Corollary 5.3.2, we can deduce that

$$
\begin{aligned}
I_6 &\triangleq J^h_{z,0}\left(u^*_k(\cdot)\right) - w_h(z,0) = \sum_{j=0}^{N-1} hL\left(y^j, s_j, u^*_{k,j}\right) + \psi\left(y^N\right) - w_h(z,0) \\
&= \sum_{j=0}^{N-1}\left[-w^k_h\left(y^j + hf\left(y^j, s_j, u^*_{k,j}\right), s_{j+1}\right) + P^k\left(y^j, s_j, u^*_{k,j}\right)\right] + \psi\left(y^N\right) \\
&\quad - \left(\sum_{j=0}^{N-1}\left[w_h\left(y^j, s_j\right) - w_h\left(y^{j+1}, s_{j+1}\right)\right] + \psi\left(y^N\right)\right) \\
&= \sum_{j=0}^{N-1}\Big\{\left[-w^k_h\left(y^{j+1}, s_{j+1}\right) + w_h\left(y^{j+1}, s_{j+1}\right)\right] \\
&\quad + \left[P^k\left(y^j, s_j, u^*_{k,j}\right) - P\left(y^j, s_j, u^*_j\right)\right]\Big\} \\
&= \sum_{j=0}^{N-1}\Big\{\left[-w^k_h\left(y^{j+1}, s_{j+1}\right) + w_h\left(y^{j+1}, s_{j+1}\right)\right] \\
&\quad + \left[P^k\left(y^j, s_j, u^*_{k,j}\right) - P\left(y^j, s_j, u^*_{k,j}\right)\right] \\
&\quad + \left[P\left(y^j, s_j, u^*_{k,j}\right) - P\left(y^j, s_j, u^*_j\right)\right]\Big\} \\
&\geq \sum_{j=0}^{N-1}\Big\{\left[-w^k_h\left(y^{j+1}, s_{j+1}\right) + w_h\left(y^{j+1}, s_{j+1}\right)\right] \\
&\quad + \left[P^k\left(y^j, s_j, u^*_{k,j}\right) - P\left(y^j, s_j, u^*_{k,j}\right)\right]\Big\} \\
&\geq -\sum_{j=0}^{N-1} 2C\left(\frac{k}{h} + k\right) = -2NC\left(\frac{k}{h} + k\right) = -2TC(1+h)\frac{k}{h^2} \\
&\geq -4TC\frac{k}{h^2},
\end{aligned}
\tag{5.3.49}
$$

and on the other hand,

$$
\begin{aligned}
I_6 = \sum_{j=0}^{N-1}\Big\{&\left[-w^k_h\left(y^{j+1}, s_{j+1}\right) + w_h\left(y^{j+1}, s_{j+1}\right)\right] \\
&+ \left[P^k\left(y^j, s_j, u^*_{k,j}\right) - P\left(y^j, s_j, u^*_j\right)\right]\Big\}
\end{aligned}
$$

$$
\begin{aligned}
&= \sum_{j=0}^{N-1} \Big\{ \Big[-w_h^k \left(y^{j+1}, s_{j+1} \right) + w_h \left(y^{j+1}, s_{j+1} \right) \Big] \\
&\quad + \Big[P^k \left(y^j, s_j, u_{k,j}^* \right) - P^k \left(y^j, s_j, u_j^* \right) \Big] + \Big[P^k \left(y^j, s_j, u_j^* \right) - P \left(y^j, s_j, u_j^* \right) \Big] \Big\} \\
&\leq \sum_{j=0}^{N-1} \Big\{ \Big[-w_h^k \left(y^{j+1}, s_{j+1} \right) + w_h \left(y^{j+1}, s_{j+1} \right) \Big] \\
&\quad + \Big[P^k \left(y^j, s_j, u_j^* \right) - P \left(y^j, s_j, u_j^* \right) \Big] \Big\} \\
&\leq \sum_{j=0}^{N-1} 2C \left(\frac{k}{h} + k \right) = 2NC \left(\frac{k}{h} + k \right) = 2TC(1+h)\frac{k}{h^2} \leq 4TC\frac{k}{h^2}.
\end{aligned}
\tag{5.3.50}
$$

By combining the lower bound given by (5.3.49) and the upper bound established in (5.3.50), we arrive at the desired conclusion stated in (5.3.46). This completes the proof. □

It is worth mentioning that in Theorem 1.7 of [16] found on pages 479–480, convergence is deduced by directly taking the limit of k to zero in an equality similar to (5.3.48). Using our functions, this equality can be expressed in the form

$$
\begin{aligned}
&\sum_{j=0}^{N-1} hL \left(y^j, s_j, u_{k,j}^* \right) \\
&= \sum_{j=0}^{N-1} \Big[-w_h^k \left(y^j + hf \left(y^j, s_j, u_{k,j}^* \right), s_{j+1} \right) + w_h^k \left(y^j, s_j \right) \Big].
\end{aligned}
\tag{5.3.51}
$$

Unfortunately, the equality given in (5.3.51) cannot be derived using the arguments presented in [16]. The reason for this is that it is acknowledged that

$$
P^k \left(y^j, s_j, u_{k,j}^* \right) = w_h^k(y^j, s_j) \quad \text{on } \Omega \ni y^j
$$

and then replace $P^k \left(y^j, s_j, u_{k,j}^* \right)$ in (4.26) by $w_h^k(y^j, s_j)$ to get (5.3.51). But in [16], $P^k \left(x_i, s_j, u_{k,j}^* \right) = w_h^k(x_i, s_j)$ is true only on node point set $G = \{x_i\}$ not on the whole Ω.

It is worth noticing that although there is no explicit time term s in the context of [16], the introduction of the time term s in our formulation does not alter the fundamental arguments. To summarize, given that we have derived (5.3.46) rather than (5.3.51), the "unique minimum assumption" discussed in [16] becomes irrelevant. The essential components leading to our conclusion are the estimates provided by (5.3.49) and (5.3.50). These estimates can also be utilized to address the limitation identified in [16].

Theorem 5.3.11 (Minimizing Sequence of Continuous Optimal Feedback Control)

Under the assumptions of Theorem 5.3.10, *let* $k = \mathcal{O}(h^{2+\gamma})$ *for some constant* $\gamma > 0$ *and* w *the viscosity solution of HJB equation* (5.3.7). *Then for any given* $z \in \Omega$,

$$J_{z,0}\left(u_k^*(\cdot)\right) \to \inf_{u(\cdot)\in\Delta} J_{z,0}\left(u(\cdot)\right) \equiv w(z,0) \quad \text{as } h \to 0+, \tag{5.3.52}$$

where $J_{z,0}$ *is defined by* (5.3.5).

Proof The proof follows a similar structure to that of Theorem 5.3.9. For the sake of completeness, we present a detailed argument here.

Consider an arbitrary point $z \in \Omega$. We begin by observing that

$$\begin{aligned}
&\left|J_{z,0}\left(u_k^*(\cdot)\right) - w(z,0)\right| \\
\le &\left|J_{z,0}\left(u_k^*(\cdot)\right) - J_{z,0}^h\left(u_k^*(\cdot)\right)\right| + \left|J_{z,0}^h\left(z,0,u_k^*(\cdot)\right) - w_h(z,0)\right| \\
&+ \left|w_h(z,0) - w(z,0)\right| \\
\triangleq\ & I_7 + I_8 + I_9.
\end{aligned} \tag{5.3.53}$$

By (5.3.46), the term

$$I_8 \triangleq \left|J_{z,0}^h\left(u_k^*(\cdot)\right) - w_h(z,0)\right| \le 4TC\frac{k}{h^2} = \mathcal{O}\left(h^\gamma\right) \to 0 \quad \text{as } h \to 0+. \tag{5.3.54}$$

The convergence of the term I_9 to 0 as $h \to 0+$ is established by Theorem 5.3.4. Therefore, our focus shifts to the term I_7:

$$\begin{aligned}
I_7 &\triangleq \left|J_{z,0}\left(u_k^*(\cdot)\right) - J_{z,0}^h\left(u_k^*(\cdot)\right)\right| \\
&\le \left|\int_0^T L\left(y_k^*(t),t,u_k^*(t)\right)\,\mathrm{d}t + \psi\left(y_k^*(T)\right) - \sum_{j=0}^{N-1} hL\left(y^j,s_j,u_{k,j}^*\right) - \psi\left(y^N\right)\right| \\
&\le \left|\int_0^T L\left(y_k^*(t),t,u_k^*(t)\right)\,\mathrm{d}t - \sum_{j=0}^{N-1} hL\left(y^j,s_j,u_{k,j}^*\right)\right| + \left|\psi\left(y_k^*(T)\right) - \psi\left(y^N\right)\right| \\
&\triangleq I_{71} + I_{72},
\end{aligned} \tag{5.3.55}$$

where $y_k^*(\cdot)$ represents the solution of the system given by (5.3.1) with the control $u_k^*(\cdot)$ defined according to (5.3.44). Here, $u_{k,j}^* = m_k^*\left(y^j, s_j\right)$ is determined by Eqs. (5.3.42)–(5.3.43), and as before, y^j denotes the solution of system (5.3.45). Analogous to the estimate provided in (5.3.33), we can derive the following estimate:

$$\left\| y_k^*(t) - y^j \right\| \leq C_T h, \quad t \in [s_j, s_{j+1}], \; \forall\, j = 0, 1, \cdots, N-1.$$

Therefore,

$$\begin{aligned} I_{71} &\leq \sum_{j=0}^{N-1} \int_{s_j}^{s_{j+1}} \left| L\left(y_k^*(t), t, u_{k,j}^*\right) - L\left(y^j, s_j, u_{k,j}^*\right) \right| \mathrm{d}t \\ &\leq \sum_{j=0}^{N-1} \int_{s_j}^{s_{j+1}} L_L \left(\left\| y_k^*(t) - y^j \right\| + |t - s_j| \right) \mathrm{d}t \leq T L_L (C_T + 1) h, \end{aligned} \tag{5.3.56}$$

and

$$I_{72} = \left| \psi\left(y_k^*(T)\right) - \psi\left(y^N\right) \right| \leq L_\psi \left\| y_k^*(T) - y^N \right\| \leq L_\psi C_T h. \tag{5.3.57}$$

The assertion (5.3.52) then follows from (5.3.53) to (5.3.57). □

5.3.4 Approximation Algorithm

We now introduce an algorithm that encapsulates the optimal feedback control approximation process outlined in the preceding subsections. It is important to clarify that only the full discretization scheme referenced as (5.3.13) is suitable for numerical computations. In contrast, the scheme denoted as (5.3.9) holds theoretical significance alone, as its space variable remains in a continuous, non-discrete form. This algorithm computes both the viscosity solution and the corresponding optimal control-trajectory pairs.

5.3.4.1 The Algorithm

Leveraging the full discretization scheme outlined in (5.3.13) and the convex linear combination approach detailed in (5.3.14), we are now ready to introduce the following algorithm.

Algorithm 8 Full discretization and convex linear combination algorithm

1: Initial partitioning based on time and space. Let $N > 0$ be a fixed integer. Define an equal partition of the interval $[0, T]$ with the length of each subinterval being $0 < h = \frac{T}{N} < 1$, denoted as $S = \{s_j = jh \mid j = 0, 1, \ldots, N\}$. For any $q \in \{0, 1, \ldots, N\}$, the notation (u^q, y^q) represents the qth optimal control-trajectory pair.
The domain $\Omega \subset \mathbb{R}^n$ is presumed to be a closed and bounded polyhedron with the property that for all sufficiently small $h > 0$ and $(x, s, u) \in \Omega \times [0, T] \times \mathbb{U}$, $x + hf(x, s, u) \in \Omega$. Let $k = Ch^{2+\gamma}$ represent the size of the space-discretization mesh, where C and γ are positive constants chosen appropriately. To proceed, we construct a regular triangulation of Ω. This triangulation comprises a finite number of simplices, denoted by $\{T_i\}$, which satisfy $\Omega = \bigcup_i T_i$ and $\max_i\{\text{diam}(T_i)\} = k$. From a numerical standpoint, the process of regular triangulation is considered complete once the size k has been specified. This triangulation serves as the foundation for further numerical computations within the domain Ω.
Denote by x_l with $l = 1, 2, \cdots, M$, the nodes of the triangulation. Set

$$G = \{x_l \mid l = 1, 2, \cdots, M\}.$$

2: Calculate approximate viscosity solution. This is achieved through the finite-difference scheme, utilizing the time-space mesh approximation as outlined in (5.3.13):

$$\begin{cases} V_i^j = hL_i^j + \sum_{l=1}^{M} \lambda_i^{l,j} V_l^{j+1}, \\ u_i^j \in \arg\inf_{u\in\mathbb{U}} \left\{Y_i^j(u)\right\}, \\ \quad \text{where } Y_i^j(u) \triangleq hL(x_i, s_j, u) + \sum_{l=1}^{M} \lambda_i^l(u, s_j) V_l^{j+1}, \\ V_i^N = \psi(x_i) \end{cases} \tag{5.3.58}$$

for $i = 1, 2, \cdots, M$ and $j = N-1, N-2, \cdots, 1, 0$, where

$$V_i^j = V(x_i, s_j), \quad L_i^j = L\left(x_i, s_j, u_i^j\right), \quad \lambda_i^{l,j} = \lambda_i^l\left(u_i^j, s_j\right)$$

and $\left\{\lambda_i^l(u, s_j) \in [0, 1] \,\middle|\, l = 1, 2, \cdots, M\right\}$ is the set of coefficients in the unique convex linear combination

$$x_i + hf(x_i, s_j, u) = \sum_{l=1}^{M} \lambda_i^l(u, s_j) x_l$$

with $\sum_{l=1}^{M} \lambda_i^l(u, s_j) = 1$.

This convex linear combination is formulated based on the vertices of the simplex within which the point $x_i + hf(x_i, s_j, u)$ is located.

More precisely, given $\{V_i^N\}_{i=1}^{M}$, when $\left\{V_i^{j+1}\right\}_{i=1}^{M}$ is known, for any $i \in \{1, 2, \cdots, M\}$, V_i^j can be determined through (5.3.58) by following these substeps:

1). Discretize the control domain $\mathbb{U}$ into a finite set denoted as $\mathbb{U}_d \triangleq \{u_l \mid l = 0, 1, 2, \cdots\}$ using a specified size of the control-space mesh $d > 0$.

2). For a given $u \in \mathbb{U}_d$, by employing a searching technique commonly used in numerical analysis, one can determine the specific simplex T_{i_0} within which the point $x_i + hf(x_i, s_j, u)$ is located. Subsequently, by formulating the convex linear combination of $x_i + hf(x_i, s_j, u)$ with respect to the vertices of the simplex T_{i_0}, one can derive the coefficients $\left\{\lambda_i^l(u, s_j)\right\}$.

3). Calculate $Y_i^j(u)$ for all $u \in \mathbb{U}_d$, and then set

$$V_i^j = \min_{v \in \mathbb{U}_d} \left\{Y_i^j(v)\right\}, \quad A \triangleq \min_{\|\cdot\|} \left\{u \in \mathbb{U}_d \;\middle|\; Y_i^j(u) = V_i^j\right\}.$$

It is obvious that $V_i^j = Y_i^j\left(u_i^j\right)$ for any $u_i^j \in A$.

Repeat the first three substeps (the first through the third) of Step 2 as described above for every $i \in \{1, 2, \cdots, M\}$ and iterate from $j = N - 1$ down to $j = 0$ to complete Step 2. This process will yield the approximate viscosity solution $G \times S$: $\left\{\left\{V_i^j\right\}_{i=1}^{M}\right\}_{j=0}^{N}$.

It is important to mention that mathematical programming techniques are not employed to address the optimization problem stated in (5.3.13).

3: Calculate the optimal feedback control-trajectory pairs $\{(u^q, y^q)\}_{q=0}^{N}$. For initial setting, let $y^0 = z$ and set $q = 0$.

1). Calculate the qth optimal feedback control u^q.

Since both y^q and $\left\{\left\{V_i^j\right\}_{i=1}^{M}\right\}_{j=0}^{N}$ are known quantities, the determination of the qth optimal feedback control $u^q = u(s_q)$ is determined by

$$\begin{cases} u^q \in \arg\inf_{u \in \mathbb{U}} \left\{ hL\left(y^q, s_q, u\right) + W\left(y^q + hf\left(y^q, s_q, u\right), s_{q+1}\right) \right\}, \\ W\left(y^q + hf\left(y^q, s_q, u\right), s_{q+1}\right) = \sum_{l=1}^{M} \mu_q^l(u, s_q) V_l^{q+1} \end{cases} \tag{5.3.59}$$

via (5.3.14), where $\left\{\mu_q^l(u, s_q) \in [0, 1] \mid l = 1, 2, \cdots, M\right\}$ is the set of coefficients in the unique convex linear combination

$$y^q + hf\left(y^q, s_q, u\right) = \sum_{l=1}^{M} \mu_q^l(u, s_q) x_l$$

with $\sum_{l=1}^{M} \mu_q^l(u, s_q) = 1$. Here, once again, the convex linear combination is generated based on the vertices of the simplex within which $y^q + hf\left(y^q, s_q, u\right)$ is situated.

Please note that in (5.3.59), the control u^q is selected based on its minimal energy consumption. It is important to mention that this minimal energy control may not be the only option, and we have the flexibility to choose any one among them.

The specifics of the calculation closely resemble those described in Step 2.

After completing these steps, the qth optimal feedback control-trajectory pair, denoted as $(u^q, y^q) = \left(u(s_q), y(s_q)\right)$, is successfully obtained.

2). Calculate the $(q+1)$-th optimal feedback trajectory y^{q+1}.

Once $u^q = u(s_q)$ is known, we proceed to solve the state equation:

$$\frac{y^{q+1} - y^q}{h} = f(y^q, s_q, u^q)$$

to obtain the $(q+1)$-th optimal feedback trajectory $y^{q+1} = y(s_{q+1})$.

3). Iterate to the next time instant. Increment q by 1 ($q = q + 1$). If q reaches N, set $u^q = u^{q-1}$ and terminate the procedure. Otherwise, return to the first substep of Step 3. Repeat this iterative process to obtain all optimal feedback control-trajectory pairs:

$$\left\{\left(u^q, y^q\right)\right\}_{q=0}^{N} = \left\{\left(u(s_q), y(s_q)\right)\right\}_{q=0}^{N}.$$

5.3.4.2 An Example

We employ the algorithm described in Sect. 5.3.4.1 to address the following optimal control problem:

$$\begin{cases} y'(t) = 2(1 - y(t))u(t), \quad t \in (0, 1], \\ y(0) = z, \\ J(u(\cdot)) = \displaystyle\int_0^1 |1 - y(t)|(1+t)^2 u(t)^2 \, \mathrm{d}t + 2|1 - y(1)|, \\ \displaystyle\min_{u(\cdot)\in\Delta^*} J(u(\cdot)), \end{cases} \tag{5.3.60}$$

where $\Delta^* \triangleq L^\infty([0, 1]; \mathbb{U})$, $\mathbb{U} \triangleq [0, 1]$, and $z \in \mathbb{R}$ is a given initial value.

This system is chosen because its optimal control can be uniquely determined and derived analytically. Therefore, we can leverage the algorithm to compute numerical results.

The HJB equation associated with (5.3.60) is expressed as follows:

$$\begin{cases} -w_s(x, s) - \displaystyle\inf_{u\in\mathbb{U}} \left\{ \nabla_x w(x, s) \cdot 2(1 - x)u + |1 - x|(1 + s)^2 u^2 \right\} = 0, \\ \qquad\qquad (x, s) \in \mathbb{R} \times [0, 1), \\ w(x, 1) = 2|1 - x|, \quad x \in \mathbb{R}. \end{cases} \tag{5.3.61}$$

According to the uniqueness theorem [242], it acknowledges the existence of a unique viscosity solution.

It is straightforward to verify that the function $w \in C(\mathbb{R} \times [0, 1])$ given by

$$w(x, s) = |1 - x|(1 + s), \quad (x, s) \in \mathbb{R} \times [0, 1] \tag{5.3.62}$$

is a viscosity solution of (5.3.61). It is worth noting that when $z = 1$, the control problem (5.3.60) simplifies to a trivial case, namely, that $y(u(\cdot), \cdot) \equiv 1$ on $[0, 1]$ and $J(u(\cdot)) = 0$ for any control function $u(\cdot) \in \Delta^*$. We restrict our attention to the scenario where $z = 1$. Under this condition, the unique optimal feedback control $u^*(\cdot)$ and its corresponding optimal trajectory $y^*(\cdot)$ for system (5.3.60) can be analytically derived as

$$u^*(z, t) = u^*\left(y^*(z, t), t\right) = \frac{1}{1+t}, \quad t \in [0, 1], \tag{5.3.63}$$

and

$$y^*(z, t) = \frac{t^2 + 2t + z}{(1+t)^2}, \quad t \in [0, 1]. \tag{5.3.64}$$

To implement the algorithm, we must select an appropriate space domain Ω that fulfills assumption (5.3.12). Specifically, this assumption requires that for all sufficiently small values of h, $x+hf(x, s, u) \in \Omega,\ \forall\ (x, s, u) \in \Omega\times[0, 1]\times[0, 1]$. In our example, where $\Omega = [0, 1]$, assumption (5.3.12) holds true provided that $h < \frac{1}{2}$. This is because for any $(x, s, u) \in [0, 1] \times [0, 1] \times [0, 1]$, we observe that

$$1 - 2hu > 0, \quad x + hf(x, s, u) = x + 2h(1 - x)u = (1 - 2hu)x + 2hu \geq 0,$$

and

$$x + hf(x, s, u) = x + 2h(1 - x)u = (1 - 2hu)x + 2hu \leq 1 - 2hu + 2hu = 1.$$

Let $h = 0.01$, and $k = d = 0.005$. By employing the algorithm, we numerically derive the viscosity solution as well as the optimal feedback control-trajectory pairs for the control problem (5.3.60). The outcomes are graphically represented in Figs. 5.14, 5.15, and 5.16 for two distinct initial values: $z = 0.30$ and $z = 0.65$. The algorithm was executed using the MATLAB programming language.

Figure 5.14 illustrates the numerical solution of the viscosity solution given by (5.3.62). Figures 5.15 and 5.16 present the numerical solutions for the optimal feedback control and trajectory functions, (5.3.63) and (5.3.64), with initial states $z = 0.3$ and $z = 0.65$, respectively. Table 5.5 provides a summary of the computed errors in the maximum norm between the numerical solutions and the analytical solutions for w, u^*, and y^*. The results demonstrate that the numerical solutions derived from our algorithm closely approximate the analytical solutions, highlighting the algorithm's effectiveness.

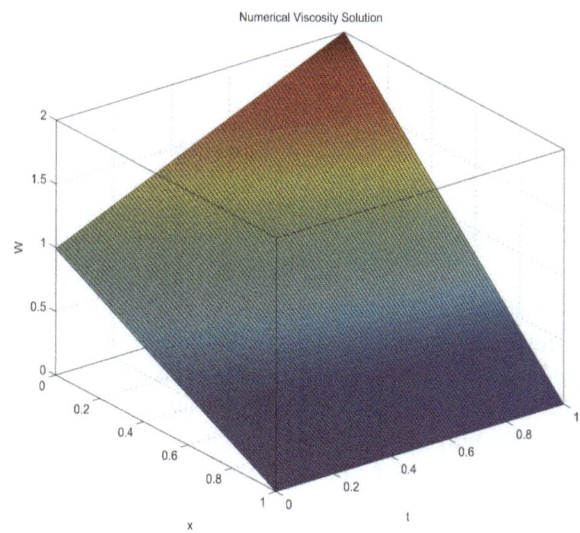

Fig. 5.14 Numerical solution of the viscosity solution

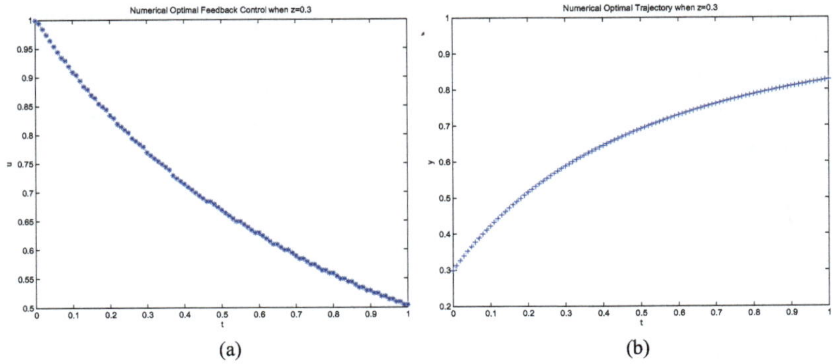

Fig. 5.15 Numerical solutions of optimal feedback control and trajectory functions with $z = 0.3$. (**a**) Control. (**b**) Trajectory

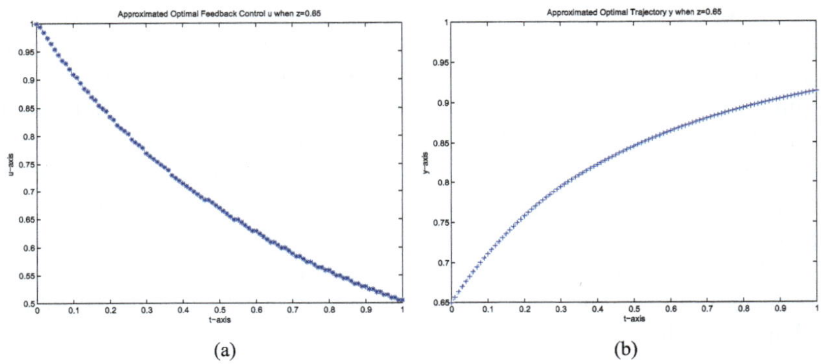

Fig. 5.16 Numerical solutions of optimal feedback control and trajectory functions with $z = 0.65$. (**a**) Control. (**b**) Trajectory

Table 5.5 The computed errors in the maximum norm between the numerical solutions and analytical solutions for w, u^*, and y^*

Error	Viscosity solution	Control	Trajectory
$z = 0.3$	7.5e-3	3.9e-3	5.0e-3
$z = 0.65$	7.5e-3	1.9e-3	5.0e-3

It should be emphasized that in this particular example, not all the conditions specified in Theorem 5.3.2 are fulfilled. Specifically, the boundedness property for f, L, ψ as stated in (5.3.8) is not satisfied. Despite this, the numerical outcome is still deemed satisfactory. The conditions imposed in Theorem 5.3.2 primarily serve the purpose of mathematical rigor in its proof. The simulation results indicate a potential broader applicability of our algorithm to a wider range of systems, suggesting its versatility and possible extension beyond the theoretical boundaries established in the theorem.

In this section, we present two discretization schemes aimed at approximating the viscosity solution of the evolutive HJB equation associated with the value function of a general continuous finite-dimensional control system having a finite horizon cost functional. One scheme relies on the time difference approximation, while the other incorporates both time and space approximations. The convergence of each scheme is established through corresponding discrete DDPs. Furthermore, we demonstrate that the optimal controls derived from these schemes exhibit "almost optimal" performance for the original continuous system.

The significance of this section lies in its ability to prove the convergence of an algorithm that approximates the optimal feedback control using dynamic programming for problems with a finite horizon cost functional, without relying on a discount factor. This is a notable advancement since existing literature primarily focuses on approximating optimal feedback controls for problems with infinite horizon cost functionals. Additionally, our approach eliminates the need for a discount factor in the cost functional, which has traditionally been a crucial component for ensuring convergence in previous studies. As a result, our work paves the way for developing simplified algorithms that can be readily applied to practical optimal control problems.

5.4 A Local Approximation Algorithm

In this section, we focus on optimal feedback control for a general continuous-time, finite-dimensional deterministic system with a finite horizon cost functional. We develop a practical algorithm that employs the dynamic programming approach to compute the numerical solution for optimal feedback control. The key features of this algorithm are twofold.

Firstly, it builds upon a convergent and constructive algorithm for the optimal feedback control law, which we introduced in Sect. 5.3. This algorithm approximates the viscosity solution of the time-space discretization scheme derived from the dynamic programming method. Secondly, the algorithm significantly reduces computational complexity by only computing values of the viscosity solution within local cones surrounding the optimal trajectory.

To demonstrate the effectiveness and efficiency of our algorithm, we present two numerical experiments. These experiments illustrate the algorithm's performance in terms of both accuracy and computational speed, highlighting its practical utility for solving optimal feedback control problems in deterministic systems with finite horizon cost functionals.

In reality, our objective is to craft a viable and practical algorithm for obtaining numerical solutions for optimal feedback control in finite-dimensional control systems with a finite horizon cost functional. This approach hinges on the principles of dynamic programming.

Our algorithm takes cue from a convergent and constructive discrete scheme for determining the optimal feedback control law, elaborated in Sect. 5.3. Furthermore, it is influenced by an approximation of the viscosity solution that assures convergence, along with a local approximation method for this viscosity solution rooted in the upwind finite-difference method (cited in [62, 185, 233]).

The main advantage of this algorithm is that it eliminates the need to compute the value function at every single point in the $(n+1)$-dimensional space-time. This aspect significantly simplifies the computational process and reduces its complexity.

Let us first revisit the problem at hand. We consider the optimal control problem (5.3.1)–5.3.2. By the dynamic programming approach, its HJB equation is (5.3.7). In this way, we know that Theorem 5.3.2 holds, providing valuable insights into the HJB equation.

It should be emphasized that our algorithm has been specifically developed for fixed-horizon problems without any state restrictions, operating under the assumption that both the viscosity solution and the optimal feedback control exist and are unique.

We proceed as follows. Firstly, we revisit the optimal feedback control law developed in Sect. 5.3, which has been adopted in our algorithm. Present a local approximation method for computing the viscosity solution along the optimal trajectory, utilizing the upwind finite-difference method. Then, we focus on the development of a new, feasible algorithm for obtaining numerical solutions to the optimal feedback control problem. Finally, we demonstrate the effectiveness and efficiency of our algorithm through two numerical experiments, including a two-dimensional problem.

5.4.1 Approximation to Optimal Feedback Control: Revisit

For each large positive integer N provided, we subdivide the interval $[0, T]$ into N equal subintervals. Define $s_j = jh,\ j = 0, 1, \cdots, N,\ h = \frac{T}{N} < 1, S = \{s_j \mid j = 0, 1, \cdots, N\}$, and let $S = \{s_j \mid j = 0, 1, \cdots, N\}$ represent this set of subdivision points. Our algorithm employs the minimal energy optimal feedback control law, which is given by

$$u_k^*(s) = u_{k,j}^* \triangleq a\left(y^j, s_j\right), \quad s \in [s_j, s_{j+1}),\ j = 0, 1, \cdots, N-1, \tag{5.4.1}$$

where $y^j \approx y(s_j)$ denotes the approximate solution at time s_j of the following time-discretized system, with the control u_k^* determined by (5.4.1):

$$\begin{cases} y^{j+1} = y^j + hf\left(y^j, s_j, u_k^*(s_j)\right), & j = 0, 1, \cdots, N-1, \\ y_0 = z, \end{cases} \tag{5.4.2}$$

and $a(x, s_j) \in A(x, s_j)$ is any one element of the nonempty subset:

$$A(x, s_j) = \left\{ v \in \mathbb{U} \;\middle|\; P^k\left(x, s_j, v\right) = \min_{u \in \mathbb{U}} P^k(x, s_j, u),\ v \text{ has minimum norm} \right\}. \tag{5.4.3}$$

In (5.4.3), P^k is defined as

$$P^k(x, s_j, u) \triangleq hL(x, s_j, u) + w_h^k\left(x + hf(x, s_j, u), s_{j+1}\right), \tag{5.4.4}$$

and w_h^k represents a known approximate solution of the viscosity solution.

It is worth noting that the selection of minimal norm for optimal control is merely one of the possible policies, as outlined in Appendix A of [16].

In Sect. 5.3, we introduce a convergent time-space discretization scheme for the HJB equation defined on $\Omega \times [0, T]$, with its solution denoted as w_h^k. It demonstrates that the optimal feedback control law u_k^* given by (5.4.1) constitutes a minimizing sequence for the original control system (5.3.1)–(5.3.3). This conclusion holds under the assumptions stated in Theorem 5.3.2 and the standard assumption ([16]) expressed in (5.4.5):

$$x + hf(x, s, u) \in \Omega, \quad \forall\ (x, s, u) \in \Omega \times [0, T] \times \mathbb{U}, \tag{5.4.5}$$

which holds for all sufficiently small h on a closed bounded polyhedron $\Omega \subset \mathbb{R}^n$.

Furthermore, a numerical experiment conducted in Sect. 5.3 yields promising results, indicating that the algorithm described therein can be applied to certain systems even when the conditions of Theorem 5.3.2 are not strictly met.

Although a convergence result exists for the approximate viscosity solution w_h^k in Sect. 5.3, the need to compute the value function at all points within a specific domain can lead to the notorious "curse of dimensionality" problem. Consequently, the primary focus of this section is to minimize computational complexity as much as possible. In the subsequent subsections, we will demonstrate that when determining the optimal feedback control, it is not necessary to compute the viscosity solution for the entire time-spatial space. Instead, computations can be confined to local cones surrounding the optimal trajectory.

5.4.2 Local Approximation of Viscosity Solution

In this subsection, we aim to devise a practical and efficient algorithm for finding the numerical solution of the optimal feedback control [108]. This approach does not rely on assumptions (5.3.8) and (5.4.5). Instead, we assume the existence and uniqueness of both the viscosity solution and the optimal feedback control.

Our algorithm incorporates two key steps to minimize computational complexity. Firstly, we confine the calculations to a local cone, denoted as R_q, around the optimal trajectory. Secondly, we avoid unnecessary rebuilds of R_q unless required. These steps significantly enhance the efficiency of our numerical method.

5.4.2.1 Build a Local Cone

It is important to note that the optimal feedback control is formulated based on the viscosity solution. Once the qth optimal state is determined, computing the qth optimal feedback control via (5.4.1) only requires the values of the viscosity solution within a local cone centered around the qth optimal state. This concept of localized approximation for the viscosity solution represents the first step in reducing computational complexity.

Now, let us introduce the local approximation method. This approach is rooted in the upwind finite-difference method [62, 185, 233]. However, there's a key distinction: We selectively choose the cones on which we numerically solve the HJB equation. Specifically, we focus on cones that surround the optimal trajectory.

In our algorithm, we simultaneously consider the calculation of the viscosity solution and the construction of optimal feedback control-state pairs. A crucial aspect of this process is that we do not always need to rebuild new local cones at each optimal state unless absolutely necessary. This significant insight can be regarded as the second step in reducing computational complexity.

Let us describe the process of local approximation for the viscosity solution. Given an initial point $y^0 = z \in \mathbb{R}^n$, the algorithm proceeds iteratively. Once a point $y^q = (y_1^q, y_2^q, \cdots, y_n^q)$, $0 \le q \le N-1$, along the optimal trajectory is known, we construct a local cone

$$R_q \triangleq \bigcup_{j=q}^{N} \left(D_j \times [s_j, s_{j+1})\right) \tag{5.4.6}$$

with

$$D_j \triangleq \prod_{p=1}^{n} \left[y_p^q - (j-q)\Delta x_p,\, y_p^q + (j-q)\Delta x_p\right], \quad j = q, \cdots, N. \tag{5.4.7}$$

We consider the following HJB equation on R_q:

$$\begin{cases} -w_s(x,s) - \inf\limits_{u\in\mathbb{U}} \left\{ \sum\limits_{p=1}^{n} w_{x_p}(x,s) f_p(x,s,u) + L(x,s,u) \right\} = 0, \\ \qquad\qquad\qquad (x,s) \in R_q, \\ w(x,T) = \psi(x), \quad x \in D_N, \end{cases} \tag{5.4.8}$$

where Δx_p is a small size of the space mesh along the x_p-axis. Observe that $D_q \subset D_{q+1} \subset \cdots \subset D_N$ where D_N represents the largest spatial domain within which we need to consider the spatial variable on R_q.

We note that the local cone R_q, which is solely determined by its vertex $\left(y^q, s_q\right)$ and the mesh size Δx_p, has been specially designed to minimize computational complexity. To facilitate numerical computations, it is necessary to discretize the continuous cone R_q. Let $\mathbf{i} = (i_1, i_2, \cdots, i_n)$, $\mathbf{i}p^+ = (i_1, i_2, \cdots, i_{p-1}, i_p + 1, i_{p+1}, \cdots, i_n)$ and $\mathbf{i}p^- = (i_1, i_2, \cdots, i_{p-1}, i_p - 1, i_{p+1}, \cdots, i_n)$. For the time-space discretization, we assume a uniform time step $h = \frac{T}{N}$, where T is the total time duration and N is the number of time steps. The discrete time points are given by $s_j = jh$ for $j = 0, 1, \cdots, N$, and

$$x_{p,i_p} = y_p^q + [i_p - (N - q)]\Delta x_p, \tag{5.4.9}$$

$p = 1, 2, \cdots, n$, $i_p = 0, 1, \cdots, 2(N - q)$.

We approximate (5.4.8) using the following upwind finite-difference scheme:

$$\begin{cases} \displaystyle\sum_{p=1}^{n}\left[\frac{1+\operatorname{sgn}\left(f_{p,\mathbf{i}}^{j+1}\right)}{2} f_{p,\mathbf{i}}^{j+1} \frac{w_{\mathbf{i}p^+}^{j+1} - w_{\mathbf{i}}^{j+1}}{\Delta x_p} + \frac{1-\operatorname{sgn}\left(f_{p,\mathbf{i}}^{j+1}\right)}{2} f_{p,\mathbf{i}}^{j+1} \frac{w_{\mathbf{i}}^{j+1} - w_{\mathbf{i}p^-}^{j+1}}{\Delta x_p}\right] \\ \qquad + \dfrac{w_{\mathbf{i}}^{j+1} - w_{\mathbf{i}}^{j}}{h} + L_{\mathbf{i}}^{j+1} = 0, \\ u_{\mathbf{i}}^{j+1} \in \operatorname*{arg\,inf}_{u \in \mathbb{U}} \left\{ \displaystyle\sum_{p=1}^{n}\left[\frac{1+\operatorname{sgn}\left(f_p(x_{\mathbf{i}}, s_{j+1}, u)\right)}{2} f_p(x_{\mathbf{i}}, s_{j+1}, u) \frac{w_{\mathbf{i}p^+}^{j+1} - w_{\mathbf{i}}^{j+1}}{\Delta x_p} \right.\right. \\ \quad \left.\left. + \dfrac{1-\operatorname{sgn}\left(f_p(x_{\mathbf{i}}, s_{j+1}, u)\right)}{2} f_p(x_{\mathbf{i}}, s_{j+1}, u) \frac{w_{\mathbf{i}}^{j+1} - w_{\mathbf{i}p^-}^{j+1}}{\Delta x_p}\right] + L(x_{\mathbf{i}}, s_{j+1}, u) \right\}, \\ w_{\mathbf{i}}^{N} = \psi(x_{\mathbf{i}}), \end{cases} \tag{5.4.10}$$

for $i_p = N - (j - q), N - (j - q) + 1, \cdots, N, \cdots, N + (j - q)$, $p = 1, 2, \cdots, n$ and $j = N - 1, N - 2, \cdots, q$, where $\operatorname{sgn}(a)$ denotes the sign of $a \in \mathbb{R}$ and

$$f_{p,\mathbf{i}}^{j+1} = f_p\left(x_{\mathbf{i}}, s_{j+1}, u_{\mathbf{i}}^{j+1}\right), \quad L_{\mathbf{i}}^{j+1} = L\left(x_{\mathbf{i}}, s_{j+1}, u_{\mathbf{i}}^{j+1}\right),$$

$$x_{\mathbf{i}} = (x_{i1}, x_{i2}, \cdots, x_{in})^{\top}.$$

It should be noted that (5.4.10) is inherently semi-Lagrangian in nature. Specifically, it represents a variation of Falcone's semi-Lagrangian scheme, as referenced in [75].

Through the application of (5.4.10), one can obtain the values of the approximate viscosity solution on the hyper-pyramidal region H_q, which represents the set of knot points within the local cone R_q as defined by

$$H_q \triangleq \bigcup_{j=q}^{N} \left(\hat{D}_j \times \{s_j\}\right), \tag{5.4.11}$$

with

$$\hat{D}_j \triangleq \left\{\left(x_{1,i_1}, x_{2,i_2}, \cdots, x_{n,i_n}\right) | i_p = 0, 1, \cdots, 2(N-q),\ p=1, 2, \cdots, n\right\} \subset D_j, \tag{5.4.12}$$

where $\{x_{p,i_p}\}$ are specified according to (5.4.9). Using piecewise linear interpolation, we can extend the values of the approximate viscosity solution from the discrete hyper-pyramidal region H_q to the continuous cone R_q. These extended values then enable us to compute the qth optimal feedback control. Importantly, the selection of Δx_p is crucial for accurate approximation. For instance, when Δx_p is chosen appropriately, such as

$$\Delta x_p \geq h \max_{u \in \mathbb{U}} \left\{\left\| f\left(y^q, s_q, u\right)\right\|\right\}, \quad p = 1, 2, \cdots, n, \tag{5.4.13}$$

then all subsequent possible states will remain within the current local cone. In other words,

$$y^{q+1}(u) = y^q + hf\left(y^q, s_q, u\right) \in D_{q+1}$$

for all $u \in \mathbb{U}$, which enables us to obtain the qth optimal feedback control u^q via (5.4.1)–(5.4.4).

From the preceding discussion, it is evident that the crucial step in determining the qth optimal feedback control u^q from the given qth optimal state y^q involves constructing a local cone R_q. This cone serves as the domain for computing the viscosity solution of the HJB equation, rather than the entire time-spatial space. This approach significantly reduces computational complexity, marking an important first step in the process.

We would like to remark that the concept of localization is not novel. From a practical perspective, a similar idea has been implemented in the work cited in [69].

To conclude this subsection, we highlight the stability of the explicit difference scheme (5.4.10) on the hyper-pyramidal region $H_q \subset R_q$. The necessary conditions for the stability of (5.4.10) are provided in [233]. Through a straightforward computation, one obtains

$$h \leq \frac{1}{\displaystyle\sum_{p=1}^{n} \frac{1}{\Delta x_p} \|f_p\|_\infty}, \tag{5.4.14}$$

where

$$\|f_p\|_\infty \triangleq \max_{(x,s,u)\in R_q\times\mathbb{U}} \left\{|f_p(x,s,u)|\right\}.$$

If $\Delta x = \Delta x_p,\ p = 1, 2, \cdots, n$, then (5.4.14) becomes

$$\Delta x \ge \sum_{p=1}^{n} \|f_p\|_\infty \cdot h. \tag{5.4.15}$$

Particularly, in the case of $n = 1$, it is indicated in [233] that (5.4.15) can be relaxed as

$$\Delta x \ge \frac{\|f\|_\infty}{2} \cdot h. \tag{5.4.16}$$

5.4.2.2 Rebuild a Local Cone if Necessary

It is important to note that our objective is to maximize the utilization of the data obtained within the local cone R_q. When the function f is bounded, the stability condition (5.4.14) ensures that (5.4.13) holds and further implies that $y^{q+i}(u) \in D_q$ for all subsequent time steps $i > 0$. However, in cases where f is unbounded, the necessity to rebuild R_q arises due to the fact that the state, driven by f, may exit the cone R_q. Technically, this occurs because f undergoes truncation during the construction of R_q (refer to Step 2 in Algorithm 9 of Sect. 5.4.3 below). To alleviate computational demands, the approach is to rebuild R_q only when it becomes necessary.

Let us explain the second step in reducing computational complexity within the aforementioned process, which entails simultaneously calculating the local approximate viscosity solution and constructing the optimal feedback control.

Notice that the $(q + 1)$-th optimal state y^{q+1} is derived from the qth optimal feedback control u^q via (5.4.2). Upon generating this new optimal state y^{q+1}, one might typically construct a fresh local cone R_q with its vertex at $\left(y^{q+1}, s_{q+1}\right)$ and repeat the aforementioned process to determine the $(q + 1)$-th optimal feedback control. While this approach is intuitive and viable, it is not always essential, as previously mentioned. In fact, whether to build a new local cone R_{q+1} at the vertex $\left(y^{q+1}, s_{q+1}\right)$ for computing the $(q + 1)$-th optimal feedback control hinges on the following observation: If all subsequent possible states reside within the current local cone R_q, that is, $y^{q+2}(u) = y^{q+1} + hf\left(y^{q+1}, s_{q+1}, u\right) \in D_{q+2}$ then we can simply set $R_{q+1} = R_q$. This situation may arise occasionally. When it does, we can leverage the existing data within the current local cone R_q to compute the $(q+1)$-th optimal feedback control, eliminating the need to recalculate the viscosity solution at this $(q + 1)$-th step. Conversely, if this condition is not met, we must construct a new local cone R_{q+1}. This insight allows us to maximize the utilization of information within each local cone once it has been established. Crucially, if a new local cone becomes necessary, any data acquired prior to this new cone becomes obsolete and can be discarded. These two steps significantly reduce computational complexity.

We are now ready to outline the process in concrete terms. Given the initial state y^0, the construction of R_0 ensures that all subsequent possible states $y^1(u) = y^0 + hf\left(y^0, s_0, u\right) \in D_1$. Therefore, we can obtain the 0th optimal feedback control u^0 using the first step of the computational complexity reduction described earlier. Similarly, we can determine the optimal state y^1 through (5.4.1) and (5.4.2), respectively.

Suppose we have already generated the qth optimal feedback control u^q within the cone R_q. Consequently, we can determine the optimal state y^{q+1} through (5.4.1) and (5.4.2). If it holds that $y^{q+2}(u) = y^{q+1} + hf\left(y^{q+1}, s_{q+1}, u\right) \in D_{q+2}$, then we simply set $R_{q+1} = R_q$. Following this principle, we can identify the maximum $q_0 \geq 0$ such that $R_{q+q_0} = R_{q+q_0-1} = \cdots = R_q$. Subsequently, we can obtain all optimal feedback controls $(u^q, u^{q+1}, \cdots, u^{q+q_0})$ and optimal states $(y^{q+1}, y^{q+2}, \cdots, y^{q+q_0+1})$ through (5.4.1) and (5.4.2). However, if there exists at least one $u \in U$ such that $y^{q+q_0+2}(u) = y^{q+q_0+1} + hf\left(y^{q+q_0+1}, s_{q+q_0+1}, u\right) \notin D_{q+q_0+2}$, we must construct a new local cone R_{q+q_0+1} and continue the process until we have found all optimal feedback control-state pairs $\{(u^q, y^q)\}_{q=0}^N$.

5.4.3 Algorithm

In this subsection, we introduce an algorithm that encapsulates the procedure for approximating optimal feedback controls, as elaborated in the preceding subsections.

Algorithm 9 The local approximation algorithm

1: Initial settings. For a large integer N, let $\Delta t = h = \frac{T}{N}, s_j = jh, y^0 = z, q = 0$.

2: Build a new local cone R_q.

Please note that R_q and its discretization H_q are fully determined by y^q and Δx_p through Eqs. (5.4.6)–(5.4.7) and (5.4.11)–(5.4.12). Given that y^q is already known ($y^0 = z$ is provided initially), the only remaining step is to select Δx_p. Set $\Delta x = \Delta x_p = hM_q$, $p = 1, 2, \cdots, n$, where M_q is to be technically selected to meet the necessary stability condition given by (5.4.15) (or (5.4.16) in the case where $n = 1$):

$$\Delta x = hM_q \geq \sum_{p=1}^{n} \|f_p\|_\infty \cdot h,$$

where

$$\|f_p\|_\infty = \sup_{(x,s,u)\in R_q\times\mathbb{U}} |f_p(x,s,u)|.$$

Suppose that

$$x_{p,i_p} = y_p^q + [i_p - (N-q)]\Delta x_p, \quad i_p = 0, 1, \cdots, 2(N-q),$$

are the $2(N-q)+1$ equal partition points of interval

$$\left[y_p^q - (N-q)\Delta x_p, y_p^q + (N-q)\Delta x_p\right]$$

on the x_p-axis, $p = 1, 2, \cdots, n$.
Please note that as h is provided, the computation cone R_q can be technically determined through an iterative process. We present one such method as follows.
Let

$$M_q(0) = \sum_{p=1}^{n} \|f_p\|_{\text{sup},0} \quad \text{with} \quad \|f_p\|_{\text{sup},0} = \sup_{(s,u)\in[s_q,T]\times\mathbb{U}} \left|f_p(y^q, s, u)\right|.$$

By setting $\Delta x(0) = \Delta x_p(0) = hM_q(0)$, we could get a local cone $R_q(0)$ via (5.4.6)–(5.4.7). Then let

$$M_q(1) = \sum_{p=1}^{n} \|f_p\|_{\text{sup},1} \quad \text{with } \|f_p\|_{\text{sup},1} = \sup_{(x,s,u)\in R_q(0)\times\mathbb{U}} |f_p(x,s,u)|$$

and set $\Delta x(1) = \Delta x_p(1) = hM_q(1)$. If it happens that $M_q(1) = M_q(0)$ (which is equivalent to $\|f_p\|_{\text{sup},1} = \|f_p\|_{\text{sup},0}$), then terminate the process. If not, repeat the entire procedure. It is important to note that this iterative method may fail to converge if the function f becomes unbounded as $\|x\| \to \infty$. In such cases, the stability conditions would not hold true.
However, as our focus lies solely on the viscosity solution within a neighboring cone of the optimal state, we can technically truncate f when x extends beyond the local cone. For instance, in the case

$$\bar{f}(x,s,u) = \begin{cases} f(-a,s,u), & x \le -a, \\ f(x,s,u), & x \in [-a,b], \\ f(b,s,u), & x \ge b, \end{cases} \tag{5.4.17}$$

in which $(s,u) \in \mathbb{U} \times [0,T]$.

3: Calculate the viscosity solution w on H_q.
For $j = N - 1$ to q, do
For $i_p = 0$ to $2(N - q)$, $p = 1, 2, \cdots, n$, do

$$\begin{cases} w_{\mathbf{i}}^N = \psi(x_{\mathbf{i}}),\ \mathbf{i} = (i_1, i_2, \cdots, i_p, \cdots, i_n), \\ u_{\mathbf{i}}^{j+1} \in \arg\inf_{u \in \mathbb{U}} \left\{ \sum_{p=1}^{n} \left[\frac{1 + \operatorname{sgn}\left(f_p(x_{\mathbf{i}}, s_{j+1}, u)\right)}{2} f_p(x_{\mathbf{i}}, s_{j+1}, u) \frac{w_{\mathbf{i}p^+}^{j+1} - w_{\mathbf{i}}^{j+1}}{\Delta x_p} \right.\right. \\ \left.\left. + \frac{1 - \operatorname{sgn}\left(f_p(x_{\mathbf{i}}, s_{j+1}, u)\right)}{2} f_p(x_{\mathbf{i}}, s_{j+1}, u) \frac{w_{\mathbf{i}}^{j+1} - w_{\mathbf{i}p^-}^{j+1}}{\Delta x_p} \right] + L(x_{\mathbf{i}}, s_{j+1}, u) \right\}, \\ f_{p,\mathbf{i}}^{j+1} = f_p\left(x_{\mathbf{i}}, s_{j+1}, u_{\mathbf{i}}^{j+1}\right),\ p = 1, 2, \cdots, n,\ L_{\mathbf{i}}^{j+1} = L\left(x_{\mathbf{i}}, s_{j+1}, u_{\mathbf{i}}^{j+1}\right), \\ w_{\mathbf{i}}^j = \left(1 - \sum_{p=1}^{n} \frac{h}{\Delta x_p} \left|f_{p,\mathbf{i}}^{j+1}\right|\right) w_{\mathbf{i}}^{j+1} + \sum_{p=1}^{n} \frac{1 + \operatorname{sgn}\left(f_{p,\mathbf{i}}^{j+1}\right)}{2} \frac{h}{\Delta x_p} f_{p,\mathbf{i}}^{j+1} w_{\mathbf{i}p^+}^{j+1} \\ \quad - \sum_{p=1}^{n} \frac{1 - \operatorname{sgn}\left(f_{p,\mathbf{i}}^{j+1}\right)}{2} \frac{h}{\Delta x_p} f_{p,\mathbf{i}}^{j+1} w_{\mathbf{i}p^-}^{j+1} + h L_{\mathbf{i}}^{j+1}, \end{cases}$$

end do i_p, $p = 1, 2, \cdots, n$;
end do j.

4: Calculate optimal feedback control-state pair. This step will only be executed when all subsequent potential states reside within the present local cone, that is,

$$y^{q+1}(u) = y^q + hf\left(y^q, s_q, u\right) \in D_{q+1}$$

for all $u \in \mathbb{U}$. We emphasize that this condition always holds for a newly constructed local cone R_q because of the specific choices made for M_q and Δx.
To compute the qth optimal feedback control using formula (5.4.1), one must first determine the values of the viscosity solution at the set of points $\{(y^{q+1}(u), s_{q+1}) \mid u \in \mathbb{U}\}$. Fortunately, these values can be easily obtained through linear interpolation of the viscosity solution values on H_q that were previously calculated. Once these values are known, the qth optimal feedback control can be derived using (5.4.1). To accomplish this, the first step is to locate the hypercube within which $y^{q+1}(u)$ resides, utilizing a searching technique. This hypercube has 2^n vertices, denoted by v_l for $l = 1, 2, \cdots, 2^n$. Next, we express $y^{q+1}(u)$ as a convex linear combination of these vertices:

$$y^{q+1}(u) = \sum_{l=1}^{2^n} \lambda^l(u) v_l \quad \text{with } \lambda^l(u) \geq 0 \text{ and } \quad \sum_{l=1}^{2^n} \lambda^l(u) = 1.$$

Discretize the control domain $\mathbb{U}$ into a finite set $\mathbb{U}_d = \{u_l \mid l = 1, 2, \cdots\}$ using a specified mesh size $d > 0$ for the control space. Subsequently, compute

$$W\left(y^{q+1}(u), s_{q+1}\right) = \sum_{l=1}^{2^n} \lambda^l(u) w(v_l, s_{q+1}),$$

$$P^k\left(y^q, s_q, u\right) = hL\left(y^q, s_q, u\right) + W\left(y^{q+1}(u), s_{q+1}\right)$$

for all $u \in \mathbb{U}_d$. With these values of viscosity solution, we can then construct the optimal feedback control. Set

$$P^k_{\min} = \min_{u \in \mathbb{U}_d} \left\{P^k\left(y^q, s_q, u\right)\right\},$$

$$A = \left\{u \in \mathbb{U}_d \mid P^k\left(y^q, s_q, u\right) = P^k_{\min},\ u \text{ has minimum norm}\right\}.$$

Choose any element a from the set A to define the qth optimal feedback control as $u^q = u(s_q) = a$. This control action drives the control system to the $(q+1)$-th optimal state $y^{q+1} = y(s_{q+1}) = y^q + hf\left(y^q, s_q, u^q\right)$.
Increment q by setting $q = q + 1$. If q reaches N, assign the value of the previous control, $u^q = u^{q-1}$, and terminate the procedure. Otherwise, proceed to Step 5.

5: Rebuild a new local cone or not?
The next step depends on whether all subsequent possible states are contained within the current local cone. If $y^{q+1}(u) \in D_{q+1}$ for every $u \in \mathbb{U}_d$, then we keep the local cone unchanged by setting $R_q = R_{q-1}$ and proceed to Step 4 to compute the next optimal feedback control and state. However, if there exists at least one $u \in U_d$ such that $y^{q+1} \notin D_{q+1}$, we return to Step 2 to construct a new local cone starting at $\left(y^q, s_q\right)$. This process is repeated until all optimal feedback control-state pairs $\{(u^q, y^q)\}_{q=0}^N$ are obtained through continuous iteration on the index q.

5.4.4 Numerical Experiments

To assess the efficiency and speed of the algorithm outlined in Sect. 5.4.3, we conduct two numerical experiments on an identical computer. We utilize both Visual C++ and MATLAB as our programming languages to implement and execute the experiments.

Firstly, we apply the algorithm to (5.3.60) in Sect. 5.3.4.2, which we have previously studied.

To do this, we need to technically truncate the function f and determine suitable values for M_q and subsequently R_q through an iteration process. This is necessary to ensure that the necessary stability condition (5.4.16) is satisfied on R_q. This step is crucial because, as stated in Step 2 of Algorithm 9, the function f is unbounded. One possible approach to achieve this is to define the truncation in a way that

$$M_q = N^{\frac{1}{2}} \sup_{u \in \mathbb{U}} \left\| f\left(y^q, s_q, u\right) \right\|$$

and truncate f with $a = b = M_q$ in (5.4.17) where $N = 40$.

By implementing Algorithm 9, we numerically obtained the optimal feedback control-state pairs for the control problem (5.3.60). The results are presented in Figs. 5.17 and 5.18, which depict the outcomes for two distinct initial values: $z = 10$ and $z = 0.99$, respectively.

Figures 5.17 and 5.18 exhibit both the numerical and analytical solutions for the optimal feedback control and trajectory functions (5.3.63) and (5.3.64), respectively. These solutions are presented for the initial states $z = 10$ and $z = 0.99$. Additionally, Table 5.6 provides a summary of the computed errors in the maximum norm between the numerical and analytical solutions:

$$\|e(\cdot)\| \triangleq \max_{t \in [0,T]} |e(t)|$$

between the numerical solutions and analytical solutions of u^* and y^*. By contrast, in Sect. 5.3.4.2, the calculation time for the case $z = 0.99$ is approximately 78 seconds. More importantly, the numerical solutions are extremely close to the analytical ones, which underscores the effectiveness and efficiency of our algorithm.

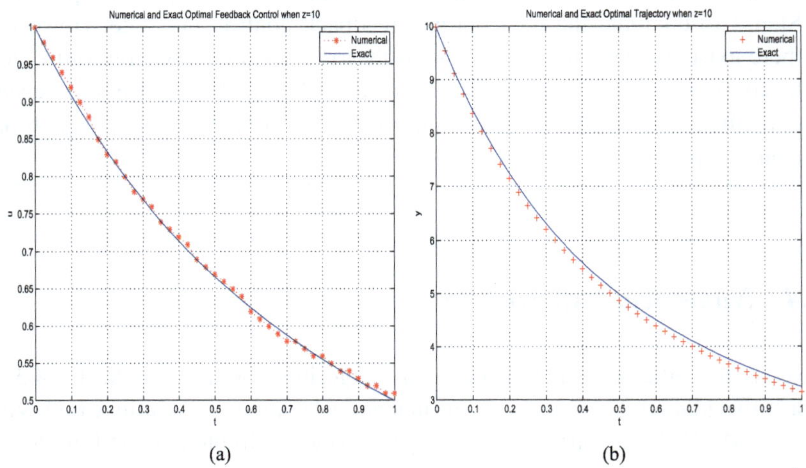

Fig. 5.17 Numerical and analytical solutions of optimal feedback control and trajectory functions with $z = 10$. (**a**) Control. (**b**) Trajectory

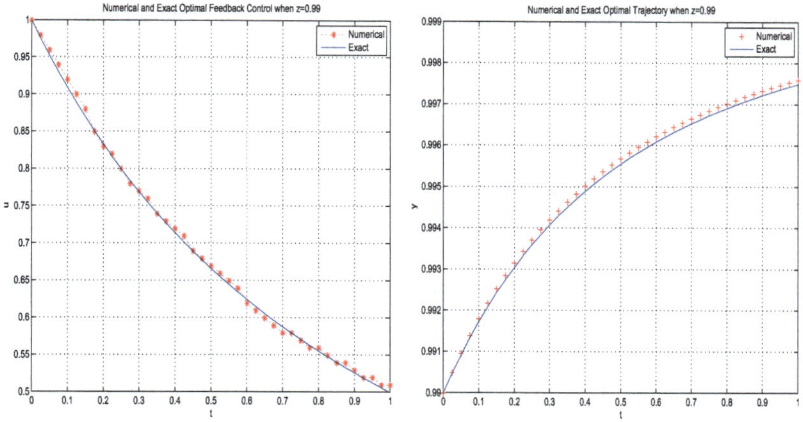

(a) (b)

Fig. 5.18 Numerical and analytical solutions of optimal feedback control and trajectory functions with $z = 0.99$. (**a**) Control. (**b**) Trajectory

Table 5.6 The computed errors in the maximum norm between the numerical solutions and analytical solutions for u^* and y^*

Error	Control u^*	Trajectory y^*
$z = 10$	3.56e-2	4.35e-2
$z = 0.99$	1.11e-2	1.31e-4

Our next example involves a two-dimensional optimal control problem, which is derived from the well-known time-optimal control problem:

$$\begin{cases} \dot{x}_1(t) = x_2, \\ \dot{x}_2(t) = u(t), \\ x_1(0) = 4, \ x_2(0) = -1, \ x_1(t_f) = x_2(t_f) = 0, \\ J(u(\cdot)) = \displaystyle\int_0^{t_f} \mathrm{d}t = t_f, \\ \min\limits_{u(\cdot)\in\Delta^*} J(u(\cdot)), \end{cases} \tag{5.4.18}$$

where $\Delta^* \triangleq L^\infty([0, t_f], \mathbb{U})$ and $\mathbb{U} \triangleq [-1, 1]$.

This problem possesses a unique solution, as detailed at Page 16 of [135].

$$t_f^* = 3\sqrt{2} - 1 = 3.242\cdots, \quad t_1 = \frac{3\sqrt{2} - 2}{2} = 1.121\cdots,$$
$$u^*(t) = \begin{cases} -1, & t \in [0, t_1), \\ 1, & t \in \left[t_1, t_f^*\right], \end{cases} \tag{5.4.19}$$

which is a bang-bang control. It is well known that the optimal feedback control of (5.4.18) can be analytically derived as

$$u^*(x_1, x_2) = \begin{cases} 1, & x_1 + \frac{1}{2}x_2|x_2| < 0, \ \text{or } x_1 + \frac{1}{2}x_2|x_2| = 0, \ x_2 < 0, \\ -1, & x_1 + \frac{1}{2}x_2|x_2| > 0, \ \text{or } x_1 + \frac{1}{2}x_2|x_2| = 0, \ x_2 > 0, \end{cases} \tag{5.4.20}$$

or it can be expressed in the simplified form as

$$u^*(x_1, x_2) = -\operatorname{sgn}\left(x_1 + \frac{1}{2}x_2|x_2|\right).$$

Let $T = t_f^* = 3\sqrt{2} - 1$ be fixed. We proceed to construct a two-dimensional optimal control problem, as outlined in Example 5.4.1 below.

Example 5.4.1 Consider the optimal control problem:

$$\begin{cases} \dot{x}_1(t) = x_2, \\ \dot{x}_2(t) = u(t), \\ x_1(0) = 4, \ x_2(0) = -1, \\ J(u(\cdot)) = -\int_0^T |u(t)|^2 \, \mathrm{d}t - \dfrac{1}{1 + x_1^2(T) + x_2^2(T)}, \\ \min\limits_{u(\cdot) \in \Delta_*} J(u(\cdot)), \end{cases} \tag{5.4.21}$$

where $\Delta_* \triangleq L^\infty([0, T], \mathbb{U})$ and $\mathbb{U} \triangleq [-1, 1]$. The optimal trajectory for the control problem (5.4.21) can be analytically found as

$$\begin{cases} x_1^*(t) = \begin{cases} -\frac{1}{2}t^2 - t + 4, & t \in [0, t_1), \\ \frac{1}{2}[t - (3\sqrt{2} - 1)]^2, & t \in [t_1, T], \end{cases} \\ x_2^*(t) = \begin{cases} -t - 1, & t \in [0, t_1), \\ t - (3\sqrt{2} - 1), & t \in [t_1, T], \end{cases} \end{cases} \tag{5.4.22}$$

and the optimal feedback control given by (5.4.20) now becomes

$$\begin{aligned} u^*(t) &= u^*(x_1^*(t), x_2^*(t)) \\ &= -\operatorname{sgn}\left(x_1^*(t) + \frac{1}{2}x_2^*(t)\left|x_2^*(t)\right|\right) = \begin{cases} -1, & t \in [0, t_1), \\ 1, & t \in [t_1, T], \end{cases} \end{aligned} \tag{5.4.23}$$

where $t_1 \triangleq \frac{3\sqrt{2}}{2} - 1$. Obviously, for any admissible control $u \neq u^*$, it holds that

$$J(u(\cdot)) > -(T + 1) = J(u^*(\cdot)),$$

where u^* is defined by (5.4.23). Therefore, the bang-bang control u^* serves as the unique solution to the optimal control problem (5.4.21).

Please note that in this example, the function f is also unbounded. Akin to the previous example in this subsection, we are technically capable of identifying one of these suitable solutions. In other words,

$$M_q = 2.2 \sup_{u \in \mathbb{U}} \left\| f\left(y^q, s_q, u\right) \right\|,$$

where $N = 159$. By employing the algorithm, we numerically obtain the optimal feedback control-state pairs for the control problem (5.4.21). The corresponding results are presented in Fig. 5.19.

Figure 5.19 compares the numerical and analytical solutions for the optimal feedback control and trajectory functions (5.4.23) and (5.4.22). Table 5.7 summarizes the computed errors in the maximum norm between the numerical solutions and the analytical solutions of u^* and $\left(x_1^*, x_2^*\right)$. Notably, for this two-dimensional control problem, the calculation took approximately 1 minute. It is evident that the numerical solutions are highly accurate, closely approximating the analytical solutions, further demonstrating the efficiency and effectiveness of our algorithm.

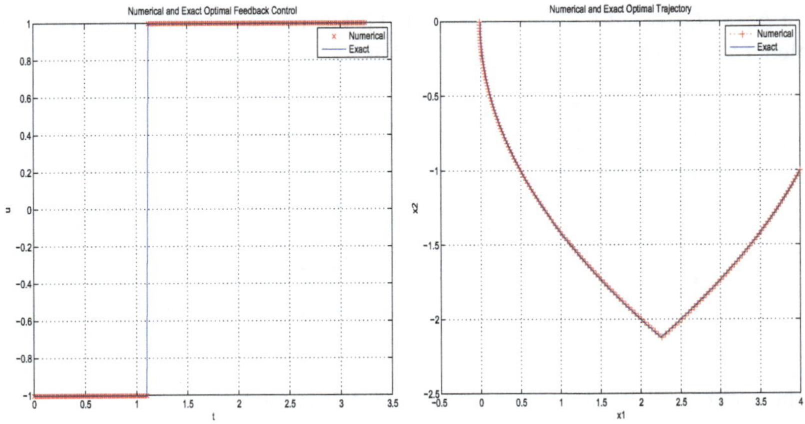

(a) (b)

Fig. 5.19 Numerical and analytical solutions of optimal feedback control and trajectory functions when $N = 159$. (**a**) Control. (**b**) Trajectory

Table 5.7 The computed errors in the maximum norm between the numerical solutions and analytical solutions for u^* and (x_1^*, x_2^*) when $N = 159$

Error	Control u^*	Trajectory x_1^*	Trajectory x_2^*
Measured on $[0, T]$	0	1.17e-2	6.96e-4

It is worth emphasizing that the analytical optimal feedback control is discontinuous and exhibits jumps at specific time points, such as t_1. In our simulation, we intentionally selected $N = 159$ to ensure that the jump point t_1 closely aligns with a time node of our discretization scheme. This strategic choice significantly minimizes all associated errors, resulting in highly accurate numerical solutions that closely match the analytical ones.

However, this is merely a technical issue in simulation. Even with a different choice of N, the results remain satisfactory. If the jump point t_1 falls within a time subinterval $I \triangleq [s_k, s_{k+1}]$ for some k, we can subdivide I into two smaller parts: $I_1 \triangleq [s_k, t_1)$ and $I_2 \triangleq [t_1, s_{k+1}]$. During simulation, the errors between the numerical and analytical solutions for the optimal trajectory (x_1^*, x_2^*) remain small. For the optimal feedback control u^*, the error can reach as high as 2 in the maximum norm over the entire time interval $[0, T]$ due to the jump property. This large error is expected because the numerical solution of the optimal control remains constant within the time subinterval I, taking either the value -1 or 1, while the analytical solution transitions from -1 on I_1 to 1 on I_2. Crucially, this significant error occurs only on a very narrow time interval I_3 which is either I_1 or I_3. On the remainder of the interval $[0, T]\backslash I_3$, the error for the optimal control remains small. Therefore, despite the presence of the jump, the numerical solution provides an accurate approximation of the optimal control and trajectory over most of the time interval.

To illustrate this situation, we employ $N = 140$ for the simulation. We make a technical choice

$$M_q = 2.05 \sup_{u \in \mathbb{U}} \left\| f\left(y^q, s_q, u\right) \right\| .$$

In this case,

$$t_1 \approx 1.1213 \in I \triangleq [s_{48}, s_{49}] = [1.1118, 1.1349],$$
$$I_1 \triangleq [s_{48}, t_1) = [1.1118, 1.1213),$$

and

$$I_2 \triangleq [t_1, s_{49}] = [1.1213, 1.1349].$$

The error of 2 occurs only on I_1. The simulation results and errors are shown in Fig. 5.20 and Table 5.8, respectively.

Drawing inspiration from the convergence results for approximating the optimal feedback control, in this section we devise a straightforward yet efficient algorithm that can be easily implemented in practice. This algorithm enables us to obtain the numerical solution of the optimal feedback control for fixed-horizon problems without state constraints, assuming that both the viscosity solution and the optimal feedback control exist and are unique.

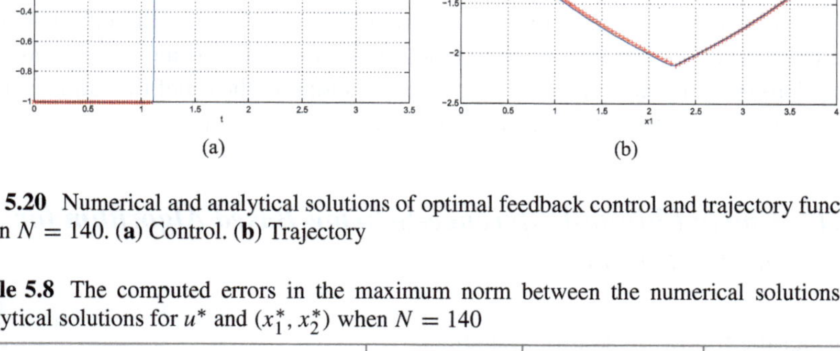

(a) (b)

Fig. 5.20 Numerical and analytical solutions of optimal feedback control and trajectory functions when $N = 140$. (**a**) Control. (**b**) Trajectory

Table 5.8 The computed errors in the maximum norm between the numerical solutions and analytical solutions for u^* and (x_1^*, x_2^*) when $N = 140$

Error	Control u^*	Trajectory x_1^*	Trajectory x_2^*
On $[0, T]$=$[0, 3.2426]$	2	2.88e-2	1.91e-2
On $I_1 = [s_{48}, t_1) = [1.1118, 1.1213)$	2	–	–
On $[0, T]\backslash I_1$	0	–	–

The core concept lies in computing the viscosity solution of the associated HJB equation solely on localized cones surrounding the optimal trajectory, rather than across the entire time-space domain. This approach maximizes the utilization of computational data while discarding it only when necessary for the computation of a new local cone.

The algorithm's efficiency and speed are demonstrated through two numerical experiments, including a 2-dimensional problem. Future exploration is warranted to investigate the convergence properties of this algorithm and its relationship with existing simpler algorithms.

5.5 Convergence of an Upwind Finite-Difference Scheme

Among the various numerical schemes for solving the HJB equation, the upwind finite-difference scheme stands out as a particularly well-suited approach that has been successfully applied to numerous examples [91, 233]. Its straightforward implementation for low-dimensional systems enables the computation of numerical solutions suitable for feedback control. Furthermore, by leveraging model reduction techniques, the upwind finite-difference scheme can be effectively applied to both finite-dimensional and infinite-dimensional systems. Additionally, not only the max-plus method but also its alternative proposed in [118] could potentially benefit from this scheme.

In fact, the upwind finite-difference scheme has been potentially used in numerous numerical solutions of the HJB equations and the optimal feedback controls. Some notable examples include [141, 144, 225].

While the upwind finite-difference scheme has proven its effectiveness in numerous applications, the crucial question of convergence remains unresolved [29, 126, 215]. Therefore, the primary objective of this section is to establish a convergence result for this scheme. Leveraging the methods introduced in [18] and [85], we aim to demonstrate that the solution obtained through the upwind finite-difference scheme converges to the value function of the optimal control problem [221]. This convergence analysis is crucial for validating the reliability and accuracy of the scheme in approximating the optimal feedback control.

5.5.1 Upwind Finite-Difference-Scheme-Based Algorithm for the HJB Equation

Consider the following control system:

$$\begin{cases} \dfrac{d}{dt}y(t) = f(t, y(t), u(t)), \quad t \in [0, T], \\ y(0) = z, \end{cases} \tag{5.5.1}$$

where $y \in \mathbb{R}^n$ is the state, $u \in \mathcal{U}[0, T] = L^\infty([0, T]; U)$ is the admissible control with $U \subset \mathbb{R}^m$ being a compact set, z is the initial value, and $f : [0, T] \times \mathbb{R}^n \times U \to \mathbb{R}^n$ is continuous.

Let a cost functional associated with system (5.5.1) be given by

$$J(u) = \int_0^T L(\tau, y(\tau), u(\tau))\, d\tau + \psi(y(T)), \tag{5.5.2}$$

where $L(t, y, u)$ represents a running cost and $\psi(y)$ denotes a terminal cost. The optimal control problem is to minimize (5.5.2) over $\mathcal{U}[0, T]$, that is, to seek an optimal control $u^* \in \mathcal{U}[0, T]$ such that

$$J(u^*) = \inf_{u(\cdot)\in\mathcal{U}[0,T]} J(u(\cdot)).$$

Drawing upon the DPP, we consider a family of optimal control problems. For each $(t, x) \in \Sigma_T = [0, T) \times \mathbb{R}^n$, minimize the cost functional

$$J_{t,x}(u) = \int_t^T L(\tau, y_{t,x}(\tau), u(\tau))\, d\tau + \psi(y_{t,x}(T)),$$

subject to

$$\begin{cases} \dfrac{d}{ds} y_{t,x}(s) = f(s, y_{t,x}(s), u(s)), & s \in [t, T], \\ y_{t,x}(t) = x. \end{cases}$$

We define the value function

$$V(t, x) = \inf_{u(\cdot)\in\mathcal{U}[t,T]} J_{t,x}(u), \quad \forall\, (t, x) \in \Sigma_T, \tag{5.5.3}$$

with the terminal value

$$V(T, x) = \psi(x), \quad \forall\, x \in \mathbb{R}^n. \tag{5.5.4}$$

If V is smooth enough, e.g., $V \in C^1([0, T] \times \mathbb{R}^n)$, then V satisfies the following HJB equation [16]:

$$\begin{cases} -\dfrac{\partial}{\partial t} V(t, x) + \mathcal{H}(t, x, \nabla_x V(t, x)) = 0, & (t, x) \in \Sigma_T, \\ V(T, x) = \psi(x), & x \in \mathbb{R}^n, \end{cases} \tag{5.5.5}$$

where

$$\mathcal{H}(t, x, \nabla_x V(t, x)) = \sup_{u \in U}\{-f(t, x, u) \cdot \nabla_x V(t, x) - L(t, x, u)\}.$$

From (5.5.5), it becomes evident that to obtain the feedback control law, we must possess knowledge of not only the value function V itself but also its gradient $\nabla_x V$. According to the viscosity solution theory, the value function V attains the status of a viscosity solution to (5.5.5) under three specific assumptions regarding the coefficients f and L [16].

Assumption 5.5.1 *The functions f, L, L_t, and L_x are bounded on $\Sigma_T \times U$.*

Assumption 5.5.2 *The function $f(\cdot, \cdot, u)$ is of class $C^1\left(\overline{\Sigma_T}\right)$, where $\overline{\Sigma_T}$ is the closure of Σ_T. In addition, there exists a constant $C > 0$ such that*

$$|f_t| + |f_x| \le C, \quad |f(t, x, u)| \le C(1 + |x| + |u|).$$

Here, the subscripts t and x represent the partial derivatives with respect to t and x, respectively. It is worth noting that the first inequality mentioned above is equivalent to the Lipschitz condition:

$$|f(s, y, u) - f(t, x, u)| \le C[|s - t| + |y - x|] \quad \textit{for all } (s, y), (t, x) \in \overline{\Sigma_T}.$$

Assumption 5.5.3 *L and ψ are continuous and satisfy the following polynomial growth condition for all $(t, x) \in \overline{\Sigma_T}$:*

$$|L(t, x, u)| \le C\left(1 + |x|^\ell + |u|^\ell\right), \quad |\psi(x)| \le C\left(1 + |x|^\ell\right)$$

with a suitable constant $\ell > 0$, where the notation $|\cdot|$ represents the operator norm.

Definition 5.5.1 Let $V \in C([0, T] \times \mathbb{R}^n)$. Then

(i) V is a viscosity subsolution to (5.5.5) in Σ_T if for every test function $\phi \in C^1(\Sigma_T)$,

$$-\frac{\partial}{\partial t}\phi\left(\bar{t}, \bar{x}\right) + \sup_{u \in U}\left\{-f\left(\bar{t}, \bar{x}, u\right) \cdot \nabla_x \phi\left(\bar{t}, \bar{x}\right) - L\left(\bar{t}, \bar{x}, u\right)\right\} \le 0,$$

at every $\left(\bar{t}, \bar{x}\right) \in \Sigma_T$ that is a maximizer of $V - \phi$ on Σ_T with $V\left(\bar{t}, \bar{x}\right) = \phi\left(\bar{t}, \bar{x}\right)$.

(ii) V is a viscosity supersolution to (5.5.5) in Σ_T if for every test function $\phi \in C^1(\Sigma_T)$,

$$-\frac{\partial}{\partial t}\phi\left(\bar{t}, \bar{x}\right) + \sup_{u \in U}\left\{-f\left(\bar{t}, \bar{x}, u\right) \cdot \nabla_x \phi\left(\bar{t}, \bar{x}\right) - L\left(\bar{t}, \bar{x}, u\right)\right\} \ge 0,$$

at every $(\bar{t}, \bar{x}) \in \Sigma_T$ that is a minimizer of $V - \phi$ on Σ_T with $V(\bar{t}, \bar{x}) = \phi(\bar{t}, \bar{x})$.

(iii) V is a viscosity solution to (5.5.5) in $[0, T] \times \mathbb{R}^n$ if it is both a viscosity subsolution and a viscosity supersoluton to (5.5.5) in Σ_T.

We now delve into the study of the unique viscosity solution to the HJB equation (5.5.5). Given Assumptions 5.5.1–5.5.3, the value function V defined in (5.5.3) and (5.5.4) is identified as the unique viscosity solution to (5.5.5), as established in [16, 85]. In the absence of a closed-form solution, we propose the utilization of an upwind finite-difference scheme to obtain a numerical solution.

For practical considerations, we confine our analysis to a finite region $\prod_{k=1}^{n} [a_k, b_k]$ rather than $\mathbb{R}^n$ for the HJB equation (5.5.5).

Let $\mathcal{V}(t, x) = V(T - t, x)$. Equation (5.5.5) becomes an initial value problem:

$$\begin{cases} \dfrac{\partial}{\partial t}\mathcal{V}(t, x) + \sup\limits_{u \in U} \left\{ -\sum\limits_{k=1}^{n} f_k(T - t, x, u) \cdot \partial_k \mathcal{V}(t, x) - L(T - t, x, u) \right\} = 0, \\ \mathcal{V}(0, x) = \psi(x), \end{cases} \tag{5.5.6}$$

where $(t, x) \in (0, T) \times \prod_{k=1}^{n} [a_k, b_k] \subset \mathbb{R}^{n+1}$, $x = (x_{(1)}, x_{(2)}, \cdots, x_{(n)})$; $f = (f_1, f_2, \ldots, f_n) \in \mathbb{R}^n$; and $\partial_k \mathcal{V}(t, x) = \partial_{x_k} \mathcal{V}(t, x)$.

Set $i = (i_1, i_2, \cdots, i_n)$, $ik^+ = (i_1, i_2, \cdots, i_{k-1}, i_k + 1, \cdots, i_n)$, $ik^- = (i_1, i_2, \cdots, i_{k-1}, i_k - 1, \cdots, i_n)$, $x_i = (x_{i_1}, x_{i_2}, \cdots, x_{i_n})$, $x_{i_k} = a_k + i_k \Delta x_k$, $i_k = 0, 1, \cdots, M_k$, with $\Delta x_k = \frac{b_k - a_k}{M_k}$, $t_j = (j - 1)\Delta t$, $j = 1, 2, \cdots, N + 1$, with $\Delta t = \frac{T}{N}$. Here M_k, $k = 1, 2, \cdots, n$, and N are given positive integers. Let

$$f_{k,i}^{j+}(u) = \max\left(f_k(T - t_j, x_i, u), 0\right), \quad f_{k,i}^{j-}(u) = \min\left(f_k(T - t_j, x_i, u), 0\right),$$

$$D_{ik}^{+}\mathcal{V} = \frac{\mathcal{V}_{ik^+} - \mathcal{V}_i}{\Delta x_k}, \quad D_{ik}^{-}\mathcal{V} = \frac{\mathcal{V}_i - \mathcal{V}_{ik^-}}{\Delta x_k}.$$

Let us consider the following upwind finite-difference scheme for the HJB equation (5.5.6):

$$\begin{aligned} \frac{\mathcal{V}_i^{j+1} - \mathcal{V}_i^j}{\Delta t} &= \sum_{k=1}^{n} \left(f_{k,i}^{j+}(u_i^j) \cdot D_{ik}^{+}\mathcal{V}^j + f_{k,i}^{j-}(u_i^j) \cdot D_{ik}^{-}\mathcal{V}^j \right) + L(T - t_j, x_i, u_i^j), \\ u_i^j &= \arg\sup_{u \in U} \left[-\sum_{k=1}^{n} \left(f_{k,i}^{j+}(u) \cdot D_{ik}^{+}\mathcal{V}^j + f_{k,i}^{j-}(u) \cdot D_{ik}^{-}\mathcal{V}^j \right) - L(T - t_j, x_i, u) \right], \end{aligned} \tag{5.5.7}$$

with

$$\mathcal{V}_i^1 = \psi(x_i) \tag{5.5.8}$$

for $j = 1, 2, \cdots, N$, $i_k = j$ to $M_k - j$, $k = 1, 2, \cdots, n$.

It is worth noting that the computational region takes the shape of a hyper-pyramidal region, which encompasses the original orthotope as a subset. This is due to the explicit nature of the discrete scheme employed. Consequently, the index i_k should range from j to $M_k - j$ instead of from 1 to M_k. This aspect has been elaborated in greater detail in [233] through a simplified case study.

Additionally, for the stability of the difference scheme, it is crucial to adhere to the Courant-Friedrichs-Lewy (CFL) condition

$$\Delta t \cdot \sum_{k=1}^{n} \frac{\|f_k\|_\infty}{\Delta x_k} \leq 1 \tag{5.5.9}$$

that should be assumed [91], where

$$\|f_k\|_\infty = \max_{(t,x,u)\in\Xi} |f_k(t, x, u)|, \quad \Xi \triangleq [0, T] \times \prod_{k=1}^{n} [a_k, b_k] \times U.$$

Condition (5.5.9) serves as a prerequisite for ensuring stability when numerically solving certain PDEs, typically hyperbolic ones, using the finite-difference method.

The aforementioned finite-difference approximation scheme was initially introduced in [233] for one-dimensional problems. Subsequently, this scheme was generalized in [91] to address n-dimensional cases. Leveraging this finite-difference scheme, we present the following algorithm to numerically solve the HJB equation (5.5.6).

Algorithm 10 Upwind finite-difference scheme based algorithm

1: Initialize the value function $\{\mathcal{V}_i^1\}_{i_k=0}^{M_k}$ and the control $\{u_i^1\}_{i_k=0}^{M_k}$, $k = 1, 2, \cdots, n$.
2: For $j = 1$ to N do
For $i_k = j$ to $M_k - j$, $k = 1, 2, \cdots, n$ do

$$\mathcal{H}_i^j(u) = \left[-\sum_{k=1}^{n} \left(f_{k,i}^{j+}(u) \cdot D_{ik}^{+}\mathcal{V}^j + f_{k,i}^{j-}(u) \cdot D_{ik}^{-}\mathcal{V}^j \right) - L(T - t_j, x_i, u) \right],$$

$$u_i^j = \arg\sup_{u\in U} \mathcal{H}_i^j(u),$$

$$\mathcal{V}_i^{j+1} = \mathcal{V}_i^j - \Delta t\, \mathcal{H}_i^j(u_i^j),$$

end do i_k;
end do j.

It is noteworthy that the efficiency and accuracy of the aforementioned method have been thoroughly validated through numerical experiments detailed in [233].

Example 5.5.2 Consider the following HJB equation:

$$\begin{cases} \dfrac{\partial}{\partial t}\mathcal{V}(t,x) + \sup\limits_{u\in[0,1]}\{-xu\mathcal{V}_x(t,x)\} = 0, \\ \qquad (t,x) \in [0,1]\times[-1,1], \\ \mathcal{V}(0,x) = -x, \end{cases}$$

which associates with the optimal control problem that minimizes $-y(1)$ subject to

$$\begin{cases} \dfrac{d}{ds}y(s) = y(s)u(s), \quad s \in [t,1], \\ y(t) = x, \end{cases}$$

and the control constraint $u \in [0,1]$. This HJB equation has the analytical solution

$$\mathcal{V}(t,x) = \begin{cases} -xe^t, & x > 0, \\ -x, & x \le 0. \end{cases}$$

The problem has been successfully solved using the above algorithm, and the numerical solution has been demonstrated to converge to the analytic solution with a first-order rate of convergence [233].

Example 5.5.3 Consider the following HJB equation:

$$\begin{cases} \dfrac{\partial}{\partial t}\mathcal{V}(t,x) + \sup\limits_{u\in[-1,1]}\{-u\mathcal{V}_x(t,x)\} = 0, \\ \qquad (t,x) \in [0,1]\times[-1,1], \\ \mathcal{V}(0,x) = -x^2, \end{cases}$$

which is deduced from the optimal control problem which minimizes $-y^2(1)$ subject to

$$\begin{cases} \dfrac{d}{ds}y(s) = u(s), \quad s \in [t,1], \\ y(t) = x, \end{cases}$$

and the control constraint $u \in [-1, 1]$. This HJB equation also has an analytical solution given by

$$\mathcal{V}(t, x) = -\left(|x| + t\right)^2 .$$

Using the same algorithm described above, the problem has been resolved, and the numerical solution has been shown to converge to the analytic solution [233].

5.5.2 Convergence of Numerical Approximation

Drawing from the works of [18] and [215], to establish the convergence of the upwind finite-difference scheme, we must demonstrate its monotonicity, stability, and consistency. The first two properties have been established in [91], and for the one-dimensional case, stability has also been addressed in [233]. However, consistency has not been addressed in the existing literature. For completeness and ease of reference, we present both the stability and monotonicity proofs here.

5.5.2.1 Monotonicity

Writing $\mathcal{V}_i^{j+1}$ in function of $\mathcal{V}_{i-1}^j$, $\mathcal{V}_i^j$, and $\mathcal{V}_{i+1}^j$, we obtain

$$\begin{aligned}
\mathcal{V}_i^{j+1} &= \mathcal{V}_i^j - \Delta t \Bigg[- \sum_{k=1}^{n} \left(f_{k,i}^{j+} \left(u_i^j\right) \cdot D_{ik}^{+} \mathcal{V}^j + f_{k,i}^{j-} \left(u_i^j\right) \cdot D_{ik}^{-} \mathcal{V}^j \right) \\
&\qquad - L\left(T - t_j, x_i, u_i^j\right) \Bigg] \\
&= \left[1 - \Delta t \sum_{k=1}^{n} \frac{f_{k,i}^{j+}\left(u_i^j\right) - f_{k,i}^{j-}\left(u_i^j\right)}{\Delta x_k} \right] \mathcal{V}_i^j \\
&\quad + \Delta t \sum_{k=1}^{n} \frac{f_{k,i}^{j+}\left(u_i^j\right) \mathcal{V}_{ik+}^j - f_{k,i}^{j-}\left(u_i^j\right) \mathcal{V}_{ik-}^j}{\Delta x_k} + \Delta t \, L\left(T - t_j, x_i, u_i^j\right).
\end{aligned} \tag{5.5.10}$$

By condition (5.5.9), it follows that

$$\begin{aligned}
& 1 - \Delta t \sum_{k=1}^{n} \frac{f_{k,i}^{j+}\left(u_i^j\right) - f_{k,i}^{j-}\left(u_i^j\right)}{\Delta x_k} \\
&= 1 - \Delta t \sum_{k=1}^{n} \frac{\left| f_{k,i}^{j}\left(u_i^j\right) \right|}{\Delta x_k} \geq 0.
\end{aligned} \tag{5.5.11}$$

In addition, since $\Delta t > 0$, $\Delta x_k > 0$, $f_{k,i}^{j+}\left(u_i^j\right) \geq 0$, and $-f_{k,i}^{j-}\left(u_i^j\right) \geq 0$, substituting (5.5.11) into (5.5.10), we conclude that $\mathcal{V}_i^{j+1}$ is a nondecreasing monotone function of $\mathcal{V}_{i-1}^j$, $\mathcal{V}_i^j$, and $\mathcal{V}_{i+1}^j$.

5.5.2.2 Stability

As demonstrated in [18], the stability of the scheme is guaranteed by the satisfaction of the CFL condition. Furthermore, according to [219], it is sufficient to verify the stability condition for the case when $L = 0$.

By setting $L\left(T - t_j, x_i, u_i^j\right) = 0$ in (5.5.10), we arrive at

$$\begin{aligned}
\left|\mathcal{V}_i^{j+1}\right| &= \left|\left(1 - \Delta t \sum_{k=1}^{n} \frac{f_{k,i}^{j+}\left(u_i^j\right) - f_{k,i}^{j-}\left(u_i^j\right)}{\Delta x_k}\right)\mathcal{V}_i^j\right. \\
&\quad \left. + \Delta t \sum_{k=1}^{n} \frac{f_{k,i}^{j+}\left(u_i^j\right)\mathcal{V}_{ik+}^j - f_{k,i}^{j-}\left(u_i^j\right)\mathcal{V}_{ik-}^j}{\Delta x_k}\right| \\
&\leq \left|1 - \Delta t \sum_{k=1}^{n} \frac{f_{k,i}^{j+}\left(u_i^j\right) - f_{k,i}^{j-}\left(u_i^j\right)}{\Delta x_k}\right| \cdot \left|\mathcal{V}_i^j\right| \\
&\quad + \Delta t \left|\sum_{k=1}^{n} \frac{f_{k,i}^{j+}\left(u_i^j\right)\mathcal{V}_{ik+}^j - f_{k,i}^{j-}\left(u_i^j\right)\mathcal{V}_{ik-}^j}{\Delta x_k}\right| \\
&\leq \left|1 - \Delta t \sum_{k=1}^{n} \frac{\left|f_{k,i}^{j}\left(u_i^j\right)\right|}{\Delta x_k}\right| \sup_i \left|\mathcal{V}_i^j\right| \\
&\quad + \Delta t \sup_i \left|\mathcal{V}_i^j\right| \sum_{k=1}^{n} \frac{\left|f_{k,i}^{j+}\left(u_i^j\right)\right| + \left|f_{k,i}^{j-}\left(u_i^j\right)\right|}{\Delta x_k} \\
&= \left(\left|1 - \Delta t \sum_{k=1}^{n} \frac{\left|f_{k,i}^{j}\left(u_i^j\right)\right|}{\Delta x_k}\right| + \Delta t \sum_{k=1}^{n} \frac{\left|f_{k,i}^{j}\left(u_i^j\right)\right|}{\Delta x_k}\right) \sup_i \left|\mathcal{V}_i^j\right|.
\end{aligned} \tag{5.5.12}$$

By the condition (5.5.9), it follows that

$$\left|1 - \Delta t \sum_{k=1}^{n} \frac{\left|f_{k,i}^{j}\left(u_i^j\right)\right|}{\Delta x_k}\right| = 1 - \Delta t \sum_{k=1}^{n} \frac{\left|f_{k,i}^{j}\left(u_i^j\right)\right|}{\Delta x_k}.$$

Thus, (5.5.12) becomes

$$\left|\mathcal{V}_i^{j+1}\right| \le \left(1 - \Delta t \sum_{k=1}^{n} \frac{\left|f_{k,i}^{j}\left(u_i^j\right)\right|}{\Delta x_k} + \Delta t \sum_{k=1}^{n} \frac{\left|f_{k,i}^{j}\left(u_{i'}^j\right)\right|}{\Delta x_k}\right) \sup_i \left|\mathcal{V}_i^j\right|$$
$$= \sup_i \left|\mathcal{V}_i^j\right|,$$

i.e.,

$$\left\|\mathcal{V}^{j+1}\right\|_\infty \le \left\|\mathcal{V}^{j}\right\|_\infty \le \left\|\mathcal{V}^{0}\right\|_\infty.$$

This step completes the proof of the stability of the difference scheme.

5.5.2.3 Consistency

Let

$$\phi_i^{j+1} = \phi_i^j - \Delta t \sup_{u\in U}\left\{ - \sum_{k=1}^{n}\left(f_{k,i}^{j+}(u)\cdot D_{ik}^{+}\phi^j + f_{k,i}^{j-}(u)\cdot D_{ik}^{-}\phi^j\right)\right.$$
$$\left. - L(T - t_j, x_i, u)\right\},$$

and define

$$f_k^+(T-s,y,u) = \max\left(f_k(T-s,y,u),0\right),$$
$$f_k^-(T-s,y,u) = \min\left(f_k(T-s,y,u),0\right).$$

Denote by δ_k an n-dimensional vector, where the kth element is 1 and all other elements are zero, namely, $\delta_k = (0,\cdots,0,1,0,\cdots,0)$, $k = 1,2,\cdots,n$. Let $\delta x_k = \Delta x_k \cdot \delta_k$. Define

$$D_k^+\phi(s,y) = \frac{\phi(s,y+\delta x_k) - \phi(s,y)}{\Delta x_k}, \quad D_k^-\phi(s,y) = \frac{\phi(s,y) - \phi(s,y-\delta x_k)}{\Delta x_k}.$$

Then

$$\phi(s+\Delta t, y)$$
$$= \phi(s,y) - \Delta t \sup_{u\in U}\left\{ - \sum_{k=1}^{n}\left(f_k^+(T-s,y,u)\cdot D_k^+\phi(s,y)\right.\right.$$
$$\left.\left. + f_k^-(T-s,y,u)\cdot D_k^-\phi(s,y)\right) - L(T-s,y,u)\right\}.$$

The consistency means that

$$\lim_{\substack{(s,y)\to(t,x)\\ \Delta t+\Delta x\to 0+}} \left\{ \frac{\phi(s+\Delta t, y)-\phi(s,y)}{\Delta t} \right.$$

$$+\sup_{u\in U}\left[-\sum_{k=1}^{n}\left(f_k^+(T-s,y,u)\cdot D_k^+\phi(s,y)\right.\right.$$

$$\left.\left.\left.+f_k^-(T-s,y,u)\cdot D_k^-\phi(s,y)\right)-L(T-s,y,u)\right]\right\}$$

$$=\phi_t(t,x)+\sup_{u\in U}\left\{-\sum_{k=1}^{n} f_k(T-t,x,u)\cdot\partial_k\phi(t,x)-L(T-t,x,u)\right\} \tag{5.5.13}$$

for every test function $\phi(\cdot,\cdot)\in C^{1,1}(\mathbb{R}^{n+1})$, where $\Delta x \triangleq \max\{\Delta x_1, \Delta x_2, \cdots, \Delta x_n\}$. To verify (5.5.13), we need only confirm the approximation of the gradient of the value function. To accomplish this, observe that for any positive integer k ranging from 1 to n,

$$\begin{aligned} &f_k^+(T-s,y,u)\cdot D_k^+\phi(s,y)+f_k^-(T-s,y,u)\cdot D_k^-\phi(s,y)\\ &\quad= f_k(T-s,y,u)\cdot\left(\partial_k\phi(s,y)+\frac{o(\Delta x_k)}{\Delta x_k}\right). \end{aligned} \tag{5.5.14}$$

Given the assumption on f, (5.5.14) implies that

$$\begin{aligned} &f_k^+(T-s,y,u)\cdot D_k^+\phi(s,y)+f_k^-(T-s,y,u)\cdot D_k^-\phi(s,y)\\ &\approx f_k(T-s,y,u)\cdot\partial_k\phi(s,y) \end{aligned}$$

for sufficiently small Δx_k. This shows the consistency of (5.5.13).

5.5.2.4 Proof of Convergence

Set

$$\begin{aligned} \mathcal{V}^*(t,x) &= \limsup_{\substack{(s,y)\to(t,x)\\ \Delta t+\Delta x\to 0+}} \mathcal{V}_{\Delta x,\Delta t}(s,y),\\ \mathcal{V}_*(t,x) &= \liminf_{\substack{(s,y)\to(t,x)\\ \Delta t+\Delta x\to 0+}} \mathcal{V}_{\Delta x,\Delta t}(s,y), \end{aligned} \tag{5.5.15}$$

for any $(t, x) \in Q_T := (0, T] \times \mathbb{R}^n$, where $\mathcal{V}_{\Delta x, \Delta t}$ satisfies the upwind finite-difference scheme for the HJB equation outlined in Algorithm 10 above, with the initial value $\mathcal{V}_{\Delta x, \Delta t}(0, x) = \psi(x), x \in \prod_{k=1}^{n} [a_k, b_k]$.

Therefore, the upper semicontinuous envelope $\mathcal{V}^*$ of $\mathcal{V}_{\Delta x, \Delta t}$ is the smallest upper semicontinuous function defined on Q_T that is greater than or equal to $\mathcal{V}_{\Delta x, \Delta t}$. Conversely, the analogous statement holds for the lower semicontinuous envelope $\mathcal{V}_*$.

Lemma 5.5.4 *Let $\mathcal{V}^*$ and $\mathcal{V}_*$ be defined by* (5.5.15). *Then $\mathcal{V}^*$ is a viscosity subsolution, and $\mathcal{V}_*$ is a viscosity supersolution.*

Proof Define

$$V^*(t, x) = \limsup_{\substack{(s,y)\to(t,x)\\ \Delta t+\Delta x\to 0+}} V_{\Delta x, \Delta t}(s, y),$$

$$V_*(t, x) = \liminf_{\substack{(s,y)\to(t,x)\\ \Delta t+\Delta x\to 0+}} V_{\Delta x, \Delta t}(s, y),$$

for any $(t, x) \in \Sigma_T$, where $V_{\Delta x, \Delta t}$ satisfies the following upwind finite-difference scheme for the HJB equation (5.5.5), similarly to scheme (5.5.7) and (5.5.8) for (5.5.6).

$$\frac{V_i^{j+1} - V_i^j}{\Delta t} = \sup_{u \in U} \left[-\sum_{k=1}^{n} \left(f_{k,i}^{j+}(u) \cdot D_{ik}^{+} V^j + f_{k,i}^{j-}(u) \cdot D_{ik}^{-} V^j \right) - L(t_j, x_i, u) \right]$$

with

$$V_i^N = \psi(x_i)$$

for $j = 1, 2, \cdots, N$, $i_k = j$ to $M_k - j$, $k = 1, 2, \cdots, n$.

On the one hand, assume that ϕ is a test function such that $V^* - \phi$ attains a maximum at $(\bar{t}, \bar{x}) \in \Sigma_T$. To demonstrate that V^* is a viscosity subsolution, we may assume without loss of generality that this maximum is strict, as stated in [85]. Consequently, there exists a sequence converging to zero, denoted by $\Delta x + \Delta t$, such that $V_{\Delta x, \Delta t} - \phi$ attains a maximum at (s_δ, y_δ) which approaches $(\bar{t}, \bar{x})$ as $\Delta x + \Delta t \to 0+$. For any

$$y \in \prod_{k=1}^{n} \left[x_{(i)} - \frac{1}{2}\Delta x_k, x_{(i)} + \frac{1}{2}\Delta x_k \right),$$

it holds

$$V_{\Delta x, \Delta t}(s_\delta, y_\delta) - \phi(s_\delta, y_\delta) \geq V_{\Delta x, \Delta t}(s_\delta + \Delta t, y) - \phi(s_\delta + \Delta t, y).$$

Observer that for any $(t,x) \in \Sigma_T$, the value function satisfies the DPP:

$$V(t,x) = \inf_{u(\cdot)\in\mathcal{U}[t,t+\Delta t]}\Bigg[\int_t^{t+\Delta t} L(\tau, y_{t,x}(\tau), u(\tau))\,\mathrm{d}\tau + V(t+\Delta t, y_{t,x}(t+\Delta t))\Bigg],$$

where $y_{t,x}(\cdot)$ is the trajectory corresponding to the control $u(\cdot)$. Therefore,

$$\begin{aligned} 0 &\le V_{\Delta x,\Delta t}(s_\delta, y_\delta) - \phi(s_\delta, y_\delta) - V_{\Delta x,\Delta t}(s_\delta+\Delta t, y) + \phi(s_\delta+\Delta t, y) \\ &\le \int_{s_\delta}^{s_\delta+\Delta t} L(\tau, y(\tau), u(\tau))\,\mathrm{d}\tau - \phi(s_\delta, y_\delta) + \phi(s_\delta+\Delta t, y). \end{aligned}$$

By dividing both sides by Δt, we proceed to let $\Delta x + \Delta t \to 0+$. Utilizing the consistency property of the finite-difference scheme, we obtain the following result:

$$-\frac{\partial}{\partial t}\phi\left(\bar{t},\bar{x}\right) + \sup_{u\in U}\left\{-f\left(\bar{t},\bar{x},u\right)\cdot\nabla_x\phi\left(\bar{t},\bar{x}\right) - L\left(\bar{t},\bar{x},u\right)\right\} \le 0.$$

Given that $\mathcal{V}(t,x) = V(T-t,x)$, it follows that $\mathcal{V}^*$ is a viscosity subsolution.

On the other hand, suppose that $V_* - \phi$ attains a minimum at $(s_\delta, y_\delta) \in \Sigma_T$. By definition, for any $\varepsilon > 0$, we can find $u(\cdot) = u_{\varepsilon,s_\delta+\Delta t}(\cdot) \in \mathcal{U}[s_\delta, T]$ satisfying

$$\begin{aligned} 0 \ge\ & V_{\Delta x,\Delta t}(s_\delta, y_\delta) - \phi(s_\delta, y_\delta) \\ & - V_{\Delta x,\Delta t}(s_\delta+\Delta t, y(s_\delta+\Delta t)) + \phi(s_\delta+\Delta t, y(s_\delta+\Delta t)) \\ \ge\ & -\varepsilon\Delta t + \int_{s_\delta}^{s_\delta+\Delta t} L(\tau, y(\tau), u(\tau))\,\mathrm{d}\tau \\ & -\phi(s_\delta, y_\delta) + \phi(s_\delta+\Delta t, y(s_\delta+\Delta t)). \end{aligned}$$

By dividing both sides above by Δt, we let $\Delta x+\Delta t \to 0+$. Once again, by applying the consistency property, we arrive at the following conclusion:

$$\begin{aligned} -\varepsilon \le\ & \frac{1}{\Delta t}\Bigg[-\int_{s_\delta}^{s_\delta+\Delta t} L(\tau, y(\tau), u(\tau))\,\mathrm{d}\tau \\ & - \phi(s_\delta+\Delta t, y(s_\delta+\Delta t)) + \phi(s_\delta, y_\delta)\Bigg] \\ =\ & \frac{1}{\Delta t}\int_{s_\delta}^{s_\delta+\Delta t}\Bigg[-L(\tau, y(\tau), u(\tau)) - \frac{\partial}{\partial t}\phi(\tau, y(\tau)) \\ & - f(\tau, y(\tau), u(\tau))\cdot\nabla_x\phi(\tau, y(\tau))\Bigg]\,d\tau \end{aligned}$$

$$\leq \frac{1}{\Delta t} \int_{s_\delta}^{s_\delta+\Delta t} \left\{ -\frac{\partial}{\partial t}\phi(\tau, y(\tau)) \right.$$

$$\left. + \sup_{u\in U} [-f(\tau, y(\tau), u) \cdot \nabla_x \phi(\tau, y(\tau)) - L(\tau, y(\tau), u)] \right\} d\tau$$

$$\to -\frac{\partial}{\partial t}\phi\left(\bar{t}, \bar{x}\right) + \sup_{u\in U} \left\{-f\left(\bar{t}, \bar{x}, u\right) \cdot \nabla_x \phi\left(\bar{t}, \bar{x}\right) - L\left(\bar{t}, \bar{x}, u\right)\right\}.$$

Therefore, $\mathcal{V}_*$ is a viscosity supersolution. □

The following assumption guarantees that $\mathcal{V}_{\Delta x,\Delta t}$ handles the initial data uniformly:

$$\lim_{\substack{(s,y)\to(0,x)\\ \Delta t+\Delta x\to 0+}} \mathcal{V}_{\Delta x,\Delta t}(s, y) = \psi(x) \tag{5.5.16}$$

uniformly for x in any compact subset of $\mathbb{R}^n$.

Theorem 5.5.5 *Let $\mathcal{V}_{\Delta x,\Delta t}(t, x)$ be a solution of the upwind finite-difference scheme. Assume that the monotonicity, stability, consistency, and* (5.5.16) *hold. Then*

$$\lim_{\substack{(s,y)\to(t,x)\\ \Delta t+\Delta x\to 0+}} \mathcal{V}_{\Delta x,\Delta t}(s, y) = \mathcal{V}(t, x)$$

uniformly on any compact subset of $\overline{Q}_T$, the closure of Q_T.

Proof Based on Assumptions 5.5.1–5.5.3, a minor adaptation of the results presented in Section 3, Chapter V of [85] demonstrates that $\mathcal{V}$ constitutes a bounded, uniformly continuous viscosity solution for the HJB equation specified in (5.5.6), subject to the prescribed initial conditions. According to Lemma 5.5.4, $\mathcal{V}^*$ qualifies as a bounded upper semicontinuous subsolution of (5.5.6), satisfying $\mathcal{V}^*(0, x) = \psi(x)$ for all $x \in \mathbb{R}^n$. A comparison theorem establishes that $\mathcal{V}_* \leq \mathcal{V}$. Analogously, it can be shown that $\mathcal{V}_* \geq \mathcal{V}$, the desired conclusion is thereby reached. □

Next, we introduce the concepts of sub- and supersolutions for the upwind finite-difference scheme. A bounded function $\mathcal{W}$ is considered a supersolution of (5.5.7) and (5.5.8) if it satisfies certain conditions specified by the scheme.

$$\begin{aligned} &\mathcal{W}(t, x) \geq \mathcal{W}(t + \Delta t, x) \\ &+\Delta t \sup_{u\in U} \left\{ -\sum_{k=1}^{n} \left(f_k^+(T - t, x, u) \cdot D_k^+ \mathcal{W}(t, x) \right.\right. \\ &\left.\left. +f_k^-(T - t, x, u) \cdot D_k^- \mathcal{W}(t, x)\right) - L(T - t, x, u)\right\} \end{aligned} \tag{5.5.17}$$

and

$$\mathcal{W}(0,x) \geq \psi(x) \tag{5.5.18}$$

for $t = t_j,\ j = 1, 2, \cdots, N,\ x = x_i,\ i_k = j$ to $M_k - j,\ k = 1, 2, \cdots, n$. In the same way, $\mathcal{W}$ is a subsolution of (5.5.7) and (5.5.8) if the inequalities are reversed in (5.5.17) and (5.5.18).

Lemma 5.5.6 $\mathcal{V}_{\Delta x,\Delta t} \leq \mathcal{W}$ *for any supersolution* $\mathcal{W}$ *of* (5.5.7) *and* (5.5.8), *and* $\mathcal{Z} \leq \mathcal{V}_{\Delta x,\Delta t}$ *for any subsolution* $\mathcal{Z}$ *of* (5.5.7) *and* (5.5.8).

Proof Since $\mathcal{V}_{\Delta x,\Delta t}(t,x)$ is the solution to the upwind finite-difference scheme (5.5.7), we have

$$\begin{aligned}\mathcal{V}_{\Delta x,\Delta t}(t,x) &= \mathcal{V}_{\Delta x,\Delta t}(t+\Delta t,x)\\ +\Delta t \sup_{u\in U}\Big\{ &-\sum_{k=1}^{n}\big(f_k^{+}(T-t,x,u)\cdot D_k^{+}\mathcal{V}_{\Delta x,\Delta t}(t,x)\\ &+f_k^{-}(T-t,x,u)\cdot D_k^{-}\mathcal{V}_{\Delta x,\Delta t}(t,x)\big) - L(T-t,x,u)\Big\}\end{aligned}$$

and

$$\mathcal{V}_{\Delta x,\Delta t}(0,x) = \psi(x)$$

for $t = t_j,\ j = 1, 2, \cdots, N,\ x = x_i,\ i_k = j$ to $M_k - j,\ k = 1, 2, \cdots, n$.

By invoking (5.5.17), (5.5.18) and leveraging the monotonicity property inherent to the upwind finite-difference scheme, we deduce that $\mathcal{V}_{\Delta x,\Delta t} \leq \mathcal{W}$. Analogously, a reversed inequality can be established for the corresponding subsolution case. □

The subsequent Theorem 5.5.7 establishes that the value function $\mathcal{V}_{\Delta x,\Delta t}$ associated with the upwind finite-difference scheme converges to the value function $\mathcal{V}$ in the limit as $\Delta x + \Delta t \to 0+$ (i.e., as the discretization becomes finer).

Theorem 5.5.7 *Let* $\mathcal{V}_{\Delta x,\Delta t}(t,x)$ *be a solution of* (5.5.6). *In addition to Assumptions 5.5.1 and 5.5.2, we suppose that* ψ *is bounded and uniformly continuous. Then,*

$$\lim_{\substack{(s,y)\to(t,x)\\ \Delta t+\Delta x\to 0+}} \mathcal{V}_{\Delta x,\Delta t}(s,y) = \mathcal{V}(t,x) \tag{5.5.19}$$

uniformly on any compact subset of $\overline{Q}_T$.

Proof Firstly, let us assume that both ψ and $\psi_{x(k)}$ for $k = 1, 2, \cdots, n$ are bounded and uniformly continuous. By invoking Taylor's formula, we can deduce that the difference quotients $D_{ik}^{+}\psi$ and $D_{ik}^{-}\psi$ are also bounded. Defined

$$\mathcal{W}(t,x) = -Kt + \psi(x),$$

where the constant K is chosen large enough so that

$$K > \left|\tilde{\mathcal{H}}(t, x, D_{ik}^{+}\psi, D_{ik}^{-}\psi)\right|$$

for all $x \in \prod_{k=1}^{n} [a_k, b_k]$, where

$$\tilde{\mathcal{H}}\left(t, x, D_{ik}^{+}\psi, D_{ik}^{-}\psi\right) = \sup_{u\in U} \left\{ -\sum_{k=1}^{n} \left(f_k^{+}(T-t, x, u) \cdot D_{ik}^{+}\psi(x) \right.\right.$$
$$\left.\left. +f_k^{-}(T-t, x, u) \cdot D_{ik}^{-}\psi(x)\right) - L(T-t, x, u)\right\}.$$

It then follows that

$$\begin{aligned} &\mathcal{W}(t+\Delta t, x) + \Delta t\, \tilde{\mathcal{H}}\left(t, x, D_{ik}^{+}\psi, D_{ik}^{-}\psi\right) \\ &= \mathcal{W}(t, x) - \Delta t\left[K - \tilde{\mathcal{H}}\left(t, x, D_{ik}^{+}\psi, D_{ik}^{-}\psi\right)\right] \\ &\le \mathcal{W}(t, x). \end{aligned}$$

This shows that $\mathcal{W}$ is a supersolution. In light of Lemma 5.5.6, we conclude that $\mathcal{V}_{\Delta x, \Delta t} \le \mathcal{W}$. Analogously, let

$$\mathcal{Z}(t, x) = Kt + \psi(x)$$

be a subsolution. Then $\mathcal{Z} \le \mathcal{V}_{\Delta x, \Delta t}$. Since

$$\begin{aligned} \left|\mathcal{V}_{\Delta x, \Delta t}(s, y) - \psi(x)\right| &\le \left|\mathcal{V}_{\Delta x, \Delta t}(s, y) - \psi(y)\right| + |\psi(y) - \psi(x)| \\ &\le -Ks + \|D\psi\||y - x|, \end{aligned}$$

we see that (5.5.16) holds uniformly for all $x \in \mathbb{R}^n$. Furthermore, given that the finite-difference scheme meets the criteria of monotonicity, stability, and consistency, Theorem 5.5.7 can be inferred directly from Theorem 5.5.5.

When ψ is bounded and uniformly continuous, we can employ a standard smoothing technique to infer that for any given $a > 0$ and a function $\tilde{\psi}$ satisfying $\left\|\tilde{\psi} - \psi\right\| \le a$, the difference quotients $D_{ik}^{+}\tilde{\psi}$ and $D_{ik}^{-}\tilde{\psi}$ are also bounded and uniformly continuous. Let $\tilde{\mathcal{V}}$ and $\tilde{\mathcal{V}}_{\Delta x, \Delta t}$ represent the respective value functions obtained when the initial cost function is replaced by $\tilde{\psi}$ instead of ψ. By definition,

$$\left\|\mathcal{V} - \tilde{\mathcal{V}}\right\| \le a, \quad \left\|\mathcal{V}_{\Delta x, \Delta t} - \tilde{\mathcal{V}}_{\Delta x, \Delta t}\right\| \le a.$$

We have thus shown that

$$\lim_{\substack{(s,y)\to(t,x)\\ \Delta t+\Delta x\to 0+}} \tilde{\mathcal{V}}_{\Delta x,\Delta t}(s,y) = \tilde{\mathcal{V}}(t,x)$$

uniformly on compact subsets of $\bar{Q}_T$. Because a can be chosen arbitrarily small, the same convergence property applies to the original functions $\mathcal{V}_{\Delta x,\Delta t}$ and $\mathcal{V}$. Consequently, the convergence stated in (5.5.19) is established. □

References

1. A.S. Ackleh, H.T. Banks, G.A. Pinter, A nonlinear beam equation. Appl. Math. Lett. **15**, 381–387 (2002)
2. R.A. Adams, J.J.F. Fournier, *Sobolev Spaces*, 2nd edn. Pure and Applied Mathematics (Amsterdam), vol. 140 (Elsevier/Academic Press, Amsterdam, 2003)
3. B.M. Adams, H.T. Banks, H.-D. Kwon, H.T. Tran, Dynamic multidrug therapies for HIV: optimal and STI control approaches. Math. Biosci. Eng. **1**, 223–241 (2004)
4. B.M. Adams, H.T. Banks, M. Davidian, H.-D. Kwon, H.T. Tran, S.N. Wynne, E.S. Rosenberg, HIV dynamics: modeling, data analysis, and optimal treatment protocols. J. Comput. Appl. Math. **184**, 10–49 (2005)
5. B.M. Adams, H.T. Banks, M. Davidian, E.S. Rosenberg, Estimation and prediction with HIV-treatment interruption data. Bull. Math. Biology **69**, 563–584 (2007)
6. A.A. Alonso, J.R. Banga, R. Perez-Martin, Modeling and adaptive control of a batch sterilization process. Comput. Chem. Eng. **22**, 445–458 (1998)
7. L.J. Alvarez-Vázquez, A. Martínez, Modelling and control of natural convection in canned foods. IMA J. Appl. Math. **63**, 247–265 (1999)
8. L.J. Alvarez-Vázquez, M. Marta, A. Martínez, Sterilization of canned viscous foods: an optimal control approach. Math. Models Methods Appl. Sci. **14**, 355–374 (2004)
9. S. Aniţa, Optimal harvesting for a nonlinear age-dependent population dynamics. J. Math. Anal. Appl. **226**, 6–22 (1998)
10. S. Aniţa, M. Iannelli, M.-Y. Kim, E.-J. Park, Optimal harvesting for periodic age-dependent population dynamics. SIAM J. Appl. Math. **58**, 1648–1666 (1998)
11. J.R. Banga, Z. Pan, R.P. Singh, On the optimal control of contact-cooking processes. Food Bioprod. Process. **79**, 145–151 (2001)
12. H.T. Banks, D.S. Gilliam, V.I. Shubov, Well-posedness for a one-dimensional nonlinear beam, in *Computation and Control IV: Proceedings of the Fourth Bozeman Conference* (Bozeman, MT, 1994). Progress Systems Control Theory, vol. 20 (Birkhäuser, Boston, 1995), pp. 1–21
13. V. Barbu, M. Iannelli, Optimal control of population dynamics. J. Optim. Theory Appl. **102**, 1–14 (1999)
14. V. Barbu, T. Precupanu, *Convexity and Optimization in Banach Spaces*, 4th edn. Springer Monographs in Mathematics (Springer, Dordrecht, 2012)
15. M. Bardi, Some applications of viscosity solutions to optimal control and differential games, in *Viscosity Solutions and Applications*, Lecture Notes in Mathematics, vol. 1660 (Springer-Verlag, Berlin, 1997), pp. 44–97

B. Sun et al., *Maximum Principle and Dynamic Programming Viscosity Solution Approach*, Systems & Control: Foundations & Applications,
https://doi.org/10.1007/978-981-96-5739-1

16. M. Bardi, I. Capuzzo-Dolcetta, *Optimal Control and Viscosity Solutions of Hamilton-Jacobi-Bellman Equations*. Systems & Control: Foundations & Applications, with appendices by Maurizio Falcone and Pierpaolo Soravia (Birkhäuser, Boston, 1997)
17. M. Bardi, P. Soravia, Hamilton-Jacobi equations with singular boundary conditions on a free boundary and applications to differential games. Trans. Am. Math. Soc. **325**, 205–229 (1991)
18. G. Barles, P.E. Souganidis, Convergence of approximation schemes for fully nonlinear second order equations. Asymptotic Anal. **4**, 271–283 (1991)
19. E.N. Barron, Application of viscosity solutions of infinite-dimensional Hamilton-Jacobi-Bellman equations to some problems in distributed optimal control. J. Optim. Theory Appl. **64**, 245–268 (1990)
20. A. Bazezew, J.C. Bruch, J.M. Sloss, Optimal (distributed or boundary) control of the vibrations of continuous systems solved numerically in a space-time domain. Engrg. Comput. **14**, 746–758 (1997)
21. A. Bazezew, J.C. Bruch, J.M. Sloss, Numerical solution of the optimal boundary control of transverse vibrations of a beam. Numer. Methods Partial Differ. Equ. **15**, 558–568 (1999)
22. M. Bergounioux, K. Kunisch, Augmented Lagrangian techniques for elliptic state constrained optimal control problems. SIAM J. Control Optim. **35**, 1524–1543 (1997)
23. M. Bergounioux, D. Tiba, General optimality conditions for constrained convex control problems. SIAM J. Control Optim. **34**, 698–711 (1996)
24. A. Bermúdez, Some applications of optimal control theory of distributed systems, ESAIM Control Optim. Calc. Var. **8**, 195–218 (2002)
25. A. Bermúdez, A. Martínez, A state constrained optimal control problem related to the sterilization of canned foods. Automatica J. IFAC **30**, 319–329 (1994)
26. C. Bernardi, Y. Maday, Spectral methods, in *Handbook of Numerical Analysis*, vol. V, ed. by P.G. Ciarlet, J.-L. Lions (North-Holland, Amsterdam, 1997), pp. 209–485
27. C. Bernardi, G. Coppoletta, Y. Maday, Some spectral approximations of two-dimensional fourth-order problems. Math. Comp. **59**, 63–76 (1992)
28. C. Bernardi, M. Dauge, Y. Maday, *Spectral Methods for Axisymmetric Domains* (North-Holland, Amsterdam, 1999)
29. O. Bokanowski, N. Megdich, H. Zidani, Convergence of a non-monotone scheme for Hamilton-Jacobi-Bellman equations with discontinuous initial data. Numer. Math. **115**, 1–44 (2010)
30. V.G. Boltyanskii, *Mathematical Methods of Optimal Control*, Translated from the Russian by K. N. Trirogoff, ed. by I. Tarnove. Balskrishnan-Neustadt Series (Holt, Rinehart and Winston, Inc., New York-Montreal, Que.-London, 1971)
31. M.E. Brandt, G.-R. Chen, Feedback control of a biodynamical model of HIV-1. IEEE Trans. Biomed. Eng. **48**, 754–759 (2001)
32. J.K. Brimacombe, Design of continuous casting machines based on a heat-flow analysis: state-of-the-art review. Can. Metall. Quart. **15**, 163–175 (1976)
33. J.C. Bruch, S. Adali, J.M. Sloss, I.S. Sadek, Maximum principle for the optimal control of a hyperbolic equation in one space dimension. II. Application. J. Optim. Theory Appl. **87**, 287–300 (1995)
34. J.C. Bruch, I.S. Sadek, S. Adali, J.M. Sloss, A maximum principle approach to optimal control for one-dimensional hyperbolic systems with several state variables. J. Franklin Inst. **343**, 480–494 (2006)
35. A.E. Bryson Jr., Optimal control - 1950 to 1985. IEEE Control Syst. **16**, 26–33 (1996)
36. J. Burger, M. Pogu, Functional and numerical solution of a control problem originating from heat transfer. J. Optim. Theory Appl. **68**, 49–73 (1991)
37. S. Butler, D. Kirschner, S. Lenhart, Optimal control of chemotherapy affecting the infectivity of HIV, in *Advances in Mathematical Population Dynamics - Molecules, Cells and Man, Series in Mathematical Biology and Medicine*, vol. 6 (World Scientific Publishing Co., Inc., River Edge, 1997), pp. 557–569
38. G.P. Cai, S.X. Yang, A discrete optimal control method for a flexible cantilever beam with time delay. J. Vib. Control **12**, 509–526 (2006)

39. P. Cannarsa, F. Gozzi, H.M. Soner, A dynamic programming approach to nonlinear boundary control problems of parabolic type. J. Funct. Anal. **117**, 25–61 (1993)
40. C. Canuto, M.Y. Hussaini, A. Quarteroni, T.A. Zang, *Spectral Methods in Fluid Dynamics* (Springer-Verlag, Berlin, 1988)
41. J. Cao, J. Zhang, X. Yang, Fully-discrete spectral-Galerkin scheme with second-order time-accuracy and unconditionally energy stability for the volume-conserved phase-field lipid vesicle model. J. Comput. Appl. Math. **406**, 113988 (2022), 18 pp.
42. I. Capuzzo Dolcetta, On a discrete approximation of the Hamilton-Jacobi equation of dynamic programming. Appl. Math. Optim. **10**, 367–377 (1983)
43. I. Capuzzo-Dolcetta, H. Ishii, Approximate solutions of the Bellman equation of deterministic control theory. Appl. Math. Optim. **11**, 161–181 (1984)
44. E. Carlini, M. Falcone, R. Ferretti, An efficient algorithm for Hamilton-Jacobi equations in high dimension. Comput. Vis. Sci. **7**, 15–29 (2004)
45. E. Casas, Control of an elliptic problem with pointwise state constraints. SIAM J. Control Optim. **24**, 1309–1322 (1986)
46. E. Casas, M. Mateos, A.Rösch, Improved approximation rates for a parabolic control problem with an objective promoting directional sparsity. Comput. Optim. Appl. **70**, 239–266 (2018)
47. E. Casas, M. Mateos, A. Rösch, Error estimates for semilinear parabolic control problems in the absence of Tikhonov term. SIAM J. Control Optim. **57**, 2515–2550 (2019)
48. E. Casas, M. Mateos, F. Tröltzsch, Necessary and sufficients optimality conditions for optimization problems in function spaces and applications to control theory. ESAIM Proc. **13**, 18–30 (2003)
49. Z.S. Chalabi, L.G. van Willigenburg, G. van Straten, Robust optimal receding horizon control of the thermal sterilization of canned foods. J. Food Eng. **40**, 207–218 (1999)
50. W.L. Chan, B.-Z. Guo, Optimal birth control of population dynamics. J. Math. Anal. Appl. **144**, 532–552 (1989)
51. W.L. Chan, B.-Z. Guo, Optimal birth control of population dynamics. II. Problems with free final time, phase constraints, and mini-max costs. J. Math. Anal. Appl. **146**, 523–539 (1990)
52. Y. Chen, F. Huang, Galerkin spectral approximation of elliptic optimal control problems with H^1-norm state constraint. J. Sci. Comput. **67**, 65–83 (2016)
53. Y. Chen, F. Huang, Spectral method approximation of flow optimal control problems with H^1-Norm state constraint. Numer. Math. Theory Methods Appl. **10**, 614–638 (2017)
54. Y. Chen, Y. Huang, N. Yi, A posteriori error estimates of spectral method for optimal control problems governed by parabolic equations. Sci. China Ser. A **51**, 1376–1390 (2008)
55. Y. Chen, N. Yi, W. Liu, A Legendre-Galerkin spectral method for optimal control problems governed by elliptic equations. SIAM J. Numer. Anal. **46**, 2254–2275 (2008)
56. Y. Chen, F. Huang, N. Yi, W. Liu, A Legendre-Galerkin spectral method for optimal control problems governed by Stokes equations. SIAM J. Numer. Anal. **49**, 1625–1648 (2011)
57. F. Clarke, M.R. de Pinho, Optimal control problems with mixed constraints. SIAM J. Control Optim. **48**, 4500–4524 (2010)
58. F.H. Clarkel, *Optimization and Nonsmooth Analysis*. Canadian Mathematical Society Series of Monographs and Advanced Texts (John Wiley & Sons, Inc., New York, 1983)
59. I.K. Craig, X. Xia, Can HIV/AIDS be controlled? applying control engineering concepts outside traditional fields. IEEE Control Syst. Mag. **25**, 80–83 (2001)
60. I.K. Craig, X. Xia, J.W. Venter, Introducing HIV/AIDS education into the electrical engineering curriculum at the University of Pretoria. IEEE Trans. Educ. **47**, 65–73 (2004)
61. M.G. Crandall, P.-L. Lions, Viscosity solutions of Hamilton-Jacobi equations. Trans. Amer. Math. Soc. **277**, 1–42 (1983)
62. M.G. Crandall, P.-L. Lions, Two approximations of solutions of Hamilton-Jacobi equations. Math. Comp. **43**, 1–19 (1984)
63. M.G. Crandall, H. Ishii, P.L. Lions, User's guide to viscosity solutions of second order partial differential equations. Bull. Amer. Math. Soc. **27**, 1–67 (1992)
64. K. Deckelnick, M. Hinze, Convergence of a finite element approximation to a state-constrained elliptic control problem. SIAM J. Numer. Anal. **45**, 1937–1953 (2007)

65. J.C. de los Reyes, K. Kunisch, A semi-smooth Newton method for regularized state-constrained optimal control of the Navier-Stokes equations. Computing **78**, 287–309 (2006)
66. Department of Demographic Census in National Bureau of Statistics of China, *China Population Statistics Yearbook* (Scientific and Technical Documents Publishing House, Beijing, 1990). In Chinese
67. Y. Ding, Z. Wang, H. Ye, Optimal control of a fractional-order HIV-immune system with memory. IEEE Trans. Control Syst. Tech. **20**, 763–769 (2012)
68. A.V. Dmitruk, A.A. Milyutin, N.P. Osmolovskii, Lyusternik's theorem and the theory of extrema. Uspekhi Mat. Nauk. **35**, 11–46 (1980)
69. D. Drake, M. Xin, S.N. Balakrishnan, A new nonlinear control technique for ascent phase of reusable launch vehicles. AIAA J. Guid. Control Dynam. **27**, 938–948 (2004)
70. N. Dunford, J.T. Schwartz, *Linear Operators*, Part I (Interscience, New York, 1958)
71. C.M. Elliott, Z. Songmu, On the Cahn-Hilliard equation. Arch. Rational Mech. Anal. **96**, 339–357 (1986)
72. R.S. Esfandiari, I.S. Sadek, Optimal open/closed-loop control of a Rayleigh beam. Optimal Control Appl. Methods **13**, 247–253 (1992)
73. L. Euler, *Methodus Inveniendi Lineas Curvas Maximi Minimive Propriatate Gaudientes, sive Solutio Problematis Isoperimetrici Latissimo Sensu Accepti*. A method for finding curved lines enjoying properties of maximum or minimum, or solution of isoperimetric problems in the broadest accepted sense (Bousquent, Lausannae and Genevae, 1744)
74. G. Fabrini, M. Falcone, S. Volkwein, Coupling MPC and HJB for the computation of POD-based feedback laws, in *Numerical Mathematics and Advanced Applications ENUMATH 2017*. Lecture Notes in Computational Science and Engineering, vol. 126, ed. by G. Fabrini, M. Falcone, S. Volkwein (Springer, Cham, 2019), pp. 941–949
75. M. Falcone, R. Ferretti, Discrete time high-order schemes for viscosity solutions of Hamilton-Jacobi-Bellman equations. Numer. Math. **67**, 315–344 (1994)
76. M. Falcone, R. Ferretti, *Semi-Lagrangian Approximation Schemes for Linear and Hamilton-Jacobi Equations* (SIAM, Philadelphia, 2014)
77. M. Falcone, T. Giorgi, An approximation scheme for evolutive Hamilton-Jacobi equations, in *Stochastic Analysis, Control, Optimization and Applications*, Systems & Control: Foundations & Applications (Birkhäuser, Boston, 1999), pp. 289–303
78. M. Falcone, P. Lanucara, A. Seghini, A splitting algorithm for Hamilton-Jacobi-Bellman equations. Appl. Numer. Math. **15**, 207–218 (1994)
79. R.S. Falk, Approximation of a class of optimal control problems with order of convergence estimates. J. Math. Anal. Appl. **44**, 28–47 (1973)
80. H.O. Fattorini, *Infinite-Dimensional Optimization and Control Theory*. Encyclopedia of Mathematics and its Applications, vol. 62 (Cambridge University Press, Cambridge, 1999)
81. H.O. Fattorini, *Infinite Dimensional Linear Control Systems. The Time Optimal and Norm Optimal Problems*. North-Holland Mathematics Studies, vol. 201 (Elsevier Science B.V., Amsterdam, 2005)
82. C. Fefferman, E.M. Stein, H^p spaces of several variables. Acta Math. **129**, 137–193 (1972)
83. J.A.M. Felippe de Souza, M.A.L. Caetano, T. Yoneyama, Optimal control theory applied to the anti-viral treatment of AIDS, in *Proceedings of the 39th IEEE Conference on Decision and Control*, Sydney (2000), pp. 4839–4844
84. W.H. Fleming, R.W. Rishel, *Deterministic and Stochastic Optimal Control* (Springer-Verlag, New York, 1975)
85. W.H. Fleming, H.M. Soner, *Controlled Markov Processes and Viscosity Solutions*, 2nd edn. Stochastic Modelling and Applied Probability, vol. 25 (Springer-Verlag, New York, 2006)
86. H. Fu, H. Rui, Adaptive characteristic finite element approximation of convection-diffusion optimal control problems. Numer. Methods Partial Differ. Equations **29**, 979–998 (2013)
87. T. Geveci, On the approximation of the solution of an optimal control problem governed by an elliptic equation. RAIRO Anal. Numer. **13**, 313–328 (1979)

88. V. Girault, P.-A. Raviart, *Finite Element Methods for Navier-Stokes Equations. Theory and Algorithms*. Springer Series in Computational Mathematics, vol. 5 (Springer-Verlag, Berlin, 1986)
89. I.V. Girsanov, *Lectures on Mathematical Theory of Extremum Problems*. Lecture Notes in Economics and Mathematical Systems, vol. 67 (Springer-Verlag, Berlin, 1972)
90. H.H. Goldstine, *A History of the Calculus of Variations from the 17th through the 19th Century*. Studies in the History of Mathematics and Physical Sciences, vol. 5 (Springer-Verlag, New York, 1980)
91. S. Gombao, Approximation of optimal controls for semilinear parabolic PDE by solving Hamilton-Jacobi-Bellman equations, in *Electronic Proceedings of 15th International Symposium on the Mathematical Theory of Networks and Systems*, ed. by D.S. Gilliam, J. Rosenthal (University of Notre Dame, South Bend, IN, August 12–16 (2002)
92. W. Gong, M. Hinze, Error estimates for parabolic optimal control problems with control and state constraints. Comput. Optim. Appl. **56**, 131–151 (2013)
93. R.G. Gottlieb, Rapid convergence to optimum solutions using a min-h strategy. J. Guidance Control Dynam. **5**, 322–329 (1967)
94. D. Gottlieb, S.A. Orszag, *Numerical Analysis of Spectral Methods: Theory and Applications*. CBMS-NSF Regional Conference Series in Applied Mathematics, vol. 26 (Society for Industrial and Applied Mathematics, Philadelphia, 1977)
95. D. Gottlieb, S.A. Orszag, *Numerical Analysis of Spectral Methods: Theory and Applications* (Society for Industrial and Applied Mathematics, Philadelphia, 1983)
96. F. Gozzi, S.S. Sritharan, A. Święch, Viscosity solutions of dynamic-programming equations for the optimal control of the two-dimensional Navier-Stokes equations. Arch. Ration. Mech. Anal. **163**, 295–327 (2002)
97. F. Gozzi, A. Święch, X.Y. Zhou, A corrected proof of the stochastic verification theorem within the framework of viscosity solutions. SIAM J. Control Optim. **43**, 2009–2019 (2005)
98. Z.-Z. Guan, J.-Q. Han, H.-S. Qin, C.-Z. Wang, S.-L. Wang, *Extremal Control and the Maximal Principle*. Modern Control System Theory Series (Science Press, Beijing, 1980). In Chinese
99. B.-Y. Guo, *Spectral Methods and Their Applications* (World Scientific Publishing Co., Inc., River Edge, 1998)
100. B.-Y. Guo, Error estimation of Hermite spectral method for nonlinear partial differential equations. Math. Comput. **68**, 1067–1078 (1999)
101. B.-Z. Guo, Riesz basis approach to the tracking control of a flexible beam with a tip rigid body without dissipativity. Optim. Methods Softw. **17**, 655–681 (2002)
102. B.-Z. Guo, B. Sun, Numerical solution to the optimal birth feedback control of a population dynamics: viscosity solution approach. Optimal Control Appl. Methods **26**, 229–254 (2005)
103. B.-Z. Guo, B. Sun, Numerical solution to the optimal feedback control of continuous casting process. J. Global Optim. **39**, 171–195 (2007)
104. B.-Z. Guo, B. Sun, A new algorithm for finding numerical solutions of optimal feedback control IMA J. Math. Control Inform. **26**, 95–104 (2009)
105. B.-Z. Guo, B. Sun, Dynamic programming approach to the numerical solution of optimal control with paradigm by a mathematical model for drug therapies of HIV/AIDS. Optim. Eng. **15**, 119–136 (2014)
106. B.-Z. Guo, J.-M. Wang, *Control of Wave and Beam PDEs, The Riesz Basis Approach*. Communications and Control Engineering Series (Springer, Cham, 2019)
107. B.-Z. Guo, T.-T. Wu, Approximation of optimal feedback control: a dynamic programming approach. J. Global Optim. **46**, 395–422 (2010)
108. B.-Z. Guo, T.-T. Wu, Numerical solution to optimal feedback control by dynamic programming approach: a local approximation algorithm. J. Syst. Sci. Complex. **30**, 782–802 (2017)
109. B.-Z. Guo, C.-Z. Yao, New results on the exponential stability of non-stationary population dynamics. Acta Math. Sci. (English Ed.) **16**, 330–337 (1996)
110. B.-Z. Guo, X. Xia, F.R. Camisani-Calzolari, I.K. Craig, A semi-discrete approach to modelling and control of the continuous casting process. Steel Res. **71**, 220–227 (2000)

111. N. Gupta, N. Nataraj, A posteriori error estimates for an optimal control problem of laser surface hardening of steel. Adv. Comput. Math. **39**, 69–99 (2013)
112. W. Hackbusch, A numerical method for solving parabolic equations with opposite orientations. Computing **20**, 229–240 (1978)
113. Y. Han, A class of fourth-order parabolic equation with arbitrary initial energy. Nonlinear Anal. Real World Appl. **43**, 451–466 (2018)
114. Y. He, Stability and convergence of iterative methods related to viscosities for the 2D/3D steady Navier-Stokes equations. J. Math. Anal. Appl. **423**, 1129–1149 (2015)
115. M. Hinze, A variational discretization concept in control constrained optimization: the linear-quadratic case. Comput. Optim. Appl. **30**, 45–61 (2005)
116. Y.-C. Ho, On centralized optimal control. IEEE Trans. Automat. Control **50**, 537–538 (2005)
117. Y.-C. Ho, D.L. Pepyne, Simple explanation of the no-free-lunch theorem and its implications. J. Optim. Theory Appl. **115**, 549–570 (2002)
118. M.B. Horowitz, A. Damle, J.W. Burdick, Linear Hamilton Jacobi Bellman equations in high dimensions, in *53rd IEEE Conference on Decision and Control*, Los Angeles, CA, Dec. 15–17 (2014)
119. F. Huang, Y. Chen, Error estimates for spectral approximation of elliptic control problems with integral state and control constraints. Comput. Math. Appl. **68**, 789–803 (2014)
120. C.-S. Huang, S. Wang, K.L. Teo, Solving Hamilton-Jacobi-Bellman equations by a modified method of characteristics. Nonlinear Anal. **40**, 279–293 (2000)
121. C.-S. Huang, S. Wang, K.L. Teo, On application of an alternating direction method to Hamilton-Jacobi-Bellman equations. J. Comput. Appl. Math. **166**, 153–166 (2004)
122. F. Huang, Z. Zheng, Y. Peng, Error estimates of the space-time spectral method for parabolic control problems. Comput. Math. Appl. **75**, 335–348 (2018)
123. A.D. Ioffe, *Variational Analysis of Regular Mappings. Theory and Applications*, Springer Monographs in Mathematics (Springer, Cham, 2017)
124. A.D. Ioffe, An elementary proof of the Pontryagin maximum principle. Vietnam J. Math. **48**, 527–536 (2020)
125. A.M. Jeffrey, X. Xia, I.K. Craig, When to initiate HIV therapy: a control theoretic approach. IEEE Trans. Biomed. Eng. **50**, 1213–1220 (2003)
126. M. Jensen, I. Smears, On the convergence of finite element methods for Hamilton-Jacobi-Bellman equations. SIAM J. Numer. Anal. **51**, 137–162 (2013)
127. H.R. Joshi, Optimal control of an HIV immunology model. Optimal Control Appl. Methods **23**, 199–213 (2002)
128. D. Kalise, K. Kunisch, Polynomial approximation of high-dimensional Hamilton-Jacobi-Bellman equations and applications to feedback control of semilinear parabolic PDEs. SIAM J. Sci. Comput. **40**, 629–652 (2018)
129. T. Kato, *Perturbation Theory for Linear Operators*. Die Grundlehren der mathematischen Wissenschaften, Band 132 (Springer-Verlag New York, Inc., New York, 1966)
130. D. Kirschner, Using mathematics to understand HIV immune dynamics. Notices Amer. Math. Soc. **43**, 191–202 (1996)
131. D. Kirschner, G.F. Webb, Immunotherapy of HIV-1 infection. J. Biol. Syst. **6**, 71–83 (1998)
132. D. Kirschner, S. Lenhart, S. Serbin, Optimal control of the chemotherapy of HIV. J. Math. Biol. **35**, 775–792 (1997)
133. D. Kleis, *Augmented Lagrange SQP Methods and Application to the Sterilization of Prepackaged Food*, Ph.D thesis, Universität Trier (1997)
134. D. Kleis, E.W. Sachs, Optimal control of the sterilization of prepackaged food. SIAM J. Optim. **10**, 1180–1195 (2000)
135. G. Knowles, *An Introduction to Applied Optimal Control* (Academic Press, New York, 1981)
136. M. Kocan, P. Soravia, A viscosity approach to infinite-dimensional Hamilton-Jacobi equations arising in optimal control with state constraints. SIAM J. Control Optim. **36**, 1348–1375 (1998)
137. K. Kohls, A. Rösch, K.G. Siebert, A posteriori error analysis of optimal control problems with control constraints. SIAM J. Control. Optim. **52**, 1832–1861 (2014)

138. D.A. Kopriva, *Implementing Spectral Methods for Partial Differential Equations. Algorithms for Scientists and Engineers*. Scientific Computation (Springer, Berlin, 2009)
139. W. Kotarski, *Some Problem of Optimal and Pareto Optimal Control for Distributed Parameter Systems* (Wydawinictwo Uniwersytetu Śląskiego, Katowice, 1997)
140. K. Kunisch, S. Volkwein, L. Xie, HJB-POD-based feedback design for the optimal control of evolution problems. SIAM J. Appl. Dyn. Syst. **3**, 701–722 (2004)
141. K. Kunisch, L. Xie, POD-based feedback control of the Burgers equation by solving the evolutionary HJB equation. Comput. Math. Appl. **49**, 1113–1126 (2005)
142. K. Kunisch, L. Xie, Suboptimal feedback control of flow separation by POD model reduction, in *Real-Time PDE-Constrained Optimization*, ed. by L.T. Biegler, O. Ghattas, M. Heinkenschloss, D. Keyes, B. van Bloemen Waanders, Computational Science & Engineering (SIAM, Philadelphia, 2007), pp. 233–250
143. E. Laitinen, P. Neittaanmäki, On numerical simulation of the continuous casting process. J. Engrg. Math. **22**, 335–354 (1988)
144. G. Lantoine, R. Epenoy, Quadratically constrained linear-quadratic regulator approach for finite-thrust orbital rendezvous. J. Guid. Control Dynam. **35**, 1787–1797 (2012)
145. U. Ledzewicz and H. Schättler, A high-order generalization of the Lyusternik theorem. Nonlinear Anal. **34**, 793–815 (1998)
146. S. Lenhart, J.T. Workman, *Optimal Control Applied to Biological Models*. Chapman & Hall/CRC Mathematical and Computational Biology Series (Chapman & Hall/CRC, Boca Raton, 2007)
147. G. Leugering, Dynamic domain decomposition of optimal control problems for networks of Euler-Bernoulli beams, in *ESAIM Proceedings, vol. 4: Control and partial differential equations (Marseille-Luminy, 1997)*. Society Mathematics Applied Industries, Paris (1998), pp. 223–233
148. G. Leugering, Dynamic domain decomposition of optimal control problems for networks of strings and Timoshenko beams. SIAM J. Control Optim. **37**, 1649–1675 (1999)
149. R.-H. Li, G.-C Feng, *Numerical Solutions of Differential Equations* (Higher Education Press, Beijing, 1996). In Chinese
150. X. J. Li, J.M. Yong, *Optimal Control Theory for Infinite Dimensional Systems*. Systems & Control: Foundations & Applications (Birkhäuser Boston, Inc., Boston, 1995)
151. D. Liberzon, *Calculus of Variations and Optimal Control Theory: A Concise Introduction* (Princeton University Press, Princeton, 2012)
152. X. Lin, Y. Chen, Y. Huang, Galerkin spectral approximation of optimal control problems with L^2-norm control constraint. Appl. Numer. Math. **150**, 418–432 (2020)
153. J.-L. Lions, *Optimal Control of Systems Governed by Partial Differential Equations*. Translated from the French by S. K. Mitter, Die Grundlehren der mathematischen Wissenschaften, Band 170 (Springer-Verlag, New York, 1971)
154. W. Liu, Optimal control of a rotating body beam. ESAIM Control Optim. Calc. Var. **7**, 157–178 (2002)
155. C. Liu, A fourth order parabolic equation with nonlinear principal part. Nonlinear Anal. **68**, 393–401 (2008)
156. W. Liu, N. Yan, A posteriori error estimates for control problems governed by Stokes equations. SIAM J. Numer. Anal. **40**, 1850–1869 (2003)
157. H. Liu, N. Yan, Global superconvergence for optimal control problems governed by Stokes equations. Int. J. Numer. Anal. Model. **3**, 283–302 (2006)
158. W. Liu, N. Yan, *Adaptive Finite Element Methods for Optimal Control Governed by PDEs* (Scientific Press, Beijing, 2008)
159. W. Liu, W. Gong, N. Yan, A new finite element approximation of a state-constrained optimal control problem. J. Comput. Math. **27**, 97–114 (2009)
160. W. Liu, D. Yang, L. Yuan, C. Ma, Finite element approximations of an optimal control problem with integral state constraint. SIAM J. Numer. Anal. **48**, 1163–1185 (2010)
161. Z. Lu, F. Cai, R. Xu, C. Hou, X. Wu, Y. Yang, A posteriori error estimates of *hp* spectral element method for parabolic optimal control problems. AIMS Math. **7**, 5220–5240 (2022)

162. S.-Y. Luo, T.-W. Chou, Finite deformation and nonlinear elastic behavior of flexible composites. ASME J. Appl. Mech. **55**, 149–155 (1988)
163. Z.-H. Luo, B.-Z. Guo, O. Morgül, *Stability and Stabilization of Infinite Dimensional Systems with Applications* (Springer-Verlag, London, 1999)
164. L.A. Lyusternik, V.I. Sobolev, *Elements of Functional Analysis* (Nauka, Moscow, 1965). In Russian
165. K. Malanowski, Convergence of approximations vs. regularity of solutions for convex, control-constrained optimal-control problems. Appl. Math. Optim. **8**, 69–95 (1982)
166. W.M. McEneaney, *Max-Plus Methods for Nonlinear Control and Estimation* (Birkhäuser, Boston, 2006)
167. W.M. McEneaney, Convergence rate for a curse-of-dimensionality-free method for Hamilton-Jacobi-Bellman PDEs represented as maxima of quadratic forms. SIAM J. Control Optim. **48**, 2651–2685 (2009)
168. D. Meidner, R. Rannacher, B. Vexler, A priori error estimates for space-time finite element discretization of parabolic optimal control problems. I. Problems without control constraints. SIAM J. Control Optim. **47**, 1150–1177 (2008)
169. D. Meidner, R. Rannacher, B. Vexler, A priori error estimates for finite element discretizations of parabolic optimization problems with pointwise state constraints in time. SIAM J. Control Optim. **49**, 1961–1997 (2011)
170. B. Meuris, S. Qadeer, P. Stinis, Machine-learning-based spectral methods for partial differential equations. Sci. Rep. **13** (2023). Article number: 1739
171. K. Miettinen, M.M. Mäkelä, T. Männikkö, Optimal control of continuous casting by nondifferentiable multiobjective optimization. Comput. Optim. Appl. **11**, 177–194 (1998)
172. K. Miettinen, M.M. Mäkelä, J. Toivanen, Numerical comparison of some penalty-based constraint handling techniques in genetic algorithms. J. Global Optim. **27**, 427–446 (2003)
173. S. Mirică, MR2316829 (2008f:49001), Mathematical review of [146]. Available at http://www.ams.org/mathscinet/, last accessed 28 Oct 2022
174. I. Neitzel, B. Vexler, A priori error estimates for space-time finite element discretization of semilinear parabolic optimal control problems. Numer. Math. **120**, 345–386 (2012)
175. I. Neitzel, J. Pfefferer, A. Rösch, Finite element discretization of state-constrained elliptic optimal control problems with semilinear state equation. SIAM J. Control Optim. **53**, 874–904 (2015)
176. H. Niu, D. Yang, Finite element analysis of optimal control problem governed by Stokes equations with L^2-norm state-constriants. J. Comput. Math. **29**, 589–604 (2011)
177. H. Niu, L. Yuan, D. Yang, Adaptive finite element method for an optimal control problem of Stokes flow with L^2-norm state constraint. Internat. J. Numer. Methods Fluids **69**, 534–549 (2012)
178. M.A. Nowak, R.M. May, *Virus Dynamics: Mathematical Principle of Immunology and Virology* (Oxford University Press, Oxford, 2000)
179. B.G. Pachpatte, *Inequalities for Differential and Integral Equations*. Mathematics in Science and Engineering, vol. 197 (Academic Press, Inc., San Diego, 1998)
180. B. Passenberg, M. Leibold, O. Stursberg, M. Buss, A globally convergent, locally optimal min-H algorithm for hybrid optimal control. SIAM J. Control Optim. **52**, 718–746 (2014)
181. A. Pazy, *Semigroups of Linear Operators and Applications to Partial Differential Equations*. Applied Mathematical Sciences, vol. 44 (Springer-Verlag, New York, 1983)
182. R. Peng, The Hahn-Banach separation theorem and other separation results. University of Chicago REU 2014, Preprint (2014)
183. A.S. Perelson, P.W. Nelson, Mathematical analysis of HIV-1 dynamics in vivo. SIAM Rev. **41**, 3–44 (1999)
184. A. Perelson, D. Kirschner, R. Deboer, The dynamics of HIV infection of $CD4^+T$ Cells. Math. Biosci. **114**, 81–125 (1993)
185. R. Peyret, T.D. Taylor, *Computational Methods for Fluid Flow* (Springer-Verlag, New York, 1983)

186. L.S. Pontryagin, V.G. Boltyanskii, R.V. Gamkrelidze, E.F. Mishchenko, *The Mathematical Theory of Optimal Processes* (Interscience Publishers John Wiley & Sons, Inc., New York, 1962)
187. W.H. Press, S.A. Teukolsky, W.T. Vetterling, B.P. Flannery, *Numerical Recipes in C++: The Art of Scientific Computing*, 2nd edn. (Cambridge University Press, Cambridge, 2002)
188. V. Radisavljevic-Gajic, Optimal control of HIV-virus dynamics. Ann. Biomed. Eng. **37**, 1251–1261 (2009)
189. T. Rees, A.J. Wathen, Preconditioning iterative methods for the optimal control of the Stokes equations. SIAM J. Sci. Comput. **33**, 2903–2926 (2011)
190. S. Richardson, S. Wang, Numerical solution of Hamilton-Jacobi-Bellman equations by an exponentially fitted finite volume method. Optimization **55**, 121–140 (2006)
191. R.T. Rockafellar, R.J.-B. Wets, *Variational Analysis*. Grundlehren der mathematischen Wissenschaften, vol. 317 (Springer-Verlag, Berlin, 1998)
192. L. Rong, A.S. Perelson, Modeling HIV persistence, the latent reservoir, and viral blips. J. Theor. Biol. **260**, 308–331 (2009)
193. L. Rong, Z. Feng, A.S. Perelson, Emergence of HIV-1 drug resistance during antiretroviral treatment. Bull. Math. Biol. **69**, 2027–2060 (2007)
194. J.E. Rubio, *Control and Optimization: The Linear Treatment of Nonlinear Problems (Nonlinear Science: Theory and Applications)* (Manchester University Press, Manchester, 1986)
195. W. Rudin, *Functional Analysis*, 2nd edn. (McGraw-Hill, Inc., New York, 1991)
196. I.S. Sadek, R.S. Esfandiari, Optimal active control of distributed-parameter systems with applications to a Rayleigh beam. IMA J. Math. Control Inform. **7**, 59–76 (1990)
197. I.S. Sadek, V.H. Melvin, Optimal control of serially connected structures using spatially distributed pointwise controllers. IMA J. Math. Control Inform. **13**, 335–358 (1996)
198. I. Sadek, J.M. Sloss, J.C. Bruch, S. Adali, Optimal control of a Timoshenko beam by distributed forces. J. Optim. Theory Appl. **50**, 451–461 (1986)
199. I.S. Sadek, J.M. Sloss, S. Adali, J.C. Bruch, Maximum principle for the optimal control of a hyperbolic equation in two space dimensions. J. Vib. Control **2**, 3–15 (1996)
200. I.S. Sadek, J.M. Sloss, S. Adali, J.C. Bruch, A maximum principle for optimal control using spatially distributed pointwise controllers. J. Vib. Control **4**, 445–462 (1998)
201. I. Sadek, M. Abukhaled, T. Abualrub, Optimal pointwise control for a parallel system of Euler-Bernoulli beams. J. Comput. Appl. Math. **137**, 83–95 (2001)
202. R.W.H. Sargent, Optimal control. J. Comput. Appl. Math. **124**, 361–371 (2000)
203. T. Sauer, *Numerical Analysis*, 2nd edn. (Pearson Education, Essex, 2012)
204. J. Shen, Efficient Spectral-Galerkin method I: direct solvers for second and fourth order equations by using Legendre polynomials. SIAM J. Sci. Comput. **15**, 1489–1505 (1994)
205. J. Shen, On fast direct Possion solver, inf-sup constant and iterative Stokes solver by Legendre-Galerkin method. J. Comput. Phys. **116**, 184–188 (1995)
206. J. Shen, T. Tang, *Spectral and High-Order Methods with Applications*. Mathematics Monograph Series, vol. 3 (Science Press, Beijing, 2006)
207. J. Shen, L.-L. Wang, Fourierization of the Legendre-Galerkin method and a new space-time spectral method. Appl. Numer. Math. **57**, 710–720 (2007)
208. J. Shen, T. Tang, L.-L. Wang, *Spectral Methods. Algorithms, Analysis and Applications*. Springer Series in Computational Mathematics, vol. 41 (Springer, Heidelberg, 2011)
209. I. Silberman, Planetary waves in the atmosphere. J. Atmospheric Sci. **11**, 27–34 (1954)
210. J.M. Sloss, I.S. Sadek, J.C. Bruch, S. Adali, Maximum principle for the optimal control of a hyperbolic equation in one space dimension. I. Theory. J. Optim. Theory Appl. **87**, 33–45 (1995)
211. J.M. Sloss, J.C. Bruch, I.S. Sadek, S. Adali, Maximum principle for optimal boundary control of vibrating structures with applications to beams. Dynam. Control **8**, 355–375 (1998)
212. J.M. Sloss, I.S. Sadek, J.C. Bruch, S. Adali, Optimal control of structural dynamic systems in one space dimension using a maximum principle. J. Vib. Control **11**, 245–261 (2005)
213. J. Song, J.Y. Yu, *Population System Control* (Springer-Verlag, Berlin, 1988)

214. E.D. Sontag, *Mathematical Control Theory: Deterministic Finite Dimensional Systems* (Springer-Verlag, New York, 1990)
215. P.E. Souganidis, Approximation schemes for viscosity solution of Hamilton-Jacobi equations. J. Differ. Equ. **59**, 1–43 (1985)
216. E.M. Stein, *Singular Integrals and Differentiability Properties of Functions*. Princeton Mathematical Series, vol. 30 (Princeton University Press, Princeton, 1970)
217. R.F. Stengel, R.M. Ghigliazza, N.V. Kulkarni, Optimal enhancement of immune response. Bioinformatics **18**, 1227–1235 (2002)
218. J. Stoer, R. Bulirsch, *Introduction to Numerical Analysis*. Texts in Applied Mathematics, vol. 12 (Springer-Verlag, New York, 2002)
219. J.C. Strikwerda, Finite difference methods for the Stokes and Navier-Stokes equations. SIAM. J. Sci. Stat. Comput. **5**, 56–68 (1984)
220. B. Sun, B.-Z. Guo, Maximum principle for the optimal control of an ablation-transpiration cooling system. J. Syst. Sci. Complex. **18**, 285–301 (2005)
221. B. Sun, B.-Z. Guo, Convergence of an upwind finite-difference scheme for Hamilton-Jacobi-Bellman equation in optimal control. IEEE Trans. Automat. Control **60**, 3012–3017 (2015)
222. B. Sun, M.-X. Wu, Maximum principle for optimal control of sterilization of prepackaged food. IMA J. Math. Control Inform. **24**, 493–505 (2007)
223. B. Sun, Z.-Z. Tao, Y.-Y. Wang, Dynamic programming viscosity solution approach and its applications to optimal control problems, in *Mathematics Applied to Engineering, Modelling, and Social Issues, Studies in Systems, Decision and Control*, vol. 200, ed. by F.T. Smith, H. Dutta, J.N. Mordeson (Springer, Cham, 2019), pp. 363–420
224. H.J. Sussmann, J.C. Willems, 300 years of optimal control: from the Brachystochrone to the maximum principle. IEEE Control Syst. **17**, 32–44 (1997)
225. G.-Y. Tang, H.-H. Wang, Successive approximation approach of optimal control for nonlinear discrete-time system. Int. J. Syst. Sci. **36**, 153–161 (2005)
226. Z.-Z. Tao, B. Sun, Galerkin spectral method for a fourth-order optimal control problem with H^1-norm state constraint. Comput. Math. Appl. **97**, 1–17 (2021)
227. Z.-Z. Tao, B. Sun, A feedback design for numerical solution to optimal control problems based on Hamilton-Jacobi-Bellman equation. Electron. Res. Arch. **29**, 3429–3447 (2021)
228. Z.-Z. Tao, B. Sun, Space-time spectral method for a fourth-order parabolic optimal control problem in three control constraint cases. Discrete Contin. Dyn. Syst. Ser. B **28**, 359–384 (2023)
229. Z.-Z. Tao, B. Sun, Error estimates for spectral approximation of flow optimal control problem with L^2-norm control constraint. J. Ind. Manag. Optim. **19**, 2020–2049 (2023)
230. The INSIGHT-ESPRIT Study Group and SILCAAT Scientific Committee, Interleukin-2 therapy in patients with HIV infection. N. Engl. J. Med. **361**, 1548–1559 (2009)
231. R.B. Vinter, *Optimal Control*. Modern Birkhäuser Classics (Birkhäuser, Boston, 2000)
232. O. von Stryk, R. Bulirsch, Direct and indirect methods for trajectory optimization. Ann. Oper. Res. **37**, 357–373 (1992)
233. S. Wang, F. Gao, K.L. Teo, An upwind finite-difference method for the approximation of viscosity solutions to Hamilton-Jacobi-Bellman equations. IMA J. Math. Control Inform. **17**, 167–178 (2000)
234. S. Wang, L.S. Jennings, K.L. Teo, Numerical solution of Hamilton-Jacobi-Bellman equations by an upwind finite volume method. J. Global Optim. **27**, 177–192 (2003)
235. G. Wang, L. Wang, Y. Xu, Y. Zhang, *Time Optimal Control of Evolution Equations*. Progress in Nonlinear Differential Equations and their Applications, vol. 92, Subseries in Control (Birkhäuser/Springer, Cham, 2018)
236. L.M. Wein, S.A. Zenios, M. Nowak, Dynamic multidrug therapies for HIV: a control theoretic approach. J. Theor. Biol. **185**, 15–29 (1997)
237. Wikipedia, Semigroup, available at https://en.wikipedia.org/wiki/Semigroup, last accessed on 28 Oct 2022
238. Wikipedia, C_0-semigroup, available at https://en.wikipedia.org/wiki/C0-semigroup, last accessed on 28 Oct 2022

239. Wikipedia, Spectral method, available at https://en.wikipedia.org/wiki/Spectral_method, last accessed on 30 Jan 2023
240. T.P. Witelski, Similarity solutions of the lubrication equation. Appl. Math. Lett. **10**, 107–113 (1997)
241. C.-Z. Yao, B.-Z. Guo, Pointwise measure, control and stabilization of elastic beams. Control Theory Appl. **20**, 351–360 (2003). In Chinese
242. J.M. Yong, *Dynamic Programming and Hamilton-Jacobi-Bellman Equations* (Shanghai Scientific & Technical Publishers, Shanghai, 1992). In Chinese
243. J.M. Yong, X.J. Li, *A Concise Lecture Note on Optimal Control Theory* (Higher Education Press, Beijing, 2006). In Chinese
244. J. Yong, X.Y. Zhou, *Stochastic Controls: Hamiltonian Systems and HJB Equations*. Applications of Mathematics (New York), vol. 43 (Springer-Verlag, New York, 1999)
245. K. Yosida, *Functional Analysis*, 6th edn. Classics in Mathematics (Springer-Verlag, Berlin, 1995)
246. D.B. Yudin, E.G. Gol'shteĭn, *Problems and Methods of Linear Programming* (Sovetskoe Radio, Moscow, 1964). In Russian
247. S.P. Yung, Multiprocess optimal control problem on Hilbert space and related Hamilton-Jacobi equation. Nonlinear Anal. **29**, 1319–1342 (1997)
248. H. Zhang, F. Liu, X. Jiang, I. Turner, Spectral method for the two-dimensional time distributed-order diffusion-wave equation on a semi-infinite domain. J. Comput. Appl. Math. **399** (2022), Paper No. 113712, 15 pp.
249. X.Y. Zhou, Verification theorems within the framework of viscosity solutions. J. Math. Anal. Appl. **177**, 208–225 (1993)
250. L. Zhou, A priori error estimates for optimal control problems with state and control constraints. Optimal Control Appl. Methods **39**, 1168–1181 (2018)
251. J. Zhou, D. Yang, Spectral mixed Galerkin method for state constrained optimal control problem governed by the first bi-harmonic equation. Int. J. Comput. Math. **88**, 2988–3011 (2011)
252. J. Zhou, D. Yang, Legendre-Galerkin spectral methods for optimal control problems with integral constraint for state in one dimension. Comput. Optim. Appl. **61**, 135–158 (2015)
253. J. Zhou, J. Zhang, X. Xing, Galerkin spectral approximations for optimal control problems governed by the fourth order equation with an integral constraint on state. Comput. Math. Appl. **72**, 2549–2561 (2016)
254. R. Zurakowski, A.R. Teel, A model predictive control based scheduling method for HIV therapy. J. Theor. Biol. **238**, 368–382 (2006)

Index

B. Sun et al., *Maximum Principle and Dynamic Programming Viscosity Solution Approach*, Systems & Control: Foundations & Applications,
https://doi.org/10.1007/978-981-96-5739-1

MIX
Papier aus verantwortungsvollen Quellen
Paper from responsible sources
FSC® C105338

If you have any concerns about our products,
you can contact us on
ProductSafety@springernature.com

In case Publisher is established outside the EU,
the EU authorized representative is:
Springer Nature Customer Service Center GmbH
Europaplatz 3, 69115 Heidelberg, Germany

Printed by Libri Plureos GmbH
in Hamburg, Germany